PHYSICS and CULTURE

PHYSICS and CULTURE

Brian Cotterell
University of Sydney, Australia

World Scientific

NEW JERSEY • LONDON • SINGAPORE • BEIJING • SHANGHAI • HONG KONG • TAIPEI • CHENNAI • TOKYO

Published by

World Scientific Publishing Europe Ltd.

57 Shelton Street, Covent Garden, London WC2H 9HE

Head office: 5 Toh Tuck Link, Singapore 596224

USA office: 27 Warren Street, Suite 401-402, Hackensack, NJ 07601

Library of Congress Cataloging-in-Publication Data
Names: Cotterell, Brian, 1934– author.
Title: Physics and culture / Brian Cotterell (University of Sydney, Australia).
Description: Singapore ; Hackensack, NJ : World Scientific, [2017]
Identifiers: LCCN 2017014320| ISBN 9781786343765 (hc ; alk. paper) |
ISBN 1786343762 (hc ; alk. paper)
Subjects: LCSH: Physics. | Culture.
Classification: LCC QC21.3 .C68 2017 | DDC 303.48/3--dc23
LC record available at https://lccn.loc.gov/2017014320

British Library Cataloguing-in-Publication Data
A catalogue record for this book is available from the British Library.

Desk Editors: V. Vishnu Mohan/Mary Simpson/Jennifer Brough/Koe Shi Ying

Typeset by Stallion Press
Email: enquiries@stallionpress.com

Printed in Singapore

To Maureen for her support and love

Preface

This book is a celebration of how physics has been an integral part of our culture concentrating on the period from the seventeenth century until the present day, though parts do go back earlier. The term physics is used loosely and includes astronomy, which always was included in what was called natural philosophy until the nineteenth century. There is a difficulty in naming the three major periods in our culture since the seventeenth century because nearly each discipline has a different definition of the terms classical, modern, and postmodern. I personally do not like these three terms because of this problem but I am forced to use them. I was brought up to understand that the classical period meant only one thing, the period of the ancient Greeks and Romans, but its use in physics now means the Newtonian period and I have taken the two terms to be synonymous. Modern physics separated from the Newtonian period at the end of the nineteenth century. Postmodernism began to be distinguished from modernism in the arts around the middle of the twentieth century and following the lead of others, I also see a difference in the character of string theory in physics from modern physics and classify it as postmodern. The first two cultural periods have been so distinct that one can talk about two revolutions but I do not think the distinction between the modern and postmodern periods is great enough to be called a revolution especially since postmodern physics has not yet been definitely defined. Occasionally the periods of physics and non-physics do not match and the book is organized on the basis of the physics on those occasions. I am British and the book is written from a British perspective but I have tried not to be too Britain centric. I have deliberately made extensive use of quotations from the works of physicists and non-physicists because I believe that those directly involved tell their own story better than anyone else.

I have always had wide interests in the humanities as well as the sciences and I have never really understood the destructive Two Cultures Debate, which raged when I was a student. To me there is a unity to human culture just as there is a unity to humans and in writing this book I hope that I demonstrate this unity.

The index is an important part of a book, in compiling this one I have tried to limit entries to significant subjects so that it is as concise and useful as I can make it. I have not used cross references because I personally find them annoying. The notes are not indexed.

I would like to acknowledge the help and encouragement I have received: from my old colleagues Gordon Williams and Tony Atkins, the composer Hèctor Parra, the postmodern artists Matthew Ritchie and Julian Voss-Andreae. My editor Mary Simpson is thanked for her patience; she has been extremely helpful and made my writing task easier. My old School of Aerospace, Mechanical and Mechatronic Engineering at the University of Sydney, especially Doug Auld, is thanked for their assistance in what was an eclectic topic for them.

Brian Cotterell
Sydney

Contents

Preface vii

Colour Plates xv

Part I Introduction **1**

Chapter 1 Outline of the Physics 3
1.1 Newtonian Physics 4
1.1.1 Nicolaus Copernicus (1473–1543) 4
1.1.2 Johannes Kepler (1571–1630) 5
1.1.3 Galileo Galilei (1564–1642) 6
1.1.4 Isaac Newton (1643–1727) 9
1.1.5 Electromagnetic force 14
1.1.6 Thermodynamics 16
1.1.6.1 The first law of thermodynamics 16
1.1.6.2 The second law of thermodynamics 19
1.2 Modern Physics 20
1.2.1 The theory of relativity 22
1.2.1.1 Special theory of relativity 23
1.2.1.2 General theory of relativity 26
1.2.1.3 Experimental verification of the general theory of relativity 29
1.2.1.4 Black holes 32
1.2.2 Quantum mechanics 34
1.2.2.1 Black body radiation 34
1.2.2.2 Planck units 35
1.2.2.3 Particle-wave duality 36
1.2.2.4 The structure of the atom 39
1.2.2.5 Heisenberg's uncertainty principle 41
1.2.2.6 Wave mechanics 42
1.2.2.7 Relativistic quantum field theory 43
1.2.2.8 The standard model of particle physics 46
1.3 Postmodern String Theory 47
1.3.1 Superstring theories 48
1.3.2 The problem with string theories 50

1.4 Concluding Remarks 52
1.5 Notes 53

Part II The Newtonian Period 59

Chapter 2 Religion 61
2.1 Newton: His Religion and Natural Philosophy 66
2.2 The Boyle Lectures 70
2.2.1 Richard Bentley (1662–1742) 72
2.2.2 Samuel Clarke (1675–1729) 76
2.2.3 William Derham (1657–1735) 77
2.3 Natural and Revealed Religion 78
2.4 The Anti-Newtonians 86
2.4.1 The Hutchinsonians 87
2.4.2 The Muggletonians 89
2.5 The Plurality of Worlds 90
2.5.1 Whewell and the Antipluralists 99
2.6 Scientific Materialism 102
2.7 Thermodynamics 105
2.8 Concluding Remarks 108
2.9 Notes 109

Chapter 3 Popularization of Classical Newtonian Physics 114
3.1 Popular Eighteenth Century Lectures and Books 114
3.1.1 Popular lectures 115
3.1.2 Popular books 121
3.1.2.1 Natural philosophy for ladies 124
3.1.2.2 Natural philosophy for children 126
3.2 Popular Nineteenth Century Institutes, Associations, Books, and Periodicals 131
3.2.1 The mechanics' institutes 132
3.2.2 The Royal Institution of Great Britain 136
3.2.2.1 Christmas lectures at the Royal Institution 142
3.2.3 The British Association for the Advancement of Science 142
3.2.4 Popular books 144
3.2.4.1 Women popular writers 149
3.2.5 Popular periodicals 154
3.3 Anti-Newtonians 155
3.4 Concluding Remarks 157
3.5 Notes 159

Chapter 4 Philosophy and Politics 164
4.1 Metaphysics and Epistemology 167
4.1.1 John Locke (1632–1704) 167
4.1.2 Gottfried Wilhelm Leibniz (1646–1716) 171
4.1.3 George Berkeley (1685–1753) 175
4.1.4 Immanuel Kant (1724–1804) 177

4.1.5 Georg Hegel (1770–1831) ... 179
4.1.6 Philosophers of science ... 181
4.1.7 The North British Group of physicists and engineers ... 186
4.1.8 Paving the way to the philosophy of modern physics ... 191
4.2 Moral Philosophy ... 194
4.3 Politics ... 202
4.3.1 The Declaration of Independence and the Constitution of the United States ... 205
4.3.2 Marxism and dialectic materialism ... 208
4.4 Concluding Remarks ... 211
4.5 Notes ... 212

Chapter 5 Imaginative Arts ... 217
5.1 The Cultural Revolution of the Seventeenth Century ... 221
5.1.1 Literature ... 222
5.1.2 Music ... 232
5.1.2.1 Harmony of the world ... 232
5.1.2.2 Compositions inspired by Kepler and other Giants of Newton ... 233
5.1.2.3 The musical intervals in Western music ... 235
5.1.3 Visual arts ... 237
5.2 The Enlightenment ... 239
5.2.1 Literature ... 240
5.2.2 Music ... 252
5.2.3 Visual arts ... 254
5.3 The Romantic Age ... 257
5.3.1 Poets and their arts ... 258
5.3.1.1 Poets receptive to science ... 259
5.3.1.2 Poets antagonistic to science ... 264
5.3.2 Visual arts ... 266
5.4 Victorian Age and Afterwards ... 274
5.4.1 Literature ... 275
5.4.2 Helmholtz and music ... 286
5.4.3 Helmholtz and painting ... 289
5.5 Concluding Remarks ... 290
5.6 Notes ... 291

Part III Modern and Postmodern Period ... 297

Chapter 6 Philosophy, Politics, and Religion ... 299
6.1 Spiritualism, Aether, Continuity, and Discontinuity ... 301
6.2 Reality, Determinism, and Causality ... 304
6.2.1 Relativity and spacetime ... 310
6.2.2 The philosophical problems of Bohr's quantum mechanics ... 314
6.2.3 Politics and quantum mechanics ... 317
6.2.4 Bohm's causal interpretation of quantum theory ... 322
6.2.5 The Age of Aquarius and quantum mechanics ... 324
6.3 Cosmology and the Big Bang ... 325

6.3.1 Inflation ... 330
6.3.2 Dark matter and energy ... 333
6.3.3 Creation ... 335
6.3.4 Theism, atheism and the Big Bang cosmology ... 338
6.4 The Anthropic Principle ... 341
6.4.1 Ratio of the electromagnetic to gravitational force N ... 343
6.4.2 Nuclear efficiency in converting mass into energy ... 345
6.4.3 The amplitude of primordial fluctuations Q ... 346
6.4.4 Number of spatial dimensions D ... 346
6.4.5 What can be learned from the anthropic principle? ... 348
6.4.5.1 No fine-tuning was necessary because as yet undiscovered laws of nature determine the constants of particle physics and cosmological constants ... 349
6.4.5.2 A Supreme Being created the universe so that it would support intelligent life ... 350
6.4.5.3 The physical and cosmological constants have the values that have enabled intelligent life to evolve by chance ... 350
6.4.5.4 Our universe is not the only one but there is a whole multiverse of universes ... 350
6.5 Conluding Remarks ... 353
6.6 Notes ... 354

Chapter 7 Popularization of Modern and Postmodern Physics ... 360
7.1 Invisible Rays ... 365
7.1.1 X-rays ... 366
7.1.2 Alpha, beta, and gamma rays ... 370
7.2 Quantum Mechanics ... 372
7.2.1 The structure of Bohr's 1913 atom ... 373
7.2.2 The new quantum mechanics after 1925 ... 376
7.3 The Theory of Relativity ... 381
7.3.1 Special theory of relativity ... 381
7.3.2 General theory of relativity ... 386
7.3.3 Anti-Einsteinians ... 389
7.4 Cosmology ... 391
7.4.1 The Steady State theory ... 395
7.5 String Theory and Postmodern Physics ... 398
7.5.1 The superstring revolutions ... 400
7.5.2 String theory atheists ... 403
7.6 Concluding Remarks ... 404
7.7 Notes ... 405

Chapter 8 Modernism and Postmodernism ... 410
8.1 Literature ... 414
8.1.1 James Joyce (1882–1941) ... 417
8.1.2 Virginia Woolf (1882–1941) ... 421
8.1.3 D. H. Lawrence (1885–1930) ... 423

8.1.4 Aldous Huxley (1894–1963) 425
8.1.5 Lawrence Durrell (1912–1990) 427
8.1.6 Vladimir Nabokov (1899–1977) 428
8.2 Visual Art 431
8.2.1 Cubism and the fourth dimension 432
8.2.2 Dada and De Stijl 437
8.2.3 The Russian avant-garde 438
8.2.4 Surrealism 440
8.2.5 Postmodernism 442
8.3 Music 444
8.3.1 Modern physics, the theory of relativity, and black holes 445
8.3.2 Postmodernism and hyperspace 447
8.4 Concluding Remarks 449
8.5 Notes 449

Epilogue 455

Bibliography 461

Index 503

Colour Plates

Plate 1. Thomas and Paul Sandby, The Great Meteor of 1783, etching and aquatint, 33 × 49.3 cm, courtesy British Museum.

Plate 2. Muggletonian System of Astronomy: The Earth is at the centre of the universe separated from heaven by a firmament within which the Sun, planets, Moon, and stars revolve, Frost (1846).

Plate 3. An experimental lecture on the power of air: a cartoon of a Royal Institution lecture on pneumatics, 1802 print by James Gillray.

Plate 4. Terence Cuneo, Sir Lawrence Bragg Giving a 1961 Christmas Lecture on Electrostatics in the Royal Institution Theatre, oil on canvas, 120 × 172 cm, with permission Science Photo Library.

Plate 5. Antonio da Correggio, Madonna with St Francis, Oil on panel, 245 × 245 cm, 1514, Gemaldegalerie, Dresden.

Plate 6. Giovanni Pittoni and the Valeriani brothers, Allegorical Monument to Isaac Newton, Oil on canvas, 229 × 139 cm, 1730, with permission Fitzwilliam Museum, Cambridge.

Plate 7. Joseph Wright, An Experiment on a Bird in the Air Pump, Oil on canvas, 183 × 244, 1768, courtesy National Gallery, London.

Plate 8. William Blake, *Newton*, Colour print, ink and watercolour on paper, 480 × 600 cm, 1805, with permission Tate Gallery.

(a) (b)

Plate 9. (a) Additive mixing of coloured primary light. (b) Subtractive mixing of coloured pigments.

Plate 10. J.M.W. Turner, *Shields, on the River Tyne*, Watercolour on paper, 15.4 × 21.6 cm, 1823, with permission Tate Gallery.

Plate 11. George Seurat, *Sunday Afternoon on the Island of La Grande Jatte*, Oil on canvas, 206 × 305 cm, 1884–1886, The Art Institute of Chicago.

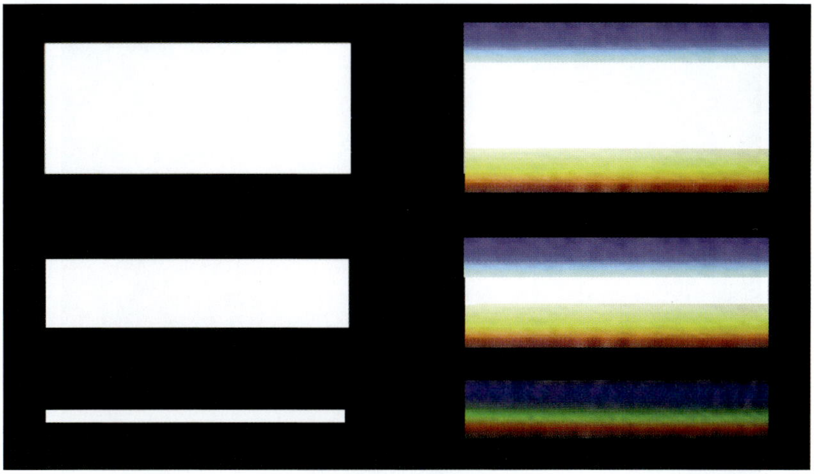

Plate 12. Colours obtained when viewing white panels on a black background through a prism, after Platts (2006).

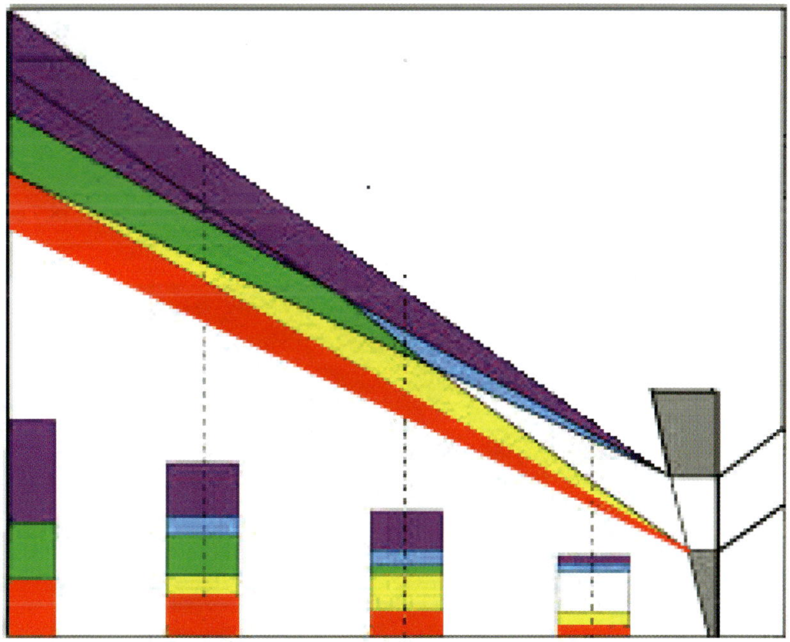

Plate 13. The colours seen when sunlight falls on a prism with light excluded except for a wide band near its centre, after Fig. 1, Plate IV, Goethe (1810).

Plate 14. J.M.W. Turner, *Light and Colour* (*Goethe's Theory*), Oil on canvas, 78.7 × 78.7 cm, 1843, with permission Tate Gallery.

Plate 15. John Constable, *View from Hampstead with a Double Rai*nbow, watercolour on grey paper, 19.7 × 32.4 cm, 1831, courtesy British Museum.

Plate 16. John Constable, *Salisbury Cathedral from the Meadows*, Oil on canvass, 38.5 × 51.1 cm, 1831, purchased by Tate Gallery in 2013.

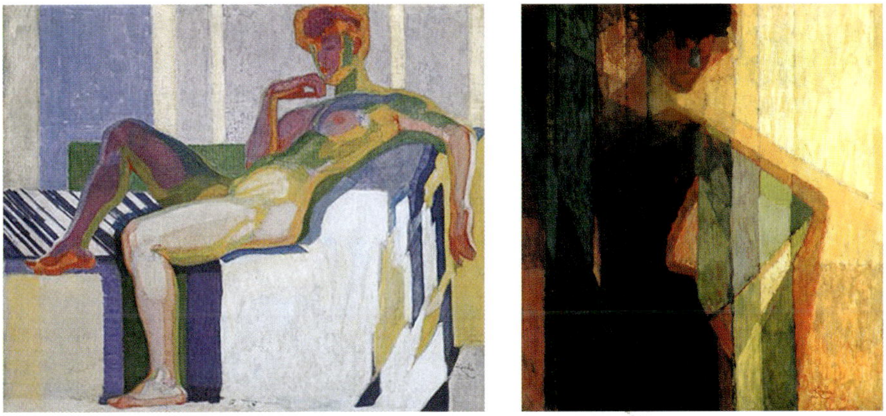

(a) (b)

Plate 17. František Kupka (a) Planes by Colours, Large Nude, oil on canvas, 150 × 180 cm, 1909–1910, courtesy Guggenheim Museum, New York. (b) Planes by Colours, oil on canvas, 109 × 99.5 cm, 1911–1912, courtesy Musée National d'Art Moderne, Paris.

Plate 18. Lithograph El Lissitzky (1923), *New Man* (*Neuer*), lithograph on paper, 51 × 43 cm, 1923, with permission Tate Gallery.

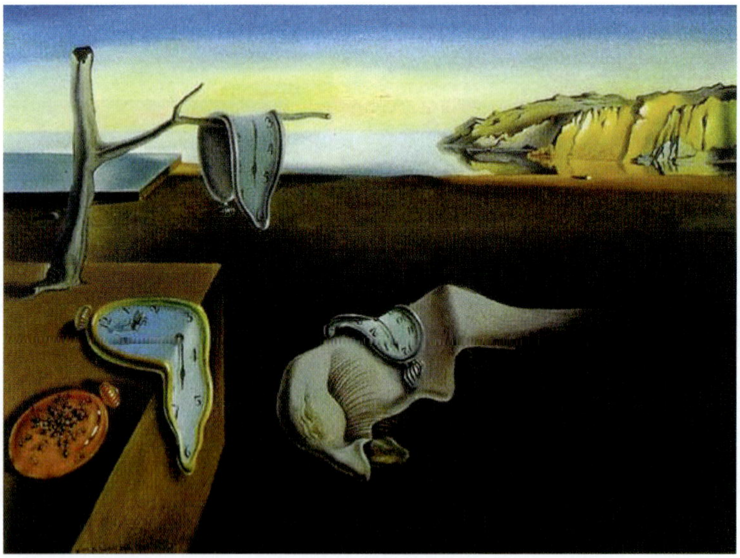

Plate 19. Salvador Dali, *Persistence of Memory*, oil on canvas, 24 × 33 cm, 1931, Museum of Modern Art, New York. © Salvador Dali, Fundació Gala-Salvador Dali/VEGAP.

Plate 20. Salvador Dali, *Corpus Hypercubicus*, oil on canvas, 194 × 123 cm, 1954, Metropolitan Museum of Art, New York. © Salvador Dali, Fundació Gala-Salvador Dali/VEGA.

Plate 21. Wolfgang Paalen, *Figure Pandynamique*, oil on canvas, 129 × 96 cm, 1940, private collection, courtesy Andreas Neufert.

Plate 22. Andy Warhol, *Moonwalk*, silk screen print, 96.5 × 96.5 cm, 1987, The Andy Warhol Museum, Pittsburgh. © Andy Warhol Foundation.

Plate 23. Matthew Ritchie, *M-Theory*, oil and marker on canvas, 208 × 279 cm, 2000, San Francisco Museum of Modern Art, courtesy Matthew Ritchie.

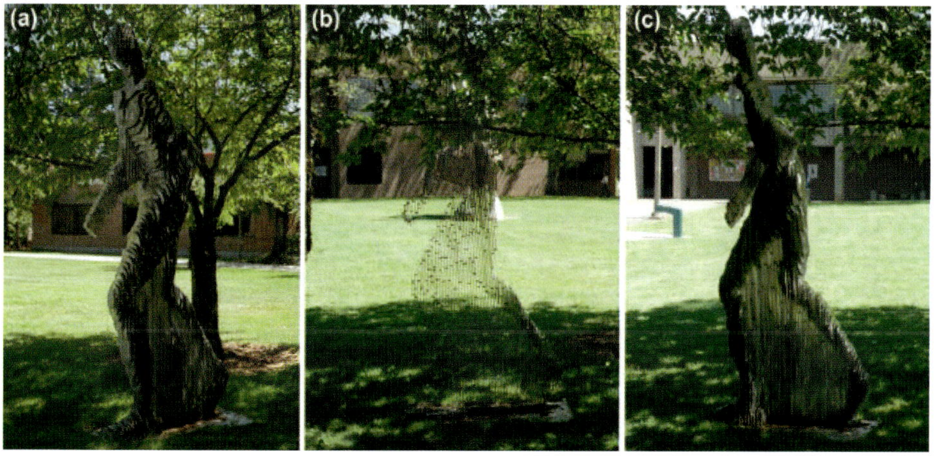

Plate 24. Julian Voss-Andreae, *Quantum Man*, stainless steel sculpture, 250 × 110 × 50 cm, 2007, private collection, Zurich, Switzerland, courtesy Julian Voss-Andreae.

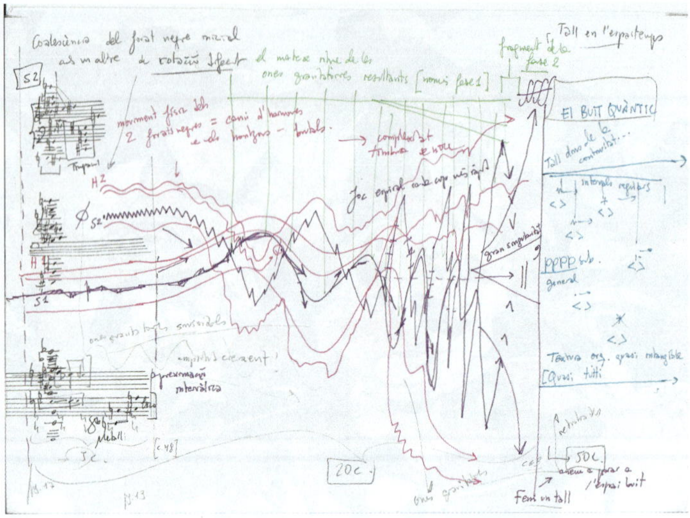

Plate 25. Hèctor Parra's sketch in preparation to writing the music to represent the collision of two black holes in *Caressant l'Horizon* first performed in 2011, courtesy Hèctor Parra.

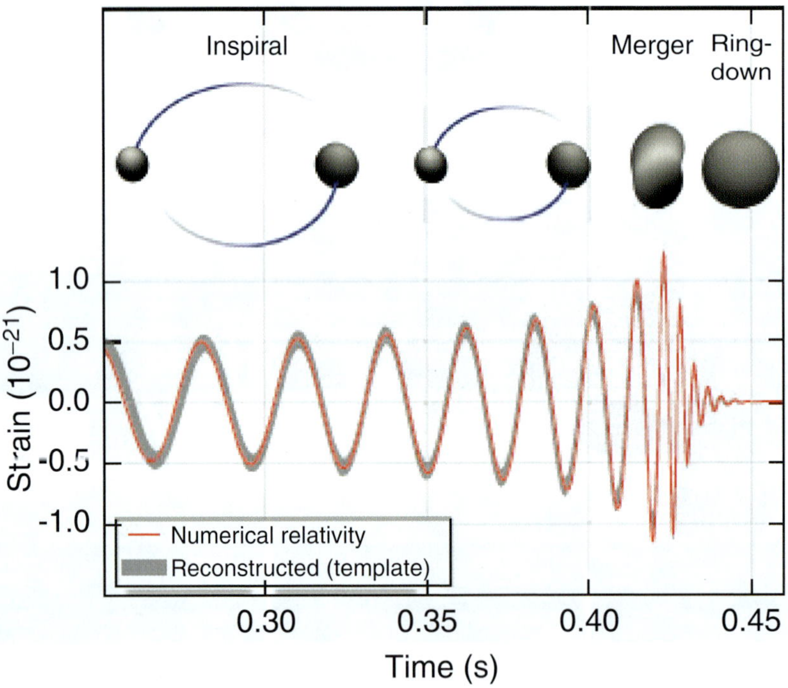

Plate 26. The modelling of the gravitational wave of spatial strain emanating from the inspiral, merger and ringdown of the first observed collision of two black holes in September 2015. Fig. 2 Abbott *et al.* (2016), courtesy American Physical Society.

Part I

Introduction

Chapter 1

Outline of the Physics

In this chapter the revolutions in physics and their main developments are outlined. There have been two revolutions in physics, which did not occur in isolation but were part of wider cultural revolutions. The first revolution in physics, which led to Newtonian physics,[1] was part of a cultural revolution that was also about the replacing of the scholastic tradition by Erasmian humanism, as has been well argued by Barbara Shapiro.[2] In the seventeenth century the new scientific ideas were shared by the educated in England and became an integral part of their culture.[3] The second cultural revolution, which began at the end of the nineteenth century, led to modernism and postmodernism. In physics, the revolution was caused by the realization that Newtonian physics, while it explained most physical phenomena on a macroscale, was not exact at either the incredibly small scale or on the very large cosmic scale. Pure physics, while an important influence on the general modernist and postmodernist culture has perhaps not played quite such large role in this second cultural revolution. However, if pure physics is combined with the applied physics of modern technology, its influence on culture has been even greater. However the effect of technology on culture, which has always been important,[4] is not considered here.

The fundamental principles of classical physics were established by Isaac Newton (1643–1727) in his epoch-making *Philosophiae Naturali Principia Mathematica*,[5] commonly known as the *Principia*, first published in 1687 and in his *Opticks*[6] published in 1704. Newtonian physics has been continually evolving since then. The world of Newtonian physics is mechanical. Every effect has its cause and if everything was known at a certain date, in theory, everything after then could be predicted. By the end of the nineteenth century it was clear that some predictions of Newtonian physics were not correct. Modern physics, which is not mechanical, is not completely deterministic.

Modern physics has similarities to the modernist movement in the arts that also began in the late nineteenth century. Both modernisms caused the certainty of the Enlightenment to be at least partially rejected. The main developments in modern physics have been the theory of relativity on the cosmic scale, which is indelibly linked to Albert Einstein (1879–1955), and quantum mechanics on the incredibly small scale.

In the late twentieth century, physics has seen the development of string theory whose foundation is not experimental and has similar characteristics to postmodernism in the arts.[7] The fundamental particles in string theory, which are point particles in quantum theory, have extension albeit extremely small. Gabriele Veneziano is considered to have written the first paper on string theory in 1968 but its origins, in the concept that some at least of the fundamental particles had extension, go back to 1943 and the S-matrix of quantum physicist Werner Heisenberg (1901–1976).[8] The postmodernism of string theory, which has not become fully accepted, has not yet caused a revolution in physics.

1.1. Newtonian Physics

Newtonian physics was the culmination of the first scientific revolution but it did not start from a vacuum. As Newton recognized himself in a letter to Robert Hooke (1635–1703) written in 1676:[9] 'If I have seen further it is by standing on the shoulders of Giants.' Newton's Giants extended back to the Ancient Greeks and even earlier. Nicolaus Copernicus (1473–1543), Johannes Kepler (1571–1630), and Galileo Galilei (1564–1642) led the revolution in science that enabled Newton to synthesize terrestrial and celestial mechanics. The *Principia* consists of three books, the first deals with forces and motion in free space, the second expands the treatment of motion to resisting media and the third book applies the first two books to the *System of the World.* At first Newton tells in the introduction to Book 3 he had originally intended to present this last book in a popular style but decided later that it would lack clarity and lead to misunderstanding and so wrote it in the same mathematical style as the first two books. However he told William Durham (1657–1735), a natural philosopher and rector at Upminister, that he made it difficult 'to avoid being baited by little Smatterers in Mathematicks'.[10]

1.1.1. *Nicolaus Copernicus* (1473–1543)

In the beginning of the sixteenth century the cosmology was that of Claudius Ptolemaeus (ca. 90–168), known in English as Ptolemy, who lived in Egypt, which was then part of the Roman Empire, and died in Alexandria. The Earth was at the centre of the universe in Ptolemy's astronomy. The fixed stars revolved around the Earth on perfect circular orbits, but the Sun and the planets moved relative to the fixed stars and could not follow exact circular orbits. In particular the superior planets like Jupiter or Saturn, which are further from the Sun than Earth, seem at times to have retrograde motion and at some periods of the year appear to move westwards relative to the stars instead of in the usual easterly direction. To account for this apparent motion, Ptolemy had the planets and the Sun follow epicyclic paths (a curve traced by a point on a circle which rolls around another circle). Even then the centre of the main circle had to be displaced from the centre of the Earth to account for the known apparent motion of the planets and the Sun. Copernicus placed the Sun at the centre of the universe and had the Earth and the planets revolve around it. He assumed that the Earth moved uniformly on a circular orbit.[11] The planets rotated around the Sun on circular orbits modified by epicycles, not to account for their retrograde motion but in order that their paths could reasonably accurately predicted since he believed, like the Greeks, the orbits had to be based on the circle because that was the perfect figure. The uniform apparent motion of the stars was explained by their being at an immense distance from the solar system. We now know that the orbits of the planets are near elliptical but, apart from Mercury the planet nearest the Sun, the orbits of the now recognized eight planets are actually not far from circular. Although Copernicus had basically finished his book, *De Revolutionibus Orbium Coelestium*, by about 1532, he did not publish it until just before his death only seeing the first copy on the day that he died. The book was dedicated to Pope Paul III and escaped official Catholic condemnation until the time of Galileo despite Scripture's central position for the Earth as expressed in Psalm 93:1 'the world ... is stablished, that it cannot be moved', because Copernicus was careful to include statements that the theory was only put forward as a hypothesis. The shifting the Earth from its central position was essential for astronomy, but Copernicus was not the first to suggest that the Earth and the planets revolved around the Sun. Anaxagoras of Samos (ca. 310–230 BC) had previously advanced the idea of heliocentricity, but his idea did not survive very long.

1.1.2. *Johannes Kepler* (1571–1630)

The German astronomer, Johannes Kepler, who established the orbits of the planets, was the next important figure to produce the ground work necessary for Newton. Kepler was the first major astronomer to adopt the heliocentric theory of Copernicus. His work on the orbits of the planets used not his own observations but those of Tycho Brahe (1546–1601) who, without the aid of a telescope, made systematic observations of the positions of the stars and planets with a median error of only 1.5 minutes of arc. However, Tycho Brahe, himself had a hybrid version of the universe where the Sun and the Moon revolved around the Earth, but the planets revolved around the Sun. Kepler briefly worked as an assistant of Brahe in the year before Brahe's death. Even before Kepler started to work for Brahe he had some ideas about the motion of the planets. Brahe would only let Kepler have access to his data on the orbit of Mars and Kepler used this data to formulate the first two of what are now known as Kepler's laws for Mars: that it revolves around the Sun on an elliptical orbit sweeping out equal areas in equal time, which were published in the *Astronomia Nova* in 1609. Kepler published his third law: that the square of the period of revolution of a planet is proportional to the cube of its mean distance from the Sun, in the *Harmonices Mundi*, published in 1619. Finally in the *Epitome Astronomia Copernicanae* completed in 1621 Kepler published all three laws generalizing the first two for all planets. There is some controversy about how widely Kepler's work was appreciated before Newton used it in his *Principia.* The young English astronomer Jeremiah Horrocks (1618–1641) certainly embraced Kepler's work and showed that the Moon's orbit was elliptical. He also predicted the transit of the Sun by Venus in his *Ephemerides* (table of astronomical positions) for the years 1629–1639,[12] Kepler had predicted the transit of the Sun by Mercury in 1631 but died before he could observe it. Kepler also predicted a transit of Venus in the same year.[13]

1.1.3. *Galileo Galilei* (1564–1642)

Galileo is one of the greatest of the founders of classical physics. Perhaps his best legacy was the understanding of the importance of acceleration in dynamics. His two major works were the *Dialogue Concerning the Two Chief World Systems*, published in 1632 and *Discourse Concerning the Two New Sciences*, published in 1638.[14] Both books were written in the form of a conversation over four days between Salviati, who had the role of the neutral observer, Sagredo, who presented the modern view of Galileo, and Simplicio, who presented the traditional view of the Ptolemy and the Aristotelians.[15] In the *Dialogue*, Galileo presented both the Ptolemaic and the Copernican systems of the Universe. Although Kepler had shown that the all the planets had elliptical orbits more than ten years before Galileo wrote the *Dialogue*, the orbits of the planets are still described as circular. Catholic opposition to the heliocentric theory arose in 1612. Galileo had been careful to avoid public statements in support of the Copernican system, but his views were well known. The Dominican Tommaso Caccini (1574–1648) denounced Galileo's opinions in 1614, from the pulpit of the Basilica of Santa Maria Novella in Florence. On the 24th of February 1616, the Holy Office of the Inquisition gave a report on the heliocentricity and found that 'this proposition is foolish and absurd in philosophy, and formally heretical since it explicitly contradicts in many places the sense of Holy Scripture.' Two days later Galileo was called before Cardinal Roberto Bellarmine (1542–1621) and he was warned that heliocentricity was erroneous and should be abandoned. Galileo acquiesced with this injunction and promised to obey. A year earlier Bellarmine had written a letter to the Carmelite, Paolo Foscarini (1565–1616), which seemed to imply that if 'a true demonstration that that the Sun is at the centre of the world and the Earth in the third heaven, and that the Sun does not circle the Earth but the Earth circles the Sun,' were found, then the Scriptures might be reinterpreted. However, Bellarmine also stated in the letter that he did not believe such a demonstration could be found because Solomon, who spoke inspired by God, had said 'the Sun also ariseth, and the Sun goeth down, and hasteth to his place where he arose.' Nevertheless Galileo may have thought that if he could give a true demonstration of heliocentricity then he might be allowed to publish his book. Galileo seems to have been convinced that a true demonstration could be obtained from an explanation of tides. It has been argued that Galileo's interest in tides, which he mistakenly believed were caused by the double motion of the Earth's diurnal rotation while it orbits the Sun, began in the mid-1590s.[16,17] Only a month before Bellarmine delivered the Holy Office's censure, Galileo wrote a long letter to Cardinal Alessandro Orsini (1592–1626) on his theory of the tides. In 1623 Cardinal Maffeo Barberini (1568–1644) was elected Pope taking the name Urban VIII. Barberini had been sympathetic to Galileo since 1610. In the year that Barberini became Pope, Galileo published *The Assayer*, dedicated to Pope Urban, which was a polemic against the treatise on the comets of 1618 by Orazio Grassi (1583–1654), a mathematician at the Jesuit Collegio Romano. Following publication of the book, Barberini became particularly friendly to Galileo and invited him to six interviews declaring him to be the Pope's beloved son. During these meetings Galileo was given permission to write about the Copernican theory, as long as he treated it as a hypothesis and so he started work on his *Dialogue*. The first three days in *The Dialogue* are devoted to a discussion of the relative merits of the Ptolemic and Copernican systems of the Universe, with the final day devoted to the theory of tides. The original title for his *Dialogue*, when it was presented to the Inquisition for approval in 1632, was *Dialogue on the Ebb and Flow of the Sea*. The Inquisition, not wishing to give approval to Galileo's theory of the tides which was the subject of the fourth day, made him change the title and add a disclaimer at the end of the book. Galileo chose Simplico to voice the disclaimer 'it would be an extravagant boldness for anyone to limit and confine the Divine.' Since Simplicio presented the traditional Ptolemaic view of the World and his views were weak compared with the Copernican view presented by Salviati. Pope Urban believed that divine omnipotence was still challenged and Galileo was called to Rome to defend his writings. In 1633 Galileo was found guilty of heresy and Galileo was required to recant; *The Dialogue* was banned, and publication of any future works was forbidden. Thus Galileo turned to the less-controversial study of the strength of materials and mechanics described in *The Discourse*. Since Galileo was banned by the Holy Office from publishing any books, *The Discourse* was published by Elsevir[18] in the Protestant Netherlands where the writ of the Holy Office was of no concern.

Galileo anticipated the first of Newton's three laws of motion, the principle of inertia, or as Newton expressed it:[5]

> 'Every body perseveres in its state of rest, or of moving uniformly straight forward, except insofar as it is compelled to change its state by forces impressed.'

He also recognized that force determines acceleration and came very near to formulating the second law of Newton: 'The change in motion is proportional to the motive force impressed.'[5] Newton's unique contribution to the laws of motion is his third one: 'To any action there is always an opposite and equal reaction.'[19] One of Galileo's major contributions was the understanding of terrestrial gravity. Galileo flirted in the early 1590s with the notion that the speed of fall of bodies is proportional to their specific gravities, but by at least 1604 he had reasoned that all bodies, regardless of their specific or actual weight, suffered constant acceleration if they fell in a vacuum and that the distance fallen is proportional to the square of the time taken.[20] However, an experimental proof of this observation was not possible until the air pump was invented in 1650 by Otto von Guericke (1602–1686). Galileo's style of the writing makes it difficult to decide how much of his mechanics was based on actual rather than thought experiments. Most of Galileo's manuscripts were published in twenty volumes by Antonio Favaro (1847–1922) over twenty years starting in 1890 but some manuscripts were not published by Favaro, since they contained only calculations or diagrams without attendant propositions or explanations. These extra manuscripts have shown that Galileo did perform extensive mechanical experiments.[21] It is unlikely that Galileo performed experiments where he dropped weights and measured the time of their fall and instead described an experiment where balls were rolled down a slope where the distance rolled in a given time is increased. Galileo stated that the velocity of the ball rolling down a slope of given vertical height is independent of the inclination of the slope and the same as that obtained in free fall, which is nearly but not quite correct. It is true that if the ball rolls down the slope without slipping, then the velocity is independent of the slope but, because the ball gathers rotational energy, its velocity is less that that during free fall. If the acceleration due to gravity is calculated according to Galileo's experiment then it would be expected to be underestimated by about 29%. Galileo does not actually give the acceleration due to gravity but, from what he wrote in *The Dialogue*, it can be inferred that his estimate was around 450 m/s^2 as compared with the actual value of 981 m/s^2, that is much less that the 29% underestimate that would be expected. What is puzzling is that Galileo, added to his own copy of the first edition, additional data which would imply that this value was about 960 m/s^2 and close to the real value which has led to an argument as to whether he actually timed free fall.[22] Whichever way he measured it, Galileo's analysis of free fall is magnificent and an enormous step forward in mechanics. Galileo described free fall as a natural motion, in Book 4 of *The Discourse* he went on to discuss the violent motion of projectiles which he had studied because of its practical importance to his patron, Cosimo II de Medici (1590–1621) the Duke of Tuscany. Before Galileo's *Discourse*, it was thought that a projectile fired horizontally would move horizontally for a while and then suddenly fall vertically. Galileo showed that the trajectory of a projectile, in the absence of air resistance, was parabolic.

1.1.4. *Isaac Newton* (1643–1727)

Newton marks the emergence of classical physics. He built on the work of Copernicus, Kepler and Galileo and many others of the seventeenth century such as Christiaan Huygens (1629–1695). What Newton bought to science was a new methodology and, as the full title of the *Principia* makes plain, the use of mathematics to deduce motions from general mathematical principles. Rene Descartes (1596–1650) wrote the *Principia Philosophiae* published in 1644 and Newton deliberately chose his title, *Philosophiae Naturali Principia Mathematica*, to reflect that his *Principia* was different. Newton wrote about natural philosophy, what we would now call physics, not general philosophy and emphasized that its principles were mathematical. What Newton strongly objected to in the *Principia Philosophiae* was Descartes' use of unsubstantiated hypotheses. Descartes was the founder of modern philosophy and his mathematical work was also very important but his scientific work was not as important. In the General Scholium, which Newton added to the second edition of the *Principia* in 1713, he made plain his methodology:[5]

> 'I do not feign hypotheses. For whatever is not deduced from the phenomena must be called a hypothesis; and hypotheses, whether metaphysical or physical, or based on occult qualities,

or mechanical, have no place in experimental philosophy. In this experimental philosophy, propositions are deduced from the phenomena and are made general by induction.'

It is application of this principle and his great mathematical ability that sets Newton apart from the other natural philosophers of the seventeenth century. The *Principia* was written in Latin, as was customary then, and an English translation by Andrew Motte (1696–1734), a one time lecturer at Gresham College, was not published until 1729. The definitive translation of the 1726 third edition is by Bernard Cohen (1914–2003) and Anne Whitman (1937–1984).[5]

Newton started the *Principia* with a series of definitions. The first definition is for the 'quantity of matter' or mass, which is defined as 'a measure of matter that arises from its density and volume.'[5] The second definition is for the quantity of motion, which is defined as 'a measure of motion that arises from the velocity and the quantity of matter jointly,' or the mass times the velocity.[5] Descartes had introduced the concept of quantity of motion, which he believed was conserved in his *Principia Philosophiae* where he wrote: [23]

'God imparted various motions to parts of matter when he first created them, and He now preserves all this matter in the same way ... and it follows [that it is] most reasonable to think that God likewise always preserves the same quantity of motion in matter.'

However it is clear from an application of his principle to colliding bodies that Descartes' quantity of motion was the product of the mass and the magnitude of the velocity and a scalar quantity. Newton in the *Principia* did not make his definition of the quantity of motion explicit, but in the Scholium to his Laws of Motion he referred to the work of John Wallis (1616–1703), Christopher Wren (1632–1723), and Christiaan Huygens published in 1688 on experimental impact between elastic and non elastic bodies. In these experiments two pendulums of equal length were used with impact at the lowest point so that on impact the velocities were collinear and they correctly took account of the direction of impact.[5] However, though their quantities of motion (scalar mass m times the velocity vector $\mathbf{v}$ or $m\mathbf{v}$) were vector quantities of motion or momentum their vectorial nature was not fully appreciated. Newton in query 31 to the second edition of the *Opticks* published in 1718 definitely used a scalar interpretation in analysing the quantity of motion in a simple example of two equal masses at the end of a light rod which was both rotating and translating writing:[6]

'tis very certain that there is not always the same quantity of Motion in the World. For if two Globes joined by a slender Rod revolve about their common Center of Gravity with an uniform Motion, while that Center moves on in a right Line drawn in the Plane of their circular Motion; the Sum of the Motions of the two Globes, as often as the Globes are in the right Line described by their common Center of Gravity, will be bigger than the Sum of their

> Motions when they are in a Line perpendicular to that right Line. By this Instance it appears that Motion may be got or lost. But ... Motion is much more apt to be lost than got, and is upon the Decay.'

If the quantity of motion, or momentum, is correctly interpreted as a vector quantity then it is easy to see that the quantity of motion is preserved. The German philosopher and mathematician, Gottfried Leibniz (1646–1716), entered the argument in 1686 with a short paper known as the *Brevis Demonstration*, in which he advanced his concept of *vis viva*, or living force, which is equivalent to the modern kinetic energy ($mv^2/2$), over Descartes' quantity of motion, starting a controversy that lasted over fifty years. The controversy was clarified by Roger Boscovich (1711–1787) in 1745 but only published by Jean-Baptiste d'Alembert (1717–1783) in 1758; both showed that *vis viva* and momentum were equally valid.[24] However discussion of *vis viva* will be left until Chapter 4.

In the *Principia* Newton followed his definitions with his famous three laws of motion but his greatest achievement was the realization that the force of gravity that caused an apple to fall from a tree on Earth was the same force that kept the planets in their orbits. According to William Stukeley (1687–1765) Newton told him in 1726 the story that that he came to the realization of universal gravity in 1665 through the fall of an apple from a tree in the garden of his family's home in the hamlet of Woolsthorpe in Lincolnshire when Cambridge University was closed because of the plague.[25] Most commentators have believed that this story is a myth but to me it has the ring of truth.[26] Descartes believed that the planets were swept around the Sun by a vortex in the aether. Newton seems to at first believed Descartes explanation at least until 1680 when he thought that a comet seen that year must have gone behind the Sun so that it was swept in the same direction as the planets by the vortex.[27] The first real evidence that Newton had begun to formulate a law of universal gravity comes in his manuscript *De Motu* which he sent to Edmund Halley (1656–1742) in 1684.[28] Halley had visited Newton earlier in that year to see whether he could find the planetary orbit that would be produced by a central force that was inversely proportional to the square of the distance to the planet. Newton said that he knew the answer to be an ellipse but he could not find the proof among his papers. The result was that Newton wrote *De Motu*, which eventually evolved into the *Principia*. It has been suggested that Halley's Comet of 1682 whose path was in the opposite direction to that of the 1680 comet and hence against the flow of any vortex may have been the key to Newton completely abandoning Descartes' vortex hypothesis.[27] Perhaps the late date of Newton's realization that Descartes' hypotheses were not well founded accounts for some of his overt hostility to

Cartesian ideas in the *Principia*. Newton gave his *System of the World* in Book 3 based on the law, which he stated in Book 1 Proposition 75:[5]

> 'Every point mass attracts every single other point mass by a force pointing along the line intersecting both points. The force is directly proportional to the product of the two masses and inversely proportional to the square of the distance between the point masses.'

Today Newton's the force of gravity, F_G, is expressed by the equation:

$$F_G = G\frac{m_1 m_2}{r^2}, \tag{1.1}$$

where $G = 6.67 \times 10^{-11}$ m^3 kg^{-1} s^2 is the universal gravitational constant, m_1, m_2 are the two point masses, and r is the distance between them. Newton also showed that if the masses were spherical, then the same law applied if r was the distance between their centres. The inverse square law for the force of gravity had been advanced by others, including Robert Hooke (1635–1703), in the late 1660s, but it was Newton who showed that an isolated planet's orbit about a Sun under this law was an ellipse. The orbits of the four inner planets, Mercury, Venus, Earth, and Mars are very near to ellipses as observed by Kepler. However, Newton knew that there was significant deviation from an elliptical orbit for the two largest planets, Jupiter and Saturn which, because of their immense mass, exerted a significant force compared to that of the Sun on each other and distorted the their orbits from a true ellipse.

Newton made it plain in the *General Scholium* to the *Principia* that he had not assigned a cause to gravity, writing:[5]

> 'I have not as yet been able to deduce from phenomena the reason for these properties of gravity... It is enough that gravity really exists and acts according to the laws that we have set forth and is sufficient to explain all the motions of the heavenly bodies and of our sea.'

The strength of Newton's approach to natural phenomena was that he explained the mechanics of the phenomena but did not attempt to explain the cause. Descartes tried to understand the cause of phenomena, which is much more problematic, and consequently made little headway in describing the mechanics of the heavenly bodies. As mathematics developed more and more mechanical problems both terrestrial and celestial have been solved using the principles given in the *Principia*. Newtonian mechanics can be used to predict the motion of bodies at scales from less than a micron to that of the solar system with, as Albert Einstein put it, 'a delicacy of detail little short of wonderful.' Only incredibly minute deviations from Newtonian mechanics can be found within the solar system such as the advance in the perihelion[29] of Mercury by 43 seconds of arc per century. A problem in Newton's theory of gravity, with which he was not aware, was that it implied instant action whereas nothing not even gravity can propagate faster than the speed of light. It takes about eight minutes for light to travel from the Sun to the Earth as Newton himself knew. If the Sun were suddenly obliterated we would not feel nor see the obliteration until after about eight minutes.

Newton's second most important publication was his *Opticks*, which he first published in 1704. Unlike the *Principia*, *Opticks* was first published in English; it was translated into Latin for the continental market in 1706 by Samuel Clarke (1675–1729). *Opticks* had its beginnings in a series of lectures to the Royal Society during the 1670s to which Newton added most of the rest of its first two books after the publication of the *Principia.* Newton delayed publication of *Opticks* until after the death of Hooke in 1703 because of his quarrel with him over the nature of light, which Newton thought was corpuscular but Hooke believed to be a wave. Because Newton had protested to Hooke he was making no hypotheses, he had to resort to the subterfuge of adding to the end of the book hypotheses thinly disguised as queries stating: 'I shall conclude with proposing only some queries in order to [facilitate] a farther search to be made by others.' In the 1706 Latin edition, Newton added two important queries, which in the second 1718 edition became queries 28 and 31. In query 28 the Cartesian notion that an aetheral medium fills the space between particles is attacked, Newton preferred an atomistic view and believed that all substances were comprised of hard impenetrable particles moving and interacting through forces at a distance. Newton expanded these ideas in query 31 which also served as the basis of the *General Scholium* of the second edition of the *Principia*, the philosophical concerns of which may be partly due to Clarke.[30]

The extraordinary scientific advances of the seventeenth century saw a complete change in the outlook of most educated people. Witch trials were common in England at the beginning of the century but after 1684 nobody was executed for witchcraft and, after the early eighteenth century, witch trials came to an end in Britain.[31] Comets were portents in Shakespeare's time, but Newton and Edmund Halley calculated the orbits of comets and showed that they obeyed the same law as planets. From the seventeenth century onwards physics began to have a significant influence on people's thought in general, though it can be argued that now in the twenty first century that this influence is unfortunately in decline despite the great technological advances due to physics.[32] Newton lectured in Cambridge on mechanics between 1684 and 1687. His lectures, like the *Principia*, were difficult to understand by anyone even university students and generally had to be explained by others. The clergyman William Whiston

(1667–1752), who attended 'one or two' of Newton's lectures, later admitted that he understood 'not at all at the time'.[33] Whiston became a senior fellow of Clare College[34] in 1693 and took up Newtonian mechanics after reading a paper by the Scottish mathematician David Gregory (1659–1708) who gave 'the most prodigious commendations' to Newton's *Principia* 'as not only right in all things, but in a manner the effect of a plainly divine genius.'[33] Gregory taught Newtonian mechanics to undergraduates at Edinburgh University, and after 1691, at Oxford, while Cambridge students were still studying Cartesian mechanics. Mathematics was a neglected subject in universities at the end of the seventeenth century and Gregory's mathematical presentation of Newtonian mechanics was not well understood. John Keill (1671–1721), who was Gregory's student at Edinburgh and followed him to Oxford, started a more easily understood course in experimental mechanics probably in 1694.[35] Whiston first met Newton in 1694 and the two became friends. Newton became Warden of the Royal Mint in 1696 and Master in 1699. These positions were meant as sinecures but Newton took them seriously and in 1701 resigned all his positions in Cambridge. Whiston became Newton's deputy Lucasian Professor and was given the full Lucasian Professorship in 1702. Although Whiston produced no original works on mechanics, he was important in spreading knowledge of Newtonian mechanics to a wide audience both in Cambridge and London. Between 1701 and 1703, Whiston delivered thirty one lectures on astronomy which were an introduction to Book 3 of the *Principia*; these lectures were published in Latin in 1707 and in English in 1715.[36] Whiston followed up the lectures on astronomy with another series between 1704 and 1708 based on a simplified version of the *Principia*; he published a Latin version of these lectures in 1710 followed by an English version in 1716.[37]

1.1.5. *Electromagnetic force*

During the second half of the eighteenth century natural philosophers began to look for a law describing the force between two electrically charged bodies and since gravity was a long range force it was assumed that the force would have a similar behaviour. Some of the first experiments were made about 1760 by Daniel Bernoulli (1700–1782), a member of the famous family of Swiss mathematicians and natural philosophers, he suspected that electric charge followed the inverse square law but little is known of his experiments. Joseph Priestley (1733–1804), a natural philosopher and a dissenting minister,[38] having been told by the American Benjamin Franklin (1716–1804) when he was in England that there was no electric charge within a charged metal cup, repeated

Franklin's experiment in 1766 and inferred that an electric forced followed the inverse square law writing:[39]

> 'May we not infer from this experiment, that the attraction of electricity is subject to the same laws with that of gravitation, and is therefore according to the squares of the distances; since it is easily demonstrated, that were the Earth in the form of a shell, a body in the inside of it would not be attracted to one side more than the other?'

However, it was not until 1785 that the forces were actually measured by Charles-Augustin de Coulomb (1736–1806), a French natural philosopher. The small electrostatic forces were measured by Coulomb using a torsion balance independently invented by him in 1777.[40] In his paper he stated that the repulsive force between two similarly charged spheres is inversely proportional to the square of the distance between the spheres. The modern form of Coulomb's law is

$$F_e = k_e \frac{q_1 q_2}{r^2}, \tag{1.2}$$

where the two point charges are q_1 and q_2, r is the distance between them, and k_e is Coulomb's constant.

It is no coincidence that in Coulomb's law, like Newton's law of gravity, the force is inversely proportional to the square of the distance. It is a consequence of the three-dimensional nature of space. Immanuel Kant (1724–1804) linked the nature of gravity's force to the dimensionality of space in his first published work in 1746–1749 stating that:[41]

> 'The three-dimensional character seems to derive from the fact that substances in the existing world act on each other in such a way that the strength of the action is inversely proportionate to the square of the distances....God could have chosen another, e.g., the inverse-cube, relation; fourthly, and finally, that an extension with different properties and dimensions would also have resulted from a different law.'

Although Kant gave no details of the lines of his thought he probably envisioned gravity, like Michael Faraday (1791–1867) did electric force, as a flux whose density decreased as the reciprocal of the square of distance because the area of a sphere about a point charge is $4\pi r^2$ so that the density is proportional to $1/4\pi r^2$ in fact in Eq. (1.2). k_e is often replaced by $1/(4\pi\varepsilon)$ where ε is called the permittivity of the medium in which the charges are immersed.

1.1.6. *Thermodynamics*

Thermodynamics, the branch of physics concerned with heat and its relation to energy and work, began with Guericke and Boyle though the name itself was not coined until the 1840s and was first used in 1854 by William Thomson (1824–1907) who was raised to the peerage with the title of Baron Kelvin in 1892. In 1660 Robert Boyle (1627–1691) wrote an account of his work with Guericke's air pump in which he stated his famous gas law[42] that the volume of a gas varies inversely with pressure that was the first stage in formulating the ideal gas law.[43] The second gas law is attributed to Jacques Charles (1746–1823), who in the late 1780s did unpublished experiments that showed that the volume of a gas at constant pressure increased linearly with temperature. The results were first published by Joseph Louis Gay-Lussac (1778–1850) in 1802 who gave credit to Charles. From his experiments on dry air Gay-Lussac calculated that absolute zero temperature at which the volume became zero to be −266.66°C, which compares well with the modern value of −273.16°C.[44] Finally in 1834 Émile Clapeyron (1799–1864) put Boyle's and Charles' laws[45] together to obtain the ideal gas law, which in modern notation is given as

$$PV = nRT, \tag{1.3}$$

where P and V are the pressure and volume of the gas, n is the number of moles,[46] R is the universal gas constant,[47] and T is the temperature in degrees Celsius measured from absolute zero, which is the modern absolute temperature and given the name kelvin in honour of Lord Kelvin. Guillaume Amontons (1663–1738) is usually attributed has having inspired the idea of measuring temperature from absolute zero,[48] but it was Thomson in a brief paper *On an Absolute Thermometric Scale* published in 1848 who first discussed how it could be obtained from the ideal gas law and be independent of the substance.[49]

The two thermodynamic laws along with the unified field theory of electromagnetism of James Clerk Maxwell (1831–1879), the Scottish theoretical physicist and mathematician, mark the beginning of the transition from classical to modern physics

1.1.6.1. *The first law of thermodynamics*

Mechanical energy is only conserved in ideal situations consequently Newton did not formulate the principle of the conservation of mechanical energy because in practice it is never achieved. In impact, mechanical energy is not conserved and this non-conservation led to the disagreement between Leibniz and his *vis viva*, or kinetic energy, and Newton with his momentum which is conserved in an impact. The main cause of the non-conservation of mechanical energy is friction. What became apparent in the nineteenth century, from the work of Julius Robert von Mayer (1814–1878), James Prescott Joule (1818–1889), Ludwig August Colding (1815–1888), and Herman von Helmholtz (1821–1894) was that the mechanical energy lost in friction reappeared as heat which is simply a different form of energy. Mayer published the first paper on the conservation of energy in 1842[50] but his work was not well known in the 1840s and it was the publications of Joule beginning in 1843 that eventually led to the first law of thermodynamics that energy is conserved.[51] Although the polymath Thomas Young (1773–1829) had introduced the modern term *energy* in his 1802 lectures to the Royal Society, that overworked word of the English language *force* was still commonly used in the sense of *energy* until the second half of the nineteenth century as Peter Guthrie Tait (1831–1901) wrote in 1876:[52] 'Perhaps no scientific English word has been so much abused as the word force.'

Mayer was a German physician who was led to consider the conservation of energy while he was a ship's doctor on a Dutch ship on a voyage to Java in 1840. Bloodletting was common in the early nineteenth century and Mayer noticed that the blood of the sailors became redder as the ship approached the tropics. The redness of the blood indicted an excess of oxygen and led Mayer to conclude that in warm climates men consumed less oxygen and generated less heat though they performed the same mechanical work. This fact led Mayer to see a correspondence between heat and mechanical work. Mayer did not consider the conservation of energy of living things in his 1842 paper but in 1845 he published a memoir on *Organic Motion in its Connection with Nutrition* in which he applied the principle of the conservation to the vital processes in plants and animals. At the time the origin, growth and energies of living things was assigned to a special agent called *vital force*. Mayer proposed that, as far as energy was concerned, living and inorganic things were no different. In 1843 Joule published three papers on the calorific effects of magneto-electricity and the mechanical equivalent of heat, followed in 1845 by a report on the mechanical equivalent of heat obtained the classic paddlewheel experiment.[53] However it was not until his paper *On Matter Living Force and Heat* delivered to the British Association at Oxford in 1847, did Joule clearly state the principle of the conservation of energy.[54] The philosophical section of Mayer's 1842 paper and that of Joule's 1847 paper are similar. Mayer argued that in a chain of causes and effects no effect can become nothing and energy is convertible and indestructible, Joule reasoned that:[55]

> 'absolute destruction of living force [energy] cannot possibly take place, because it is manifestly absurd to suppose that the powers with which God has endowed matter can be destroyed ... We have reason to believe that the manifestations of living force [energy] on our globe are, at the present time, as extensive as those which have existed at any time since its creation, or, at any rate, since the deluge'

No details were given by Joule on the measurement of the mechanical equivalent of heat but he favoured measuring the rise in temperature induced in water by paddles in his 1847 paper. However Mayer gave a neat method using the specific heats of a gas at constant pressure and constant volume to obtain the mechanical equivalent of heat.

The work of the Danish engineer Colding published in Danish during the 1840s was also not well known. His first treatise *Theses concerning Force* was presented to the Royal Society of Copenhagen in 1843 and was followed by more results presented to the Danish Association of Natural Philosophers in 1847.[56] Colding in his 1843 paper called the conservation of energy 'the principle of the perpetuity of energy'.[56]

Helmholtz's main role in the formulation of the principle of the conservation of energy was showing the generality of the principle in a masterful paper *On the Conservation of Force* published in 1847.[57] During the 1870s the Royal Society awarded the Copley Medal to Joule (1870), Mayer (1871), and Helmholtz (1873) for their work on the principle of the conservation of energy.

In 1850 the German physicist and mathematician Rudolf Clausius (1822–1888) put the principle of the conservation of energy on a sound foundation[58] and traced its beginnings to the French military engineer Sadi Carnot (1796–1832), who gave the principles of the ideal heat engine in *Réflexions sur la Puissance Motrice du Feu* (*Reflections on the Motive Power of Fire*) published in 1824. Carnot had made use of the impossibility of perpetual motion to show that the work obtained from the ideal Carnot cycle was the greatest that could be obtained from a heat engine and thus in effect made use of the conservation of energy though he used the discredited caloric theory of heat.[59]

Prior to the development of modern physics, mass was thought to be conserved but it is now known that mass is another form of energy and now comes under the conservation of total energy. The *Principia* opens with the definition that 'the quantity of matter [mass] is a measure of matter that arises from its density and volume jointly' that, as Ernst Mach (1838–1916) pointed out, contains a circular argument.[60] Galileo's measure of matter was weight, but Newton knew that a body's weight varied depending upon its position and he wanted a measure of matter that was inherent to a body. Newton assumed, without explicit statement, that the mass of a body was constant. The law of mass conservation came much later from the study of chemical reactions and was first clearly stated by Antoine-Laurent de Lavoisier (1743–1794) in *Traité Élémentaire de Chemie* published in 1789.

The first law of thermodynamics that total energy is conserved first formulated in the 1840s is a fundamental law of nature, which will always remain valid, though virtual particles that exist for extremely short period sometimes need to be invoked to preserve it in modern physics. Its apparent violation of the first law in 1950 led to the discovery of the neutrino.

1.1.6.2. *The second law of thermodynamics*

As well as effectively using the first law of thermodynamics in his *Reflections on the Motive Power of Fire,* Carnot also tacitly assumed the second law when he showed that the work that could be extracted from an ideal engine depended on the difference in temperature between the heat source and the heat sink. To extract work from heat, a minimum of two heat sources are necessary one at a higher temperature than the other. Because heat cannot be completely prevented from dissipating, the efficiency of a real engine is always less than that of the ideal Carnot engine. The ideal Carnot engine is reversible, that is it can transfer the same heat from the lower source to the higher if same work is done on it as was extracted but real processes can only at best approximate to a reversible process. No work can be extracted from a single source of heat. Such a process would not violate the first law of thermodynamics but violates what became known as the second law. Although work cannot be obtained from a single source of heat, but heat can be obtained from the dissipative mechanical work that is called friction, which is an irreversible process.

The second law of thermodynamics was obtained independently by Rudolf Clausius and William Thomson in the early 1850s. Clausius introduced the name *entropy* and the symbol S in 1865 for the thermodynamic state variable with which the second law of thermodynamics can be defined mathematically.[61] For a reversible process the change in entropy dS in system that receives an infinitesimal heat input of dQ is given by[62]

$$dS = \frac{dQ}{T}. \tag{1.4}$$

For an irreversible process

$$dS > \frac{dQ}{T}. \tag{1.5}$$

The entropy of a real system always increases. There was no rigorous proof of the second law; it is based on experience. After considering finite isolated systems throughout his ninth memoir, Clausius took the great leap of applying his results to the universe as a whole and stated that:[61]

> '1 The energy of the universe is constant
> 2 The entropy of the universe tends to a maximum.'

Expressed in classic thermodynamics terms the second law of thermodynamics does not seem to be Earth shattering but they are perhaps the most important of the physical laws. Strictly the second law can only be applied to a macro-state but its implications when extended to the universe as a whole are that eventually it will reach a uniform temperature and no life will be possible.

The kinetic theory of gases, developed by Clausius and Maxwell in the late 1850s, led to an attempt in 1872 by the Austrian physicist Ludwig Boltzmann (1844–1906), one of the founders of statistical mechanics, to prove the second law. Boltzmann introduced the H-theorem. The function H for a dilute gas composed of particles is a function of the number of particles in a particular small volume, which have a particular momentum distribution, summed over all the possible positions and momenta. For a gas in equilibrium, Boltzmann showed that the entropy was proportional to H and suggested that this was also true for a gas not in equilibrium. Boltzmann's H-theorem was not seen as verifying the second law of thermodynamics but his work widened the meaning of entropy to statistical mechanics. This approach to the second law leads to a definition of entropy that implies that the disorder of a system increases with time. It also links entropy with the arrow of time popularized by Arthur Eddington (1882–1944) in his book *Nature of the Physical World* published in 1928. Eddington described time's arrow as only being 'determined by that incongruous mixture of theology and statistics known as the second law of thermodynamics.'[9] In Newtonian mechanics time can run equally well backwards as forwards, it is entropy that dictates it can only go forwards:[9]

> 'If as we follow the arrow we find more and more of the random element in the state of the world, then the arrow is pointing towards the future; if the random element decreases the arrow points towards the past.'

1.2. Modern Physics

The first evidence that physics was changing came with the gradual realization that there were mysterious invisible rays. Their discovery started with Michael

Faraday who performed experiments in the 1830s where he passed a current through a partially evacuated glass tube which created a glow around the cathode. These rays were named cathode rays in 1879 by the German physicist Eugen Goldstein (1850–1930), they are invisible but produce visible light when they collide with atoms. Cathode rays were shown by Joseph John Thomson (1856–1940), better known as simply J. J. Thomson, to be a stream of electrons[63] in 1897. Further discussion of invisible rays will be left until Chapter 7.

In the nineteenth century it was thought that light and electromagnetic waves needed to propagate in aether,[64] a subtle massless medium that permeated space, like sound waves, which propagate in air or another gas. Because the Earth is in motion around the Sun, it was expected that the flow of aether across the Earth should produce a detectable aether wind and that the speed of light would be dependent on the direction of propagation relative to that wind. It was known that the speed of light was of the order of 300×10^6 m/s so very accurate methods of measuring time differences were necessary to detect any changes due to the direction of propagation relative to any aether wind. Albert Michelson (1852–1931) designed an interferometer that could detect changes in the phase of light that had been split into two beams with a half silvered mirror. The beams of light travelled at right angles to each other along tubes of equal length and were then reflected back and combined to detect any interference fringes. The first instrument was not sensitive enough and Michelson collaborated with Edward Morley (1838–1923) to build a better instrument. With this new instrument they could detect about 1/100th of an interference fringe and Michelson and Morley showed in 1887 that the difference detected in the speed of light propagating in different directions was negligible, being within the limits of experimental error. Lasers have since enabled experiments similar to those of Michelson-Morley to be repeated with extreme accuracy with the same result. The Michelson-Morley experiments led to Einstein's special theory of relativity.

The idea that matter is composed of tiny particles or atoms that cannot be divided is as old as the Greek philosopher Democritus (ca. 460–370 BC). The English chemist and physicist John Dalton (1766–1848) put the concept on surer foundations in the early 1800s. To Dalton, atoms were quanta of matter and matter contained a great number of atoms.[65] In 1901 Max Planck (1858–1947) introduced a similar concept for energy to account for blackbody radiation where heat is radiated from a hot body in discrete quanta not continuously.[66] Although we now know that atoms are made up of much smaller fundamental quanta, these twin concepts of Dalton and Planck mark the beginning of the idea of quantum mechanics.

In 1928 Eddington gave a parable of two tables in the introduction in his popular book *Nature of the Physical World* on modern physics at both the atomic and the cosmic scale, which neatly demonstrated the difference between classical and modern physics.[9] Eddington's first table was an ordinary substantial table of deterministic Newtonian physics with which we are familiar in ordinary life; his second scientific table is that of modern physics and mostly empty space,

> 'scattered in that emptiness are numerous electric charges rushing about with great speed; but their combined bulk amounts to less than a billionth[67] of the bulk of the table itself. The second table supports writing paper just as satisfactorily since little electric particles with their headlong speed keep on hitting the underside, so that the paper is maintained in shuttlecock fashion at a nearly steady level.... [T]here seems to be nothing to choose between the two tables for ordinary purposes; but when abnormal circumstances befall, then my scientific table shows to advantage... Modern physics has by delicate test and remorseless logic assured me that my second ... table is the only one which is really there – wherever "there" may be. On the other hand ... modern physics will never succeed in exorcising that first table – strange compound of external nature, mental imagery and inherited prejudice – which lies visible to my eyes and tangible to my grasp.'

1.2.1. *The theory of relativity*

Although Einstein's paper on the theory of special relativity in 1905[68] marks the emergence of relativity it was not without precedents. As Newton wrote in the Scholium to the definitions in the *Principia*, 'although time, space, place, and motion are very familiar to everyone, it must be noted that these quantities are popularly conceived solely with reference to the objects of sense perception.' In classical Newtonian mechanics time and space are absolute and the laws of motion, as defined by Newton, hold in any inertial reference frame, which is any reference frame that is moving at constant speed in a straight line.[69] In physics, the principle of relativity is the requirement that the equations, describing the laws of physics, have the same form in all admissible frames of reference. Einstein in 1905 announced the *principle of relativity* stating that no experiment can distinguish one inertial frame from another and hence all physical laws must reflect this invariance. This concept can be traced back to Galileo's discussion of motion viewed both from the deck of a ship and from land, a consequence the transformation of the coordinates from one inertial frame of reference to another in Newtonian physics is called a *Galilean transformation*. Newton's laws of motion are invariant under a Galilean transformation which, if there are two inertial reference frames Σ and Σ' where the reference frame Σ' has a velocity, v, along the x axis, is given by

$$x' = x - vt, \tag{1.6}$$
$$t' = t.$$

Until the second half of the nineteenth century there was no conflict with observed phenomena for relativity based on a Galilean transformation.

1.2.1.1. *Special theory of relativity*

The special theory of relativity applies where there is only a very weak gravitational field or accelerating inertial reference frame.[70] Einstein, using the Michelson-Morley experiments, also saw that the velocity of light must be invariant of the inertial frame. Since the equations of motion are linear the transformation must also be linear in all inertial frames. With no loss in generality we can consider the linear transformation of a one dimensional space and time (x, t) from a reference frame Σ to Σ' whose velocity in the x direction is v. Under the Galilean transformation the velocity of light depends upon the speed of the observer and time is absolute, but in the special theory of relativity a general linear transformation becomes

$$\begin{aligned} x' &= \phi(v)[x - vt], \\ t' &= \psi(v)t + \chi(v)x, \end{aligned} \tag{1.7}$$

where ϕ, χ, and ψ are functions of the velocity. The necessary forms of these functions that satisfy a constant velocity of light, c, in any inertial frame of reference, and the requirement that the transformation must degenerate to the Galilean when the velocity v is small, can be found by simple algebra. The resulting transformation is known as the Lorentz transformation, after the Dutch physicist Hendrik Lorentz (1853–1928), who first published them in 1904 unbeknown by Einstein.[71] Symmetry dominates modern physics and, whereas the Galilean transformation given by Eq. (1.6) is asymmetrical, the Lorentzian transformation is symmetrical. This symmetry is enhanced when time is given the same dimensions as length by multiplying it by the speed of light giving $\tau = ct$; it is also convenient to express the velocity v by $\beta = v/c$, and the Lorentz transformation is given by

$$\begin{aligned} x' &= (x - \beta\tau)/\sqrt{1-\beta^2}, \\ \tau' &= (-\beta x + \tau)/\sqrt{1-\beta^2}. \end{aligned} \tag{1.8}$$

There is one crucial difference between Einstein's transformation and Lorentz's: whereas Lorentz assumed that the reference frame was at rest with respect to the aether, Einstein only considered the relative motion of two reference frames; the concept of the aether was superfluous.

1.2.1.2. *General theory of relativity*

The general theory of relativity did not come easy to Einstein; in 1912 he began working intensely on his gravitational theory and from then, until he published a detailed and self-contained exposition in 1916, the history is complex.[76] Einstein wrote to the German physicist Arnold Sommerfeld (1868–1951) that 'compared to the problem [of general relativity] the original [special relativity] theory is child's play.' No attempt will be made to present the mathematics involved in the general theory of relativity and the treatment is based on Bertrand Russell's (1872–1970) short popular book, *The ABC of Relativity*, first published in 1925, which is still one of the best non-mathematical approaches.[77]

A blindfolded stationary observer in an inertial frame of reference can know nothing about his velocity, but if the frame of reference is accelerating he feels a force. This is a common experience in a lift, when it starts or stops we feel an acceleration or deceleration force in our stomach, but once we are travelling at constant speed we cannot tell that we are moving. Einstein thought about this problem of an observer in a frame of reference undergoing constant acceleration and whether the observer could tell he was accelerating. In the lift we know we are accelerating because we know that nothing else could have happened. However, if we were suddenly transported blindfolded in our sleep to another much larger planet and then awoke, the gravitational force on us would be much larger than on Earth, would we know we were on a much larger planet or would we think we were in an accelerating lift or rocket on Earth? Einstein concluded that there was no way we could tell whether the increase in force was due to acceleration or to gravity and hence a field of uniform gravitation is physically equivalent to uniform acceleration. This result is known as the *equivalence principle*.

Our common experience is that we live in a three-dimensional space with time, a separate variable, marking events in that space and that is the basis for classical physics. However in Einstein's theory of relativity, space and time are inseparable; time becomes a fourth dimension. As in the special theory of relativity the equations of general relativity are more symmetric if the time dimension is multiplied by the speed of light so that all the coordinates have the dimension of space. Just as the Newtonian gravitational law was a consequence of Euclidean geometry, so too is the general theory of relativity a geometric property but of four dimensional non-Euclidean spacetime not three dimensional Euclidean space. In the presence of large gravitational fields near stars, or in non-inertial frames of reference with large accelerations, spacetime is warped or distorted. It is impossible to visualize the distortion of spacetime though this distortion can be described mathematically and we can only resort to analogy to understand it. We perceive ordinary objects in an undistorted three-dimensional Euclidian space but we have to resort to two-dimensions to understand how even space can be warped. At school we are taught two-dimensional plane geometry that was studied by Euclid where we draw our objects on flat paper. However a flat sheet is not the only possible form of a two-dimensional surface, the surface of the sheet can be curved and either open or closed.[78] The surface of a sphere is a two-dimensional closed surface. We define a straight line in a flat plane as the shortest distance between two points; the line defining the shortest distance between any two points on a curved surface are called a *geodesic*. On a spherical surface, like the Earth, the geodesics are great circles. Lines of longitude are great circles but, except for the equator, lines of latitude are not. A small area on the surface of a sphere is almost flat and the angles of a triangle on add up to 180°, but the angles of a large triangle, on a sphere or the Earth, formed by the intersection of the equator by lines of longitude 45° W and 45° E at the north pole add up to 270°. On a sphere the sum of the angles of a triangle vary from 180° for the smallest to 540° for the largest triangle. The two-dimensional geometry of any surface that cannot be formed from a sheet of paper, like a cylinder or cone can, is non-Euclidean. The Earth cannot be depicted on flat maps without some distortion or warping. The geometry of spacetime is warped and non-Euclidean. A small area of the Earth can be mapped without warping and is in the limit Euclidean, so too is the geometry of a small volume in three or more dimensions. What Einstein did in his general theory of relativity was to show that matter warps spacetime; the larger the mass the bigger is the warping. In the absence of non-gravitational forces, bodies travel on the geodesics in spacetime that take the longest times as measured by clocks moving with them, which Russell called the law of cosmic laziness.[77] A light ray travels on a geodesic, which means that rays of light are bent when passing near a massive body like the Sun. Warped spacetime is a special case of the geometry of multi-dimensional non-Euclidean surfaces, which the German mathematician Bernhard Riemann (1826–1866) had developed in 1854 in a purely mathematical thesis that then had no application to physics, but was applied by Einstein to general relativity. Physics cannot be completely understood without mathematics. Newton had to develop calculus, which he called the *method of fluxions*, to be able to write his *Principia*. The seventeenth century was as much a revolution in mathematics as science and mathematics has advanced as rapidly, or even faster, than science since then and it is uncanny that much of what was first developed in mathematics has found later application in physics like Riemann surfaces.

When Einstein presented his theory of general relativity in 1915 the general concept of the universe was much the same as that in the time of Newton. In the first of his series of four letters to Richard Bentley (1662–1742) in December 1692 Newton wrote that:

> 'if the matter of our sun and planets and all the matter in the universe was evenly scattered throughout all the heavens, and every particle had an innate gravity towards all the rest, and the whole space throughout which this matter was scattered was but finite: the matter on the outside of this space would, by its gravity: tend toward all the matter on the inside and, by consequence, fall down to the middle of the whole space and there compose one great spherical mass. But if the matter was evenly diffused throughout an infinite space, it would never convene into one mass; but some of it convene into one mass and some into another, so as to make an infinite number of great masses scattered at great distances from one another throughout all that infinite space.'

Before 1915 the universe was seen to be essential static as Newton had imagined. However, by the end of the nineteenth century it was realised that the extent of the universe could not be infinite because the pull on any given star would be infinite in all directions. The gravitational potential of a star is the scalar field generated and if the number of stars were infinite the gravitational potential could not be defined because it would be the difference of infinite quantities. Thus it was realized that the universe could not be infinite. To reconcile a finite number of stars and a steady state universe, Carl Neumann (1832–1925) in 1896 introduced a cosmological constant into the equation for the Newtonian gravitational potential whose effect was to balance the attractive force of gravity with a repulsive cosmological force.[79] The modification to the equation for gravitational potential was purely an empirical ad hoc measure. Einstein realised that his field equations for the general theory of relativity led to a dynamic universe, similar to that of a finite Newtonian universe, in which spacetime was not static but expanding or contracting. Since such a universe was contrary to the current astronomy, Einstein added a cosmological constant, Λ, similar to that of Neumann, which can be interpreted as the energy density of the vacuum of space.[80] All the visible stars and nebulae were thought to be part of the Milky Way when Einstein presented his theory of general relativity, but in 1922–1923 Edwin Hubble (1889–1953) proved that the nebulae were much too distant to be part of the Milky Way and were entire galaxies outside our own. In 1929 Hubble showed that the galaxies were receding from us at speeds that increased with their distance showing that the universe is not static but expanding.[81] Einstein is reported to have said that the cosmological constant was his biggest blunder, though the statement does not appear in any of his writings and is probably apocryphal[82]; he removed the constant in 1931.[83] Viewed at a sufficiently large scale, the universe looks the same in all directions from any point within it. This fact, which was initially assumed to be true by the Russian physicist Alexander Friedmann (1888–1925),[84] makes necessary an expanding universe where the speed at which two galaxies are moving apart is proportional to the distance between them exactly as found experimentally by Hubble. In Friedmann's model, with a zero cosmological constant, the universe initially expands but at a decreasing rate until the pull of gravity causes the universe to contract. Friedmann's assumptions also led to another model where the expansion of two galaxies decreases with time, but tend to a steady speed of separation which can be zero. However, in 1998 observations of very distant supernovae in the Hubble Space Telescope showed that, though the expansion of the universe did initially slow, it is now expanding at an accelerating rate. This observation combined with the orbiting of spiral galaxies, like our Milky Way, have led to the postulation of *dark energy* and *dark matter*, which together, using the Einsteinian energy/mass equivalence, are thought to constitute 95% of the total content of the universe.[85]

1.2.1.3. *Experimental verification of the general theory of relativity*

All theories are approximations to reality that provisionally hold until a repeatable experiment shows them to be incorrect in some way. Einstein devised three experimental tests of the theory of general relativity that are now known as the classic tests. The only one that could be verified in 1915 when Einstein announced his theory was the advance in the perihelion of Mercury to which reference has already been made. Because of the presence of all the other planets, classical physics does not predict that the orbit of Mercury is constant but, even allowing for the change in the position of the perihelion caused by these interactions; there remains an unexplained advance of 43 seconds of arc per century. Einstein's theory exactly accounted for this difference. There are even more minute shifts in the perihelion of other planets and modern measurements of these shifts for Venus and Earth by radio telescopes are also accurately predicted by the general theory of relativity.

The next of Einstein's tests was the deflection of light from a star when it passes close to the Sun. Since stars generally cannot be observed when they are close to the Sun, observations have to be made during a solar eclipse. Einstein's paper on general relativity published in Germany in 1916 was not generally available to the Allied Nations during World War I. However Eddington did receive knowledge of the theory from Willem de Sitter (1872–1934) in the Netherlands, which was neutral, and published a short article on it in *Nature* in 1916.[86] Cambridge was very active in the field of relativity and it was another

Cambridge man, Ebenezer Cunningham (1881–1977) who in 1914 wrote the first English book-length account of the special theory of relativity.[87] Eddington, a Quaker, was a pacifist[88] and at the end of the war the Quakers did what they could to alleviate the sufferings to the populations of both the Allies and the Central Powers. Seven hundred thousand starved to death in Germany because of the post-armistice blockade and the Emergency Committee of the Quakers shipped food and humanitarian supplies to Germany in defiance of the blockade. Eddington saw that he could do much repair the connections of the German scientific community if he could verify Einstein's predicted deflection of light.[89] A suitable eclipse of the Sun was due in 1919 and Eddington and Frank Dyson (1868–1939), the Astronomer Royal, organized an expedition under the auspices of the Royal Society's Joint Permanent Eclipse Committee to observe it in Sobral in North Brazil and also on the island of Principe off the west coast of Africa. Einstein's theory of general relativity predicted that light from stars near the Sun should be deflected by 1.75 seconds of arc. Earlier in 1911, Einstein using his principle of equivalence without the full equations of general relativity had predicted exactly half this value for a classical prediction where light is assigned a mass as a consequence of its energy. Newton himself had in fact suggested the effect of gravity on light in his *Opticks* but made no calculations. It was left to Henry Cavendish (1731–1810) to make the first calculation of the deflection of light passing close to the Sun; Johann von Soldner (1776–1833) published this result in 1801.[90] Thus the Eddington-Dyson expedition gave three possibilities for the starlight passing close to the Sun: no deflection, a classical deflection of 0.87″, or an Einsteinian deflection of 1.75″.[91] The results of the observations were announced at a special joint meeting of the Royal Society and the Royal Astronomical Society in November 1919. Not surprisingly difficulties were encountered in the observations and the results from the Sobral astrographic plates were thought to be unreliable and the Principe observations were partly obscured by cloud. The results reported, namely 1.98 ± 0.12″ for the Sobral observations and 1.61 ± 0.3″ for the Principe observations, essentially confirmed Einstein's predictions. In 2004 the deflection by the Sun of radio waves from quasars measured by long base line radio interferometry has verified Einstein's prediction to an accuracy of 0.008%.[92] Eddington's announcement of the results of the measurements created great excitement and the observations by Eddington and Dyson brought almost universal acceptance of Einstein's theory of relativity not only by scientists, but by the population at large.[93] In 1933 Ernest Rutherford (1871–1937), the father of nuclear physics, was asked why the Eddington-Dyson expeditions had brought such acclaim to Einstein's theory of relativity and he replied:[94]

'The war had just ended; and the complacency of the Victorian and the Edwardian times had been shattered. The people felt that all their values and all their ideals had lost their bearings. Now, suddenly they learnt that an astronomical prediction by a German scientist had been confirmed ... and, indeed, prepared for already during the war, by British astronomers. Astronomy had always appealed to public imagination; and an astronomical discovery, transcending worldly strife, struck a responsive chord.'

The third test suggested by Einstein, examined the equivalence principle by testing local position invariance which means the independence from the spacetime location of the reference frame, when observing results of local non-gravitational test experiments. A gravitational redshift experiment covering this last of the original tests was suggested by Einstein but was not performed until 1959 when Robert Pound and Glen Rebka placed a radioactive isotope of iron that emitted gamma rays at the centre of a loudspeaker cone near the roof of the Jefferson Physical Laboratory at Harvard University.[95] A similar radioactive isotope of iron in the basement had a scintillation counter beneath it. By vibrating the speaker at different frequencies, the gamma source moved with varying speed creating varying Doppler shifts similar to the change in the note of a train's horn as it passes you. When the Doppler shift cancelled out the gravitational blueshift the detector experienced a drop in the number of gamma rays received. The experiment confirmed the local position invariance to an accuracy of 10% which in later tests was decreased to 1%. Using a caesium atomic fountain clock with a hydrogen maser, Andreas Bauch and Stefan Weyers have since confirmed the local position invariance to an accuracy of 0.002%.[96] Despite the fact that Einstein thought that this test was a crucial test of the general theory of relativity it is really only a test of the equivalence principle.[92]

Another prediction by Einstein from his general theory of relativity in 1916 was gravitational waves, which are transverse waves similar to electromagnetic waves but they act on matter and make ripples in spacetime. Since the waves are extremely small, Einstein thought that they would never be detected. In science, one should never say never; the first direct evidence of gravitational waves was found in 2015.[97] The source of these waves was two black holes of mass 29 and 36 times that of the Sun spiralling towards each other about 1.3 billion years ago and merging to form one black hole of 62 times the mass of the Sun. The waves were detected by two interferometers in the USA one at Hanford, WA, and the other at Livingston, LA. Predictions from the general theory of relativity of the spatial strain caused by the collision of these two black holes are shown in Plate 27 and they closely match the detected wave forms.

1.2.1.4. *Black holes*

John Wheeler (1911–2008) is credited with the first use of the term *black hole* during a public lecture on the universe held at the Goddard Institute of Space Studies in 1967 to describe region of space from which nothing, not even light can escape. Stars of mass smaller than Chandrasekhar's limit, of about one and a half times that of our Sun, eventually collapse into stable dwarf stars whose size is comparable to that of our Earth. Larger stars collapse unstably to form black holes where the gravity field is so strong that nothing can emerge. At a distance from a black hole its effect is simply that of the mass of the star and matter is drawn towards it in exactly the same fashion as for any other star of the same mass. However, a black hole has what is described as an event horizon close to the hole in spacetime where the gravitational pull becomes so great that nothing, not even light can escape.

By their very nature black holes cannot be directly observed, how then is it known that they are a real? The first evidence for black holes was obtained from the study of binary stars where two stars orbit about their common centre of gravity. With some binary stars only one is visible though observation shows that the one that can be seen is part of a binary system. In most of these binaries the unseen secondary is less massive than the primary observed star and the secondary is not observed because it is not bright enough in comparison with the primary star. In a small number of cases the secondary unobserved star is the heavier. Gary Gibbons and Stephen Hawking argued in 1971 that there was evidence of black holes in a few of these binaries.[98] Some of these binary systems emit X-rays like the one called Cygnus X-1, which was discovered by Riccardo Giacconi in 1964, and is located about six thousand light years from us in the constellation Cygnus of our galaxy the Milky Way. The X-rays are emitted as matter crosses the event horizon of a black hole and is drawn into it. Although Cygnus X-1 was found in the sixties, its status as a black hole was not firmly established for a long time. Stephen Hawking, who in 1975 thought that it Cygnus X-1 was most likely to be a black hole, made a bet to avoid tempting fate with Kip Thorne of the California Institute of Technology that it was not one; he conceded the bet in 1990.[99] The mass of the black hole for Cygnus X-1 is estimated to be 8 to 16 times that of our Sun.[100] Supermassive black holes are now believed to exist in the centre of galaxies; the one at the centre of our galaxy, the Milky Way, has a mass estimated to be an almost unbelievable size of about four million times that of our Sun and is the second smallest that has been observed.[101]

The existence of black holes was predicted almost immediately after Einstein's theory of general relativity was published. The German physicist Karl Schwarzschild (1873–1916) was at the Russian front in 1916 when he obtained the first exact solution to Einstein's ten field equations which contain six independent unknowns; he sent this solution to Einstein who communicated it to the Prussian Academy and wrote back 'I had not expected that one could formulate the exact solution of the problem in such a simple way.' In his second paper from the front, Schwarzschild showed that if the mass to radius ratio (*M/R*) of a star exceeded a critical value given by

$$\frac{M}{R} = \frac{c^2}{2G}, \tag{1.14}$$

where *G* is the universal gravitation constant, it would collapse to form what is now called a black hole. Tragically Schwarzschild died from a skin disease in the same year that he wrote these two important papers on the theory of relativity, he was only 42 years old. Rather surprisingly Schwarzschild's result had been obtained at the end of the eighteenth century from Newtonian mechanics, where the "particles of light" are assumed to be corpuscles subject to gravitational interactions with massive bodies, exactly as any other particle, for what was termed *dark bodies*, which were not quite the same as black holes. John Michell (1724–1793) presented a paper to the Royal Society in 1783 where he hypothesized that: [102]

> '[If the] diameter of a sphere of the same density as the sun were to exceed that of the sun in the proportion of 500 to 1, a body falling from an infinite height towards it, would have acquired at its surface a greater velocity than that of the light, and consequently, supposing light to be attracted by the same force in proportion to its *vis inertiæ*, with other bodies, all light emitted from such a body would be made to return towards it, by its own proper gravity.'

Pierre-Simon Laplace (1749–1827) in his *Exposition du Système du Monde*, published in 1796 gave an algebraic proof of the argument for dark bodies that leads to Eq. (1.14).[103] The velocity of light is assumed to be slowed by gravity as any other particle. Light can escape from a dark body, but falls back if the escape velocity is less than the velocity of light. As Giovanni Preti pointed out in 2009, since the Michell-Laplace argument is based on the concept of escape velocity, the critical *M/R* ratio is bound to be identical to the general relativistic Schwarzschild one.[103] However, dark bodies are not the same as black holes. Nothing can leave a black hole and it cannot be observed, but a dark body is luminous and can be seen by a close observer but is too dim to be observed by a far one.[103]

1.2.2. *Quantum mechanics*

The general theory of relativity breaks down at the sub-atomic scale. Quantum mechanics describes the interaction of the smallest particles in the universe such as electrons and quarks, which are considered point particles with no spatial dimensions. Brian Greene explains in his book *The Elegant Universe*, written in an attempt to make the forefront of physics research understandable, that whereas the concepts of the theory of relativity can be grasped by the average person, quantum mechanics is so far from our normal everyday experience that its concepts are almost unbelievable.[104] Classical mechanics and relativistic mechanics are deterministic, exact values can be found for all the variables. However as Wolfgang Pauli (1900–1958) said in his Nobel Lecture in 1946[105]:

> 'the statements of quantum mechanics are only dealing with possibilities not actualities. They have the form *this is not possible* or *either this or that is possible*, but they can never say *this will happen then and there*.'

It is the uncertainty central to quantum mechanics that led Einstein, who was important in the early development of quantum theory, to write in 1926 to Max Born (1882–1970) the often-quoted words: 'quantum ... theory yields much, but it hardly brings us close to the Old One. In any case I am convinced he does not play dice.'[106]

1.2.2.1. *Black body radiation*

The fundamental principle of quantum mechanics, that energy is not emitted continuously but in small quanta, was first advanced by Max Planck in 1901 to explain black body radiation. A black body is one that absorbs all the radiation that falls upon it, and because no radiation is reflected, it appears black when cold; the inside of a camera is painted black so that stray light that falls on it is absorbed rather than reflected. Jožef Stefan (1835–1893) and Ludwig Boltzmann (1844–1906) formulated the fundamental law of black body radiation for the energy flux density, Ψ, from experiment and theory between 1879 and 1884, which is given by

$$\Psi = \sigma T^4, \tag{1.15}$$

where T is the absolute temperature in degrees kelvin, and σ is a physical constant now known as the Stefan-Boltzmann constant,[107] The Stefan-Boltzmann law says nothing about the frequency distribution of the radiant energy. In order to obtain a distribution in energy that was compatible with the Stefan-Boltzmann law, Planck had to assume that the energy, E, was emitted in small quanta that were proportional to the frequency, f, and inversely proportional to its wavelength, λ,[66] so that

$$E = hf = \frac{hc}{\lambda}, \tag{1.16}$$

where the constant of proportionality, h, is what is now termed *Planck's constant* a fundamental physical constant that Planck calculated from the Stefan-Boltzmann constant to be 6.55×10^{-34} Js (the modern value is 6.63×10^{-34}). Since it has become customary in quantum mechanics to use a frequency based on radians per second rather than rotations, a reduced Planck's constant $\hbar = h/2\pi = 1.05 \times 10^{-34}$ Js is now more usual. The quanta of energy are so incredibly minute that the discrete energy steps can not be discerned.

1.2.2.2. *Planck units*

General systems of scientific units, like the Système International (SI), are arbitrary. In 1899 Planck introduced what he called natural units (subsequently called Planck units in his honour) based on universal physical constants that provide a scale for any universe. Originally Planck chose four constants to which a fifth was added later to form his natural units. The five constants used today and their associations are shown in Table 1.1. The actual values of these physical constants have no significance except as yardsticks. In our physiology and that of all living things, motion is slow relative to the Earth on which we live but the Earth is orbiting the Sun at over 100,000 km/hr, which is fast but still only 0.01% of the speed of light. None of these speeds have any physical significance, only the speed of light has such significance that it is can be an independent yardstick. In 1899, using dimensional analysis, Planck obtained 'units of length, mass, time, and temperature, which are independent of special bodies or substances, which necessarily retain their significance for all times and for all environments, terrestrial and human or otherwise, and which may therefore be described as *natural units*, which are given in Table 1.2 together with their values in SI units.[108] Except for mass, the Planck units are either very small or very large. They are not practical as everyday units of measure but most importantly they provide physically meaningful scales.

Table 1.1. The five universal Planck units.

Physical constant	Symbol	Association
Speed of light	c	Electromagnetic waves and special relativity
Gravitational	G	Newtonian gravity and general relativity
Reduced Planck constant	$\hbar$	Quantum mechanics
Boltzmann's constant	$k_B = R/N_A$	Thermodynamics and statistical mechanics
Coulomb's constant	k_e	Electrostatics and quantum mechanics

Table 1.2. Planck units and their SI values.

Planck Unit	Planck unit in terms of the five physical constants	Value of Planck unit in SI units
Length	$l_{Pl} = \sqrt{\frac{\hbar G}{c^3}}$	1.616×10^{-35} m
Mass	$m_{Pl} = \sqrt{\frac{\hbar c}{G}}$	2.177×10^{-8} kg
Time	$t_{Pl} = \frac{l_P}{c} = \sqrt{\frac{\hbar G}{c^5}}$	5.391×10^{-44} s
Temperature	$T_{Pl} = \sqrt{\frac{\hbar c^5}{G k_B^2}}$	1.417×10^{32} K
Charge	$q_{Pl} = \sqrt{\frac{\hbar c}{k_e}}$	1.876×10^{-18} °C

1.2.2.3. *Particle-wave duality*

The concept that electromagnetic radiation propagates in discrete quanta was used by Einstein to explain the photoelectric effect,[109] in which high frequency electromagnetic radiation can cause electrons to be emitted from metals and an electric current to flow. The energy of the emitted electrons only depends upon the frequency of the light and not on its intensity. The intensity of the light simply increases the number of electrons that are emitted. In the early nineteenth century Thomas Young and Augustin-Jean Fresnel (1788–1827) had advanced the wave theory of light which became accepted at the expense of Newton's corpuscular theory. However, the wave theory predicts that the energy of the emitted electrons should depend upon the light intensity not the frequency. Einstein returned to a particulate theory of light based on what is now called the photon. The photon, the electromagnetic force particle, travels at the speed of light so, if it had even a tiny rest mass, its energy would be infinite by Eq. (1.12), hence the photon must be massless but it possesses momentum p which is given by

$$p = \frac{E}{c} = \frac{hf}{c} = \frac{h}{\lambda}. \tag{1.17}$$

Table 1.3. Particle-wave duality.

Particle-wave	Wave length (m)	Mass (kg)	Velocity (m/s)	Momentum (kg m/s)
Cricket ball	10^{-34}	0.14	30	4.2
Alpha particle	7×10^{-15}	6×10^{-27}	1.5×10^{7}	9×10^{-20}
Electron	6×10^{-12}	10^{-30}	10^{8}	10^{-22}
γ-rays	10^{-12}–10^{-11}	massless	3×10^{8}	6×10^{-23}–10^{-22}
X-rays	10^{-10}–10^{-8}			6×10^{-26}–10^{-24}
Visible light	3.8–7.5×10^{-7}			9–17×10^{-28}
Infra red	10^{-6}–10^{-3}			6×10^{-31}–10^{-27}
Microwaves	10^{-2}			6×10^{-32}
Radio and TV	0.3–10			6–200×10^{-35}

Light has a wave-particle duality because of the inadequacy of the classical concepts of a wave and a particle to describe their behaviour completely. It is not at all surprising that the debate as to whether light was a wave as proposed by Descartes and Huygens or a particle as proposed by Newton should have continued for two centuries until their duality was finally recognized. This duality is central to quantum mechanics not just for light. Louis de Broglie (1892–1987) showed the wave-like characteristics of all matter in 1923.[110] In Table 1.3 the wave length and momentum of different electromagnetic waves and some particles are given. The electromagnetic waves all propagate at the speed of light and are massless. The particles that have mass must travel at less than the speed of light.[111] Although according to quantum mechanics there is a duality between all particles and electromagnetic waves it is only over the central part of the table where this duality has any real meaning. A cricket ball has a wave whose length is of the order of Planck's length and waves of this length have no real meaning; conversely radio and TV waves have an incredibly small momentum and virtually no particle-like character. George Thomson (1892–1975) and Clinton Davisson (1881–1958) received the Nobel Prize for their experimental discovery of electron diffraction where electrons behave like waves, which led to the transmission electron microscope developed during the 1930s.

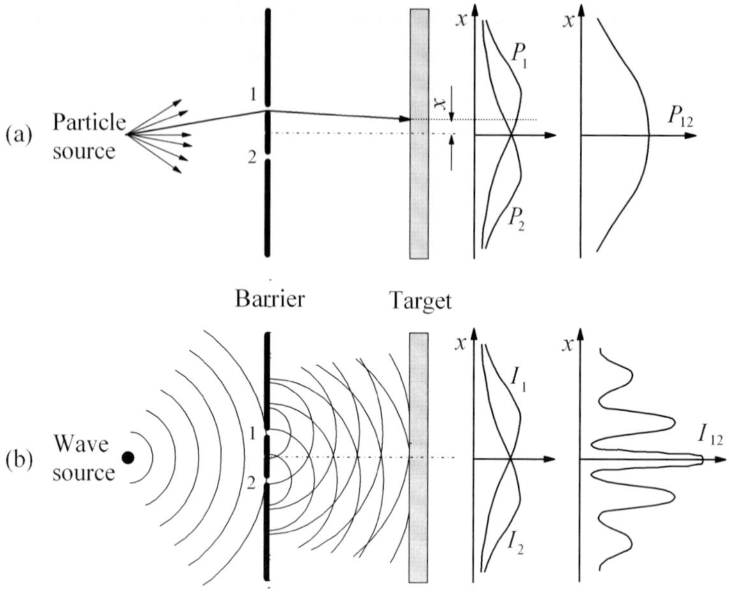

Fig. 1.2. The two slit experiment with particles and waves, after Feynman (2011).

The particle-wave duality causes an apparent paradox in the classic two slit experiment a form of which was first performed using light by Thomas Young in 1803. Richard Feynman (1918–1988) began his third volume of his famous *Lectures on Physics* on quantum mechanics, first published in 1964, with an explanation of the two slit experiment, which is much more detailed than the simple summary I can give here.[112] Briefly if nanoparticles (containing more than about 1000 atoms or more massive than about 10^{-22} kg) are emitted at a constant rate at a random angle, or waves from a single source impinge on a barrier containing two slits their arrival at an absorbing target would be expected to be different (see Fig. 1.2). In the case of particles we can speak about the probability, $P_1(x)$, of a particle passing through slit 1 and arriving at a point x on the target and similarly the probability, $P_2(x)$, of particle passing through slit 2 and arriving at the target (see Fig. 1.2(a)). The total probability, $P_{12}(x)$, of a particle passing through either slit 1 or slit 2, shown in Fig. 1.2(a) is then given by the simple sum of the probabilities,

$$P_{12}(x) = P_1(x) + P_2(x). \tag{1.18}$$

The intensity of the of a wave arriving at the target though slit 1, with slit 2 blanked off, $I_1(x)$, and similarly the intensity of the wave passing through slit 2, $I_2(x)$, are similar to the probability of particles as shown in Fig. 1.2(b). However, if both slits are open the waves passing through the two slits interfere with one another since the waves have travelled different distances and arrive out of phase, the combined intensity, $I_{12}(x)$, oscillates and

$$I_{12}(x) \neq I_1(x) + I_2(x). \tag{1.19}$$

The intensity I_{12} is shown in Fig. 1.2(b). However if the particles are small, say electrons, then Eq. (1.18) does not hold and the probability $P_{12}(x)$ is similar to the intensity $I_{12}(x)$ obtained using a wave with the right wave length (or, from Table 1.3, 6×10^{-12} m) as the particle-wave duality predicts. How can this result occur if an electron can only pass through one of the slits and not both at the same time? The quantum mechanics answer is the unexpected one that the electron can go through both slits at the same time. Now, is it possible to observe an electron with a split personality going through both slits at the same time? If we devise any method of observing through which slit the electron passes, we affect the experiment and the interference pattern is lost and the probability $P_{12}(x)$ is once again given by Eq. (1.18).

Now in Table 1.3 a wave length is assigned even to a cricket ball, why should we not get interference even for such a large particle? Feynman answered this question.[112] Look at the wave length of a cricket ball 10^{-34} m. The interference pattern would be extremely fine. 'So fine, in fact, that no detector could distinguish the separate maxima and minima [and we see] only a kind of average, which is the classical curve' given by Eq. (1.18). The ability to produce an interference effect in the two slit experiment is a sure sign of quantum behaviour. The largest particle so far that has proved to behave according to quantum mechanics is the huge molecule with 810 atoms whose chemical formula is $C_{284}H_{190}F_{320}N_4S_{12}$ which has a mass of 1.69×10^{-23} kg.[113]

1.2.2.4. *The structure of the atom*

Prior to 1925 a quantum mechanics picture of the atom was built on the model of 1911 proposed by Ernest Rutherford where a small positively charged nucleus is surrounded by a cloud of orbiting electrons similar to planets orbiting the Sun, which he obtained from analysis of the scattering of alpha particles by a very thin gold foil. Niels Bohr (1885–1962) at the University of Copenhagen developed a very different theory of the atom, in a series of papers from 1913 to 1918, where electrons orbited a central nucleus but not like planets around the Sun.[114] In Bohr's model the electrons do not obey the usual laws of electrodynamics and can only jump from one stationary state or shell to another when a quanta of energy was emitted or absorbed according to the relationship

$$E' - E'' = hf, \tag{1.20}$$

where E' and E'' are the energy of the two states, h is Planck's constant, and f is the frequency. For the simplest atom, hydrogen, which consists of a nucleus of a single proton with a single electron, the energy levels in electron volts[115] of the different shells are given by

$$E_n = -\frac{13.6}{n^2}\ \text{eV}, \tag{1.21}$$

where n is the *principal quantum number* of the electron, which in the Bohr atom defines the possible orbit shells of the electron with $n = 1$ being the closest possible orbit. Bohr used classical mechanics and Coulomb's law to calculate Eq. (1.21). The Bohr model of the hydrogen atom, based on these assumptions, accurately predicted the spectral lines in the spectrum of hydrogen. However, as pointed out by James Jeans (1877–1946) at the British Association meeting in 1913:[116]

> 'The only justification at present put forward for these assumptions is the very weighty one of success.'

Although energy is absorbed or emitted in quanta as an electron changes its quantum number, if n is large there is little difference between a classical and a quantum mechanics interpretation. This conclusion gave rise to the *correspondence principle* of Bohr which states that in the limit of large quantum numbers classical and quantum mechanics are the same.

In 1921 Bohr deduced that in elements of higher atomic number than helium the electrons are arranged in shells, which can contain no more than $2n^2$ electrons, which are filled from the inner shell outwards; thus the inner shell can have a maximum of 2 electrons, next 8 electrons and so on.[117] Bohr knew that the electrons in each shell other than the first were apportioned between sub-groups, but he got the distribution wrong. We now know that the atomic structure is governed by three other quantum numbers besides the principal one: the *azimuthal quantum number* which gives the orbital angular momentum and determines the sub-group of the electron with the integer numbers $0 \le l \le (n-1)$, the *magnetic quantum number* which takes the integer values $-l \le m_l \le l$, and lastly the *spin quantum number*, m_s, which determines the spin of the electron and takes the values of −½ or ½. Bohr recognized that a fundamental problem with his model was the question why all the electrons did not collapse into the innermost shell which would be the lowest energy state. The answer was provided in 1925 by Wolfgang Pauli (1900–1958) who formulated the *exclusion principle* which states that no two electrons can have the same four quantum numbers with this principle electrons cannot collapse into the lowest energy state.[118] Electrons are a class of fundamental particles called fermions which obey the exclusion principle.

The exclusion principle prevents the collapse of stars smaller in mass than Chandrasekhar's limit. As the atoms in a star are compressed by gravity, the electrons repel each other. The closer the electrons are compressed, the faster they oscillate to repel one another. However, the speed of the electrons is limited to the speed of light and when a star's gravity forces create a pressure greater than can be resisted by this limit, the star collapses into a black hole.

1.2.2.5. *Heisenberg's uncertainty principle*

Prior to 1925 quantum theory relied heavily on classical mechanics and as Heisenberg wrote in his famous 1925 paper:[119]

> '[contained] relationships between quantities that are apparently unobservable in principle... It seems more reasonable to try to establish a theoretical quantum mechanics, analogous to classical mechanics, but in which only relationships between observable quantities occur.'

In June 1925 a heavy attack of hay fever had forced Heisenberg to leave Göttingen for the island of Helgoland where there is no grass. The hay fever lessened and, in the comparative isolation of Helgoland, Heisenberg saw the necessary relationships that must hold if the correspondence principle was to be true for large principle quantum numbers. The key idea was in the kinematic interpretation of the location of an electron which was replaced by a quantum quantity based on the transition amplitudes from one quantum state to the next. Heisenberg did not realize at the time that the resulting set of equations representing the motion were matrix equations, but Born did. With the help of his student Pascual Jordan (1902–1980), Born reinterpreted Heisenberg's symbolic multiplication as matrix multiplication.[120] Ordinary algebraic multiplication is commutative, that is $pq = qp$, but matrix multiplication in general is not and if $\boldsymbol{p}$, $\boldsymbol{q}$ are the matrices[121] that represent the momentum and the position of electrons for transitions between quantum states $\boldsymbol{pq} \neq \boldsymbol{qp}$. Born and Jordan calculated the difference between the non-commutative product of the matrices and showed that energy was conserved in the transformations and Bohr's frequency condition given by Eq. (1.20) was met. They also extended the new quantum mechanics, which became known as *matrix mechanics* to electrodynamics. In the final paper of the trinity, Heisenberg, Born and Jordan further developed general quantum mechanics extending the theory to systems having more than one degree of freedom.[122] All the three papers which established general quantum mechanics were written between June 1925 to November 1925 — a remarkable advance in such a short time.

Heisenberg, who had moved to the University of Copenhagen in 1926, realized in 1927 that the non-commutativity of the product of $\boldsymbol{p}$ and $\boldsymbol{q}$ implies that the momentum and the position of an electron cannot both be known to high accuracy.[123] This necessary impreciseness is now known as the *uncertainty principle*.[124] Heisenberg did not precisely define the degree of uncertainty and it was Earle Kennard (1885–1968), an American physicist on sabbatical leave from Cornell University at the University Copenhagen who in 1927 gave the relationship between the standard deviations of the momentum, $\sigma(p)$, and the position $\sigma(q)$[125]

$$\sigma(p)\sigma(q) \geq \frac{h}{4\pi}. \tag{1.22}$$

Recently it has been shown that Eq. (1.22) is not valid in all circumstances and a universal uncertainty principle has been established.[126]

With the departure of the old quantum theory that relied on classical mechanics there was a loss of visual imagery. In 1927 Heisenberg argued that the ordinary view or intuition [gewöhnliche Anschauung] based on visual imagery of phenomena that have been seen had to be replaced by imaginative visualization [Anschaulichkeit] of the intrinsic attributes of sub atomic particles that cannot be perceived such as an electron's spin.[123]

1.2.2.6. *Wave mechanics*

Erwin Schrödinger (1887–1961) at the University of Zurich took the view that electrons:[127]

> 'consist of, or are nothing, but, wave-systems. This extreme conception may be wrong, indeed it does not offer as yet the slightest explanation of why only such wave-systems seem to be realized in nature as corresponding to mass-points of definite mass and charge.'

De Broglie had tried to introduce relativistic notions into the wave theory and Schrödinger at first attempted to develop this work, but he quickly decided to deal with non-relativistic wave mechanics. In a few short months in 1926 Schrödinger established wave mechanics in a convincing form in four classic papers in *Annalen der Physik*.[128] Schrödinger showed that from a purely mathematical point of view that his wave theory was identical to the matrix mechanics of Heisenberg, Born, and Jordan.[129]

Born gave a probability interpretation of Schrödinger's wave function.[130] The interpretation, Born stated, arose from a remark of Einstein on the relationship between photons and wave fields. Einstein had said that waves only served to indicate the path of the particles and spoke of the "phantom field" [Gespensterfeld]. The wave field gives the probability that a photon should take a given path. An electron wave is therefore interpreted as giving the probability of the particle being in a particular position and it is more likely to be where the magnitude of the wave is large. The idea that wave mechanics gives only the probability of the position of an electron is in complete accord with Heisenberg's uncertainty principle.

1.2.2.7. *Relativistic quantum field theory*

Relativistic quantum field theory combines the special theory of relativity with quantum mechanics. In 1928 Paul Dirac (1902–1984) developed the relativistic equation for the wave function of the electron, which explained the spin of an electron.[131] Two solutions of the equation correspond to the two spin states of the electron. The other two solutions predicted the possible existence of an electron with a positive charge, *e*, and a negative energy, which Dirac predicted in 1931 to be:[132]

> 'a new kind of particle, unknown to experimental physics, having the same mass and opposite charge to an electron. We may call such a particle an anti-electron.'

In 1932 the American physicist, Carl Anderson (1905–1991), observed such a particle which he called the *positron*. It is now known that there is an antiparticle with the same mass and opposite electric charge corresponding to all fundamental particles. In the same paper that Dirac predicted the positron he also showed that there is 'a symmetry between electricity and magnetism quite foreign to current views' and predicted that just as there are single electric charges so too are there isolated magnetic charges or monopoles. Dirac only had to wait a year for his prediction of the positron to be confirmed by experiment but the existence of the magnetic monopole was not shown experimentally until 2013.[133]

The first attempt at producing a relativistic quantum theory of electrodynamics, now known as *quantum electrodynamics* (QED), was made in 1929 and 1930 by Heisenberg and Pauli.[134] A major problem was that the self-energy of an electron, which is its total energy in free space when isolated from other particles, was infinite because in quantum theory the electron is a point particle. This problem did not arise in the classic theory of Lorentz because he assumed that the electron had a finite size.[135] However there are difficulties in reconciling finite sized electrons with relativity. Attempts were made to formulate quantum field theories where electrons have finite size and are consistent with basic physical principles but they all failed. The problem of the infinite self-energy of point fundamental particles was not solved until the 1940s and the work of Richard Feynman, Julian Schwinger (1918–1994), and Sin-Itiro Tomonaga (1906–1979) who showed that *renormalization* could obtain unambiguous answers from electrodynamics in spite of the infinities. Renormalization consists of changing the values of the mass and charge of an electron so that the physical values are different from the mathematical parameters that occur in the equations of QED. Dirac objected to the process of renormalization because the changes have to be infinitely great though he admitted that it gave a theory which agreed with observation.[136] In particular QED predicts the small difference between two energy levels of the hydrogen atom[137], known as the *Lamb shift*, which were the same in Dirac's 1928 theory.[131] Dirac was unhappy with renormalization until the end of his life and one of his last seminars given to an Orbis Scientiae meeting in 1983 was on his opposition to renormalization. In this talk he said:

> 'These rules of renormalisation give surprisingly, excessively good agreement with experiment. Most physicists say that the working rules are, therefore, correct. I feel that is not an adequate reason. Just because the results happen to be in agreement with observations does not prove that one's theory is correct.'

Karl Popper (1902–1994), the great philosopher of science, agreed fully with Dirac's comments.[138]

In 1979 Feynman gave a series of lectures on QED in New Zealand at the University of Auckland. The book, *QED: The Strange Theory of Light and Matter*, first published in 1985, was based on these lectures and provides a marvellous non-technical introduction to QED.[139] The QED is constructed on the foundation of classic Maxwell-Lorentz electrodynamics. Because QED essentially gives the probability of what happens, electromagnetic waves take all possible paths and can propagate slower or faster than the speed of light and only on average propagate at the speed of light. QED emerged in two forms. Tomonaga and Schwinger presented a formalized QED, whereas Feynman gave his intuitive visualization [Anschaulichkeit] using formal diagrams based on mathematical rules for such processes as an electron emitting a photon. Initially Feynman's diagrams were met with scepticism but they have proved to be very useful in bookkeeping quantum interactions. Toichiro Kinoshita at Cornell University has led the way in using Feynman diagrams to calculate detailed properties of electrons to high precision.[140] Perturbation methods based on Feynman diagrams can be used to calculate the anomalous magnetic moment of an electron. The theoretical expression for the electron anomaly, $a(e)$, can written in terms of a series expression with decreasing terms in the electron charge e

$$a(e) = C_1\left(\frac{\alpha}{\pi}\right) + C_2\left(\frac{\alpha}{\pi}\right)^2 + C_3\left(\frac{\alpha}{\pi}\right)^3 + C_4\left(\frac{\alpha}{\pi}\right)^4 + \cdots, \tag{1.23}$$

where $\alpha = 2\pi e^2/hc$ is a non-dimensional parameter which is about 0.0073. The first term is represented by a Feynman diagram of one closed loop and can be relatively easily found but C_2, C_3, and C_4 require 7, 72, and 891 Feynman diagrams respectively and took Kinoshita some 25 years to complete. The theoretical value of $a(e)$ agrees with the experimental value to within less than 0.00001%; at the moment the biggest error comes from uncertainty in the precise value of α. Agreement between this theoretical calculation and the real world is truly incredible.

An analogous approach to QED has been used to develop a quantum mechanics understanding of the weak and strong forces which are called the *quantum electroweak theory* and *quantum chromodynamics* respectively. Electromagnetic and weak forces are very different at low energy levels but until very shortly after the Big Bang, when the temperature of the universe was greater than about 10^{15} degrees kelvin, they were the same. The force particles that carry the weak force are called the weak gauge bosons. The weak gauge bosons, unlike photons, have mass and they are more than a hundred thousand times as massive as an electron and even more massive than an atom of iron. However, the particles have a very short life of only about 3×10^{-25} s. The symmetry of *gauge theories* predicts that the weak gauge bosons are massless, they gain mass as a result of the *Higgs mechanism* named after the English theoretical physicist Peter Higgs. The Higgs mechanism requires a fundamental particle, the *Higgs boson*, a particle which was found by use of the new CERN Large Hadron Collider in 2011 and 2012. These experimental results cannot be distinguished from the Standard Model predictions for the Higgs boson.[141] In quantum chromodynamics there are three types of charge named after the three primary colours: red, green and blue. The strong force is carried by massless gluons.

Table 1.4. The three quantum fields.

Quantum field theory	Force	Force particles
Electrodynamic	Electromagnetic	Photons
Electroweak	Weak	W and Z bosons
Chromodynamic	Strong	Gluons

The relativistic three quantum fields and forces are summarized in Table 1.4. All force particles are bosons and unlike the fundamental particles of matter, which are called *fermions*, they can occupy the same quantum state. Bosons have integer spin in contrast to fermions which have a half integer spin. Not all bosons are force particles. Bosons do not obey the exclusion principle.

1.2.2.8. *The standard model of particle physics*

The standard model of particle physics began with the unification in 1960 of electromagnetic and weak interactions by Sheldon Glashow.[142] By 1967 Steven Weinberg[143] and Abdus Salam[144] (1926–1996) had given the unification of the electromagnetic and weak forces its modern form and all three received the Nobel Prize for Physics for this work in 1979. Although high-energy physics experiments have confirmed the Standard Model, it is not complete because it does not include gravitation and other descriptions. The standard model is based on the symmetry of gauge theories. For example the phase of the wave functions describing a quantum field can be changed by a constant arbitrary amount without altering the theory. An electromagnetic force is necessary to keep this symmetry with a phase change that varies over spacetime. The Standard Model is a gauge theory of the electromagnetic, weak and strong interactions. It needs some 26 freely adjustable parameters whose values are obtained experimentally and which cannot be determined from first principles. A very brief taste of the Standard Model is given here that hopefully allows its flavour to be savoured; a popular account that is more substantial is *The Theory of Almost Everything* by Robert Oerter.[145]

Table 1.5. The three families of fundamental particles (masses given relative to the electron).

Family 1		Family 2		Family 3	
Particle	Mass	Particle	Mass	Particle	Mass
Electron	1	Muon	207	Tau	3478
Electron-neutrino	$< 10^{-5}$	Muon-neutrino	< 0.3	Tau-neutrino	< 5
Up-quark	8.7	Charm quark	2435	Top quark	335000
Down-quark	1.37	Strange quark	204	Bottom quark	8219

Atoms are built out of the smaller fundamental particles shown in Table 1.5. Between the 1930s and 1968 atoms were thought to consist of a nucleus of protons and neutrons surrounded by a cloud of electrons. The proton has a mass of about 1.67×10^{-27} kg and a positive electric charge equal to an electron and a neutron has almost the same mass as a proton but no charge. However, protons and neutrons were found to consist of smaller fundamental particles called *quarks* by Murray Gell-Mann after the line 'Three quarks for Muster Mark' in James Joyce's *Finnegans Wake*.[146] These particles were also independently predicted by George Zweig who called them *aces*. In the first family of fundamental particles there are two types of quarks: an up-quark of charge $2e/3$ and a down quark of charge $-e/3$; a proton consists of two up and one down quark giving a net charge e and a neutron consists of one up and two down quarks with no net charge. There is also a fourth fundamental particle, the *neutrino*, which was postulated Wolfgang Pauli in 1930 to preserve the conservation of energy and momentum in beta decay.[147] The neutrino a tiny particle with no electric charge was difficult to find because they only rarely interact with other matter and they were not detected until 1956.[148] Even an average energy neutrino passes readily through our world. Most of the neutrinos we encounter are the result of nuclear reactions in our Sun.

1.3. Postmodern String Theory

Quantum mechanics accounts for most of the physical behaviour of atoms and their fundamental particles but it does not explain everything. Einstein was particularly concerned that gravitational and electromagnetic forces were separate and spent the last thirty years of his life unsuccessfully searching for the so-called unified theory of everything. Three of the fundamental forces have been reconciled with relativity but gravity has not. Matter warps spacetime according to the theory of general relativity but at large scales it remains smooth. However it loses its smooth character at subatomic scales because of quantum fluctuations, to become what John Wheeler has described as *quantum foam*.[149] General relativity cannot be combined with the highly fluctuating quantum foam. Clearly quantum mechanics does not model reality completely.

Nature and the physical laws are very symmetrical and string theory has much possible symmetry that seems capable of reconciling behaviour at the extremely small scale with relativity. In the original string theory the fundamental particles of physics, which in quantum theory are point particles, are modelled by strings that can be in the form of tiny closed loops or open strings of the order of Planck's length. According to string theory, the properties of the fundamental particles are the vibrational modes of the strings. However for consistency strings need to vibrate in a spacetime of more than three space dimensions. Despite an initial success, the string theory for strong force interactions was shown to conflict with later high-energy experiments and the success of quantum chromodynamics caused the initial enthusiasm for string theory to wane. However, when in 1984 Michael Green and John Henry Schwarz showed that the conflicts of string theory with quantum field theories could be overcome by supersymmetric *superstring theory* interest again began to wax.[150] As Edward Witten has stated:[151]

> 'string theory forces general relativity upon us, whereas standard quantum field theory apparently makes it impossible to incorporate general relativity. And string theory leads in a remarkably simple way to a reasonable rough draft of particle physics unified with gravity.'

However, string theory is not without problem as Peter Woit has pointed out.[7]

1.3.1. *Superstring theories*

Since 1984 string theories have been based on *supersymmetry*.[150] Symmetry has always played a large part in physics and in the 1970s a possible new symmetry, called supersymmetry, was proposed between fermions, or the fundamental particles of matter, and bosons, or the force particles. In supersymmetry there is a symmetric boson for every fermion and vice versa. Thus for example the electron has a symmetric boson called the *selectron* and the photon has a symmetric fermion called the *photino*. The partner particles tend to cancel one another and the supersymmetric standard model does not rely on the numerical adjustments needed in the standard model. Unfortunately none of the supersymmetric partners have been discovered. Supersymmetry also enables a unification of the fundamental forces. The four fundamental forces have a huge range in strength. The strong force is the largest, being a thousand times stronger than the electromagnetic force. The weak force is about a 10^{13} times weaker than the strong force. Gravity, the weakest force, is about 10^{40} times weaker than the strongest. This huge difference between gravity and the other fundamental forces is known as the *hierarchy problem*. All the four fundamental forces were unified up until about the Planck time after the Big Bang when the temperature was in excess of 10^{27} degrees kelvin; the gravitational force separated from the other forces at about this time.[152] The strong force separated next followed by the splitting of the electroweak forces into electromagnetic and the weak forces moments later at 10^{15} degrees kelvin and 10^{-10} seconds. When Ugo Amaldi and his co-workers re-examined the extrapolations from the data necessary to show unification of the forces they found that without supersymmetry the three non-gravitational forces did not quite become one; supersymmetry removed the discrepancy.[153] Thus supersymmetry, if even if it not real, is a good approximation to reality. Supersymmetry enabled three of the four fundamental forces to be unified assuming dimensionless point fundamental particles, but gravity still could not be accommodated.

It was found that there was not one superstring theory satisfying all the possible symmetries but five and that they all needed not just the four dimensions of Einstein, but ten: one dimension of time and nine of space. These facts presented difficulties. Not only was there not a unique theory but, in addition to the three space dimensions we can see, there were six other dimensions. The possible existence of more than three space dimensions was not new. Theodor Kaluza (1885–1954) had suggested four space dimensions in 1919 and Oskar Klein (1894–1977) showed in 1926 that it was possible for the fourth space dimension to be real but tightly curled up so that it could not be seen. The extra equations due to the fourth space dimension led to not only Einstein's equations of general relativity but also Maxwell's equations of electromagnetism; the weak and strong forces were not known in 1919. However, it was found that the theory did not predict the experimental data. The predicted relationship between the mass and charge of an electron were very much different to that observed. The difficulty of the superfluous space dimensions has been explained by assuming that they are compacted or curled up so that they are smaller than Plank's length (1.6×10^{-35} m) and cannot be detected. A rope or wire is usually given as an illustration of a two dimensional space which if the diameter is very small appears to have only a single dimension. However, it was shown that the six extra dimensions of superstring theory could only be compacted in a special way into what are called Calabi-Yau spaces.[154] The fact that there were five possible superstring theories was an initial embarrassment, but in the second superstring revolution, which dates from the conjectures made by Witten during the Strings Conference at the University of Southern California in 1995, an all embracing theory, which Witten called M-theory, where 'M can stand for either "magic", "mystery", or "matrix", according to taste,' was proposed that encompassed the five existing superstring theories.[155] M-theory is more complex than the previous superstring theories having vibrating *p*-branes or objects with dimensions varying from zero, point particles, up to nine dimensions which is the maximum possible with ten space dimensions because a *p*-brane vibrates in a *p*+1 space. The *p*-branes with dimensions greater than one are massive and have little effect. M-theory has, as Witten has stated, 'the remarkable property of predicting gravity — that is it requires the existence of a massless spin-2 particle, [the graviton], whose couplings at long distance are those of general relativity.'[156] Witten has also shown that M-theory enables gravitational force to be merged with the other three fundamental forces at very small distances. Stephen Hawking, though his book *The Grand Design* seemed in 2010 to indicate that M-theory had the potential to be a theory of everything but was less certain earlier in 2002 when he had the view that perhaps it is impossible to formulate a theory of the universe in a finite number of statements just as Krut Gödel (1906–1978) had shown it impossible to formulate all of mathematics from a finite system of axioms.[157]

1.3.2. *The problem with string theories*

String theory began some forty years ago yet the theory part of its name is still open to question. Is string theory a scientific theory at all or is it almost in a similar category as *intelligent design*?[158] As Karl Popper emphasized, a scientific theory must be capable of being falsified.[159] However, the means of potentially falsifying a scientific theory do not necessarily have to be available before a scientific hypothesis is advanced but it must be capable of being falsified in principle. If a narrower view of what constitutes science was taken, then experiment would have to precede every theory and progress in science would be difficult if not impossible; theory itself often suggests the experiments for its verification even if those experiments cannot yet be performed. Newton's theory of gravity explained the empirical data already gathered by Kepler. On the other hand, though Einstein's theory of general relativity explained the hitherto anomaly in the precession of Mercury, it also made further predictions which were not verified until later. Intelligent design is not scientific because it is not based on falsifiable facts and does not suggest experiments that would be capable of its falsification at any time in the future. String theory is not metaphysics because linear accelerators have too small an energy to probe the structure of strings by a factor of some 10^{16} but in practice such tests would be too costly if performable at all.

The literal interpretation of fundamental particles as strings is not necessary. Just as in quantum theory the imaginative visualization of fundamental particles as point particles means that they behave as if they were point particles not that they are in fact point particles. Again there are two ways of conceiving quantum theory as a particle or a wave and as Popper discussed 'there is ...no symmetry or duality between particles and waves; the waves describe the dispositional *properties* of the particles' not the actual particles.[138] However, string theory has other questions to be answered before it can be considered a true scientific theory. Tempers run high among the protagonists in the argument over the legitimacy of string theory. A definite verdict will only come with time which cannot be influenced by the effort put into string theory nor by the brilliance of its adherents; intemperate derogatory remarks, like the calling of those who question string theory's legitimacy the *popperazzi*, do not help. However the legitimacy of postmodern string or other postmodern theory may yet be fully established. Never say never.

The question of what causes even a scientific theory to be abandoned must also be briefly addressed. First it must be realized that no scientific theory explains the whole truth but simply explains particular phenomena that we perceive to be part of the real world. Almost certainly we will never be able to know everything about the real world and a scientific theory can only ever be an approximate explanation of it. Ptolemy's cosmology centred on the Earth was a scientific theory of its period that explained the motion of the planets and the stars but not the cause for that motion. The Copernican heliocentric system, based on circular orbits, was no more scientific than that of Ptolemy at the time but just simpler; even Kepler's elliptical orbits were only a simpler method of explaining cosmology in the absence of Newton's theory of gravity. Conventionalists such as Henri Poincaré (1854–1912) and Pierre Duhem (1861–1916) believed that the criterion of simplicity could be used to choose theories, but like aestheticism or beauty these attributes cannot be defined and are merely of a personal nature. However, Newton demonstrated that the Earth and the planets revolve around the Sun on approximately elliptical orbits because of gravity, which applied equally to bodies on this Earth, and gave scientific validity to Kepler's system and the Ptolemaic and the Copernican system of circular and epicyclic orbits were abandoned. With the passage of time Newton's theory of gravity itself was shown by Einstein not to be the whole truth. The falsification of the Newtonian prediction of the precession in the perihelion of Mercury did not lead to the abandonment of Newtonian mechanics but only to the realization that there were limits to its application. Newton himself knew that there was a problem with gravity acting at a distance and that the theory enabled the motion of the planets to be predicted, but did not explain how the force of gravity was communicated across vast distances. But then Einstein's general theory of relativity and the warping of spacetime while being a more accurate description of cosmology still does not answer why spacetime is affected by objects in its environment. The discovery of the theory of everything and an end to physics will never happen. Falsifiability is a necessary attribute of a scientific theory, but falsification of a prediction of a particular phenomenon does not necessarily lead to its abandonment, provided there are sufficient valid predictions, but establishes the limitations to that theory. As has been seen quantum theory enables some predictions to be made with extraordinary precision, but not all predictions seem true. In 1998 observations of a supernova showed that the universe was accelerating faster than previously thought implying the existence of a cosmological constant in Einstein's theory of general relativity.[160] However, quantum theory predicts a cosmological constant which is an incredible 120 orders of magnitude too large.[161] This conflict between theory and observation has caused intense questioning but has not led to the abandonment of quantum field theory. Singular problems of the same order with string theory need not lead to its abandonment.

There is absolutely no direct evidence that the fundamental particles have extension but neither is there for the point like particles of quantum mechanics. However what is promising is the graviton behaviour of some of the excitations of the particles in string theory. Whereas quantum mechanics resists gravity, superstring theory embraces it.[162] However superstring theory requires the folding of the superfluous dimensions that apparently has existed from the Big Bang and the beginning of time. However the greatest weakness is the 'lack of directly testable experimental predictions that would signal *string physics*,' though this weakness is apologetically termed only a shortcoming are there really testable predictions either indirect or direct?[163]

1.4. Concluding Remarks

Physics has always been part of our culture and has had a much wider influence than its direct effect on science and technology. We evolved through technology that enabled our hominid ancestors to become successful. In particular the exploitation of flaking to form stone tools enabled efficient butchering of carcasses and the increased nutrition led to an increase in the ratio of brain to body weight in the early hominids and eventually to human intelligence.[164] Benjamin Franklin called man 'a tool-making animal.' Although tool-making is not exclusive to man, he is the only animal to have relied upon tool making and technology for his survival and development. Mechanics was the first of the physical sciences to be developed. Initially mechanics was a practical art but the Greeks began to understand its principles. Many early developments in physics were not only for practical reasons, though the Ancient Egyptians used the helical rising of Sirius to predict the annual Nile flood, much of the early study of astronomy was for religious reasons, curiosity and wonder played their part too. Up until Newton the development of physics was piecemeal; he was the first to produce a unified approach. In this very brief survey of physics and astronomy a few of the most important milestones in classical, modern, and postmodern physics have been discussed. It is the wider cultural influence of physics that is the subject of the rest of this book.

1.5. Notes

[1]*Newtonian physics* is frequently called *classical physics*, especially in comparison with modern or postmodern physics, and I have treated the two expressions as synonymous. There are problems with using the adjective *classical* since historically it was reserved as an attribute of the ancient Greeks or Romans and it still retains that use in most cases. Thus the term *classical period* can only mean the ancient Greek and Roman period. I have tended to use the term *Newtonian* in those instances where I believe the use of *classical* is confusing.
[2]Shapiro (1991).
[3]Shapiro (1969).
[4]Technology was influencing the culture of our hominid precursors long before they evolved into human and could handle abstract ideas. Cotterell and Kamminga (1992). See also §1.4.
[5]Newton (1726).
[6]Newton (1718).
[7]Woit (2006).
[8]Rickles (2014).
[9]Eddington (1928).
[10]Snoblen (1998).
[11]Although the Earth's orbit is elliptical its eccentricity is quite small. The aphelion, or maximum distance from the Sun, is 1.52×10^8 km and the perihelion, or minimum distance from the Sun is 1.47×10^8 km.
[12]Published in 1630, after the death of Kepler, by his son-in-law Jacob Bartsch (ca. 1600–1633).
[13]Kepler calculated that the transit of Venus in 1631, while visible in America, would not be visible in Europe, which was the case.
[14] Galilei (1632, 1638).
[15] Salviati and Sagredo are named after Galileo's friends Filippo Salviati (1582–1614), and Giovanni Francesco Sagredo (1571–1620). Simplicio is named after Simplicius of Cilicia, a sixth-century commentator on Aristotle.
[16]Naylor (2003).
[17]Kepler had an idea that the Moon caused the tides, but his view was dismissed by Galileo as a 'useless fiction'. Finocchiaro (1989).
[18]The modern publishing company Elsevier take their name from this early publishing house, but Elsevir ceased publishing in 1712 and Elsevier has no connection other than name with this famous publishing company.
[19]Newton foresaw that his third law might be misinterpreted and also included a second version: In 'other words the actions of two bodies on each other are always equal and always opposite in direction.'
[20]In *The Discourse* Galileo went on to prove a corollary that the distances fallen during successive equal time intervals are in proportion to the series of the odd numbers 1:3:5:7: …
[21]Drake (1973).
[22]Lindberg (1965); Naylor (1973).
[23]Descartes (1644).
[24]Iltis (1971).
[25]McKie and de Beer (1951).
[26]The physicist Richard Feynman was famous for his habit of writing quantum theory in strip clubs

and, though I do not equate myself with either Newton or Feynman, when I was working on my doctorate, I regularly used to solve knotty problems in programming in the late 1950s while cycling between the Engineering Department and the Cavendish, where Cambridge's computer EDSAC 2 was located. I strongly believe that people do get some of their best ideas when they are relaxed and not specifically working, though I would not recommend thinking too hard while cycling in the traffic of the twenty-first century in Cambridge or anywhere else. Now I find my best ideas come when I wake in the still of the night.

[27]Kollerstrom (1999).

[28]See Cohen's *Brief History of the Principia*. Newton (1726).

[29]The perihelion is the point in a planet's orbit that is nearest to the Sun.

[30]Gascoigne (2004).

[31]Even at the end of the seventeenth-century many well educated men did still believe in witches. Joseph Glanvill (1636–1680) a member of the Royal Society wrote *Saducismus Triumphatus*, a book on witchcraft which was published, with material added by another Royal Society man, Henry More (1614–1687), posthumously in 1681.

[32]Mooney and Kirshenbaum (2009).

[33]Snobelen (2004).

[34]Fellows could not be married and Whiston's Fellowship came to an end in 1699 when he married Ruth Antrobus. Snobelen (2004).

[35]The first use of experiments in physics teaching was by Jacques Rohault (1620–1672) in his lectures on Cartesian natural philosophy given in France around 1670.

[36]Whiston (1715).

[37]Whiston (1716).

[38]Better known for his discovery of the gases which are now called: oxygen, sulphur dioxide, hydrogen chloride, and nitrous oxide.

[39]Priestley (1767).

[40]The invention of the torsion balance is commonly credited to the English geologist John Michell in ca.1750.

[41]Kant (1746-1749).

[42]Boyle did not publish an explicit statement of his law. In 1660 he published a book, *New Experiments Physico-Mechanical Touching the Spring of Air and its Effects*, but it was in answering a critic Francis Linus that he made an explicit statement of the law. In continental Europe this law is known as Mariotte's law but Edme Mariotte did not publish his work until 1676.

[43]Boyle (1660).

[44]Gay-Lussac (1802).

[45]Clapeyron referred to them as the Mariotte and Gay-Lussac laws.

[46]A mole of any gas contains Avogadro's number (6.022×10^{23}) of molecules.

[47]Clapeyron (1834).

[48]Chang and Li (2005).

[49]Thomson (1848).

[50]Mayer (1842).

[51]Earlier in 1798 Benjamin Thompson (1753–1824), better known as Count Rumford had demonstrated how effective mechanical work was at generating heat in his cannon boring experiments.

[52]Tait (1876a).

[53]Joule (1843, 1845).

[54]Reproduced in Watson (1947).

[55]Watson (1947).

[56]Colding (1864).
[57]Cahan (2012).
[58]English translation, Clausius (1851).
[59]See §3.2.2.
[60]Mach (1883).
[61]Clausius (1865).
[62]The sign convention used for the heat input to the system is the one usually used today and was used by Clausius in his earlier memoirs but in the ninth he reversed it.
[63]At the time J. J. Thomson called them corpuscles.
[64]Although the more usual modern spelling is *ether* I have chosen to use the rather archaic spelling *aether* so that there is no confusion with the organic chemical ether.
[65]Dalton (1808).
[66]Planck (1901).
[67]Eddington used the English definition of a billion that is 10^{12}. Apart from where it occurs in quotations, I have used the American definition 10^9 because it has become the norm in the twenty-first century.
[68]Einstein (1905a).
[69]Speed in any direction is a scalar quantity. Speed measured in a particular direction is velocity a vector quantity.
[70]A body in a gravitational field is indistinguishable from one that is accelerating. An accelerating frame of reference, like gravity, can be determined because a force will be detected on any mass.
[71]Lorentz (1904).
[72]A very good brief account of Minkowksi diagrams, but with more detail than I could justify here, is contained in Chang (1993).
[73]Chosen to give simple dimensions in the Minkowski diagram.
[74]He considered the motion of an electron, but remarked that the results are also valid for particles.
[75]Translation of a letter from Einstein to Lincoln Barnett, 19 June 1948.
[76]Einstein (1916a).
[77]Russell (2009).
[78]The surface of a sphere is two dimensional; it is also a closed finite surface whereas a flat sheet is open and can be of infinite extent. Since astronomy has shown that the universe originated with the Big-Bang, its spacetime is of finite extent. Finite three-dimensional space cannot be visualized but by analogy with the surface of a sphere the finite spacetime of the universe can be appreciated.
[79]Neumann (1896).
[80]Einstein (1917).
[81]Hubble (1929).
[82]Harvey and Schucking (2000).
[83]Einstein (1931).
[84]Friedmann (1922).
[85]Spergel (2015).
[86]Eddington (1916).
[87]Cunningham (1914).
[88]Cunningham, a Congregationalist, was also a pacifist whom I knew in the late 1950s but he was a modest man and I never knew at the time of his early work in relativity. His pacifism was first expressed during the Boer War and in the First World War he worked on the land rather than join the army and found it difficult to keep up with the advances in the theory of relativity. Unfortunately he never did any more major research after the First World War.
[89]Stanley (2003).
[90]Will (1988).

[91]Dyson *et al.* (1920).
[92]Will (2006).
[93]There were a very small number of scientists and the general public who refused to recognize Einstein's theories.
[94]An after dinner remark at Trinity College, Cambridge in 1933, Chandrasekhar (1987).
[95]Pound and Rebka (1959).
[96]Bausch and Meyers (2002).
[97]Abbott *et al.* (2016).
[98]Gibbons and Hawking (1971).
[99]Hawking (1988).
[100]Caballero-Nieves *et al.* (2009).
[101]Bender and Saglia (2007).
[102]Michell (1784).
[103]Preti (2009).
[104]Greene (1999).
[105]Pauli was awarded the Nobel Prize for Physics in 1945, but because of the war his lecture was delayed until December 1946.
[106]Jammer (1999).
[107]John Tyndall had found that the radiation of white hot platinum wire was 11.7 times that when it was dull red. He estimated that the temperatures of the dull red and white hot platinum wire to be 525°C and 1200°C respectively in 1879 Stefan noticed that 11.7 is equal to the fourth power of the ratio of the absolute temperatures. However Tyndall's estimates of the temperature were not very accurate and actually radiation from polished platinum is highly selective and the ratio is more nearly to the fifth power of the ratio of the absolute temperatures.
[108]Planck (1899).
[109]Einstein (1905b). The Nobel Prize for Physics awarded to Einstein in 1921 was 'for his services to theoretical physics, and especially for his discovery of the law of the photoelectric effect.'
[110]De Broglie was the first person to receive a Nobel Prize for a doctoral thesis.
[111]If the speed of the particles is comparable to the speed of light a relativistic mass must be used to calculate the momentum.
[112]Feynman (2011).
[113]Eibenberger *et al.* (2013).
[114]Bohr (1913, 1914, 1918).
[115]1 eV = 1.602×10^{-19} joules. The energy level is relative to that of an electron infinitely far from the nucleus.
[116]Jeans (1914).
[117]Bohr (1921).
[118]Pauli (1925).
[119]Heisenberg (1925).
[120]Born and Jordan (1925).
[121]Matrices are two dimensional arrays. The components of the matrix $\boldsymbol{p}$ are $p_{n,m}$ where m and n each can take the values of 1, 2, 3, ..., N or M, N and M being the dimensions of the matrix. In this context the dimensions of the matrices are infinite and n and m are the principal quantum numbers.
[122]Born *et al.* (1925).
[123]Heisenberg (1927).
[124]Heisenberg did not use the expression 'uncertainty principle' which is now the accepted term, but used the term 'Ungenauigkeit' or inaccuracy in his 1927 paper.
[125]Kennard (1927a,b).
[126]Ozawa (2003).

[127]Schrödinger (1926a).
[128]Schrödinger (1926b).
[129]Schrödinger (1926c).
[130]Born (1926).
[131]Dirac (1928a, b).
[132]Dirac (1931).
[133]Ray *et al.* (2014).
[134]Heisenberg and Pauli (1929, 1930).
[135]Lorentz (1904).
[136]Dirac (1978).
[137]Lamb and Retherford (1947).
[138]Popper (1982a).
[139]Feynman (1985a).
[140]Kinoshita (1990a, b).
[141]Khachatryan *et al.* (2015).
[142]Glashow (1961).
[143]Weinberg (1967).
[144]Salam (1968).
[145]Oerter (2005).
[146]Gell-Mann (1964).
[147]In an open letter to a group of nuclear physicists who were to meet at a conference in Tübingen, Germany.
[148]Cowan *et al.* (1956).
[149]Wheeler (1957).
[150]Green and Schwartz (1984).
[151]Witten (2005).
[152]Georgi and Glashow (1974); Georgi *et al.* (1974).
[153]Amaldi (1991).
[154]Candelas *et al.* (1985). Calabi-Yau spaces were conjectured by Eugenio Calabi in 1957 and proved by Shing-Tung Yau in 1977.
[155]A form of symmetry in the mathematics of the five superstring theories, called duality, was found whereby they were not independent of each other.
[156]Witten (1996).
[157]Hawking (2002); Hawking and Mlodinow (2010); Gödel (1931).
[158]Smolin (2006); Woit (2006).
[159]Popper (1959a).
[160]Riess *et al.* (1998).
[161]Sahni (2002).
[162]'t Hooft (2013).
[163]de Haro (2013).
[164]Cotterell (2010).

Part II

The Newtonian Period

Chapter 2

Religion

Science has had an ambivalent influence on religion. It has been used to bolster religious beliefs and condemned where it has been seen to contradict religious dogma. Anaxagoras (ca. 500–428 BC) was banished from Athens in about 450 BC for the impiety of believing that the Sun was a red-hot stone and in the third century BC Aristarchus (ca. 310–230 BC) was persecuted by Cleanthes for teaching that the Sun not the Earth was the centre of the Universe. In the Christian Church the use of natural philosophy to understand the world that God had created began with the church fathers, especially St Augustine (354–430). The distinction between reason and revelation was emphasized, if there was only one truth which could be gained from reason or revelation it was safer to go from revelation to reason. After the fall of the Roman Empire in the fifth century most of the Greek works on science were lost to the Western World, but by the thirteenth century these works, especially those of Aristotle, and commentaries by medieval Islamic philosophers such as Ibn Sina Avicenna (ca. 980–1037) and Ibn Rushd Averroes (1126–1198) had been rediscovered. At the same time universities were forming in the West. Aristotle's works on logic, known collectively as the *Organon*, were already available and had caused no problems with the Church. Aristotle's natural philosophy was more problematic. The new mendicant orders led by the Dominicans began to establish themselves in the University of Paris. The secular professors at first welcomed the mendicants, but they did not respect the University's privileges and autonomy. The teaching of the natural philosophy of Aristotle was condemned in 1210 but the autonomy of the University protected the professors. In 1254 the great Dominican, St Thomas Aquinas (1225–1274) studied at the University of Paris from 1245 to 1248 returning again in 1252. Proselytization by the mendicant orders at the University led to a confrontation with the secular professors during 1250–1256. Aquinas, who was appointed a professor in theology in 1256, defended the mendicants. From 1265 to 1274 Aquinas wrote his famous *Summa Theologiæ* producing a comprehensive philosophical doctrine that blended Catholic orthodoxy with Aristotelian philosophy despite the 1270 condemnation in Paris of some of

Aristotle's natural philosophy and the Averroist interpretations. Aquinas died in 1274 before his *Summa Theologiæ* was completed. In 1277 there was a further condemnation of aspects of Aristotelian natural philosophy including some that had been incorporated by Aquinas in his works. It is believed by some that the condemnations had a positive effect on natural philosophy by liberating it from Aristotle's strangle hold.

Traditionally Martin Luther (1483–1546) initiated the Protestant Reformation in 1517 by nailing his ninety-five theses on the door of All Saints' Church in Wittenberg.[1] The Reformation was a revolt against Catholic authority. Luther brought a new era where he believed we would recover the knowledge of the external world that was lost through the fall of Adam. For Luther the reality of experience was open to rational thought and not a subject of faith. Thus his influence on science was generally benign. From the end of 1520, Luther was accustomed to hold regular discussions for his young followers at his home and much has been made of Luther's derogatory remarks he made at one in 1539 on Copernicus' heliocentric hypothesis, which though not yet published was known.[2] However science was allowed to flourish unhindered in Protestant countries and even the Catholic Church took no direct action against Copernicus' hypothesis for seventy years after the publication of *De Revolutionibus*.

Although the more traditional natural philosophers still believed in the Ptolemaic System at the beginning of the seventeenth century, by about 1633 nearly all astronomers and mathematically competent philosophers believed in heliocentricity.[3] The Holy Office condemned heliocentricity in 1616, but by about 1630 the effect of the condemnation had weakened. Kepler was a Lutheran but, as an Imperial Mathematician to the Holy Roman Emperor Rudolf II, he was allowed to live in the Catholic city of Prague from 1600–1612. Rudolf II abdicated in 1611 and the Emperor Matthias, Rudolf's brother, appointed him the district mathematician in Linz where he lived until 1628. Until 1633 there was no attempt to identify and discipline anyone with Copernican leanings in Catholic cities and Kepler was unhindered in his work on the motion of the planets. In Protestant countries the condemnation of heliocentricity by the Holy Office naturally had no effect. Even in Catholic countries after the condemnation of Galileo in 1633 there was no attempt to identify and condemn anybody professing heliocentricity. Galileo's condemnation did cause self censure by some, particularly Descartes who abandoned publishing *Le Monde* which was a Copernican publication even though he was living in Holland and beyond the reach of the Holy Office.[3] In fact the French ecclesiastical authorities did not promulgate the decree of the Holy Office that the Copernican system was heretical and after about ten years the condemnation was almost forgotten outside of Rome and there were no prosecutions of those who quietly propagated the Copernican system. By 1644 Descartes felt safe in publishing his *Principia Philosophiae*, which contained his heliocentric vortex system, in Amsterdam. Finally in 1744 permission was given for the publication of Galileo's *Dialogues* in Rome itself.[3] However, a formal apology for all the mistakes committed by Catholics, including the trial of Galileo, was only issued by Pope John Paul II in 2000, though in 1939 Pope Pius XII, in his first speech to the Pontifical Academy of Sciences, had included Galileo among the 'most audacious heroes of research… not afraid of the stumbling blocks and the risks on the way, nor fearful of the funereal monuments.'[4]

In the seventeenth century natural philosophy began to be used as an argument in support of Christianity and as a unifying force in Protestant England. Puritanism began with those who thought that the Church of England restored by Elizabeth I was not radical enough. With the accession of James I in 1603 the Puritans again pressed for reform of the Church of England. Disenchanted with the Hampton Court Conference of 1604 some separated but most remained within the Church. Under Charles I (1625–1649) there was less tolerance of Puritans and the high church William Laud, who became Archbishop of Canterbury in 1633, viewed Puritans as a schismatic threat. During the Civil War (1642–1651) the Long Parliament abolished episcopacy in the Church of England and replaced it with a Presbyterian structure which in effect caused a breakdown in religious uniformity and allowed independent sects to flourish. With the Restoration of Charles II the Act of Uniformity was reinstated in 1662 and the majority of clerics conformed. After the Restoration many churchmen were indifferent to the particular forms of church worship or government and fostered an inclusive approach to the church and became known as the Latitudinarians the term arising from a tract written by Simon Patrick (1626–1707), Bishop of Ely in 1662 entitled *A Brief Account of the New Sect of Latitude-Men together with some Reflections upon the New Philosophy*.[5]

The Latitudinarians developed what they called natural theology or religion often using natural philosophy in their reasoning. The movement had its beginnings in what became known in the nineteenth century as the Cambridge Platonists whose most important members were Henry More (1614–1687) and Ralph Cudworth (1617–1690). The Cambridge Platonists rejected Aristotelian philosophy and were well acquainted with the new developments in science of Galileo and Descartes which they used in their development of natural theology but, while they welcomed the new natural philosophy, they became worried by its possibilities. However, in general latitudinarians believed that revelation could be supplemented by natural philosophy. The Cambridge Platonists were primarily clergymen and relied on the scientific ideas of others. Other latitudinarians such as John Wilkins (1614–1672) and Robert Boyle, who were both founders of the Royal Society, did study science for its own sake. Of course, in the seventeenth century the professions were not distinct. Every fellow of the universities had to take holy orders. Wilkins, who was first Master of Wadham College, Oxford and then Trinity College, Cambridge during the interregnum, became Bishop of Chester near the end of his life after the Restoration. Robert Boyle, a son of the Earl of Cork the wealthiest man in Britain, did not have to obtain any office and did not take holy orders, but he was a very devout Christian. In 1663 Boyle published a book: *Some Considerations Touching the Usefulness of Experimental Natural Philosophy*, in the first part of which he discussed the religious value of study of the natural world. It is significant that Wilkins' main theological work: *Of the Principles and Duties of Natural Religion*, published posthumously in 1675, while showing his mathematical and scientific background in his style of writing with the use of postulates, definitions and axioms, relied mainly on reason for his arguments and only touched on science very occasionally to stiffen his arguments. Thomas Burnet (ca. 1635–1715) joined Clare College, Cambridge, which was Cudworth's College, in 1651 and followed him when he moved to Christ's College where Burnet spent twenty-one years as a fellow teaching the new philosophy. In 1681 Burnet published the first two books of his *Sacred Theory of the Earth* in Latin, which was translated into English in 1684.[6] Burnet expanded on the account of the creation and form of the Earth given by Descartes in his *Principia Philosophiae*. In the first book Burnet suggests that the antediluvian Earth must have been smooth since it was impossible to flood the whole present Earth to the tops of all the mountains as described in the biblical account of Noah's Flood. The second book deals with the state of the primaeval Earth, and paradise. With the *Sacred Theory of the Earth*, Burnet started a debate on the proper use of natural philosophy to comment on the Bible. Many latitudinarians believed in the coming of the millenarian paradise[7] and the third and fourth books of Burnet's *Sacred Theory of the Earth*, published in Latin in 1689[8] dealt with the coming conflagration of the world and the creation from it of a new heaven and a new Earth. In his third book Burnet, while not denying the possibility of miracles, saw that not all so-called miracles were evidence of God intervening in the laws of nature and wrote 'the first Rule concerning miracles is this, That we must not flie to miracles, where Man and Nature are sufficient.' This ambivalent approach to miracles was adopted by Newton and his followers. Burnet's *Sacred Theory of the Earth* was very popular and there were seven editions, the last being published in 1759.

Burnet's four books used unsubstantiated hypotheses and his *Sacred Theory of the Earth* attracted considerable criticism from both theologians and natural philosophers. Burnet sent a prepublication copy of his book to Newton in late 1680 and Newton read the book and comments on it in a letter which, apart from a section quoted by Burnet in his reply, has been lost.[9] Part of the fragment reads:[10]

> 'Ye heat of ye Sun rarefying that side of ye Chaos that ley next it, or by ye pressure of ye vortex or of ye Moon upon ye Waters, some inequalities might bee made in ye Earth, & then ye waters flowing to those lower parts or cavities would make ye seas there, & ye upper parts of ye Earth towards ye poles which they flowd from, would bee dry land. And all this might ye rather bee, because at first wee may suppose ye diurnal revolutions of ye Earth to have been very slow, soe that ye first 6 revolutions or days might containe time enough for ye whole Creation.'

This fragment is interesting because the first sentence shows that Newton, though advancing the idea of universal gravity with the phrase *pressure of the moon*, had not yet completely given up Descartes vortex theory. The second sentence shows that Newton considered that the Genesis account of the creation of the world out of chaos in six days was probably not literally true. In his reply to Burnet's second letter, Newton advances a hypothesis of how the days at the beginning of creation could have been much longer.[9] He supposed that the Earth, once created to become more or less its present form after the second day of creation, did not at first rotate, but by the application of a constant torque it started rotating. If the 'first revolution was done in one of our years, in the time of another year there would be three revolutions, of a third five, of a fourth seven, etc. and of the 183rd year 365 revolutions,' which follows from the concept of a constant torque being applied to the Earth. The tenor of Newton's letter is that there are many possible hypotheses for the creation of the Earth and Noah's flood but that it is impossible to determine which if any of them are true. As he wrote: 'All this I write not to oppose you, for I think the main part of your hypothesis as probable as what I have here written, if not in some respects more probable,' and as he stated in his last sentence: 'I have not set down any thing I have well considered, or will undertake to defend.' What one sees is a forerunner of Newton's famous statement in the *Principia* that 'I do not feign hypotheses.' Other critics were not so restrained as Newton, John Keill, a young follower of Newton at Oxford, wrote a blistering attack on Burnet's *Theory* in his *Examination of Dr Burnet's Theory of the Earth* (1698) where he used mechanical analysis to show that many of Burnet's hypotheses were shallow.[11] The influential French mathematician, cosmologist and naturalist of the eighteenth century, Georges-Louis Leclerc, Comte de Buffon (1707–1788), in 1749 described the *Sacred Theory of the Earth* as 'an elegant romance, a book which may be read for amusement, but cannot convey any instruction.'[12]

The latitudinarians strongly believed that natural philosophy had to be consistent with revelation. This synthesis was opposed by freethinking deists and atheists who either thought that the nature of a non-interventionist supreme creator of the universe could be determined from natural world without the help of revelation, or that a creator was unnecessary. In the seventeenth century the freethinkers' philosophers were Thomas Hobbes (1588–1679) and Baruch Spinoza (1632–1677) who were popularly thought to be atheists. Hobbes' great work was the *Leviathan* published in 1651. The *Leviathan*, particularly the last part, had an anti-theological tone. His views on religion gained him many enemies; he was attacked by contemporaries such as Henry More and Ralph Cudworth and again later in the eighteenth century. Spinoza was born a Jew in Amsterdam and lived all his life in Holland. His main work was *Ethics* which was only published posthumously. The Jews excommunicated Spinoza in 1656. Spinoza was, what the Irish deist John Toland (1670–1722) later called a pantheist, that is, he believed God and Nature or the Universe, were identical. All was God, individual souls and matter were not individual things but simply attributes of an infinite Divine Being. Spinoza was naturally abhorred by orthodox Christians, though he was buried in sanctified ground in the churchyard of the Nieuwe Kerk in The Hague. Both Hobbes and Spinoza were seen as the faces of atheism in the seventeenth century.

Although on the Continent Descartes' system of natural philosophy contained in his *Principia Philosophiae*, was widely followed until the middle of the eighteenth century, in England Newton's system was quickly adopted by most to explain natural religion. Newton, while he was vitally interested in religion and wrote many theological papers, published none. It is appropriate to begin a discussion on the interaction between natural philosophy and religion with Newton's own view.

2.1. Newton: His Religion and Natural Philosophy

There is a tendency to over analyze the writings of famous men such as Newton. Inconsistencies and contradictions in letters and manuscripts, which Newton himself did not publish are not surprising. Even the ideas of great men are not always fruitful and in my view most emphasis should be given to their published works not their letters and manuscripts which can contain their prior attempts at understanding problems. Newton published little on the relation of natural philosophy to religion. In the first edition of the *Principia* there is only one mention of God and one to Holy Scripture. However in a letter to Richard Bentley written in answer to Bentley's questions when he was preparing his Boyle Lectures for publication in 1692, Newton wrote of the third book of the *Principia*:[13]

> 'When I wrote my treatise about our system [of the world], I had an eye upon such principles as might work with considering men, for the belief of a deity.'

Newton's published remarks on the relationship between natural philosophy and the nature of God are contained in the General Scholium[14] to the *Principia* which he added to the 1713 second edition and in the *queries* at the end of the second edition of *Opticks*.[15] In the General Scholium, Newton repeats the substance of his remark about the relation between religion and natural philosophy writing that 'to treat of God from phenomena is certainly a part of natural philosophy.' Also in query 28 of the *Opticks* Newton discussed the main purpose of natural philosophy which he saw to be 'to argue from phenomena without feigning hypotheses for explaining all things mechanically, and to deduce causes from effects till we come to the very first cause, which certainly is not mechanical.' He continued this theme in query 31, 'it's unphilosophical to seek for any other origin of the world [other than God], or to pretend that it might arise out of a chaos by the mere laws of nature.' The same view is expressed in the General Scholium to the *Principia*:

> 'all these regular motions [of the planets] do not have their origins in mechanical causes, since comets go freely in very eccentric orbits and in all parts of the heavens... This most elegant system of the Sun, planets and comets could not have arisen without the design and dominion of an intelligent and powerful being.'

These quotations form part of Newton's design argument for the existence of God. However, Newton thought that God might have to intervene at times to ensure the stability of the world. The quotation from query 31 given above continues:

> 'though being once form'd, it [the world] may continue by those laws for many ages. For while comets move in very excentrick orbs in all manner of positions, blind fate could never make all the planets move one and the same way in orbs concentrick, some inconsiderable irregularities excepted, which may have risen from the mutual actions of comets and planets upon one another, and which will be apt to increase, till this system wants a reformation.'

This passage prompted Leibniz to complain in a letter written to Caroline of Ansbach (1683–1737), the Princess of Wales that: [16]

> 'Sir Isaac Newton, and his Followers, have also a very odd Opinion concerning the Work of God. According to their Doctrine, God Almighty wants to wind up his Watch from Time to Time: Otherwise it would cease to move. He had not, it seems, sufficient Foresight to make it a perpetual Motion. Nay, the Machine of God's making, is so imperfect, according to these Gentlemen; that he is obliged to clean it now and then by an extraordinary Concourse, and even to mend it, as a Clockmaker mends his Work.'

Newton also thought about the stability of the universe as a whole. In his letter to Bentley on the 10th December 1692, he stated that if the universe were finite then gravity would eventually:[9]

> 'fall down into the middle of the whole space and there compose one great spherical mass. But if the matter was evenly dispensed throughout an infinite space, it could never convene into one mass, but some of it would convene into one mass and some into another so as to make an infinite number of great masses scattered at great distances from one to another throughout all that infinite space.'

Newton does not state whether he believed the universe to be finite or infinite, but appears to have favoured a finite universe since David Gregory reported in 1694 that Newton observed:[17] 'that a continual miracle is needed to prevent the Sun and the fixed stars from rushing together through gravity.'

It has been suggested that another interpretation to the phrase in query 31: 'till this System wants a Reformation' that Newton did not mean that God would intervene to establish the regular motion of the solar system but that the Reformation was in fact the Apocalypse predicted in the Bible.[18] Newton was definitely a millenarian and he made many estimates of the date of the Apocalypse, ranging from 1869 to 2370.[19] However in the *Principia*, Newton stated: 'that the motions of the planets in the heavens may subsist an exceedingly long time.' Even if Jupiter was revolving in a medium as dense as the upper air on Earth, he stated in the *Principia* that it 'would not lose by the presence of the medium the 1,000,000th part of its motion in 1,000,000 years', so he could not think that the seven hundred years given by his latest biblical estimate could be due to the laws of nature. Newton was a convinced, albeit unorthodox, Christian, some of his theology may very well have been influenced by natural philosophy. His antitrinitarianism is very obvious in the General Scholium; Christ is very rarely called God in the Bible and Newton points out that 'the word "god" is used far and wide to mean lord.'[20] However, his religion came from his biblical studies not natural philosophy. If Newton's natural philosophy had played a large part in his theology then would he not have paid more attention to possible physical causes of the end of the world such as the instability of the solar system or the collapse of a finite universe due to gravity? Newton clearly believed that natural philosophy proved the existence of a deity who was:[21]

> 'eternal and infinite, omnipotent and omniscient, that is he endures from eternity to eternity, and he is present from infinity to infinity: he rules all things and he knows all things that happen or can happen,'

but without revelation this only leads to deism. Although many natural philosophers of the eighteenth century were deists, Newton certainly was not. Thus, though Newton saw that natural philosophy was not inconsistent with Christianity and provided evidence in support of the existence of the 'Lord God *Pantokrator* [all-powerful]'; his theology, like the quotation, came from the Bible.

The idea that there were laws of nature imposed by God on matter grew out of the natural philosophy of the seventeenth century. The first use of the term *laws of motion* was by John Wallis (1616–1703) and Christopher Wren (1632–1723) in 1669.[22] Newton frequently described the mechanical relationships in his *Principia* as laws. Although Newton did not use the more general term *laws of nature* in the *Principia* he did in his *Opticks*. These laws were believed to be independent of time and place. There must have been a conflict in Newton's mind because the General Scholium makes it plain that he believed God was omnipotent and could do anything even alter the laws of nature and he believed in miracles. In a letter written to John Locke (1632–1704) in 1691, Newton stated that 'Miracles of good credit' occurred in the first two or three years of the Christian era; contemporary miracles were Catholic and denied by the Church of England.[9] He believed in prophesies of the Bible and also, as indicated above, thought that God might have to intervene to correct the motion of the planets, all of which require miracles. However as the Scottish philosopher David Hume (1711–1776) wrote in the middle of the eighteenth century: [23]

> 'A miracle is a violation of the laws of nature; and as a firm and unalterable experience has established these laws, the proof against a miracle, from the very nature of the fact, is as entire as any argument from experience can possibly be imagined.'

If it is argued that the laws of nature are incompletely understood then what seems to be a miracle may not contravene the real laws of nature. On the other hand if it is denied that there are immutable laws of nature then everything is a miracle and the word has no meaning. John Wilkins would have agreed with Hume's definition of a miracle and Boyle wrote of God's laws of nature:[18]

> 'he has not bound up his own hands by them, but can invigorate, suspend, overrule, and reverse any of them as he thinks fit.'

However, Newton realized that there were problems with such a definition of a miracle. Now just as Thomas Huxley (1825–1895) was known as Darwin's Bulldog for his advocacy of Charles Darwin's (1809–1882) theory of evolution, Newton's Bulldog for his natural philosophy was Samuel Clarke. Newton and Clarke fell back on the St Augustine's definition of a miracle as:[18]

> 'anything which appears arduous or unusual, beyond the expectation or ability of the one who marvels at it.'

Thus in the published version of his 1705 Boyle Lectures , Clarke wrote[24]

> 'the true definition of a miracle... is a work effected in a manner unusual or different from the common and regular method of Providence, by the interposition either of God himself, or some intelligent agent superior to man.'

Leibniz disagreed strongly with Newton's mathematical view of natural philosophy. The dispute between Leibniz and the Newtonians had begun in 1705 over who had first invented calculus. The dispute broadened. In 1710 Leibniz criticized Newton theory of gravity on the grounds of the impossibility of force acting at a distance. Roger Cotes[25] (1682–1716) answered Leibniz's criticism in his preface to the second edition of the *Principia*.[26] The dispute over the theological implications of Newton's natural philosophy started with the letter Leibniz wrote to Caroline, the Princess of Wales, quoted above, which started an exchange between him and Clarke during 1715–1716. In his reply to the first of Leibniz's letters, Clarke affirms the view of the Newtonians that God's specific intervention affirms his all powerful nature and wrote:[16]

> 'The notion of the world's being a great machine, going without the interposition of God, as a clock goes without the assistance of a clockmaker; is the notion of materialism and fate, and tends... to exclude providence and God's reality out of the world'

Leibniz counters this statement with:[16]

> 'God has foreseen everything, he has provided a remedy for everything before-hand; there is in his works a harmony, a beauty, already pre-established.'

2.2. The Boyle Lectures

Robert Boyle established a series of lectures in his will, which became known as the Boyle Lectures, where a clergyman was elected:

> 'to preach Eight Sermons in the year, for proving the Christian Religion, against notorious Infidels, *viz.* Atheists, Theists, Pagans, Jews, and Mahometans, not descending lower to any Controversies, that are among Christians themselves... to answer such new Objections and Difficulties as may be started, to which good Answers have not been made.'

The Glorious Revolution of 1688 had secured the position of the Church of England but the Toleration Act of 1689 allowed freedom of worship to those Nonconformists who had pledged to the Oaths of Allegiance and Supremacy but not to Catholics. Boyle understood that only a broad natural theology could enable the Church to formulate an acceptable Protestantism.[27] The four trustees named to administer the Boyle Lectures were the diarist John Evelyn (1620–1706), the cleric Thomas Tenison (1636–1715), Bishop of Lincoln and later Archbishop of Canterbury, the politician Henry Ashurst (1645–1711), and the lawyer John Rotherham (ca. 1630–1708). Of these trustees Evelyn and Tenison were latitudinarians. Not enough is known about the other two trustees to determine their views definitely; certainly a significant number of the original Boyle Lecturers appointed by these trustees, and all those who used natural philosophy in their lectures, were latitudinarians.

The original series of Boyle Lectures ran from 1692 to 1732 and then continued spasmodically until 1905; in 2004 the lectures were restarted. Originally the Boyle Lectures consisted of eight sermons delivered throughout the year. The first Boyle Lecturer, Richard Bentley took as his topic in 1692 *The Folly and Unreasonableness of Atheism* and made extensive use of the *Principia* in his arguments and these sermons contain the first popular presentation of Newton's *System of the World*.[28] Bentley also delivered the Boyle Lectures for 1694 but this series was not published and the manuscript is lost. The failure to publish his 1694 Boyle Lectures displeased Tenison.[29] Other Boyle Lecturers who made substantial use of the natural philosophy of the *Principia* were Samuel Clarke and William Derham. All three of these lecturers presented in part the design argument for the existence of God. The published versions of their Boyle Lectures were very popular and widely read during the eighteenth century. John Harris gave the Boyle Lectures for 1698 and while he did not directly use the *Principia* in his sermons, he did in his fifth lecture refute Toland's concept of that motion was essential to matter.[30] William Whiston, Newton's successor to the Lucasian professorship, gave the Boyle Lectures in 1707. As Newton probably had a hand in his appointment as a Boyle Lecturer, one might have expected that Whiston would have made use of Newtonian natural philosophy in his sermons, but instead he chose to lecture on the accomplishment of scriptural prophecies.

2.2.1. *Richard Bentley* (1662–1742)

Richard Bentley was neither a natural philosopher nor a mathematician and it is unclear what motivated him to undertake to refute atheism making use of the *Principia*. He probably studied some natural philosophy and mathematics and may have attended lectures by Newton when a student at St Johns College, Cambridge during 1676–1680; though the curriculum at Cambridge in the seventeenth century was vague and still based on the medieval trivium of grammar, rhetoric, and logic and the quadrivium of arithmetic, geometry, astronomy, and music, it also contained the three philosophies: natural, normal, and mental. However, there is no record of Bentley having an interest in natural philosophy between 1680 and July 1691 when Newton sent him a reading list to enable him to understand the *Principia*. Since this list dates to before Boyle's death on New Year's Eve 1691 it could not have been associated with the Boyle Lectures and must mark the beginning of an interest in natural philosophy.[31] There is no concrete evidence that Newton influenced the appointment of Bentley as the first Boyle Lecturer, but there is a consensus that he did have a hand in it, which may be because of Bentley's recent correspondence with him. Bentley was at a disadvantage when he wrote his sermons using Newtonian precepts because the General Scholium of the *Principia*, and the queries of the *Opticks* had not been written. When Bentley was preparing his sermons for publication he did write to Newton asking his advice on a number of points unfortunately, while all four of Newton's replies exist,[32] only the last of Bentley's letters written on the 18th February 1693 survives.

It was in his fourth sermon, first published in 1693, that Bentley made use of natural philosophy for the first time writing that:[28]

> 'without gravity, the whole universe... would have been a confused-chaos, without beauty or order, and never stable and permanent in any condition. Now it may be prov'd in its due place, that this gravity, the great basis of all mechanism, is not it self mechanical; but the immediate fiat and finger of God.'

It would appear that this published version of the fourth sermon was written before Bentley could have received any letters from Newton, if this evocative statement is solely the work of Bentley, it is remarkably similar to the sentiment of Newton's statements in the *Principia* and the *Opticks*. While the statement does not prove that because the mechanism of gravity was unknown that God must be the cause, it was good supporting evidence. Bentley may in his original sermons have assumed that gravity was essential to matter, which contradicts his statement that it 'is not it self mechanical', since in his second letter to Bentley, written on the 17th January, Newton rebukes Bentley writing:

'you sometimes speak of gravity as essential and inherent to matter. Pray do not ascribe that notion to me; for the cause of gravity is what I do not pretend to know.'

In his published sermons, while he never explicitly stated that gravity was essential to matter, Bentley allowed the possibility that it was to show that the existence of God could still be proved. However, in his seventh sermon Bentley attempted to prove that gravity cannot be innate and essential to matter.

It was the last three sermons that gave 'the proofs of a deity from the inanimate part of the world' that naturally made the most use of Newton's *Principia*. Bentley first set out to prove that:

'it impossible that the parts of our world, the Sun and the planets with their regular motions and revolutions, should have subsisted eternally in the present or a like frame and condition.'

Bentley saw the necessity of proving that the Earth and the solar system did not always exist because the Bible describes the creation of the world and the Irish Archbishop James Ussher (1581–1656) had calculated from its chronologies that the Earth was created in 4004 BC. Newton's first letter of the 10th December 1692 did discuss a possible natural philosophical argument, which Bentley had queried, that a finite universe would ultimately collapse and hence the solar system could not always have existed. Possibly Newton's letter was not received before the publication of the sixth sermon and Bentley, unsure of his natural philosophy, tried instead to prove the impossibility of the world always having existed from mathematical arguments based erroneously on the assumption that all infinities are equal as Newton wrote in his second letter on the 17th January 1693. Bentley's second proposition in the sixth sermon 'that neither matter universally and abstractly consider'd, nor motion as its attribute and property, can have existed from all eternity, [was a] metaphysical speculation... [and] of far less moment and consequence than the rest [of the propositions],' and he used only metaphysical arguments for it.

In his seventh sermon Bentley tried to prove his third proposition:

'that, though we should allow the atheists, that matter and motion may have been from everlasting; yet if (as they now suppose) there were once no Sun, nor stars, nor Earth, nor planets; but the particles, that now constitute them, were diffused in the mundane space in manner of a chaos without any concretion or coalition; those dispersed particles could never of themselves by any kind of natural motion, whether call'd fortuitous or mechanical, have conven'd into this present or any other like frame of heaven and Earth.'

The only existing letter (18th February 1693) from Bentley to Newton and the last two letters from Newton (11th and 25th February) are concerned with this seventh sermon. Newton wrote in his letter of the 11th February:[32] that he

> 'had considered... [the proposition in this sermon] very little before your letters put me upon it. [Newton then wrote:] The hypothesis of matter's being at first evenly spread through the heavens is, in my opinion, inconsistent with the hypothesis of innate gravity, without a supernatural power to reconcile them, and therefore it infers a deity.'

Since at the time of Bentley's sermons the *Principia* was only understood by a few, he started this sermon with an outline 'of the most principal and systematical phænomena, that occur in the world now that it is formed.' He started this outline with a discussion of gravity and, *inter alia*, stated that:

> 'since gravity is found proportional to the quantity of matter, there is a manifest necessity of admitting a vacuum, another principal doctrine of the atomical philosophy.'

Newton wrote very little about corpuscles or atoms. In Corollaries 3 and 4 to Proposition 6 of Book III of the *Principia* a corpuscular theory of matter is used to justify the existence of a vacuum, which was denied by the Cartesians; Newton argued that 'all spaces could not be equally full' because then the 'extreme density of matter' would cause air be as dense as 'quicksilver or gold.' Newton implied in Corollary 4 that the corpuscles that make up matter are of the same density and that the difference in different materials is the ratio or pores or vacuum to corpuscles in a material. Newton wrote a little more about corpuscles in query 31 to the *Opticks* but, of course, that was not available to Bentley. The seventeenth century saw a revival of atomism to which Robert Boyle contributed and Bentley used Boyle as well as Newton in this section of his sermon. Bentley calculated the ratio of space to matter within the orbit of the Earth, correcting his original estimate of its diameter using Newton's letter,[33] but not the calculation whose numerical value is unimportant. Having shown how isolated every corpuscle of matter would be if the all the matter in the universe were uniformly dispersed from which 'a right conception and imagination of the supposed chaos' prior to the formation of the universe, Bentley concluded that it was impossible for the world as we know it being formed by mechanical means out of chaos. There is also the problem that Newton had seen with gravity acting at a distance. In his letter to Bentley,[33] Newton wrote:

> 'That gravity should be innate, inherent and essential to matter, so that one body may act upon another at a distance thro' a *vacuum*, without mediation of anything else, by and through which this action and force may be conveyed from one to another, is to me so great an absurdity, that I believe no man who has in philosophical matters a competent faculty of thinking can ever fall into. Gravity must be caused by an agent acting constantly; but whether this agent be material or immaterial, I have left to the consideration of my readers.'

Bentley conveniently forgot the last sentence and paraphrased the first sentence as a proof that there was a divine agent.

In his eighth and last sermon Bentley undertook to prove that:

> 'That the order and beauty of the inanimate parts of the world, the discernible ends and final causes of them, ... , or a meliority above what was necessary to be, do evince by a reflex argument, that it is the product and workmanship, not of blind mechanism or blinder chance; but of an intelligent and benign agent, who by his excellent wisdom made the heavens and Earth: and gives rains and fruitful seasons for the service of man.'

Clearly a luminescent Sun is necessary for life on Earth and Bentley obviously had questioned Newton in his first letter[34] as to why some heavenly bodies were luminous and some opaque because Newton replied that 'I do not think [it] explicable by meer natural causes, but am forced to ascribe it to the counsel and contrivance of a voluntary agent.' In his last sermon Bentley made use of Newton's letter and argued that it was

> 'the contrivance and choice of a wise and good being; that the central Sun should be a lucid body, to communicate warmth and light and life to the planets around him.'

Bentley at this point in his sermon did not limit life to the Earth.

The next two points that Bentley made in his last sermon were that the regularity of the motion of the planets orbiting the Sun in the same direction within the plane of the ecliptic could not have occurred by chance. Newton had also made this point in his first letter.[34] At the time these facts seemed a strong argument for a divine creator, Laplace developed the nebular hypothesis of the formation of the solar system in 1796.[35] The nebular hypothesis, that the solar system condensed from a swirling mass of gas, had its roots in earlier works by Emanuel Swedenborg (1688–1772) and Immanuel Kant.[36,37] After the Sun condensed, the nebula contracted and its angular rotation speeded up, as required by Newtonian mechanics, and the planets formed from the nebula. Laplace's hypothesis explained why the planets all rotated in the same direction and weakened the argument for divine creation. Laplace is said to have replied to a remark of Napoleon that his *Exposition du Système du Monde* contained no reference to God that 'I had no need of that hypothesis.' The present model of the evolution of the solar system is still similar to Laplace's nebula hypothesis.[38] The rest of the points in this sermon concern the uniqueness of the Earth and its suitability for life and are not directly related to the *Principia*. Having admitted that there could possibly be life on other planets, Bentley argues that, since the heat from the Sun varies in proportion to the square of the distance from it, the atmospheres of the inner planets, Mercury and Mars would be too hot and that of the outer planets too cold for life. This point and the rest in the sermon simply beg the question. We have life as we know it on Earth because of its position in the solar system, its diurnal rotation, and the inclination of its axis to the ecliptic plane, not that the Earth is like it is so that it can support life.

Although Bentley in no way proved the existence of a deity within the limitations in knowledge of natural philosophy of the day, he did develop some arguments from the *Principia* and may have started Newton thinking about adding the General Scholium discussing the implications of natural philosophy to it.

2.2.2. *Samuel Clarke* (1675–1729)

Samuel Clarke gave the Boyle Lectures in 1704 and was reappointed in 1705 because, as Tenison said, 'persons of such abilities in theology, philosophy, and mathematics are not to be commonly found.'[29] Clarke had good credentials in natural philosophy though he was never elected to the Royal Society.[39] He was a scholar of Gonville and Caius College, Cambridge from 1691 to 1695 and his disputation for his bachelor's degree was a Proposition from the *Principia* and helped to earn him a college fellowship. Clarke continued his interest in natural philosophy. Newton asked him to translate his *Opticks* into Latin and Clarke defended the theological interpretations of the Newtonians from Leibniz's attacks, but he made no contribution to it.

The title of Clarke's first Boyle Lecture series in 1704 *was A Demonstration of the Being and Attributes of God: More Particularly in Answer to Mr Hobbs, Spinoza, And their Followers.*[40] As Clarke wrote in the published version of his second Boyle Lecture series, his arguments in his first lecture series were partly metaphysical and partly from the discoveries made in natural philosophy; the more major of the examples drawn from natural philosophy are discussed here.[24] The published versions of Clarke's lectures were not laid out as eight sermons but as a series of propositions. In the first lecture series Clarke builds up his argument from Proposition I where he sought to prove that: 'Something always Was: Otherwise the Things that Now are must have arisen out of Nothing, absolutely without Cause' using an argument that can be traced back to Parmenides (ca. 515–450 BC) that nothing comes from nothing, through Proposition II where he made it clear that 'one Immutable Being' has always existed from eternity and Proposition III where he argued: 'That the Material World cannot possibly be the First, Original and Independent Being,' until he got to Proposition XII that stated: 'That the Supreme Cause and Author of all Things must of Necessity be a Being of infinite Goodness, Justice and Truth.'

The first use of the *Principia* occurs in Proposition III. In the same year that Clarke delivered his Boyle Lecture, the deist Toland published his *Letters to Serena* to which was added two letters which purported to show that motion was essential to matter.[41] What obviously caught the attention of Newton's Bulldog was Toland's attempt to show that this statement was in keeping with Newton's *Principia*. Toland also purported to show that the necessary existence of matter precluded the existence of a vacuum. Clarke countered both of these ideas in Proposition III. In Proposition IX Clarke counters Spinoza's assertion of the absolute necessity of God's Nature, in part using the same argument that Newton later used in the General Scholium to the second edition of the *Principia* in 1713 namely:

> 'the direction of all their motions, both of the primary and secondary planets, uniformly from west to east, when by the motion of the comets it appears there was no necessity but that they might easily have moved in all imaginable directions: is evident proof that these things are the effect of wisdom and choice.'

In his argument 'That the Self-Existent Being and Supreme Cause of all Things must of Necessity have Infinite Power' in Proposition X, Clarke used as an example gravity whose cause, as Newton later stated in the General Scholium, was unknown, writing

> 'since it is ever proportional, not at all to the surfaces of bodies ... but entirely to the solid content of bodies; 'tis evident it cannot be caused by matter acting upon the surfaces of matter, which is all it can do; but must be caused by something which continually penetrates its solid surface.'

Proposition XI 'That the Supreme Cause must of Necessity be Infinitely Wise' cited the motion of the planets and contained a quote from the *Principia*.

The second series of Boyle Lectures that Clarke delivered in 1705 made only passing reference to natural philosophy and is not discussed here.

2.2.3. *William Derham* (1657–1735)

From 1689 William Derham was rector at Upminster in Essex. He was an amateur scientist interested in natural history, meteorology, and astronomy and was elected to the Royal Society in 1703. Derham, who in many ways seems to have been an earlier version of Gilbert White (1720–1793) of Selborne, was appointed a Boyle Lecturer in 1711. Tenison was obviously impressed by Derham's sermons since he reappointed him again in 1712. Derham thought that Bentley and Clarke had not presented the design argument as strongly as they might have done and he set out to give a systematic survey of the Earth its form and motion and the details of all life, which was published as *Physico-Theology*.[42] It is not clear what the form the original sermons were, since the published version is a single work that, apart from the concluding Book IX: *Practical inferences from the forgoing survey*, has little reference to religion and was mainly descriptive. Derham made use of the *Principia* and *Opticks* in a number of places. However, some of his conclusions using concepts of natural philosophy are questionable: for example he suggested that gravity prevented the Earth breaking up under the centrifugal force due to the Earth's rotation. He correctly stated that the surface speed of the Earth at the equator was about 1,000 mph, which was of course an almost unbelievable speed in the eighteenth century, but the centrifugal acceleration at the equator is only about a hundred and fiftieth part of the acceleration due to gravity. However, he did point out that the geocentric Ptolemaic system implied that the speed of the Sun to be greater than 30 million miles per hour and remarked 'but what is this to the [implied] rapidity of the stars.' Derham could have estimated the necessary speed of the nearest fixed stars since Huygens had estimated the distance of the star Sirius, one of the nearest stars to us, from its parallax when observed at across a diameter of the Earth's orbit[43] which translates to some 3×10^{12} km, a factor of about ten times too small, but even with this estimate the implied speed based on the Ptolemaic system would be of the order of 8×10^{11} km/s or 700 times the speed of light! Derham's book *Physico-Theology* was an outstanding success going through three editions in the first year alone.

Whereas Bentley's and Clarke's Boyle Lectures were tightly argued proofs the existence of God, Derham's followed the Leibnizian thesis that God's existence is proved because we live in the best of all possible worlds[44] which Voltaire[45] later parodied in *Candide*.

2.3. Natural and Revealed Religion

In the seventeenth and eighteenth century most Christians believed in a literal interpretation of the Bible. As already discussed, attempts to reconcile natural philosophy with biblical accounts of the system of the world were made by Burnet and others before the publication of the *Principia*, but the natural philosophy of the *Principia* was a much more developed system that explained much of the known universe making a reconciliation with the Bible more difficult. However on the other hand the *Principia* gave details of the mechanics of the universe for the first time, which for many Christians provided a design argument for the existence of a divine creator. For the first eighty or so years the

Royal Society actively sought to promote the design argument to demonstrate the existence of God and it was during this period that many natural philosophers wrote works that attempted to show that Newtonianism supported Christianity. However this period came to an end when the deist Martin Folkes (1690–1754) became President of the Royal Society in 1741.[46] From that time Folkes propagated:[47]

> 'The infidel System with great assiduity, and made it even fashionable in the Royal Society, so that when any mention [was] made of Moses, of the deluge, of religion, scriptures, &c., it generally was received with a loud laugh.'

In 1748 Hume published his essay *Of a Particular Providence and a Future State* in which he argued that the design argument was only speculative.[46] The Newtonian Christian apologists correspond to the earlier period of the Royal Society.

Along side works that used Newtonian natural philosophy to show that the Biblical account of the development of the world were true, there were others who used natural history either as a supplement or on its own. Thus, though natural history is outside the scope of this book, passing reference is necessary to some of these. John Woodward (ca. 1665–1728), a physician and natural historian, who was very interested in fossil shells and one of the first in the modern world to believe that they were the fossilized remains of actual marine creatures, rather than freaks of nature, wrote about the development of the Earth in 1695.[48] Woodward stated his belief that it was the Great Deluge that caused shell fossils to be found inland from the sea.[48] However, he made some silly remarks such as 'these marine bodies are now found lodged in those strata according to the order of their gravity, those which are heaviest lying deepest in the Earth, and the lighter sorts ... shallower or nearer to the surface', which was not true of deposits. Buffon quite reasonably remarked that 'of this author, it may be said that he wanted to build an immense edifice upon foundations less firm than sand.'[12]

William Whiston was one of the first to try to explain the biblical account of creation, Noah's Flood, and the apocalyptic end of the world with Newtonian natural philosophy. Whiston had been ordained in the Church of England in 1693 when he became a senior fellow at Clare College and started studying the *Principia*. In 1696 he published his *A New Theory of the Earth*.[49] He called his theory *new* to distinguish it from Burnet's Cartesian *Sacred Theory of the Earth*, which Whiston had defended in his BA examination in 1684. At the time of writing his *New Theory*, Whiston was a reasonably orthodox Anglican. He promised to 'keep sufficiently close to the Letter of Moses, and yet be far from allowing what contradicts Divine Wisdom, Common Reason, or Philosophick Deductions.' One of his motivations for writing the *New Theory* was similar to that of Burnet that is to show that what the ordinary person saw as a miracle could be often explained by natural philosophy to be a law of nature. Whiston's second postulate was: 'That which is clearly accountable in a natural way, is not, without reason to be ascrib'd to a Miraculous Power.' His first postulate stated that: 'The Obvious or Literal Sense of Scripture is the True and Real one, where no evident reasons can be given to the contrary.' In Book I, Whiston summarized Newtonian natural philosophy concentrating mainly on gravitation. However, unlike Newton who professed not to know the cause of gravity, Whiston described the 'force of Gravitation [as] being so plainly above, besides, and contrary to the Nature of Matter [being incapable of acting at a distance] ... must be the Effect of a Divine Power and Efficacy.'

Whiston made it plain that he believed that the Mosaic history of creation described only the creation of Earth and not that of the universe or of all matter. Genesis clearly states that 'God made two great lights; the greater light to rule the day, and the lesser light to rule the night: he made the stars also' on the fourth day,[11] However, Whiston interpreted this verse as God made visible the Sun, Moon and the stars stating that 'the universal Gravitation of Matter contradicts and overthrows this fancy of the heavenly Bodies having been originally included in, and at the Creation extracted from *Chaos*.'[49] As Keill pointed out, Whiston often seemed to contradict his first postulate.[11]

The first of his hypotheses was that 'THE Ancient *Chaos*, the Origin of our Earth, was the Atmosphere of a Comet.' Although he believed in the account of creation in Genesis, Whiston found like Newton that it was difficult to believe that the Earth was literally created out of chaos in just six days. In Whiston's interpretation, the Earth started orbiting the Sun at the beginning of creation, but did not rotate on its axis until after the fall of Adam which occurred on the seventh day of creation. Thus a day during the creation of the Earth, according to Whiston, was a year. Although this accounts for a primitive day lasting a year it was necessary for the Sun and the planets to rise in the west and set in the east; Herodotus and Plato were quoted in support of this phenomenon. Keill pointed out that the cold during the six month night would be unbearable to man and beast and rejected this hypothesis.[11] Whiston believed that the probable cause of the Earth starting to rotate was an oblique impact on the Earth by a comet.

Whiston also believed that the Deluge was caused by a comet. In 1680 a comet[50] was observed that Newton and Halley had calculated to have a period of greater than five hundred years and observed that it was probably the same comet that had been reported in 1106, 531 or 532 and 44 BC which would give it a period of some 575 years. In the second edition of his book, Whiston improved upon his calculations of the date of the Deluge, using his earlier book that gave chronologies derived from the Old Testament, and he calculated that it occurred 4072 years before 1724 or 4028 years before the last appearance of the comet in 1680 which is seven periods of the comet.[51] Whiston hypothesized that the tail of the comet contained vapours or mists and when the Earth passed through the tail 'it must acquire a large Cylindrical Column of Vapours', and these would have caused 'violent Rain for the Space of forty Days and as many Nights' as described in Genesis. Whiston believed

> 'The other main Cause of the Deluge, was the breaking up of the Fountains of the great *Abyss*, or causing such Chaps and Fissures in the upper Earth, as might permit the Waters contained in the Bowels of it ... to ascend,'

which Burnet had believed to be its main cause. Whiston was not alone in hypothesizing that the Great Deluge was caused by a comet. Halley in 1694 read two papers to the Royal Society in which he suggested that the Deluge was caused by the 'choc[52] of a comet.'[53] In discussing the Great Deluge, Whiston referred frequently to Woodward's work. Keill did not agree with the mechanisms Whiston invoked to account for the flood but was quite taken by the coincidence of the comet's approach to Earth at the same time as the Great Deluge calculated from Biblical chronologies.[11] Whiston republished his theory for the Great Flood as a pamphlet entitled *The Cause of the Deluge Demonstrated*, in 1714 and this caught the eye of the newly formed Scriblerus Club, a group of literary wits including Alexander Pope (1688–1744) and Jonathan Swift (1667–1745) who ridiculed falsity in learning, they took Whiston's theory one step further and postulated that a comet merely passing by the Earth would create unusually strong tides and produced fictional tide tables for such events.[54]

Whiston wrote little on the final conflagration of the Earth apart from describing the Biblical account. However, he did speculate that:

> 'if [a Comet] came near enough, and were itself big enough, it would so much retard the Earth's annual Motion, and oblige it to revolve in an *Ellipsis* so near to the Sun in its *Perihelion*, that the Sun itself would scorch and burn ... it ... and this Combustion being renew'd every Revolution would render the Earth a perfect *Chaos* again.'

The *New Theory* was very popular and went through six editions. It was praised by Locke for its novel hypotheses, but strongly criticized by Keill immediately after publication and later by Buffon who remarked that 'he treats this matter more like a polemical divine than a philosopher.'[12] Whiston answered

Keill's criticisms in 1698.[55] The main justified comment that Whiston made on his criticisms was that Keill had 'a peculiar fondness ... for the Introduction of unaccountable Miracles on all occasions.' However, Whiston showed inventiveness in his own interpretations of scripture so that they would fit his theory.

By about 1708 Whiston came to doubt the doctrine of the Trinity and to believe in Arian Christology which saw the Father as the only true God. Whiston's view of the Trinity was essentially the same as that held by Newton. However, whereas Newton kept his beliefs private, Whiston let it be known that publicly that he believed the Athanasian Trinitology to be a corruption of the beliefs of the early Church and wrote to the Archbishops of Canterbury and York asking for their counsel. By 1710 Whiston's Arian beliefs had become too widely known to be ignored and he was dismissed from his Lucasian Professorship. Although Whiston did not directly expose Newton's private religious beliefs, he did drop hints of them which brought his friendship to Newton to an end. Although Whiston had lost Newton's friendship, and it is probable that Newton blocked Folkes' proposal of Whiston for a fellowship of the Royal Society in 1716, Whiston still in 1717 dedicated his *Astronomical Principles of Religion* to Newton and the Members of the Royal Society. In the *Astronomical Principles*, Whiston made use of Newton's *System of the World* to develop the principles of natural religion and to show that Jewish and Christian revelation relied upon the laws of nature for their foundation.[56] His purpose was to demonstrate that 'Philosophy and Mathematics are on the Side of both Natural and Reveal'd Religion' and 'to shew what is properly the Religion of a genuine and considering Astronomer; or what are properly the Astronomical Principles of Natural and Reveal'd Religion.' He promised 'to Consider what [the] Degree of Evidence is, which ought to be here insisted on; without which we are not, and with which we are obliged to acquiesce in Divine Matters.' As in his *New Theory*, Whiston started by summarizing the laws of matter and motion and describing the system of the universe often quoting directly from the *Principia*. Whiston then gives the principles of natural religion that he believed can be inferred from this knowledge. He stated that the unity of God can be learnt from the true system of the universe with its one law of gravity[57] and the same laws of motion everywhere. In this section he further amplified his concept of gravity as being a direct divine action. According to Whiston, God exercises continual providence over the World and does interpose his immaterial and immediate power over gravity. Whiston believed that Jewish and Christian revelation relied upon the laws of nature for their foundation and astronomy and other mathematical sciences confirm the accounts of scripture. The *Astronomical*

Principles was not as popular as Whiston's *New Theory* and there were only two editions.

Whiston understood the *Principia* and made genuine attempt to show that it agreed with Christianity as well as natural religion. After Whiston attempted to reconcile science with Christianity he shifted somewhat from Newtonian natural philosophy and astronomy to natural history, geography, and the newly emerging science of geology. Deist and atheist natural philosophers either ignored a supreme being like Laplace, or just acknowledged his role in the creation of the universe but believed that 'in physical subjects, we ought to avoid having recourse to supernatural causes,' as Buffon, who was a materialist, did.[12] Those Christians who wrote about the theory of the Earth either mostly opposed Newtonian natural philosophy, like the Hutchinsonians and Muggletonians, or ignored it and often took their ideas from Burnet. One of the latter, John Whitehurst (1713–1788), a clock and scientific instrument maker and a geologist, was elected to the Royal Society for the publication of *An Inquiry into the Original State and Formation of the Earth* in 1778.[58] Whitehurst was a Newtonian, but it is doubtful that he understood much of the *Principia.* His chapters on the formation of the Earth are vague and are written so as conform to the spirit of the account in Genesis rather than be based on Newtonian natural philosophy. However, the appendix where he identified and examined the geological strata of Derbyshire produced important geological generalizations and it was for this work that he was elected to the Royal Society. Another writer even less qualified than Whitehurst who published a theory of the Earth was William Drysdale (1745–1803) a schoolmaster.[59] Although written in a pseudo *Principia* style with propositions etc., one of which was that 'Air is the Primum Mobile; or the First Moving Principle; in material and universal Creation', this theory of the Earth, 'which is thickly interspersed with Hebrew and Greek quotations, is sublimely unintelligible' as Eneas Mackenzie (1777–1832) stated in 1827.[60]

Thomas Dick (1774–1857) was in a different category to Whitehouse and Drysdale. On the 18th August 1783, the eight year old Thomas saw a famous meteor pass near his home in Dundee on its passage along the whole of the eastern coast of Scotland and England finally dissipating over south western France.[61] This meteor's apparent diameter was half that of the Moon but considerably brighter and led the young Thomas to make his own telescope and study astronomy.[62] The meteor was also seen by Tiberius Cavallo (1749–1809), an Italian natural philosopher, and his five companions from terrace at Windsor Castle. One of Cavallo's companions was the artist Thomas Sandby (1721–1798), who with his brother Paul Sandby (1731–1809) produced a print of the event shown in Plate 1; there are three successive superimposed views of the meteor with the third image showing the meteor beginning to break up. Dick's family was of modest means and Dick only managed to enter the University of Edinburgh in 1794 by supporting himself with private teaching and left without taking his MA because that involved extra fees. Dick came from a religious family who belonged to one of the churches that had seceded from the Established Church of Scotland and in 1804 he became a moderator of the Stirling Presbytery. However Dick got his servant pregnant and he was expelled from the Presbytery. In 1807 Dick then became a parish schoolmaster in Methven, Perthshire and continued with his astronomical observations at night.[62] Dick also started a people's library in Methven, which had some of the features of a mechanics' institute, and in 1814 he wrote articles in the *Monthly Magazine* advocating the general establishment of similar institutes.[63] Throughout his life Dick remained a committed Christian and all his writings praised God the creator of all. His most famous work first published in 1823 was *The Christian Philosopher* which connected science and philosophy with religion.[64] Dick believed:[64]

> 'that the visible manifestations of the attributes of the Deity are too frequently overlooked by Christians ... [and] it is a most dangerous and delusive error to imagine, that Reason and the study of the material world, ought to be discarded from science and religion.'

Chapter 2 of *The Christian Philosopher* gave the main sciences, including natural philosophy and astronomy that in Dick's view were related to religion and Christian theology, though he claimed that: [64]

> 'all the moral arts and sciences which have been invented by men – every department of human knowledge, however far it may, at first sight, appear to be removed from religion – may be considered as having a direct bearing on Theology.'

Dick's books had a great influence in their day; David Livingstone (1813–1873) regarded Dick's *The Philosophy of a Future State* published in 1828 as his second most important influence after the Bible and had made him resolve to be a Christian missionary.[65] In 1837 the London Religious Tract Society started its Monthly Volume Series of cheap books[66] on secular subjects with religious comments.[67] About 15% of these books were on science and religion and some of these were best-sellers including Dick's *Solar System* first published in 1846 which sold over 25,000 copies in the first five years.[68] The Religious Tract Society did not preach Christian values too obviously for then only believers would buy the book.[69] In the *Solar System* Dick stated his case for believing that the 'evidence of a self-existent and eternal Being whose wisdom is inscrutable' in the introduction, but then limits himself to remarks on how the solar system evidences God's existence at the end of some of the sections to the chapters.[70]

The will of Francis Henry Egerton (1756–1829), eighth Earl of Bridgewater, directed that £8,000 be used to publish a work *On the Power, Wisdom, and Goodness of God, as manifest in the Creation* to demonstrate natural religion. This bequest had almost as much influence in the nineteenth century as Boyle's did at the end of the seventeenth century. Eight prominent scientists of the day were appointed under the terms of the will and wrote treatises that became known as the Bridgewater Treatises. The third treatise on natural philosophy *Astronomy and General Physics Considered with Reference to Natural Theology* was written by William Whewell (1794–1866), an eminent Cambridge scientist who had a strong Christian belief.[71] The treatise has three books: Terrestrial Adaptations, Cosmical Arrangements, and Religious Views. Apart from the introduction the religious remarks are mostly left to the third book. One of the problems of applying natural philosophy to religion is the difference between laws of nature and moral laws. Whewell recognized this problem and starts his treatise with a discussion of laws of nature which all inanimate objects obey and moral laws which man can choose to obey or deny. He comes back to this theme in Book 3 where showed that as nature's laws led to good governance of the universe so the moral laws lead to good governance of man. Whewell's treatise could be read as a religious work or as a popular account of natural philosophy. Whewell sent a copy of his treatise to the London Mechanics' Institute and it likely that it, as well as the other Bridgewater treatises, were used by the mechanics' institutes to aid their understanding of their lectures on natural philosophy.[72] Certainly Whewell included some of the most up to date aspects of natural philosophy. For example in discussing whether there was a vacuum in interplanetary space or a resisting medium he introduced the orbit of Encke's Comet. The German astronomer, Johann Encke (1791–1865), did not discover this comet, whose period of about three years is the shortest of any comet, but did calculate its orbit in 1818. The comet returned in 1822 but its arrival did not exactly coincide with Encke's calculations and he concluded that the observations could only be explained on supposing that there was a resisting medium in interplanetary space. As Whewell noted Encke's comet is subject to disturbing action by the planets and calculations showed that some of the discrepancy was due to this cause but there still remained a residual discrepancy which, at the time that Whewell was writing, amounted to about two days. It was concluded that this evidence confirmed the existence of a resting medium. However, it is now known that Encke's comet loses mass at its perihelion when it is quite close to the Sun. Sometimes this loss is dramatic as in 2007 when a

NASA space probe filmed an unusually-large release of plasma from the solar corona, called a coronal mass ejection, tearing off part of the comet's tail.[73] The Bridgewater Treatises were very successful; each treatise going through many editions and the series totalled more than 60,000 copies in print by 1850.[72]

By the nineteenth century biology started to be used in the design argument. In 1802 the theologian William Paley (1743–1805) published his *Natural Theology* presented the design argument using mainly biological arguments but began with the watchmaker argument for an intelligent watchmaker and by analogy he extended it to an intelligent maker for plants and animals.[74] The importance of astronomy to the watchmaker argument was discounted with Paley writing:[74]

> 'my opinion of Astronomy has always been, that it is *not* the best medium through which to prove the agency of an intelligent Creator; ... [even though] it shews ... the magnificence of his operations.'

Paley was not the first to use the watchmaker argument and his work advanced no new ideas. Possibly Paley reworked the watchmaker argument in *Natural Theology* because of Hume's criticism of teleological arguments in *Dialogues Concerning Natural Religion* posthumously published in 1779.[75] He also wanted to refute the proto-evolutionary ideas of Buffon and Erasmus Darwin (1731–1802), the grandfather of Charles. *Natural Theology* had a great influence on many during the nineteenth century including Charles Darwin. William Thomson, who in 1863 calculated the age of the Earth from the time it would take for a molten Earth to cool to its present temperature to be within 20 to 400 million years and therefore not old enough for evolution by natural selection, concluded his 1871 Presidential Address to the British Association for the Advancement of Science by calling William Paley's *Natural Theology* 'that excellent old book' and asserted that 'overpoweringly strong proofs of intelligent and benevolent design lie all around us ... [and] all living beings depend on one ever-acting Creator and Ruler.'

2.4. The Anti-Newtonians

Newtonians generally believed that natural philosophy aided the understanding of both natural and revealed religion, but such a belief was not held by all. Opinions about the relationship between Newtonian natural philosophy and religion were not limited to the educated intelligentsia; the self-educated also had opinions. Many believed that the Bible contained all there was to know about natural philosophy and that Newtonian natural philosophy and its system of the world were contrary to Scripture. The main anti-Newtonian in the eighteenth century was John Hutchinson (1674–1737). In the nineteenth century the most significant attack on the Newtonian System came from Isaac Frost (1793–1858) a Muggletonian.[76] Whereas the Hutchinsonians were High Church Tories, the Muggletonians arose from the religious fervour of the Civil War. In February 1652, John Reeve (1608–1658), a tailor, heard Jesus speak to him on three successive mornings. Jesus told him that he was given understanding of the Scriptures 'above all men in the world' and that he was chosen as his last messenger and that his cousin also a tailor, Lodowicke Muggleton (1609–1698) was to be his mouth piece. Reeve and Muggleton saw themselves as the two prophets foretold in the Book of Revelation. The Holy Scriptures of the Muggletonians, that they became known as, consisted of the Three Commissions: the Old Testament (except those books traditionally ascribed to Solomon, but including the apocryphal Book of Enoch), the New Testament, and the Third Commission writings of Reeve and Muggleton. Since a rejection of any of the commissions led to damnation, they were originally not a proselytizing sect. The Muggletonians did not believe in the Trinity, for which blasphemy Reeve and Muggleton were imprisoned in 1653, but believed that God was the man Jesus. They were an anti-intellectual sect; one of their doctrines was that of the two seeds of man: the good seed of faith and the bad seed of reason. Despite their naive ideas about religion and science, Muggletonianism survived until well into the twentieth century, the last surviving trustee, Philip Noakes, did not die until 1979.

2.4.1. *The Hutchinsonians*

Hutchinson was influenced by Woodward, who was the Duke of Somerset's physician, when Hutchinson became Somerset's chief steward. During his business trips for Somerset, Hutchinson helped Woodward's search for minerals and fossils but broke with him when he found that Woodward was claiming his findings as his own.[77] In 1724 Hutchinson published the first two volumes of the Mosaic account of the Great Deluge, which he deliberately called *Moses's Principia*.[78] Hutchinson believed that the Old Testament revealed all that there was to know about natural philosophy and that study of the Bible, not experimental or mathematical science, was the key to understanding it. In all there were twelve volumes of Hutchinson's *Principia*, three of which were published after his death.[79] Hutchinson's writing was abstruse and difficult and at first his works did not make much impression. From the mid 1730s Hutchinson did have a following in Oxford but it was not until after his death and the publication of his collected works in 1748 that it can be said that this following became almost fashionable for a time.[80,79] Hutchinsonianism was the first but not the only anti-Newtonian movement based on religious objections.

Hutchinson thought that the Old Testament had not been read correctly because the pointing system of the Hebrew text was a relatively recent innovation and did not exist at the time when it was believed that Moses wrote Genesis. By removing the pointing system from the text, the self taught Hutchinson believed that access could be obtained to Moses' true meaning and his natural philosophy.[79] The concept that Moses' used Genesis to convey the true God-given natural philosophy seems very strange to us today as it also did to those who wrote criticizing Hutchinson in the eighteenth century. Samuel Berington in 1750 neatly caught the crux of the disbelief in Hutchinson's ideas, writing that 'it is imposing on the divine writings of Moses, to assert that he wrote the history of creation to establish a system of philosophy.'[81]

Whereas Newton did not believe in the Trinity, Hutchinsonians saw many precursors of the concept of the Trinity in the Old Testament. Fire, light and air or spirit was the agent of the philosophy which gave motion to the planets. They also believed that that the word for God used in Genesis, Elohim, which they wrote as Elahim – a legitimate interpretation of an un-pointed Hebrew text – signified the three persons of the Trinity. The identification of Elohim, which is a plural noun, with the Trinity had been made by the early Fathers of the Church, including St Augustine (354–430) in his sermons against Arianism,[82] but Hutchinson went further and stated that Elahim derived from Elah, which means to imprecate and thus Elahim meant that the Trinity imprecated the sacrifice by one of them for man.

Hutchinson was contemptuous of Newtonian natural philosophy especially the concept of gravity acting at a distance 'the aim of the hypotheses of the allies, so much commended by our undertaker, centring there: and none, no not one, for the agents mentioned in Revelation.'[83] Further on Hutchinson stated that:

> 'as long as gravity stands, Moses cannot be explain'd. So, if he'll shew, once in a month, how by any law ... the heavy nodules and light shells could both settle out of the water, and lodge together near the surface of the strata of chalk.'

Here Hutchinson correctly recognized that Woodward was completely wrong in stating that 'these marine bodies are now found lodged in those strata according to the order of their gravity.' Hutchinson then roundly condemned Newtonian natural philosophy stating 'if gravity or properties in matter once fall, there cannot an atheist stand.'

After Hutchinson's death, his followers softened the attack on Newton and his natural philosophy. Robert Spearman (1703–1761) a co-editor of the posthumous collected works of Hutchinson made it clear that he did not question Newton's status, but gave a Hutchinsonian interpretation of his natural philosophy.[79] George Horne (1730–1792) went further claiming that Newton was a Hutchinsonian.[84]

2.4.2. *The Muggletonians*

Isaac Frost and his co-Muggletonian brother Joseph had a successful brass foundry business in Derby, and when they moved to London began to finance the publication of Muggletonian books. Lectures on astronomy using orreries and planetariums were reasonably common during the nineteenth century and from the 1820s astronomy was an important subject at mechanics' institutes. The fact that many of these lectures on the Newtonian *System of the World* were given by clergymen infuriated Frost who wrote: 'it is taught to be death eternal to those who despise God's holy truth, as it is spoken in the Scriptures.'[76] By the nineteenth century Muggletonians, while they still did not proselytise, were more open to society and to counter these lectures, Frost gave his own ones to audiences who were not only Muggletonians.[85] In 1846 Frost finally published his *Two Systems of Astronomy* where he discussed the Newtonian and the Muggletonian Systems.[76] *Two Systems* was beautifully printed in quarto size and it was aimed at the general public as well as Muggletonians;[86] there were eleven colour plates printed by George Baxter, including *The System According to the Holy Scriptures* (the Muggletonian version) shown in Plate 2.[85] The book starts with a critique of the Newtonian System, the problem was that Frost did understand it having probably obtained most of his knowledge from attending popular lectures some of which at least, if his comments in *Two Systems* are to be believed, were of a poor standard. Frost recounted his experience at one such lecture where the forces keeping the Earth in its orbit round the Sun were discussed. Frost was asked his opinion by the lecturer stating that:

> 'Sir Isaac Newton has only given two proofs in support of his theory of the Earth revolving round the Sun ... the power of attraction, to keep the Earth to the Sun; [and] ... the power of repulsion, by virtue of the centrifugal motion of the Earth ... but really he has only half done his work. The Earth traverses round the Sun on an inclined plane; the consequence is, there are FOUR powers required to make his system complete:
>
> power of 1st. The ATTRACTION.
> 2ndly. The power Of REPULSION.
> 3rdly. The power of ASCENDING the inclined plane.
> 4thly. The power of DESCENDING the inclined plane.'

The lecturer supposedly commented that 'it certainly was necessary to have these FOUR points cleared up before the system could be said to be complete.'[76] There were many similar ludicrous remarks in Frost's critique of the Newtonian System.

The Muggletonian System was quite extraordinary; not only did it have to agree with the Old Testament but also with the Third Commission and other writings of Muggleton. The Earth was stationary and the centre of the universe. The Sun and the Moon were no larger than what they appeared to be and no higher than six miles above the Earth. The stars were even closer as is shown in Plate 2. Although this extraordinary system required absolute belief in the Muggletonian Third Commission, Frost was astute enough to seize on one paradoxical observation reported by the astronomer James South (1785–1867) in 1832 who observed a red star apparently advance across the face of the Moon for a short distance before vanishing. Here was the evidence that Frost needed that the stars were closer to the Earth than the Moon. South tried to suggest that the phenomenon was caused because the star was red so that its light would have a different refractive index to the Moon. In 1859 George Airy (1801–1892) tried to explain the phenomenon, which he had observed with white as well as red stars but his interpretation, that it was in some way connected with the wave theory of light, was as wrong as South's.[87] Not until 1900, when the English-born American surveyor and astronomer George Davidson (1825–1911) published a long paper giving details of the projection of stars on the limb of the Moon and other similar phenomena, was the true explanation given:[88]

> 'in the occultation of stars by the moon, a spurious and factitious limb of the moon can be formed only and solely by the unsteadiness of our atmosphere more immediately surrounding the station of the observer.'

Davidson also noted that Newton knew of the problem of observations made when the atmosphere was disturbed writing about telescopes in his *Opticks* that:[15]

> 'they cannot be so formed as to take away that confusion of the Rays which arises from the Tremors of the Atmosphere. The only remedy is a most serene and quiet air, such as may perhaps be found on the tops of the highest Mountains above the grosser Clouds.'

2.5. The Plurality of Worlds

The Dominican Giordano Bruno (1548–1600) believed that the universe was infinite, the stars also possessed planets like our Sun and were inhabited by intelligent beings. He was also a pantheist that is he believed that the universe was God, and God was the universe. For these and other unorthodox views he was burnt at the stake in the Campo de' Fiori in Rome. Eighty-six years later Bernard le Bovier de Fontenelle (1657–1757) revived the extraterrestrial debate with the publication of *Entretiens sur la Pluralité des Mondes* which described the inhabitants of the Moon and the planets.[89] Fontenelle was neither a natural philosopher nor an astronomer but was a writer and a sceptical man of the world. *Entretiens* was written as a rather light hearted conversation between a philosopher and a marquise who criticizes his description of life on the Moon as being 'a great deal of ignorance based on a very little science.' The book was a sensation and introduced the term *plurality of worlds*. *Entretiens* may have been accepted because it did not raise any religious issues but Fontenelle was also criticized for ignoring the religious aspects of the plurality of worlds. Aphra Behn (ca. 1640–1689), that amazing lady writer of the seventeenth century who published the first English translation of *Entretiens* in 1688, complained in her preface that Fontenelle 'ascribes all to Nature, and says not a Word of God Almighty ... so that one would almost take him to be a Pagan.' However after Fontenelle, most of those who believed in the plurality of worlds emphasized that God created them to be populated.

The classical world was divided over the question of extraterrestrials with Aristotle arguing against them. With the re-discovery of classical works the debate was rekindled. With Copernican dethronement of the central position of the Earth, the plurality of worlds began to become widely believed even by Christians. However Fontenelle's *Entretiens* was quickly placed on the Catholic Index in 1687.[90] Burnet believed in the plurality of worlds suggesting that David was musing upon them in his 8th psalm when he wrote:

> 'When I consider thy heavens, the work of thy fingers, the moon and the stars, which thou hast ordained; what is man, that thou art mindful of him?'

and he commented that:[6]

> 'he truly supposes the celestial bodies and the inhabitants of them, much more considerable than we are, and reckons up only terrestrial things as put in subjection to man. Can we then be so fond as to imagine all the corporeal universe made for our use?'

The Irish Franciscan, Francis Walsh (1654–1754), published his theory of the Earth in 1734, which closely following Burnet except in the question of the plurality of worlds where he gave rational reasons why the Moon and other planets could not be inhabited.[91]

Newton, like Descartes, never explicitly embraced the plurality of worlds. The nearest he went to it was in the General Scholium where in addition to recognizing that the solar system was evidence 'of an intelligent and powerful being' he went on to write: 'And if the fixed stars are the centres of similar systems, they will all be constructed according to a similar design and subject to the dominion of One;' and in query 31 of the *Opticks* writing that God could 'make Worlds of several sort in several Parts of the Universe. At least, I see nothing of Contradiction in all this.' Newton's followers mostly accepted the plurality of the worlds even though he was reluctant to do so fully.

At the end of his life, Huygens wrote *Cosmotheros*, where he attempted to use scientific knowledge of the day to discuss the possibility of other life in the solar system.[43] He thought life on all the planets was possible but some hesitancy can be gathered from his use of double negatives as when he wrote 'that the planets have those elements of earth, air, and water, as well as we, I have already made not unlikely.' As for the moons of the solar system Huygens thought these to be barren based on his knowledge of our own Moon which he deduced had no seas, rivers, clouds, air, or water. *Cosmotheros* is basically a scientific work but Huygens thought that it was not contrary to Christianity writing that 'God had no design to make a particular enumeration, in the holy scriptures, of all his works of creation.' Eminently sensible John Keill thought that:[92]

> 'those who affirm that God created these great bodies [that is stars] only to give us a small dim light must have a very mean opinion of the divine wisdom. It is more reasonable to suppose that every Sun is surrounded with a company of planets.'

He implied that these planets probably had inhabitants writing 'all the worlds or system of the worlds, are as so many theatres, in which he displays his divine power, wisdom and goodness.'

William Derham followed up his successful *Physico-Theology* with *Astro-Theology*, which was published in 1719.[93] In this book, he advanced his *New System* of the universe which proposed that the fixed stars were like our Sun and 'there are many other systems of Suns and planets besides that in which we have our residence' and that these planets were 'Worlds or places of habitation' though he did not speculate as to the nature of the inhabitants. Derham thought that his scheme gave a 'more extensive, grand, and noble view of God's works,' he had no doubts about this new system, but thought 'it to be far the most rational and probable of any.' He gave a number of reasons for believing in the new system, the first of which was:

'because it is far the most magnificent of any; and worthy of an infinite CREATOR: whose power and wisdom are without bounds and measures ... as myriads of systems are more for the glory of God and more demonstrate his attributes than one.'

Derham's book was also very successful and a major source for the physico-theological poets.[94] One such was Edward Young (ca. 1683–1765) who wrote in the Ninth Night of his mammoth *Night-Thoughts on Life, Death, and Immortality*, composed between 1742 and 1746:

An undevout astronomer is mad.
True; all things speak a God; but in the small,
Men trace out Him; in great, He seizes man;
Seizes, and elevates, and wraps, and fills
With new inquiries, mid associates new.
Tell me, ye stars! ye planets! tell me, all
Ye starr'd, and planeted, inhabitants! what is it?.

The first line in the above extract from Young's poem was given almost the stature of a verse in the Bible and was widely quoted.

Kant's early works had a scientific rather than philosophic character. His most important work of scientific nature was his *Universal History and Theory of the Heavens*.[37] This work relied largely on analogy which he thought had the same value as formal proofs if they agreed with observation. However, arguments from analogies can often suggest interesting hypotheses but there is no means of knowing which the fruitful ones are. Kant viewed the fixed stars as:

'just so many Suns, the centres of similar systems, in which everything may be set up in just as large and orderly way as in our system and the infinite space teems with cosmic systems, whose number and excellence have a relationship to the infinite nature of their Creator.'

His comments on inhabitants of planets were limited to our solar system where he had the rather peculiar notion that the inhabitants of the furthest planets were the best. Kant was influenced by Thomas Wright (1711–1786), an amateur astronomer, who in 1750 wrote *An Original Theory or a New Hypothesis of the Universe*. Wright identified nebula with island universes which consisted of a 'vast infinity of worlds crowded full of Beings, all tending through their various States to a final Perfection,' and he speculated that 'Man may be a very inferior Class.'[95] The arguments of Wright and most of those in the second half of the eighteenth century, for populated worlds, were largely teleological. God would have created the stars and their attendants for a purpose, if they were not inhabited why should they exist? Would God have simply created them as the equivalent of Christmas tree lights? However such arguments are foreign to Newtonianism, which attempted strictly to deduce knowledge from phenomena, a statement of a purpose for a phenomenon is of necessity an unsubstantiated hypothesis. Francis Bacon (1561–1626) had earlier rejected teleological arguments, while not denying that nature could demonstrate a divine purpose, which is the basis of the design argument.[3] However, teleology was a pervading explanation of the romantic era in the second half of the eighteenth century used by scientists, theologians, and poets such as Coleridge and Wordsworth.

In the eighteenth century many self-educated men like Wright were drawn to study astronomy because they saw that the night sky revealed the magnificence of God. James Ferguson (1710–1776) was another such man, who first studied the stars while working as a shepherd. In 1754 Ferguson published a short book: *An Idea of the Material Universe, Deduced from a Survey of the Solar System*, where he described his rapture at observing the stars:[96]

> 'What an august! What an amazing Conception, if human Imagination can conceive it, does this afford of the Creator! Thousands of thousands of Suns ... accompanied with ten thousand times ten thousand worlds ... and these worlds peopled with Myriads of intelligent Beings, all Candidates for Heaven ... how great, how wise, how good, must He be who made and guides the whole.'

One of the leading astronomers of the eighteen century, William Herschel (1738–1822), who changed his career from musician to astronomer and catalogued more than two thousand nebulae and star clusters, was greatly influenced by Ferguson and absorbed his passionate belief in the plurality of worlds.[97] Johann Heinrich Lambert (1728–1777) pushed the teleological argument further to state that celestial bodies always avoid one another because the preservation of their inhabitants is the aim of the Creator.[98]

By the second decade of the nineteenth century almost every astronomical body in the night sky had been populated by somebody. Lambert populated comets despite Newton's estimate of the extremely high temperatures attained at the perihelion for the 1680 comet.[99] Herschel proposed that the satellites of Jupiter and Saturn were inhabited. Both William Herschel and Johann Elert Bode (1747–1826) believed that even the Sun was inhabited. All of these conclusions were reached from teleological arguments. It was common belief that God had populated or was in the process of populating his entire universe, but there were some who took a deistic or an atheistic view of the plurality of worlds. Laplace, whose nebular hypothesis gave support to the probability that stars were the centres of many planetary systems and considered that extraterrestrial intelligent beings were most likely,[35] was a deist and may have become an atheist. Jérôme

Lalande (1732–1807), who wrote the influential *Traité d'Astronomie*, first published in 1764, included a section on the plurality of the worlds in his 1792 third edition, had definitely become an atheist by 1800.[97] Deprived of a teleological argument by his atheism, Lalande used an analogy to argue for the plurality of worlds.

The radical Tom Paine (1737–1809) attended the astronomical lectures of James Ferguson and Benjamin Martin (1705–1782) who both espoused the plurality of worlds. The *Age of Reason* published in 1793 by Paine, was a sensational attack on Christianity and all other religious institutions and argued that 'a plurality of the worlds at least as numerous as what we call stars, renders the Christian system of faith at once little and ridiculous and scatters it in the mind like feathers in the air.'[100] However Paine was not an atheist. The *Age of Reason* was written while Paine was in France as Secretary for Foreign Affairs to the American Congress. He was disturbed by the irreligious attitude of the French revolutionaries and gave his religious belief on the second page of the *Age of Reason*: 'I believe in one God and no more, and I hope for happiness beyond this life.' To Paine the only revelation was in nature:[100]

> 'THE WORD OF GOD IS THE CREATION WE BEHOLD: And it is *this word*, which no human invention can counterfeit or alter, that God speaketh universally to man.'

In the late eighteenth century others were drawn to reject Christianity for the same reason as Paine including Horace Walpole (1717–1797), son of Robert Walpole (1676–1745) the longest serving prime minister of Britain. He wrote that reading Fontenelle's *Plurality of Worlds* 'rendered me an infidel. Christianity, and a plurality of worlds, are in my opinion irreconcilable.'[101] Like Paine that did not mean that he was an atheist for he continued:

> 'Atheism I dislike. It is gloomy, uncomfortable; and in my eye unnatural and irrational. It certainly requires *more credulity to believe that there is no God than to believe there is*. This fair creation, those magnificent heavens, the fruit of matter and chance! Oh! impossible!'

Christians fought back against Paine's rejection of Christ. The Methodist Adam Clarke (ca. 1762–1832) seized on the use of the plural forms of heaven and world in the Bible to justify a Christian belief in the plurality of worlds in his sermons and commentary on the Bible.[97] A Church of Scotland minister, Thomas Chalmers (1780–1847), preached in Glasgow on the relation of Christianity to the plurality of the worlds over a series of seven sermons in 1815 and published them in his extremely successful *Astronomical Discourses* in 1817 which went through many editions in Britain and America in the first year alone.[102] The astronomical objection to Christianity caused by belief in the plurality of worlds, which Chalmers almost took for granted, was the focus of his *Discourses*. The Christian church at the beginning of the nineteenth century was under attack by both intellectuals and the artisans of the industrial revolution who had gained some education. Although Chalmers lists the Bible verses he has used in each of his discourses, he makes no direct appeal to them in his argument whose main feature is an appeal to ignorance. 'How do infidels know that Christianity is set up for the single benefit of this Earth and its inhabitants? How are they able to tell us, that if you go to other planets, the person and the religion of Jesus are unknown to them? Chalmers' *Discourses*, a considered discussion on the plurality of worlds, was very influential.

Thomas Dick was a convinced believer in the plurality of the worlds. He thought:[64]

> 'that matter exists solely for the sake of sensitive and intelligent beings, and that the Creator made nothing in vain; it is a conclusion to which we are necessarily led, that the planetary globes are inhabited by various orders of intelligent beings.'

The Bible, according to Dick, contained many references to other inhabited worlds and gave a number of examples such as in the Book of Nehemiah 9:6:

> 'thou has made heaven, and the heaven of heavens, with all their host ... and thou preservest them all.'

These arguments, according to Dick:[64]

> '[combined] with the wisdom and other attributes of the Deity ... amount not only to a high degree of probability [that there are other inhabited worlds] but to something approaching to a moral demonstration.'

Dick's belief in the plurality of worlds intensified as he got older. In *The Christian Philosopher* Dick wrote about the probability of there being inhabited other worlds and rejected some of the evidence quoted by other writers for the Moon being inhabited though he did think it was probably inhabited. In the *Celestial Scenery* first published in 1837 Dick wrote 'I have assumed the position that they [the planets and their satellites] are all peopled with intelligent beings' and devoted one chapter to the plurality of the worlds in which he gave five arguments for his position.[103] In 1840 Dick expanded his book *Celestial Scenery* in *The Sidereal Heaven* and advanced further arguments in support of the plurality of worlds 'both from reason and revelation.'[104]

Of course the pluralist debate suffered from not having any direct evidence of the existence of extraterrestrial inhabitants. Richard Locke (1800–1871), a New York journalist, wrote a series of articles in 1835 for the New York Sun supposedly supplying this evidence. In the articles Locke stated that the astronomer John Herschel (1792–1871), son of William, had observed animals on the Moon with a twenty-four foot diameter reflecting telescope, using extracts from a paper in the *Edinburgh Journal of Science* supposedly written by Herschel.[97] The animals reportedly seen through this unbelievably large telescope included an intelligent orangutan-like creature. Sixty-thousand copies of the series in pamphlet form were sold and other newspapers also carried reports of the discovery. Obviously the deception could not continue and came to an end when another reporter approached the Sun to get a copy of the nonexistent Edinburgh journal. The incident, which is very reminiscent of the Orson Welles 1938 radio production of *The War of the Worlds* broadcast in the form of news reports, was described as a hoax, but Michael Crowe convincingly has demonstrated that it was a failed attempt at satire, which may say something about the gullibility of Americans.[97]

Newtonian natural philosophy essentially added nothing to further the pluralists' debate after the beginning of the nineteenth century but a development of Newtonian optics did. Young's 1802 wave theory of light enabled the development of spectroscopy. In the same year William Wollaston (1766–1826) observed seven dark lines in the spectrum obtained from a prism using sunlight shining through a narrow slit whose width was 1/20 in (1.27 mm).[105] Newton in his famous experiments described in *Opticks* had mostly observed the spectrum obtained with a prism using sunlight shining through a circular hole of the order of one third of an inch (8.5 mm) though he did also use a narrow slit 'but a tenth or twentieth part of an inch [2.54–1.27 mm] broad' and did not report seeing any dark lines.[106] Wollaston did no further research on the dark lines in the spectrum which we now know are due to light coming through the Sun's photosphere being absorbed at specific wavelengths by elements in the solar atmosphere. The dark lines are now known as Fraunhofer lines after Joseph von Fraunhofer (1787–1826) who observed them in 1814 not knowing about the previous work of Wollaston. The explanation for the dark lines in a spectrum of light was first given in 1859 by Gustav Kirchhoff (1824–1887) and Robert Bunsen (1811–1899). Kirchhoff and Bunsen also noticed that that several of the Fraunhofer lines coincided with the bright emission lines in the spectra of heated gases; incandescent solids or liquids do not produce emission lines but only a continuous spectrum. This interpretation opened a means of determining the elements present in stars by combining a spectroscope with a telescope by William Huggins (1824–1910) in 1862. Fraunhofer had previously obtained the spectra for a few of the brightest stars in 1823 but could not interpret their meaning. Huggins in collaboration with the chemist William Miller (1817–1870) made a spectroscopic study of nearly fifty stars which showed 'the similarity of essential constitution ... [that exists] among the stars, and between them and our Sun,' this finding gave substance to the pluralists argument that the stars were like the Sun.[107] However they also examined the spectra of a few nebulae, mainly those classified as planetary nebula by William Herschel, because they somewhat resembled the planet Uranus which he had discovered, though they have nothing to do with planets.[108] For these nebulae bright line spectra indicated that they were composed of glowing gases rather than clusters of stars. In 1868 Huggins published a further paper on about seventy further nebulae stating that 'of these seventy nebulae about one-third give a spectrum of bright lines.'[109] Despite two thirds of the nebulae examined by Huggins having continuous spectra, these results contributed to the rejection of the island universe concept until it was renewed in the 1920s when Edwin Hubble identified that several spiral nebulae such as those in the constellations of Andromeda and Triangulum were galaxies or island universes far outside our Milky Way.

The analogy argument for the plurality of worlds was weak and, before Darwin published his *On the Origin of the Species*, there was no real scientific evidence for or against it and it is curious why the belief persisted especially among Christians for whom it needed some mental contortions to reconcile it with Christian revelation. Darwin showed that new species arose through natural selection taking advantage of favourable mutations, or *sports* as Darwin termed them. The wastage involved in evolution undermined the teleological argument for the plurality of worlds. However, Darwin provided a scientific explanation of how intelligent extraterrestrials might evolve. Darwin himself did not enter the plurality debate and only referred to the Creator seven times in the first edition of the *Origin*. However, there were more references to the Creator in his two earlier manuscripts of 1842 and 1844 which became part of the *Origin*.[110] In these manuscripts Darwin voiced his own concept of the Creator writing that:

> 'It is derogatory that the Creator of countless systems of worlds should have created each of the myriads of creeping parasites and worms which have swarmed each day of life on land and water on this one globe.'

Enthusiasm for the plurality of worlds, which was widespread at the middle of the nineteenth century, waned during the second half of the century. However, from about 1880 astronomers started to become convinced that they could observe canals on Mars which indicated the presence of intelligent beings. The most influential proponent of Martian canals was the American Percival Lowell (1855–1916) but by 1910 observations with higher resolution telescopes showed that the canals were illusory and belief in intelligent life within the solar system came to an end.[97] It was not until the 1960s that belief in extraterrestrials outside of the solar system began to become respectable once more but now based on exobiology. The first confirmed discovery of a planet outside the solar system was not made until 1988. The Church of Scientology, established in 1953, believes in reincarnation of extraterrestrials as people on Earth.

The plurality of worlds is unfalsifiable and it not surprising that the antipluralists were greatly outnumbered by the pluralists and driven to their position by religion not science.

2.5.1. *Whewell and the Antipluralists*

Not all believed in the plurality of the worlds. In 1699 Thomas Baker (1656–1740), a fellow of St John's College, Cambridge, anonymously published *Reflections upon Learning*, which warned against complete reliance in learning because 'Men are fond of Learning almost to the loss of Religion' and contained a discussion of the plurality of worlds in the chapter on astronomy.[111] Baker rightly points out the weakness in the chain of reasoning often used:

> 'we first observe Seas and Rivers, in the Moon, and if such be there, there must be Plants that they water, and if Plants, there must be likewise Animals to feed upon them, and all these are design'd for the service of Men. The reason is easily carry'd further, for if the Moon be a World, by parity of Reason, so must the other Planets be also, and if all the visible Planets are carried about in the Vortex of the Sun, which is no better than the other Stars, no doubt, the other fix'd Stars, have their attending Planets, as well as the Sun, and so we have a Plurality of Worlds.'

If the chain is broken so is the argument for the plurality of worlds. He also objected to the teleological argument:

> 'These World-mongers are always objecting the improbability of God's framing so many vast and glorious Bodies, only for the sake of this Earth, so inconsiderable a portion of the whole ... they that argue thus, seem to measure things by their Bulk, which is a false way of reasoning.'

It is not surprising that the Hutchinsonians and Muggletonians with their literal interpretations of the Holy Scriptures were antipluralists. Alexander Catcott (1725–1779) was a theologian and geologist, who learned his Hutchinsonianism from his father Alexander Stopford Catcott (1692–1749). In 1756 he wrote a criticism of the book *A Vindication of the Histories of the Old and New Testament* by Robert Clayton (1695–1758) the Bishop of Clogher, who was an antitrinitarian friend of Samuel Clarke and among other things supported the plurality of the worlds.[112] Catcott expressed the antipluralists thinking of Hutchinsonians who believed that:

> 'the earth and all the heavenly host were created and formed but *just* before man ... [and] that the whole was intended only as a *grand palace*, fitly furnished and richly decorated, for the future inhabitant of earth, MAN.'

Although the Hutchinsonians did not believe that the stars were closer to Earth than the Moon, they did not believe that they were at so vast a distance as calculated by the Newtonians. To show that the stars were much closer to Earth, Catcott first tried to show that the Earth is considerably closer to the Sun than given by calculations from the parallax of the Sun. To do this he relied on a naïve method proposed by Walmsley the Vicar of St Margaret's Church in Angmering, Sussex. Walmsley proposed that the Earth rolled around the Sun like a wagon or an early surveyor's wheel. From this absurd proposal, Catcott calculated the mean distance of the Earth from the Sun to be 1,454,526 miles (2,340,831 km); Newton calculated the distance at between seventy and eighty-one million miles (110–130 million km), the modern value is 149,597,890 km.[113] Catcott also doubted that the fixed stars emitted their own light and opined that they were 'only so many *specula* [mirrors]' reflecting the light of the Sun. Compared to the Muggletonians, the Hutchinsonians were men of science. Isaac Frost, whose concept of the universe required a great imagination, described the plurality of worlds as:[76]

> 'the imaginative branch, which is introduced into this science [of astronomy]; such as imagining that the planets are worlds, and the whole infinite space is an innumerable mass of worlds, revolving round Suns, and those worlds peopled like unto this globe.'

He considered that such belief amounted to wickedness and those passages in the Bible like the verse in Hebrews XI: 'through faith we understand that the worlds were framed by the word of God', which was used in a sermon in Westminster Abbey as evidence for the plurality or worlds, was a mistranslation.[94] Other clergymen such as John Wesley (1703–1791), one of the founders of Methodism, who argued that the possibility of there being many worlds did not mean that they were probable, were more rational in their objection.[94]

William Whewell was the most eminent scientist to question the plurality of the worlds prior to the publication of Darwin's *On the Origin of the Species* in 1859. Although initially supportive of the plurality of worlds, Whewell came to believe that it was incompatible with Christianity. Such was the entrenched general belief that Whewell published his rebuttal *Of the Plurality of the Worlds* anonymously in 1853 being fully aware of the consternation it would produce.

However, despite the anonymity of the publication, Whewell was immediately identified as the author and subject to surprised criticism that he should question the general perceived wisdom. Whewell anticipated this criticism in his preface writing:[114]

> 'It will be a curious, but not a very wonderful event, if it should now be deemed as blameable to doubt the existence of inhabitants of the Planets and Stars, as, three centuries ago, it was held heretical to teach that doctrine.'

Whewell's position was that there was 'no insuperable difficulty in conceiving other parts of the Universe to be tenanted by animals' but refused to believe in intelligent extraterrestrials. Whewell's belief in the possibility of extraterrestrial animals was influenced by Chalmer's discussion of microscopy in *Discourses* where he stated that astronomy 'led me to see a system in every star ... [microscopy] leads me to see a world in every atom.'[102] Chalmer's *Discourses* was seen by Whewell as the major argument against his position and he referred to it often in the first four chapters of his own work which discussed astronomy. Whewell's antipluralist position came from his strong Christian belief but his arguments were scientific.

The eighteenth century scientific arguments in discussions on the plurality of worlds were based on astronomy and Newtonian natural philosophy. By Whewell's time geology had become a developed science – Charles Lyell (1797–1875) published the first volume of his influential *Principles of Geology* in 1830 – and not surprisingly Whewell used geological arguments in his work. While geology is beyond the scope of the present book some discussion of Whewell's geological argument is necessary to understand his position. Geology had changed the belief of most thinking Christians:[114]

> 'The doctrine of the successive creation of species [being] firmly established among geologists ... [show that God] has made [man's] period ... only a moment in the ages of animal life, [and] the only period of intelligence, morality, religion... He has done this ... contrary to our conceptions of His greatness and majesty [and] it is plain that our conceptions are erroneous.'

Thus Whewell argued against the conception that God created the planets of the Sun and stars for the purpose of intelligent and moral beings.

Whewell also examined the astronomical arguments. He minimized the number of potential worlds by questioning whether all the nebulae, which had recently been discovered, were separate galaxies each with stars as numerous as our own Milky Way. He also showed that not all of the stars in the Milky Way can have a planetary system. Next he examined our solar system. The Moon has no atmosphere and cannot sustain life. The outer planets beyond Mars are so far from the Sun making the light and heat reaching them very small; their densities are small so they may not be solid; the gravity on the giant Jupiter and Saturn is much greater than on Earth. Whewell saw clearly that all these facts made life on the outer planets very unlikely but he was not so sure that Mars could not support life. Of the inner planets, the surface of Venus could not be seen because as we know now there is a dense opaque cloud and Mercury is much closer to the Sun and 'receives seven times as much light and heat as the earth.'[115] Whewell also made a special pleading for the non-existence of life on Venus and Mercury stating that:

> 'the zodiacal light[116], a nebulous appendage to the Sun, as Herschel conceives, extending beyond the orbits of Mercury and Venus ... [shows that] these planets ... have not yet fully emerged from the atmosphere in which they had their origin. Their chaotic state being not yet ended, we need not wonder, if we find no evidence of their having inhabitants, and some evidence to the contrary.'

Whewell turned to biology to show that the teleological argument was weak:

> 'Similar visible structures do not [necessarily] answer a similar purpose ... The paps and nipples of a male animal ... look like the terrestrial breast of Nature: but are they really nursing breasts? ... Or are they mere images of such breasts? Male teats, dry of all nutritive power? Sports, or rather overworks of nature ... It cannot be wise or right, to prop up our notions of God's greatness, by physical doctrines which will not bear discussion. God's greatness has no need of man's inventions for its support.'

Whewell's book provoked an intense reaction, with most commentators defending the plurality of worlds. Belief in the plurality of the worlds from a purely Newtonian or astronomical perspective did not die until the beginning of the twentieth century.

2.6. Scientific Materialism

Materialism has its roots in the Greek atomists. The soul according to Democritus was composed of atoms; thought was a physical process. The Aristotelians rejected atomism. In the seventeenth century atomism was revived by Pierre Gassendi (1592–1655) a French priest and Thomas Hobbes. Gassendi made atomism compatible with Christian doctrine by asserting that atoms are not eternal but have been created by God whereas Hobbes was a thorough-going materialist writing at the beginning of the Leviathan that 'life is but a motion of limbs' and 'that all automata (engines that move themselves by springs and wheeles as doth a watch) have an artificial life.'[117] Newton himself recoiled from materialism when others suggested that gravity was a fundamental force of matter and certainly he did not believe that universe could arise from atoms 'out of a chaos by the mere laws of nature.'[15]

Early in the nineteenth century it began to be suggested by the French naturalist Jean-Baptiste Lamarck (1744–1829) and others that life originated as part of the process of the evolution of the universe as described by the nebular hypothesis. The most popular exposition of Lamarck's ideas in Britain was given in *Vestiges of the Natural Creation* published anonymously by Robert Chambers (1802–1871) in 1844.[118] *Vestiges* caused an outcry from many because it stood the design argument of its head by linking it to materialism. The ideas of Lamarck on how different species came into being were wrong but they paved the way for Darwin. Although Darwin did not believe in the Bible as divine revelation or that Christ was the Son of God, he was not a materialist but his *Origin of the Species* promoted materialism.

John Tyndall (1820–1893) created considerable concern when, as President of the British Association,[119] he delivered his Presidential Address to the Belfast meeting in 1874 on scientific materialism. He had presented similar lectures before, for example his address as president of the Mathematical and Physical Section of the British Association in Norwich in 1868, without creating a storm of protest. The problem was that, as President of the British Association, Tyndall represented science and an attack on revealed religion was unacceptable to many Christians. Newspapers both in Britain and America reported Tyndall's speech in full. While some people applauded Tyndall's stand that science demanded materialism, many others attacked him, some even demanding that he be tried for blasphemy.[120] The Effect of Tyndall's Belfast Address was long lasting. George Bernard Shaw (1856–1950) in his 1903 play *Man and Superman* had Mrs Whitefield say 'Nothing has been right since that speech that Professor Tyndall made at Belfast.' A professor of natural philosophy at the Royal Institution in London, Tyndall was a member of the X-Club, formed by Huxley in 1864 to promote Darwinism. Most of the nine members of the X-Club were in the life sciences; Tyndall, initially the only physicist of the club, was to promote Darwinism through the physical sciences.[121] Although Tyndall promoted scientific materialism, he was not a thorough-going materialist and in his address said that materialism could not be 'a complete philosophy of life.'[122] The romantic idealism of Johann Wolfgang von Goethe (1749–1832) and especially that of Thomas Carlyle (1795–1881) had a large influence on Tyndall's thought and metaphysics.[120] Carlyle's new religion, that life contains elements of wonder that cannot be explained by physical science, softened Tyndall's science based materialism. The result was what Tyndall called scientific materialism others saw as pantheism.[120] Tyndall quoted Carlyle 'Did I not believe that an Intelligence is at the heart of things, my life on earth would be intolerable' noting that Carlyle 'is not, in my opinion, rendered less but more noble by the fact, that it was the need of ethical harmony here, and not the thought of personal happiness hereafter, that prompted his observation.'[123]

Tyndall based his physical argument for scientific materialism on atomism, evolution, and the conservation of energy, which he thought was more important than evolution. In his section on atomism Tyndall relied heavily on the *History of Materialism* written by the German philosopher[124] Friedrich Lange (1828–1875) first published in German in 1865.[125,126] We are told that 'Newton and Boyle lived and worked happily' under the Strasburg clock models of the universe, without having to enquire about the prime mover. Tyndall does not mention Newton's query 31 to the second edition of his *Opticks* where he discussed atomism writing 'that God in the beginning form'd matter in solid, massy, hard, impenetrable particles' because Newton specifically rejected the materialistic conception of the universe being organized out of atoms by the laws of nature.[15] Newton's atomism lacks the rigor of his mechanics and optics. In his historical development of atomism, Tyndall jumps from Newton to Dalton ignoring important developments in the eighteenth century especially those of Roger Boscovich, and Joseph Priestley. Boscovich first published his *Theoria Philosophiae Naturalis* in Vienna in 1758. In his theory, Boscovich replaced the 'solid massy, hard, impenetrable particles' of Newton by interacting point particles of no extension.[127] This reinterpretation removed some of the problems associated with the solid model. The translation of the full title of Boscovich's book is *A Theory of Natural Philosophy reduced to a Single Law of the Forces existing in Nature* and what he attempted was to reduce all the interactions of matter, from the smallest to the largest scale, to a single law of the forces between two particles. For very tiny distances between particles the force is repulsive it then alternates between attraction and repulsion until at large distances the force becomes gravity.[128] The particle theory of Boscovich is purely kinematic involving only the dimensions of length and time and his particles do not have a mass assigned to them.[129] Boscovich was not a materialist but Priestley was. However, though Priestly was a materialist, he saw 'nothing nothing inconsistent with Christianity and the conception of the materiality of the human and divine soul.' Boscovich's particle theory was of great interest to Priestley who used it to further his materialism and gained the enmity of Boscovich.[130] Perhaps the reason why Tyndall did not mention Priestley was because Lange opined that Priestley was for 'the history of materialism, as a whole, of but slight importance.'[125]

2.7. Thermodynamics

There have been few topics in physics that have been more exploited in religious debate than thermodynamics. The laws of which were used both to confirm and deny traditional religion though there have been some like the French physicist and historian Pierre Duhem who was a sincere Catholic believer but strongly saw his physics as completely separate from his belief. The religious import of thermodynamics has been acknowledged by most physicists except Clausius. In Joule's general exposition of the principle of the conservation of energy he wrote: 'order is maintained in the universe – nothing is deranged, nothing ever lost,' and then he quoted from Ezekiel 1:16: [131]

> '"wheel may be in the middle of wheel" and every thing may appear complicated and involved in the ... intricacy of an almost endless variety of causes, effects, conversions, and arrangements, yet is the most perfect regularity preserved – the whole being governed by the will of God.'

Colding wrote in 1856:[132]

> 'I have further endeavoured to prove that the same forces that God from the first beginning placed in the world along with matter have later been continuously active and will continue to be so and to be equal to themselves in magnitude, activity, and abundance, without ever disappearing or ceasing to exist, ... not only have existed from the world's first beginning, but that these same forces have been continually active in developing the world toward the goal that was given at the creation itself.'

And in 1863 Colding also wrote:[132]

> 'Divine reason manifests itself in nature through the activity of nature's forces. If those forces could lose their ability to act, then all further changes and developments in nature would come to an end.'

Thomson, Maxwell, and Tait were also impressed by the religious message in thermodynamics.[133]

Although the first law of thermodynamics was used to express a God given order of the universe, it did not provide any indication of how it was created. The second law, for those who believed as Clausius did that it implied that the entropy of the universe tended to a maximum, could be used to argue for a religious belief in a final end to the universe as Pope Pius XII did in an address to the Pontifical Academy of Sciences in 1951:[4]

> 'this fatal destiny, ... which ... comes from positive scientific experience, [and] eloquently postulates the existence of a necessary Being.'

Others like Tyndall who believed in materialism, tended to ignore the second law. In lectures at the Royal Institution in 1862, published the year after as *Heat a Mode of Motion*, Tyndall discussed the conservation of energy at length but said nothing about the second law.[134] William Thomson was a deeply committed Christian who believed that religion and science to be mutually supportive; he was part of the theological tradition of voluntarism where God maintains a framework of absolute laws of nature.[135] Thomson conceived the second law of thermodynamics as a law of natural history rather than of natural philosophy.[135] Thomson and Tait gave an account of the universe in a Church of Scotland magazine *Good Words* in 1862, which was particularly directed against the views of Tyndall.[136] In their article they described how the second law would lead to the heat death of the universe consistent with passages in the Bible such as Psalm 102:26 where the heavens and Earth 'shall wax old as doth a garment'.[137] Maxwell made similar use of the second law in a letter in 1868 to Mark Pattison (1813–1884), Rector of Lincoln College, Oxford where he wrote that our experience of irreversible processes codified in the second law, 'leads to the doctrine of a beginning and an end instead of cyclical progression for ever,' a prospect that confirmed his evangelical belief in a world created by God.[138] Franz Brentano (1838–1917), an important but not well-known German philosopher, who was ordained as a Jesuit priest in 1864 but lost his faith in 1870 and gave up his priesthood, delivered a series of lectures on the existence of God at the Universities of Würzburg and Vienna between 1868 and 1891 in which he used thermodynamic arguments that were similar to those held by Thomson and Maxwell.[139]

Herbert Spencer (1820–1903), who was to become the most popular of the Victorian philosophers, embarked on a monumental work on his *System of Synthetic Philosophy* in the 1860s which was a major part of his work for the rest of his life. Spencer's synthetic philosophy was a form of monism where everything was subject to scientific laws including religion. He published his first volume of this work *First Principles* in 1862 in which he attempted to reconcile the *Unknowable* and with the *Laws of the Knowable*. Spencer believed 'the basis of reconciliation must be ... [the] deepest, widest, and most certain of all facts.'[140] The deepest of the knowable facts was the 'Persistence of Force', using the expression for the conservation of energy suggested by Huxley, which he took to be a 'primordial truth which underlies all knowledge'. Spencer's interpretation of the first law of thermodynamics was very broad:[140]

> 'By the persistence of Force, we really mean the persistence of some Power which transcends our knowledge and conception. The manifestations, as occurring either in ourselves or outside of us, do not persist; but that which persists is the Unknown Cause of

these manifestations. In other words, asserting the persistence of Force, is but another mode of asserting an Unconditioned Reality, without beginning or end.'

In common with many non-physicists in the 1860s Spencer had no real conception of the second law of thermodynamics. In a letter Bertrand Russell sent to Beatrice Webb (1858–1943) in 1923, he wrote 'I don't know whether he [Spencer] was ever made to realise the implications of the second law of thermodynamics; if so, he might well be upset.'[141] Thus Spencer had no science which could imply the necessity of creation. In his search for monistic relationship between the knowable and the unknown he could only conclude that one had to see that the 'antithetical conceptions of Spirit and Matter; the one is no less than the other to be regarded as but a sign of the Unknown Reality which under lies both.'

A form of monism was common in those who sought to use thermodynamics to support non-traditional religion. The Latvian born Wilhelm Ostwald (1853–1932), who received a Nobel Medal for Chemistry in 1909, was a more extreme monist than Spencer. Ostwald had his epiphany in 1890 in Berlin's *Tiergarten* when he became suddenly aware of the unifying power of Energetics, all was energy.[142] The two laws of thermodynamics are at the core of Energetics. The monism that Ostwald adopted was that of the biologist Ernst Haekel (1834–1919) and was built on the law of substance, which was simply the law of the combined conservation of matter and energy. Haekel's monist church had the trappings of traditional religions there were Sunday sermons and a winter festival to replace Christmas complete with monist carols. The modern God was science. Everything was seen to derive from Energetics even basic ethical principles.

Not all scientists at the turn of the twentieth century were against traditional religions. Pierre Duhem, a physicist with a wide range of interests, had conservative beliefs, he denied the reality of atoms and atomic particles and he did not believe the laws of thermodynamics could be applied to the universe. In an article *Physics of a Believer* first published in 1905 Duhem stressed that 'theoretical physics has no part to play in metaphysical or theological discussions' and he thought the argument that the biblical version of the creation and final end of the universe was supported by thermodynamics was scientifically and theologically invalid.[133] In contrast William Ralph Inge (1860–1954), Dean of St. Paul's Cathedral, London, was a very conservative churchman with a good knowledge of thermodynamics which he thought was relevant to theology.[143] In 1933 Inge published his lucid Warburton Lectures on the prophecy of modern science about the ultimate fate of the world, delivered in Lincoln's Inn Chapel, during 1931–1933, as *God and the Astronomers*.[144] He stated his view of the relevance of physics in the first two sentences of the preface:

> 'A philosopher or theologian who wishes to write on cosmology – on the relation of God to the universe – must in these days acquire some knowledge of modern astronomy and physics, two closely allied sciences in which new discoveries and new theories are being published almost every year. It is no longer possible to brush these researches aside as irrelevant to metaphysics or to theology.'

Inge saw that the second law of thermodynamics confirmed the traditional Christian belief in an external Creator, who made a world which had a beginning in time and will have an end in time. The *New Götterdämerung* is what Inge called the heat death predicted by the second law borrowing the term used by the German historian Oswald Spenger (1880–1936) in the first volume of his *Death in the West* published in 1918. One fact commented on by Inge was the blind eye turned in the 1860s to the heat death of the world by:

> 'Herbert Spencer, Charles Darwin, and a host of others [who] were indulging in exultant paeans about the ineluctable law of progress, which was destined, beyond a peradventure, to lead the human race onward and upward to a condition of absolute perfection.'

Although Inge had some extremely right wing views about society, such as his belief in eugenics, *God and the Astronomers* was well written with valid comments on science and religion.

2.8. Concluding Remarks

In the seventeenth century knowledge had not been compartmentalized; natural philosophy, while separated from religious control in Protestant countries, was not wholly independent of religion. Newton clearly saw that the third book of the *Principia* could be used in natural theology and at first Newtonian natural philosophy was enthusiastically used to justify the belief in a deity. However despite the warnings of Bacon against mixing the Scriptures with natural philosophy, Newtonians began to use natural philosophy to explain the biblical account of creation, the flood, and the apocalyptic end of the world. Not surprisingly attempts at explaining the inexplicable biblical accounts led to some rejecting Newtonian natural philosophy while Newtonian deists rejected the Bible.

Belief that God created the universe for a religious purpose led to the revitalization of the idea of the plurality of worlds as astronomy developed and in the eighteenth century the plurality of worlds became almost synonymous with astronomy. There is only the most tenuous mention of the possibility of other worlds in the Bible and it is remarkable that most educated Christians believed in the plurality of the worlds. Atheists and deists, rejecting the teleological argument, seized on the plurality of worlds as an argument against either a personal God or the total concept of a Divine Being. Before the middle of the nineteenth century there was almost no evidence either for or against the plurality of worlds. In the end it was biology not Newtonian natural philosophy that gave the strongest evidence for the plurality of worlds. However the plurality of the worlds is not science, despite the very many scientists who believe in it, since it is unfalsifiable. The Search for Extraterrestrial Intelligence (SETI) is searching for extraterrestrial signals of intelligent life, but the time span, during which similar signals could be sent from Earth before our civilization destroys itself, is extremely short compared with astronomical time, and it is likely to be about the same time for any extraterrestrial civilization sending similar signals to us, thus, even if extraterrestrial civilizations exist, the probability of such signals reaching us is almost zero. That billions of dollars have been spent so far on SETI shows a still almost desperate need for humans to believe in the plurality of worlds.

Until the second law of thermodynamics the laws of nature contained no arrow of time and physics could not comment on the question of creation or the final end of the universe. The situation changed in the mid-nineteenth century but the validity of applying the second law to the entire universe was questioned and some never grasped the concept of entropy.

This chapter on the influence of Newtonian physics on religion was chosen to be the first of the chapters on the influence of physics because of the amazing gusto with which so many people from all walks of life used physics on all sides of the debate on religion from the seventeenth to the end of the nineteenth century. The enthusiasm with which ideas were developed from physics and used either to justify revealed religion or atheism has not in my mind been matched in any other application of physics to our culture.

2.9. Notes

[1] Although the Reformation was precipitated by the public burning of the papal bull, *Exsurge domine*, along with the books of canon law at Wittenberg in 1520.
[2] Kobe (1998).
[3] Russell (1989).
[4] Sorondo (2003).
[5] Patrick (1662).
[6] Burnet (1684).
[7] A future thousand-year age of peace and righteousness associated with the Second Coming of Christ.

[8] An English translation was published in 1690.
[9] Vol. 2 Newton (2008).
[10]The common abbreviations that Newton used in his letters: (yt for that, wch for which) have been replaced by the words to make the letter more readable.
[11] Keill (1698, 1699).
[12] Buffon (1785).
[13] It has been suggested that this statement is contradicted by the first of a set of seven Statements on Religion found among his papers, 'That religion and Philosophy are to be preserved distinct. We are not to introduce divine revelations into Philosophy nor philosophical opinions into religion', it being supposed that Philosophy refers to natural philosophy, Austin (1970). None of the other six statements have any connection to natural philosophy and Newton did distinguish between natural philosophy and philosophy so the suggestion does not seem likely.
[14] In this section the quotations from the General Scholium are taken from the translation by Cohen and Whitman (1999) of the 1726 third edition of the *Principia.*
[15] Newton (1718).
[16] Clarke (1717).
[17] Vol. 3 Newton (2008).
[18] Harrison (1995).
[19] Snobelen (2003).
[20] Snobelen (2007).
[21] Burtt (1951).
[22] Wallis and Wren (1669).
[23] Hume (1748).
[24] Clarke (1706).
[25] Also known as Coates.
[26] Newton (1726).
[27] Jacob (1976).
[28] Bentley (1699).
[29] Dahm (1970).
[30] Harris (1698).
[31] Bentley was elected to the Royal Society in 1695 and he served on its council in 1697 but, on the evidence of the minutes, was not a very active member.
[32] Newton (1756).
[33] 25th February 1693. Newton (1756).
[34] 10th December 1652. Newton (1756).
[35] Laplace (1796).
[36] Swedenborg (1734).
[37] Kant (1755).
[38] Montmerle *et al.* (2006).
[39] His eldest son also named Samuel was appointed to the sinecure of King's Waiter at the Port of London and was elected to the Royal Society" in 1728.
[40] Clarke (1705).
[41] Toland (1704).
[42] Derham (1713).
[43] Huygens (1698).
[44] Leibniz (1710).
[45] The nom de plume of François-Marie Arouet (1694–1778).
[46] Force (1984).
[47] Force and Popkin (1990).

[48] Woodward (1695).
[49] Whiston (1725a).
[50] First observed by telescope on the 14th November 1680 by Gotfried Kirch was one of the brightest comets of the seventeenth century even being visible during daylight.
[51] Whiston (1702).
[52] A variant in the spelling of shock which comes from the French choc.
[53] The two papers were not published in 1694 because Halley wanted to avoid 'censure of the Sacred Order' and only published in 1724. Nicolson and Rousseau (1968).
[54] The Scriblerus Club, who also ridiculed Whiston for his scheme to determine longitude at sea, is discussed in Chapter 5.
[55] Whiston (1698).
[56] Whiston (1725b)
[57] Whiston uses the term 'law of gravity' here though later he distinguishes between laws and powers.
[58] Whitehurst (1788).
[59] Drysdale (1798).
[60] Mackenzie (1827).
[61] Beech (1989).
[62] Astore (2004).
[63] Tylecote (1957).
[64] Dick (1826).
[65] Blaikie (1880).
[66] They were usually 192 pages and in fancy paper covers they cost sixpence. In cloth board covers they cost tenpence.
[67] Brown (2001).
[68] Secord (2000).
[69] Fyfe (2005).
[70] Dick (1846).
[71] Whewell (1833).
[72] Topham (1992).
[73] Vourlidas *et al.* (2007).
[74] Paley (1802).
[75] O' Flaherty (2010).
[76] Frost (1846).
[77] Mandelbrote (2004).
[78] Hutchinson (1724).
[79] Aston (2004).
[80] Hutchinson (1748–1749).
[81] Berington (1750).
[82] Contra Sermones Arianorum.
[83] Hutchinson (1748–1749). Here Hutchinson referred to Newton as 'our undertaker', a few lines later he referred to 'an undertaker' seeming by context to mean Woodward.
[84] Horne (1756).
[85] Reid (2005).
[86] The book was very expensive costing sixteen shillings, discounted to twelve for Mugletonians. Reid (2005) suggests that colour prints, which summarized the Muggletonian system, were used because many Muggletonians were illiterate, but such persons would not have been able to afford even the discounted price.
[87] Airy (1859).

[88] Davidson (1900).
[89] Fontenelle (1686).
[90] It was removed from the Index in 1825, but again placed on the Index in 1900. Crowe (1986).
[91] Walsh (1743).
[92] Keill (1721).
[93] Derham (1719).
[94] Fara (2004a).
[95] Wright (1750).
[96] Ferguson (1754).
[97] Crowe (1986).
[98] Lambert (1761).
[99] Newton estimated that the heat of the comet at its perihelion reached 'about 2000 times greater than the heat of red-hot iron.' Newton (1726).
[100] Paine (1794).
[101] Walpole (1798).
[102] Chalmers (1817).
[103] Dick (1838).
[104] Dick (1844).
[105] Wollaston (1802).
[106] Newton (1718). It is not surprising that Newton did not see any dark lines because using a comparatively wide slit of 1.27 mm they are difficult to see. Charles Babbage, who repeated the experiments of experiments of Joseph von Fraunhofer that gave dark lines in a spectrum, also could not see them.
[107] Huggins and Miller (1864a).
[108] Huggins and Miller (1864b).
[109] Huggins (1868).
[110] Darwin (1909).
[111] Baker (1700).
[112] Catcott (1756).
[113] An accurate estimate of the distance of the Earth from the Sun can be made by timing the start and end of a transit of Venus across the Sun from widely spaced locations on Earth. The first such pair of transits after the publication of the *Principia* occurred in 1761 and 1769. One of the reasons for Captain Cook's first voyage to the Pacific was to observe the transit in Tahiti in 1769. In 1771 Jérôme Lalande used data from both transits to obtain a value of 153 million km for the distance.
[114] Whewell (1853).
[115] We now know that, though Venus is further from the Sun than Mercury its temperature is higher than Mercury because of the greenhouse effect of its blanketing clouds.
[116] We now know that the zodiacal light is caused by sunlight scattered by space dust.
[117] Hobbes (1651).
[118] Chambers authorship was not revealed until 1884.
[119] Now known as the British Science Association.
[120] Barton (1987).
[121] William Spottiswoode, another member of the X-Club was a mathematician but after 1871 he became more interested in physics.
[122] Tyndall (1896).
[123] President's Address to the Mathematical and Physical Section of the British Association on *Scientific Materialism* in 1868. Tyndall (1896).
[124] Tyndall was much influenced by German Philosophers since he did his doctorate at the University of Marburg in Hesse because their course was relatively inexpensive.

[125] Lange (1879–1881).
[126] Tyndall only had volume 1 of the two volume German edition when he wrote the Belfast address. Barton (1987).
[127] Whyte (1958) writes that to call the particles, which Boscovich named *puncta* or *particula*, atoms is misleading; his particles were primary constituents of matter not chemical atoms.
[128] Boscovich's forces or *vires* are not Newtonian forces but kinematic accelerations. Whyte (1958).
[129] Whyte (1958).
[130] Schofield (2004).
[131] Watson (1947).
[132] Caneva (1997).
[133] Kragh (2008).
[134] Tyndall (1863).
[135] Smith (1976).
[136] Smith and Norten (1989).
[137] Thomson had earlier used this biblical quotation in the draft of his *Dynamical Theory of Heat*. Smith (1977).
[138] Kragh (2004).
[139] Brentano (1987).
[140] Spencer (1867).
[141] Webb (1926).
[142] Hakfoort (1992).
[143] Hiebert (1966).
[144] Inge (1933).

Chapter 3

Popularization of Classical Newtonian Physics

The *Principia* is one of the most important books in the English language, but few read it and Newton deliberately made it difficult to read. However, Newton's disciples proclaimed its importance and soon there was a clamour to understand it. The earliest popularizations of the *Principia* were contained in the Boyle Lectures, especial those by Bentley, and Whiston's *The New Theory of the Earth*, which have already been dealt with in the previous chapter. The great Samuel Johnson (1709–1784) is said to have learnt much that he knew about science from the published Boyle Lectures.[1] Newton's fame became widespread among the population even if most understood little of his natural philosophy. Paul Chamberlen (1635–1717), one of the seventeenth and early eighteenth century Chamberlen family of medical practitioners, made use of Newton's celebrity status to advertise an anodyne[2] necklace, costing the then considerable sum of five shillings, which he claimed would ease the cutting of children's teeth, the labour pains of women, and headaches. Clearly if such a charm worked it was an example of action at a distance like Newton's gravity. To advertise the anodyne necklace, Chamberlen wrote a spurious pamphlet *A Philosophical Essay Upon Actions on Distant Subjects. Wherein are clearly Explicated According to the New Philosophy and Sir Isaac Newton's Laws of Motion*, which was given away free at the sign of the Sugar-Loaf, a confectioners shop near the Royal Exchange.[3]

3.1. Popular Eighteenth Century Lectures and Books

The coffee-houses of London, especially those around the Royal Exchange, were the equivalent of today's open universities during what has become known as the Augustan Age from the end of the seventeenth century to the middle of the eighteenth century. In fact in Oxford, where coffee-houses first originated in Britain during the early 1650s, they were known as penny universities since a coffee cost a penny. Lectures on mathematics and natural philosophy in coffee-houses started in the 1690s. One of the first lecture series on mathematics by

John Harris held first in Southwark in 1698 and later in the Marine Coffee-House near the Royal Exchange, was initially free under the patronage of Charles Cox (1660–1729), a brewer and Whig Member of Parliament. By 1704 Cox's support for the lectures became uncertain and Harris broadened his lectures to include natural philosophy using a public subscription to finance them.[4] The usual subscription fee for a lecture series was from two to three guineas. Such fees effectively limited the audience to aristocrats, gentry, and a few of the more affluent traders. After a while it also became common for the lecture series to be given at the lecturer's house or shop if he was an instrument maker.

3.1.1. *Popular lectures*

Popular lectures on Newtonian natural philosophy began to proliferate at the end of the seventeenth century. The lecturers were mainly drawn from those who had learnt natural philosophy at university or through apprenticeships, like James Hodgson (ca. 1678–1755) who was apprenticed to the astronomer John Flamsteed (1646–1719), and scientific instrument makers, like Francis Hauksbee, the elder, (ca. 1660–1713). In 1704 Hodgson, who had completed his apprenticeship and had been elected to the Royal Society in 1703, was engaged by Hauksbee to give *A Course of Natural and Experimental Philosophy and Astronomy* at the Hand and Pen Coffee-House using Hauksbee's apparatus.[5] The lectures were moved first to the Queen's Head Tavern in Fleet St. and then to the Hauksbee's new instrument shop in Wine Office Court in 1707.[4,6] Hauksbee, who had been elected to the Royal Society in 1705 and became their curator of experiments, did not lecture himself until 1709 when Hodgson obtained an appointment at the Royal Mathematical School, Christ's Hospital. Hauksbee published his *Physico-Mechanical Experiments* based on his lectures in 1709.[7] Nothing is known about the details of the course given by Hodgson apart from the advertisement for the course given in Halley's *Miscellanea Curiosa*, but Hauksbee's course was a series of experiments many based on what today would be called a vacuum pump with very little comment on the results though some of the apparatus was important in its own right,[8] and the description of Hodgson's course, 'for the benefit of ... curious and inquisitive gentlemen' in his advertisement would equally apply to Hauksbee's course.[9] The audience at either lectures series would have learnt very little about Newtonian natural philosophy. Hauksbee was primarily a skilled instrument maker who, working with Newton, had made innovative apparatus. Newton treated Hauksbee as a technician, though he was regarded as a natural philosopher on the Continent and his book was translated into Italian, Dutch, and French.[10] Hauksbee's nephew, Francis Hauksbee the younger (1688–1763), was also a scientific instrument maker and in 1712 opened an instrument shop in Crane Court very near to the new premises of the Royal Society.[6]

After Whiston was dismissed from his Lucasian Professorship in 1710 he went to London where he first gave lectures on mathematics. In 1712 he and the mathematician Humphrey Ditton (1675–1714), gave lectures on natural philosophy with both the Hauksbees demonstrating the experiments in the younger Hauksbee's premises in Crane Court. The elder Hauksbee died in 1713 and Whiston carried on lecturing with a similar course of mechanical, optical, hydrostatic, and pneumatic experiments in conjunction with the younger Hauksbee, which continued until 1731.[11] Richard Steele (1672–1729) the politician, and co-founder of *The Spectator* and *The Guardian*, arranged for Whiston and Hauksbee to give a course on astronomy starting in 1714 at Button's Coffee House, in Russell St., Covent Garden, which was transferred to Will's Coffee House on the opposite side of Russell St. in 1715.[12] Steele arranged for the course to be given at a coffee-house rather than his newly opened theatre, the *Censorium*, because he was not certain that Whiston would stick to astronomy but make remarks concerning religion as well. Only the course outlines are available, but it is clear even from these that Whiston's lectures were more systematic than those of the elder Hauksbee. Clearly Whiston coming straight from lecturing at Cambridge brought an improvement in the intellectual standard of these lectures but still retained a fascinating array of natural philosophy experiments. By demonstrating his continuing commitment to publicizing Newton's work, Whiston was still hoping for reconciliation with Newton and election to the Royal Society by the encouragement he received from Newton who sent him a copy of the second edition of the *Principia* in 1713.[4]

John Theophilus Desaguliers (1683–1744) was the French born son of Huguenots refugees who fled France for the British Isles in 1690 and moved to London in 1694. In 1705 Desaguliers joined Christ Church, Oxford where he studied under Keill graduating in 1709.[13] Desaguliers, a protégé of Isaac Newton, was a Newtonian much after the style of Keill. After graduating, Desaguliers took over Keill's position as a lecturer in experimental philosophy at Hart Hall, now Hertford College, Oxford University. On moving to London, Desaguliers briefly continued the elder Hauksbee's lectures for his widow in 1713. Desaguliers was proposed as a fellow of the Royal Society by Newton and elected in 1714 and two years later he took over the elder Hauksbee's position of curator of experiments for the Royal Society. From about 1715 Desaguliers gave lectures on Newtonian mechanics with copious experiments at his house in

Channel Row, Westminster and he published the 'minutes of my lectures for the use of such gentlemen as have been my auditors' in 1717.[14] By 1719 Desaguliers was also giving his lectures in French, probably for Huguenot refugees, and in 1724 published a bilingual synopsis[15] He was ordained and became chaplain in 1716 to James Brydges (1673–1744), who was created the Duke of Chandos in 1717, but Desaguliers became more important to Brydges' entrepreneurial projects as a scientific advisor than as a chaplain. Desaguliers interest in technology grew over the years and came to occupy a considerable portion of his public lectures that were originally designed for the entertainment of the nobles and gentry.

Popular lectures on natural philosophy were not limited to London. Desaguliers boasted in his preface to his *Course of Experimental Philosophy* in 1734 'that of eleven or twelve persons, who perform experimental courses at this time in England, and other parts of the world, I have had the honour of having eight of them for my scholars.'[16] Three of the most active popularisers of Newtonian mechanics in the 1750s were: Erasmus King (1690–1760), Benjamin Martin (1704–1782), and James Ferguson (1710–1776).[17] King lectured from 1736 to 1756 in some twenty different coffee-houses or taverns,[18] and he also conducted experiments for the Royal Society Fellow, Stephen Hales (1677–1761), these experiments featured in King's lectures after about 1741.[19] One new topic in King's first lecture was the ascent of sap in plants, though it was not until the end of the nineteenth century that the biophysical processes for the ascent of sap in tall trees began to be really understood and a definitive review of the process was not written until 1981.[20] Nevertheless the elements of this biophysical process were first given some 250 years earlier by Hales in 1727.[21,22] Hales crucial experiment is shown here in Fig.3.1. A small diameter tube c was packed with wood ash and the lower tube and the spherical vessel were filled with water. At the start of the experiment the end x was placed in a vessel filled with mercury; the top tube was also filled with mercury from a to b. The ashes sucked up the water causing the mercury to rise to z in the lower tube. The tube c, in Hales first experiment was three feet (0.91 m) long and after five or six days the mercury in the lower tube rose seven inches (178 mm) which is equal to raising water to about eight foot (2.42 m). Hales commented:[21]

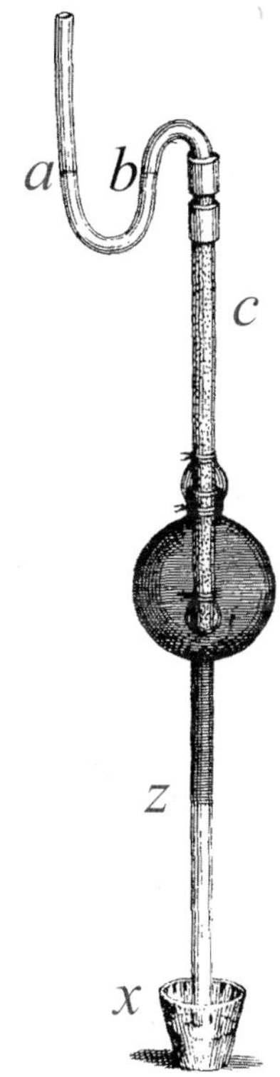

Fig. 3.1. Hales' ascent of sap experiment (Hales, 1727).

> 'by the same principle it is, that we see in the preceding experiments plants imbibe moisture so vigorously up their fine capillary vessels; which moisture, as is carried off in perspiration, (by the action of warmth) thereby gives the sap vessels liberty to be almost continually attracting of fresh supplies.'

Thus Hales correctly recognized what the biophysical process by which sap rises is due to the sap in the xylem vessels of transpiring plants flowing under tension caused by evaporation from the leaves. Under normal conditions atmospheric pressure can support a column of water just over 10 m high or about 760 mm of mercury. This implies that in trees taller than about 10 m must have a negative pressure, or tensile stress, in their sap. Tensile stress can, under suitable conditions, be sustained by sap and other liquids which is why trees can grow taller than 10 m. Hales did not explain how cohesive forces could suck up sap in tall trees, but did refer to Newton's query 31 in his *Opticks*. Newton explained that atmospheric pressure could support the[23]

> 'standing of quick-silver (mercury) in the barometer at the height of 50, 60 or 70 inches (1.27 to 1.78 m) or above when ever it is well purged of air and carefully poured in [a barometer], so that its parts be every where contiguous both to one another and to the glass. The atmosphere by its weight presses the quick-silver into the glass, to the height of 29 or 30 inches [737 or 762 mm]. And some other agent raises it higher, not by pressing it into the glass, but by making its parts stick to the glass, and to one another. For upon any discontinuation of the parts, made by either bubbles or by shaking the glass, the whole quick-silver falls down to the height of 29 or 30 inches.'

Nothing is known of what King said in his first lecture when he was discussing attraction and cohesion but it is probable that he did correctly account for the ascent of sap in plants and may have possibly noted how sap could rise in trees taller than 10 m. The ascent of sap in plants was pure science; in his eleventh lecture he discussed a technological invention of Hales and Desaguliers to measure the depth of the sea without a sounding line.[24] In 1739 King had joined Lord Baltimore's voyage to St Petersburg hoping to demonstrate his natural philosophy experiments to Empress Anna (1693–1740) and while on the voyage King had tested the invention of Hales and Desaguliers, which accurately measured the depth of the sea up to 288 feet (73.2 m).[25] In the eighteenth century King was well-known but has since been forgotten.[18]

Benjamin Martin, a ploughboy turned natural philosophy lecturer and instrument maker, gave experimental philosophy lectures including astronomy in southern England from about 1740.[26] In 1742 he moved to Reading where he published *A Course of Lectures in Natural and Experimental Philosophy*.[27] His lectures were similar to those given earlier in the eighteenth century but in the preface he emphasized the usefulness of the subjects of his lectures writing for example: 'In gunnery, the engineer finds the whole theory of his art founded in philosophy, viz. in the doctrine of projectiles.' Martin further emphasized the usefulness of natural philosophy in his *Panegyrick on the Newtonian Philosophy* published in 1749 where he showed Newtonian philosophy's to be of: [28]

> 'absolute necessity to the perfection of human nature; improvement of arts and sciences, the promotion of true religion, the increase of wealth and honour, and the completion of human felicity.'

This emphasis on the usefulness of natural philosophy went with a decrease in the cost of lectures from two to three guineas in the early eighteenth century to twelve shillings for Martin's twelve lectures.[29] Competition probably caused a necessary widening of the audience by attracting artisans, with the usefulness of the courses, who could not afford high prices. Apart from a brief period in London in 1745, Martin spent his most of his time lecturing in southern England and the Midlands until he settled in London in 1756.[30] Martin published prolifically on natural philosophy, in 1747 he published the *Philosophia Britannica* in two volumes,[31] increased to three volumes in the 1754 edition, which were extended versions of his two volume book and intended to be independent of his lectures.[30] In 1751 Martin published *A Plain and Familiar Introduction to the Newtonian Philosophy*, written mainly as a relatively cheap book, costing three shillings, for the use of those who attended his lectures. Martin continued to lecture on natural philosophy until the early 1770s.

James Ferguson was born a few kilometres from the village of Keith in Banffshire, Scotland to a farm worker who was literate. He was a child prodigy who deduced the mechanics of levers and winches when he was only seven or eight years old.[32] With the help of sympathetic employers and their staff, Ferguson educated himself in mechanics and astronomy. In 1743 Ferguson moved to London taking with him an orrery he had constructed, which he sold to the Solicitor-General Dudley Ryder (1691–1756). After being disappointed in not obtaining a position as master in a school of mathematics because the post required a bachelor and he was married, he supported himself by painting miniature portraits in Indian ink. Ferguson's first advertisement for lectures was for use of the globe and astronomy in 1746.[33] There was an eclipse of the Sun visible in London on the 14th July 1748, and Ferguson took this opportunity to give lectures on the eclipse in the days immediately before its occurrence. These lectures must have been successive since afterwards he continued lecturing with the aid of one of his orreries that he had made (see Fig. 3.2).[33] Ferguson then gave up portrait painting and employed himself 'in the much pleasanter business of reading lectures on mechanics, hydrostatics, hydraulics, pneumatics, electricity,

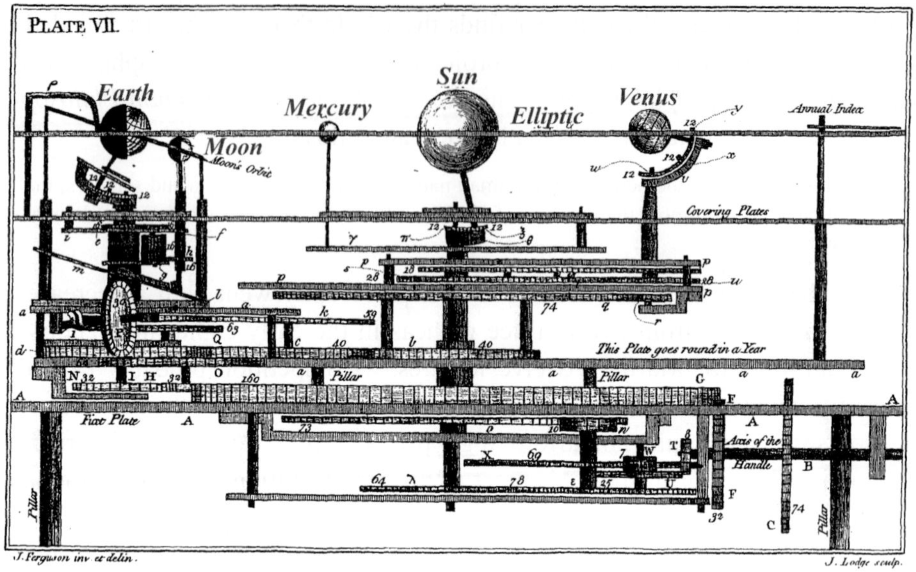

Fig. 3.2. Ferguson's orrery (Ferguson, 1773).

and astronomy.'[34] In 1756 Ferguson published *Astronomy Explained and Made Easy*[35] an expanded version of his earlier short book *An Idea of the Material Universe, Deduced from a Survey of the Solar System.*[36] From 1757 to 1760 Ferguson gradually changed his lectures from astronomy to experimental philosophy and in 1760 he published *Lectures on Select Subjects in Mechanics, Hydrostatics, Pneumatics, and Optics.*[37] In these lectures he also included the science of sundials, learnt as a young man from Alexander Cantley, the butler of Thomas Grant of Achoynaney.[34] Ferguson described the self taught Cantley as:[34]

> 'the most extraordinary man that I ever was acquainted with, or perhaps shall ever see; for he was a complete master of arithmetic, a good mathematician, a maker of music on every known instrument except the harp, understood Latin, French, and Greek, could let blood extremely well, and could even prescribe as a physician upon any urgent condition.'

After lecturing on the transit of Venus which occurred on the 6th June 1761, Ferguson resumed his itinerant lectures over a wide area of Britain from Plymouth to Edinburgh until 1774 when he was sixty-four years old.[33] In 1763 Ferguson was elected to the Royal Society.

Although independent lectures on natural and experimental philosophy continued after the middle of the eighteenth century, the growth in competition from formal institutions offering utilitarian lectures on mechanics and increased competition from other leisure activities, they became less important in spreading Newtonian philosophy.

3.1.2. *Popular books*

Apart from books explaining Newtonian natural philosophy to university and other college students, such as Humphrey Ditton's *The General Laws of Nature and Motion*,[38] many of the first popularizations were written in conjunction with lecture courses. One of the first popular books not associated with lecture courses was *A View of Sir Isaac Newton's Philosophy* by Henry Pemberton[39] (1694–1771) published in 1728 which covered both the *Principia* and *Opticks*. Although Pemberton's biographer in the *Oxford Dictionary of National Biography* opined that 'it made no great mark,' the book went through a second printing in Dublin in its first year but it had strong competition.[40] In 1734 Desaguliers published the first volume of his substantial *Course of Experimental Philosophy*.[16] The distinction between popular exposition and scientific discourse was not clear cut; as Desaguliers wrote in the preface to the first volume of his book:[16]

> 'a great many persons get a considerable knowledge of natural philosophy by way of amusement; and some are so well pleas'd with what they learn that way, as to be induced to study mathematicks, by which they at last become eminent philosophers.'

However, aiming at a wide audience, Desaguliers avoided mathematics. Desaguliers' second volume was not published until 1744 only shortly before his death.[41] The explanation for this delay was given in the preface to the second volume:

> '[the] delay was occasion'd by the promise I made in my first book to decide the question about the force of bodies in motion ... [with those in] Germany, Italy and Holland measuring that force by the product of the mass into the square of the velocity of the body; and those of France and England measuring that force by the product of the mass into the simple velocity.'

Desaguliers finally came to the correct conclusion 'that the whole was only a dispute about words; the contending parties meaning different things by the word FORCE.' In the second volume Desaguliers did not include optics, as he had earlier planned, but instead substituted a long section on water mills, water machines, and steam engines for raising water because of popular demand. His lectures which were originally designed for the entertainment of the nobles and gentry became more orientated towards technology.

A feature of eighteenth century life that fascinated the common people, nobles and King George I alike was the strongman. Desaguliers explained how their feats of strength were accomplished in his *Course of Experimental Philosophy*. The famous strong men of the first half of the eighteenth century that performed in London were William Joy[42] (1675–1734) from Kent, the German Johan Karl von Eckenberg (1684–1748) and the Londoner Thomas Topham (1702–1749). In about 1716 Desaguliers showed many of his experiments to King George I who asked him whether the feats of strength that he had been shown half a year earlier, possibly by Eckenberg, were genuine.[16] Desaguliers described how: [16]

> 'he had a Frame of Wood made to stand in [as shown in Fig. 3.3] and with a Girdle and Chain lifted an Iron Cylinder made use of to roll the Garden [weighing some 860 kg], sustaining it easily when once it was up. Some Noblemen and Gentlemen who were present 'try'd the Experiment afterwards, and lifted the Roller some with Ease, and some with more Difficulty than [he] had done.'

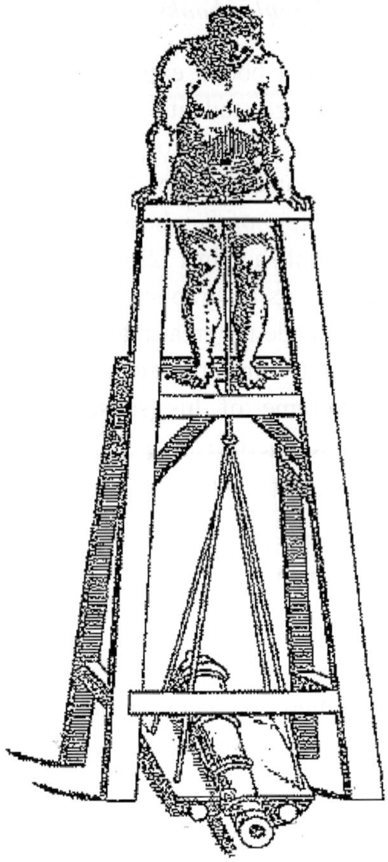

Fig. 3.3. Desaguliers' frame for lifting with a girdle and chain (Desaguliers, 1734).

The strongmen found out by practice without any mechanical theory the postures of the body which enabled them to perform their feats of strength. Some, as Desaguliers observed, discovered more advantageous postures than others. In about 1720 Eckenberg, by using the right position, resisted the pull of two horses but eight years later when Topham repeated the feat he resisted the pull of one horse but, when he tried to do the same with two horses, he was pulled up from his posture and his knees severely damaged when they struck against the support.[16] In his first volume of his *Course of Experimental Philosophy* Desaguliers explained using simple mechanics how the strongmen performed their seemingly impossible feats and stated that ever since his lectures none 'of these Sampsons have attempted ... to impose upon People in the same manner in or near London.'[16]

Voltaire, the nom de plume of François-Marie Arouet (1694–1778), was a writer of poems and plays prior to his exile to England in 1726. He had been introduced in France to Newtonianism in the early 1720s by the Tory Lord Bolingbroke (1678–1751), who was there in exile after supporting the 1715

Jacobite rebellion. In 1726 Voltaire had an argument with a French nobleman who then got his retainers to beat him up, on challenging the nobleman to a duel, the nobleman obtained a *lettre de cachet* and had Voltaire imprisoned in the Bastille. To get out of prison, Voltaire offered to go into exile to England. By 1726 Bolingbroke had been pardoned and was back in England and introduced Voltaire to his circle. However Voltaire did not confine his acquaintances to Tories and was also introduced to Whig figures such as Samuel Clarke and other Newtonians but he did not meet Newton himself who died the year after he arrived. Voltaire found London to be a culture shock. In France one could not say or print anything that might be taken as impugning the king or the nobility whereas in England there was comparative freedom. At Newton's funeral in Westminster Abbey, Voltaire was astounded to see his coffin carried by peers of the realm and buried next to kings. Voltaire was only in England for less than three years before he was allowed to return to France, but those years changed him and he became a convinced Newtonian. He propagated Newtonianism through two books: *Letters Concerning the English Nation*,[43] which consisted of 25 letters, with four about Newton and Newtonianism, first published in London in English in 1733; the second book was a popular account of Newton's natural philosophy *Elémens de la Philosophie de Neuton* published in London in an authorized version in 1738 and an English translation *The Elements of Sir Isaac Newton's Philosophy*[44] published in the same year.[45]

In his letter on Descartes and Newton, Voltaire began by characterizing the difference between Cartesianism and Newtonianism:[43]

> 'A Frenchman who arrives in *London*, will find Philosophy, like every Thing else, very much chang'd there. He had left the World a *Plenum*, and now he finds it a *vacuum*, At Paris the Universe is seen, compos'd of Vortices of subtle Matter; but nothing like it is seen in London.'

In the next letter on attraction, Voltaire gave an outline of Newton's *System of the World* contrasting it to the Cartesian System. Letter 16 is on Newton's optics, though it contains some errors, this letter does explain reasonably well the discoveries of Newton.[46] Voltaire remarks that Descartes 'hop'd to discover in the Stars, by the Assistance of Telescopes, Objects as small as those we discern upon Earth,' but that Newton realized that because of the differences in refraction of the colours, there was a limit to the resolution of a refracting telescope and invented the reflecting telescope to overcome this difficulty.[47] Voltaire's last letter on Newton is on infinites in geometry and Newton's chronology of the Ancient Biblical Kingdoms. The discussion on the problem of infinity and Newton's fluxions is short and not very illuminating.

In 1733 Voltaire began a long lasting relationship with Émilie du Châtelet (1706–1749), who was married to the Marquis Châtelet. Unusual for the time, Émilie du Châtelet as a child had been encouraged by her father to study mathematics and science and she had continued with these studies.[48] Émilie was more knowledgeable in science than Voltaire and she helped him greatly in the writing of *Elements of Sir Isaac Newton's Philosophy* in the introduction of which Voltaire wrote:[49]

> 'We shall endeavour to make these elements easy and intelligible to those, who know no more of Newton and Philosophy than their name,'

Throughout the book Voltaire was at pains to reject notions of Descartes.

In 1649, the year which she died from childbirth,[50] Émilie du Châtelet, completed a translation of Newton's *Principia* into French. Her translation was not published until ten years after her death.

3.1.2.1. *Natural philosophy for ladies*

During the eighteenth century ladies were potential audiences and readers for popular natural philosophy. Desaguliers in his *A Course of Experimental Philosophy* remarked in the preface:[16]

> 'I have with great Pleasure seen the Newtonian Philosophy so generally received among Persons of all Ranks and Professions, and even the Ladies, by Help of Experiments'

and as with most subscribed books on natural and experimental philosophy there were quite a few lady subscribers. Martin also wrote in his preface in 1743:[27]

> 'why should not the Ladies understand and study the Sciences of Humanity, of which Philosophy holds the First Place? Knowledge is now become a fashionable Thing, and Philosophy is the Science a la Mode.'

By 1747 Martin assumed in his preface that ladies as well as gentlemen would attend his course[31] and in 1751 the title of his course is *A Plain and Familiar Introduction to the Newtonian Philosophy Design ... for the Use of Such Gentlemen and Ladies as Would Acquire a Competent Knowledge of this Science.*[51] These changes in the reference to ladies attending Martin's courses suggest that their number among his audience increased and by 1751 they were not an insignificant portion of his audience.

Fontenelle's *Entretiens sur la Pluralité des Mondes*, was based on Cartesian not Newtonian natural philosophy, and as he explained in his preface: [52]

> 'I have introduced a fair lady to be instructed in philosophy ... by this fiction, I shall render my work more agreeable, and to encourage the fair sex ... by the example of a lady who had no supernatural character, and who never goes beyond the bounds of a person who has no tincture of learning ... And why should this imaginary lady have the precedency of all the rest of her delicate sex? Or do they believe they are not as capable of conceiving that which she learned with so much facility?'

The use of a dialogue between a male philosopher and a lady was natural for Fontenelle who had a familiarity with the salons of Paris where science was discussed.[53] This style of writing became a common device in the eighteenth century to popularize Newtonian natural philosophy for both men and women. The adoption of Fontenelle's style to sell popular science is not surprising since *Entretiens* was most successful with numerous editions and translations published during the eighteenth century. John Harris used Fontenelle's style in his *Astronomical Dialogues between a Gentleman and a Lady* in 1719 writing his preface that:[54]

> 'the reader will easily see that the conversation in these dialogues is feigned, and in imitation of those of the excellent Mr Fontenelle, On the Plurality of Worlds. And that the digressions and turns of wit are introduced to render those notions pleasing and agreeable, which perhaps without such a kind of dress, would appear too crabbed and abstracted.'

Whereas Harris did not specifically address his book to women despite using Fontenelle's style, in 1737 the Italian Francesco Algarotti (1712–1764) did, as his title *Il Newtonianismo Per le Dame ovvero Dialoghi Sopra la Luce e i Colori*[55] made plain but he wanted to attract both sexes. The book described some of Newton's optical experiments in a series of six dialogues between a cavaliere and a marchesa and was dedicated to Fontenelle. Although optics is the theme that runs through the dialogues, the cavaliere touches on many other aspects of philosophy in a way that would interest a cultured marchesa or anyone else. It is not surprising that an Italian should have written a book on Newtonian natural philosophy specifically for ladies since Italy was the one country in Europe where women could teach in universities.[56] The most famous was Laura Bassi (1711–1778) who was a professor of physics and mathematics at the University of Bologna from 1732 until her death. In the first dialogue the cavaliere read a part of a canzone[57] addressed to Bassi to celebrate her graduation from the University of Bologna, which had been published by Algarotti in 1732.[58] To Algarotti Newton's optics was an example of what was best in natural philosophy or as the cavaliere stated:

> 'There cannot be a better pattern of true philosophy than his treatise of *Opticks*, it was the product of thirty years application and study. One single experiment of his gives more

satisfaction and information, than one could formerly have received from all their [Descartes etc.] best digested and most ingenious systems.'

The imagery that Algaretti used in his dialogues was interesting and vivid such as his quotation, in the fourth dialogue, from the sixteenth century epic poem *Jerusalem Delivered* by Torquato Tasso in the cavaliere's discussion of the spectrum of colours:

The jolly peacock spreads not half so fair
The eyed feathers of his pompous train;
Nor golden Iris so bends in the air
Her twenty-coloured bow, through clouds of rain;

which he then followed up with the comment:

'Innumerable degrees of intermediate colours insensibly connect and unite the seven primary ones. Neither Corregio Titian, nor his rival Rosalba, did ever unite and shade their mezzo tinteos with so much exactness to form the oval of a face.'

To make the book more popular, Algarotti was mildly erotic in some places knowing, like advertising men, that sex sells. The popularity of *Il Newtonianismo Per le Dame* was such that it went through numerous Italian editions throughout the eighteenth century and as well as being translated into English, French, German, and Dutch.

Despite the apparent light hearted nature of Algarotti's work, it had a thinly veiled hidden text that variously promoted Lockean philosophy and exhibited his association with freemasonry. Because of this hidden text, *Il Newtonianismo Per le Dame* was put on the *Index of Forbidden Books* in 1739 – the only popularization of Newtonian natural philosophy to attract the wrath of the Holy Office.[58] In 1746 Algarotti published a modified edition of his work entitled: *Newtonianism, or Dialogues on Light, Colours and Attraction*; he further modified the text in 1750 and 1752 changing the title to *Dialogues on Light, Colours and Attraction*.[58] None of these later editions were put on the *Index*.

3.1.2.2. *Natural philosophy for children*

Natural philosophy had by the middle of the eighteenth century become firmly entrenched as part of popular culture and it is not surprising that books written specifically for children started to appear. Prior to Newton and the *Principia* natural philosophies had not influenced popular culture to any extent and no books had been written specifically for children. The first children's book on natural philosophy *The Newtonian System of Natural Philosophy, Being the Substance of Six Lectures Read to the Lilliputian Society, by Tom Telescope* was written anonymously by John Newbery (1713–1767), a publisher who specialized in children's books, and published in 1761. Newbury had previously published *A Museum for Young Ladies and Gentlemen* in 1750 which was cobbled together from a variety of sources but not written specifically for children and did not stand up to his motto *Delectando monemus* (instruction with delight), whereas *The Newtonian System of Natural Philosophy* certainly lived up to the motto. In this book, the young ladies and gentlemen are on holiday at the home of the Countess of Twilight when a discussion arose as to how they should amuse themselves; various children's games and cards are suggested. Master Telescope is aroused by the mention of cards 'Playing at cards for money, says he, is so nearly allied to covetousness and cheating, that I abhor it.' Instead Tom Telescope suggests natural philosophy and some other intellectual pursuits which he thinks are very entertaining and they agree to have Tom lecture them. The lectures are given at the residence of the Marquis of Setstar, who is a natural philosopher and has 'the proper instruments' for such lectures. Tom's first lecture is on matter and motion, which he illustrates with children's toys.[59] The second lecture on astronomy is held in the Marquis' observatory. Not surprisingly the plurality of worlds is touched upon but, though the planets of the stars are 'filled with animals, plants, and minerals, all perhaps different from our [Earth], but intended to magnify the Almighty Architect' there was no suggestion that there are intelligent beings on these planets. Tom made little reference to religion but he did continue at this point in his lecture:

> 'who weighed the mountains in his golden scales, who measured the ocean in the hollow of his hand, who drew out the heavens as a curtain, who maketh the clouds his chariot, and walketh upon the wings of the wind.'

which is an allusion to Isaiah 40:12 in the Bible. Parts of lectures 3 and 6 are also on Newtonian natural philosophy but the rest are mostly on natural history where the science in some of them is out of date. In lecture 3 on air, atmospheric effects and earthquakes there is a remnant of Greek scholasticism. The lecture begins with a discussion of the four elements fire, air, earth, and water which Empedocles had established in the fifth century BC despite their having been overthrown a century earlier by Boyle in his celebrated treatise *The Sceptical Chymist*, which in part earned him the title of the *Father of Chemistry*.[60] However, some topical events are covered such as the great Lisbon earthquake of 1755. Here Newbery can be forgiven for not using the very recent work of John Michell on earthquakes published only the year before where, though Michell attributed the prime cause of earthquakes to the generation of fracture by the high pressure of water vaporized by fire, he did understand that shock waves travelling at the 'rate of more than twenty miles per minute [533 m/s]'caused the damage.[61] The third lecture is also noteworthy for its social comment when the Duke of Galaxy caused social consternation by entering the observatory at the beginning of the lecture. Tom says:

> 'There is nothing betrays a mean spirit and low education so much as this ridiculous awe and dread which some people shew in the company of their superiors ... Birth and fortune are merely accidental, and may happen to be the portion of a man without merit; but the man of genius and virtue is ennobled, as it were by himself.'

In the last lecture on the five senses of man and his understanding there are more social and political topics. Cruelty to beasts and man is discussed. The Ambassador of Bantam joins this lecture and expresses horror at practices in processing animals for the table when he hears Sir Harry say:

> 'to gratify their palates ... they must roast lobsters alive, and whip young pigs to death to make them tender.'

Later on Tom contrasts the action of those who are kind to animals but cruel to their fellow man:

> 'Sir Thomas has compassion ... but I am afraid it is only for himself. He loves his dogs and horses, because his dogs and horses give him pleasure; but to creatures that afford him none, he is absolutely insensible. I have seen him, even at Christmas, feed his pupps ...with delicacies; but rave, at the same time, in a merciless manner at poor children who were shivering at his gate and send them away empty handed.'

Another neighbour of the same disposition is kind to his dogs and horses but is engaged in the slave trade. Addressing Lady Caroline, Tom asks:

> 'Is this humanity, Madam? But above all is this Christianity? And arte these the blessed effects of the liberty we boast of?'

The lecture finished with a discussion, based on Plutarch's *Life of Lycurgus*, of the golden age of Sparta and its fall

> 'when these great and virtuous people of Sparta had conquered Athens, and from thence introduced wealth and luxury into their own country, they lost their virtue, dwindled to nothing and were themselves enslaved.'

The book ends with a poem written by Oliver Goldsmith (1728–1774) especially for Newbery:

Man in that Age no rule but Reason knew,
And with a native bent did good pursue;
Unaw'd by punishment, and void of fear,
His words were simple and soul sincere.
By no *forc'd laws* his passions were confin'd,
For *conscience* kept his heart, and calm'd his mind;
Peace o'er the world, her blessed sway maintain'd,
And e'en in Deserts smiling Plenty reign'd.

Tom Telescope's *The Newtonian System of Natural Philosophy* was extremely successful. It went through nine editions, selling some 25,000–30,000 copies, during the eighteenth century alone and remained on book shelves for eighty years, and was translated into Dutch, Swedish, and Italian.[62] As with many other successful books it suffered revisions over the years that greatly altered its character. In 1806 Tom was forced to grow up and become Mr Thomas Telescope and a fully fledged philosopher since it was thought inappropriate to have a child lecturing children, the aristocrats were taken out in 1812 and the Lockean structure was removed in 1820.[62]

Textbooks for school children were written in the eighteenth century such as *The Young Man's Book of Knowledge* by Daniel Fenning (1714–1767) published in 1764 which contained sections on natural philosophy and astronomy but these are outside of the scope of this book. Ferguson wrote *The Young Gentleman and Lady's Astronomy Familiarly Explained* published in 1768.[63] This book is written as a dialogue between Neander, a young man just come down from Cambridge, and his sister Eudosia, who he is instructing. Although the title implies that the book is for young people of both sexes, it may have been more aimed at young ladies since in the first dialogue Edosia indicates that she would like her brother to instruct her in astronomy and says 'But shall I not be laughed at for attempting to learn what men say is fit only for men,' with Neander replying 'Never, by any man who thinks right; and I hope you are above minding what those say who think wrong.' The book seems to be aimed at young people in their late teens considerably older than Tom Telescope's young friends. Unlike Newbery's *The Newtonian System of Natural Philosophy*, Ferguson's book gave religious justification in the first dialogue for undertaking the study but there is no further mention of religion in the other dialogues; the plurality of worlds, discussed in Ferguson's other books on astronomy, is not mentioned.

Olinthus Gregory (1774–1841), whose main education came from studying at Richard Weston's (ca. 1732–1806) boarding school in Leicester for ten years, published *Lessons, Astronomical and Philosophical, for the Amusement and*

Instruction of British Youth in 1793 when he was only nineteen years old.[64,65] This work appears to have been designed as both a popular book and also a potential textbook for schools. The contents of the book are very similar to those of Newbery's *The Newtonian System of Natural Philosophy*; it was intended to advance the design argument for the existence of a divine creator but the religious content is mainly confined to the introduction and concluding lesson. The plurality of worlds is advanced with a quotation from Martin[31] and a paragraph using language that is very similar to Ferguson's.[35] The work was successful going through six editions by 1824.

Stephen Jones (1763–1827) wrote a very curious book for children published in 1793: *Rudiments of Reason; or the Young Experimental Philosopher.*[66] Perhaps wisely Jones did not name himself as the author but only gave his initials. Jones was primarily a journal editor and compiler of reference books and his outlook was greatly influenced by his enthusiasm for freemasonry which is very apparent in his book. The book, in three volumes, is more or less on natural philosophy (astronomy is not treated) and written in the form of a family conference between the father, Sir Thomas Howard, his wife Lady Caroline, and their ten children. What makes it different to the previous children's books discussed is that it has a very large number of experiments or natural phenomena to be explained by the children at the end of each conference. Jones, though he was educated at St Paul's School, London, had no formal contact with natural philosophy after leaving school and did not apparently gain any knowledge of it by self education as others such as Ferguson had. Some of the main text is accurate but very many of the explanations of natural phenomena, are simply not scientific and the book would have filled children with many nonsensical ideas. Surprisingly the book went through a second edition in 1805 which was 'carefully revised and enlarged by the Rev. Thomas Smith.' If it is thought that this criticism of Jones' book is an imposition of twenty first century science on the eighteenth century, the review of the second edition in the *Eclectic Review* in 1806 makes the same criticism.[67] The review opened with

> 'we can scarcely account, on any principle sufficiently respectful to the Reverend gentleman, whose name is prefixed to this volume, for its antiquated and erroneous doctrines.'

The anonymous reviewer gives a number of examples of the erroneous explanations of natural phenomena. What is puzzling is that *Rudiments of Reason* was published by Elizabeth Newbery, the widow of Francis, John's nephew, who took over John's publishing business after the death of John's son who was also called Francis. John Newbery's *The Newtonian System of Natural Philosophy* is a famous book. Obviously the talent in the Newbery family had deteriorated in the thirty two years between it and Jones' book.

The Sequel to Mentoria or the Young Ladies Instructor is interesting because it was written by a woman, Ann Murry (ca. 1755–1816) for young ladies.[68,69] The book is in the form of a dialogue between a female instructor, Mentoria, and two young ladies, Mary and Louisa. The contents of the book are again similar to those in Tom Telescope's *The Newtonian System of Natural Philosophy*. Murry does not feel the need felt earlier to justify writing a book on natural philosophy and astronomy specifically for young ladies. She does, however, feel it necessary to justify its religious relevance and included the phrase: *Being principally intended to enlarge the Ideas and inspire just Conceptions of the Deity, from the Contemplation of the general; System of the Universe* in the book's full title and again emphasized how the book would promote religious belief in her preface and interspersed moral reflections throughout the dialogues. She referred to William Herschel on the plurality of worlds and wrote:

> 'It is highly probable there are millions of suns, attended by planetary orbs, inhabited by myriads of rational beings; all tending to the glory of their great Creator, and uniting to fulfil the general purposes ordained by Divine Wisdom.'

Herschel was her main reference on astronomy and her book is well up to date describing comparatively new aspects of astronomy such as the Milky Way Galaxy. *The Sequel to Mentoria* was a fine book but there do not appear to have been any more editions.

3.2. Popular Nineteenth Century Institutes, Associations, Books, and Periodicals

The two most important British institutes that gave popular lectures on physics during the nineteenth century were the mechanics' institutes and the Royal Institution. Both originally were designed to cater for the working class though it was the mechanics' institutes that developed the concept of being either managed partly or completely by the working class themselves. The Royal Institution was from the outset run by the elite initially for both the working class and the people of fashion but soon the working class was seen as too unfashionable and eased out of the Institution.

In the early nineteenth century the lack of opportunity for scientists to meet together, combined with the feeling that science in Britain was declining led to the formation of the British Association for the Advancement of Science (BAAS) along the lines of the Deutscher Naturforscher Versammlung formed in Leipzig in 1822. Physics has always been one of the key areas covered by the BAAS.

Books were expensive at the beginning of the nineteenth century when the process of producing a book differed little from that when printing was first introduced in Europe. In addition in Britain there was a paper tax which was not abolished until 1861. Changes in paper making technology and the use of water and then steam power brought the cost of paper down as the nineteenth century progressed. By the middle of the nineteenth century power driven cylindrical printing presses were developed, though at first they were used more for newspapers than books. With the reduction in cost, the number of books published increased dramatically and by 1850 there were plenty of books available at 2s.6d, with many costing only a shilling.[70] Periodicals, which increased in frequency from quarterly, to monthly, and weekly over the first half of the nineteenth century, also fell in price. The weeklies of the 1840s and 1850s typically cost a penny and even *Nature*, when it began publication in 1869 primarily as a popular journal for the general public, cost only fourpence.[71] As the cost of books and periodicals decreased, the number published increased as did the number of popular scientific writers. In the 1871 British census 2,500 people listed their employment as writing; the proportion books published on pure science at that time was about 5% which suggests that there were over a hundred active science writers.[70,72]

3.2.1. *The mechanics' institutes*

The working class emerged as an important section of the lower orders of society during the industrial revolution in the late eighteenth century. Many of the workers and some of their employers saw a need for their education both generally and in science. Even after the decrease in the cost of subscriptions for popular lectures on mechanics and astronomy, they were still too expensive for most of the working class. Books too were expensive in early nineteenth century Britain costing about half a guinea compared with five shillings on the Continent.[73]

In 1799 George Birkbeck (1776–1841), a Quaker, became the first professor of natural philosophy at the newly formed Andersonian Institution in Glasgow. He became interested in educating the working classes when, finding that there were no instrument makers in Glasgow, he had to explain the apparatus he needed to the men in workshops. As Birkbeck noted 'I had frequent opportunities of observing the intellectual curiosity of the *unwashed artificers*, to whose skill I was often obliged to have recourse.'[73] When a model of a centrifugal pump was being tested in one workshop, Birkbeck was entranced by the 'inquisitive countenances' of the workers who had gathered round it and he resolved to offer instruction to workmen in science.[73] Birkbeck started an evening mechanics' class at the Anderson's Institution in 1800.[74] The classes were a success with 75 workers attending the first course and growing to 500 by the fourth.[73] In these early courses the mechanics studied the principles of natural philosophy. However, the classes ended in 1804 when Birkbeck resigned from Anderson's Institution in protest at being asked to make the course bring in more money.[75] The workers decided to continue their classes after Birkbeck left Glasgow and they were given by Andrew Ure (1778–1857), a chemist who succeeded Birkbeck as professor of natural philosophy.[73] Birkbeck went to Birmingham after leaving Glasgow and started similar classes for workers there before he moved to London in 1806 when he got married.[76]

There were spasmodic attempts at forming mechanics' institutes in the years after Birkbeck's experiment. These attempts were short lived may be because they were instigated not by workers but by professional men. In 1821 Leonard Horner (1785–1864) founded *The School of Arts of Edinburgh*, which is often referred to as the first mechanics' institute; but this too was controlled not by the workers but by the subscribing patrons.[73] However it was successful and finally became, after a number of name changes, *The Heriot-Watt University* in 1966. The burghers of Edinburgh were quite specific in what they wanted taught at their school of arts, it was not natural philosophy but, as the full title of the school spelt out, 'the instruction of mechanics in such branches of physical science as are of practical application in their several trades.' As with other mechanics' institutes, the mechanics quickly discovered that they lacked sufficient knowledge of arithmetic or geometry to follow the classes. To remedy this lack of knowledge the students organized their own remedial classes with their own teachers the first two being a joiner and a cabinet maker.[73] In the twenty first century the level of mathematics taught in English schools has deteriorated to such an extent that many universities, like London's Imperial College, have had to offer remedial mathematics classes for students enrolling in science and engineering courses even if they have obtained an A-grade in mathematics in the Advanced Level General Certificate of Education.

The Directors of Anderson's Institution had little interest in the mechanics' classes and disputes over the ownership of the library and the apparatus for the classes led to the workers withdrawing from the Anderson's Institution and forming their own independent Glasgow Mechanics' Institution in 1823.[73] The stated purpose of this institution was 'the instruction of artisans in the scientific principles of arts and manufactures.' The management of the institution was by the workers themselves and initially they sought no financial aid from the more affluent members of society. The first lecture courses were given by John Steele[77] on mechanics and chemistry and in 1824 natural history and mathematics were added. Whether the mechanics course contained elements of natural philosophy or was only a practical mechanics course on machinery is not known. However, Steele did not repeat his course 'owing to some difficulties with the class' and moved in May 1824 to the newly formed Mechanics' Institution in Aberdeen where he lectured on natural philosophy and chemistry and so his lectures in Glasgow probably had at least an element of natural philosophy.[78] After Glasgow other towns quickly formed their mechanics' institutes; by May 1825 there were nearly thirty operating mechanics' institutes in Great Britain and Ireland which by 1851 had increased to over 700.[81] Birkbeck, now in London, was the moving force behind the founding of the London Mechanics' Institution in November 1823.[79] The other important figures in the creation of the London Institution were Joseph Robertson (1788–1852), editor of the London Mechanics' Magazine, and the middle class radicals: Thomas Hodgskin (1787–1869) and Francis Place (1771–1854).[75] Birkbeck and Place believed that complete workers control over mechanics' institutes was not a viable option; nevertheless the constitution of the London Institution required that two-thirds of the management committee be mechanics. This provision was probably strongly urged by Robertson since in November 1824 he wrote a strong editorial in the London Mechanics' Magazine when he thought that:[80]

> 'the working classes had been deprived of the number of votes in the Committee which these laws professed to secure to them, and that one of the extra gentlemen thus thrust into the management, had accomplished the trick, by passing himself off as a *common workman*, though, in reality, a master tradesman at the head of a large establishment.'

The two-thirds provision was adopted by many of the other early institutes but some management committees, like that of the Manchester Mechanics' Institution, were entrusted to directors chosen by and among the honorary members who had paid large subscription fees. However, some commentators have considered that there was little practical difference between the types of management committees stating that 'the control and uplift rhetoric of the supposedly worker-controlled Glasgow Mechanics' Institution is little different from the bourgeois-controlled Edinburgh School of Arts.'[81] Henry Brougham (1778–1868), the Whig politician who in 1830 became Lord Chancellor and wrote the very influential *Practical Observations on the Education of the People* published in 1825, believed that it was essential that the committee of management consisted of at least two-thirds working men.[82]

It is not easy to find exactly what was taught at the mechanics' institutes. The early institutes seem to have concentrated satisfying the workers desire for education and knowledge rather than providing instruction in subjects directly related to their employment. Even the institutes like the Gas Workmen's Institution in Glasgow did not at first offer trade-related courses. In an article on mechanics' institutes in Tait's Edinburgh Magazine for 1838 'the standing dishes or staple commodities' in the seven leading institutes[83] surveyed were natural philosophy and chemistry. Astronomy was absent from the curriculum of the early institutes and does not seem to have been often offered in later years. The reason for this omission is not clear; a possible reason may have been that many popular works on astronomy were used to advance the design argument for natural theology and religion was expressly excluded from the mechanics' institutes because of the large variety of persuasions of the members. Natural philosophy and chemistry continued to be staple commodities until about 1850. James Hudson, who wrote *The History of Adult Education* published in 1851, opined that natural philosophy, which had been offered by many institutes up until then, was being dropped from the curriculum. Of Liverpool Mechanics' Institution, Hudson wrote that: [84]

> 'The classes of natural philosophy and chemistry have been long discontinued owing to the persons attending the establishment representing a different grade of society. The working-mechanics who generally prefer sterner studies have given place to others who attend dancing, and essay discussions.'

Hudson was been rightly criticized in recent years for his view that the institutes had by 1850 reached their pinnacle and were declining and losing the support of the workers but his opinion about classes in natural philosophy seem to be better founded. Of course the term *natural philosophy* was starting to become archaic in the 1850s and replaced by physics except for a few British institutions such as the Royal Institution, Oxford, Cambridge, and Glasgow Universities where there have been a professors of natural philosophy up until the present time.[85] However it is true that 1851 was a watershed for the mechanics' institutes. The Great Exhibition of 1851, organized in part by Prince Albert, was primarily a showcase of British manufacturing which was beginning to have strong competition from France and Germany. The emergence of France and Germany as competition to Britain was in no small part due to their superior technical education as was recognized by Brougham as early as 1824.[86] After the Great Exhibition, apart from the continuance of elementary classes, the courses in the mechanics' institutes became more orientated to industry and outside of the scope of this book.

The mechanics' institutes spread to Australia, Canada and the eastern states of America.

3.2.2. *The Royal Institution of Great Britain*

The Royal Institution of Great Britain was the conception of the American born Benjamin Thompson (1753–1824), Count Rumford of the Holy Roman Empire, and was founded in 1799 'for diffusing the knowledge, and facilitating the general introduction, of useful mechanical inventions and improvements; and for teaching, by courses of philosophical lectures and experiments, the application of science to the common purposes of life,' and much of the first funding came from the Society for Bettering the Conditions and Improving the Comforts of the Poor. There were 58 original Proprietors of the Institution who each subscribed fifty guineas, which then was a large sum of money. The Institution had strong links to the new Board of Agriculture which had been formed in 1793 with nearly half the Proprietors and fourteen out of nineteen of the Managers and Visitors of the Institution being members of the Board.[87] In keeping with the original purpose of the Institution, the first professorship was in natural philosophy and chemistry. The first professor appointed in 1799 was Thomas Garnett (1766–1802) who was a chemist and a physician, who was previously professor of natural philosophy at the Anderson's Institution in Glasgow. The professors at the Royal Institution in the early days did not have much academic freedom and were quite rigidly controlled by the Managers. In fact the conditions seem very similar to those at research institutes in present day Singapore. Garnet did not submit the syllabus for his 1801 lecture course to the Managers for approval and the ensuing arguments led to him resigning in that year.[88] Thomas Young, who at 28 years old was already a Fellow of the Royal Society, was quickly appointed to the natural philosophy and chemistry professorship on the recommendation of Joseph Banks.[88] In 1802 Young delivered fifty lectures on mechanics, hydrodynamics, astronomy, and physics repeating an extended version of the same lectures the following year.[88] These lectures, which were published in 1807, contained significant new research but unfortunately Young was not a good lecturer and his audience slowly diminished the main problem was that he pitched his lectures at too high a level. Part of the problem was also that Humphry Davy (1778–1829), who had become an assistant lecturer in chemistry in 1801 and promoted to the first professor of chemistry in 1802, was an excellent popular lecturer who devised exciting experiments all suited to his fashionable audience. Davy's good looks also were a magnet for the young ladies.

In 1802 Rumford quarrelled with some of the other Managers over the purpose of the Institution and left England; the school for mechanics was discontinued and Rumford formally resigned his position in 1804. Among Young's other duties was editorship of the Journals and being the Superintendent of the House, and when in 1803 his request to the Managers for a rise in salary and the employment of some librarians was refused he resigned. The Royal Institution was not the place for Young, he was a brilliant researcher but he did not have the personality nor the inclination to amuse fashionable society some of whom can be seen in the foreground of the 1802 cartoon by James Gillray, shown in Plate 3, which represents Garnett lecturing on pneumatics. This cartoon predates the change in the policy at the Institution and members of the less fashionable society can be seen at the left of the cartoon.

After Young there was no professor of natural philosophy but in 1815 John Millington (1779–1868) was engaged to give a course of lectures on natural philosophy and he continued lecturing until 1829. He was appointed professor of mechanics in 1817 but he received no salary and was simply paid three guineas a lecture. In 1831 William Ritchie (ca. 1790–1837) was appointed professor of natural philosophy. Richie's commitment at the Institution was not very onerous and entailed him delivering six lectures a year and his salary of only £50 a year (only a sixth of Young's salary) was commensurate.[89] Since Richie delivered so few lectures at the Institution he did not make much of an impact. His research was mainly on electromagnetic induction. When Ritchie died of typhoid in 1837; he was succeeded by William Brande (1788–1866) who had previously held the chair of chemistry. John Tyndall the next professor of natural philosophy was not appointed until after Brande resigned in 1852.

Michael Faraday, who only been educated at a common day-school, became fascinated by chemistry while an apprentice bookbinder. A chance gift of tickets for lectures by Davy at the Royal Institution led eventually to Faraday becoming his assistant and, after Davy had resigned his professorship, going on a scientific tour of Europe with him. On Faraday's return to London in 1815 he became a laboratory assistant at the Royal Institution working mainly under Brande, who was then the new professor of chemistry.[90] In 1821 Faraday became the acting House Superintendent at the Royal Institution and while in this position he started his famous experiments on electro-magnetism. In 1822 he published a paper that showed that a magnet could be made to rotate around a current carrying wire or the wire made to rotate around a magnet.[91] His apparatus is illustrated in Fig. 3.4 where on the left side a magnet poised at the base of a glass cup of mercury and rotates around a wire that dips into the mercury and on the right hand side a magnet is fixed into a glass cup and a wire dipping into the mercury rotates around it. Thus Faraday's interest in electricity and magnetism began, though when he was appointed a professor of the Royal Institution in 1833 it was in chemistry, this post he held for the rest of his life. Brande became increasingly involved in his duties at the Royal Mint, and during the 1820s developments at the Institution were largely due to Faraday. In 1826 Faraday introduced the tradition of Friday evening discourses for members. Although these were for members of the Royal Institution, they were widely reported in newspapers and were very influential in spreading the knowledge of the latest scientific developments in the non-scientific community. Faraday delivered 123 Friday evening discourses.[90]

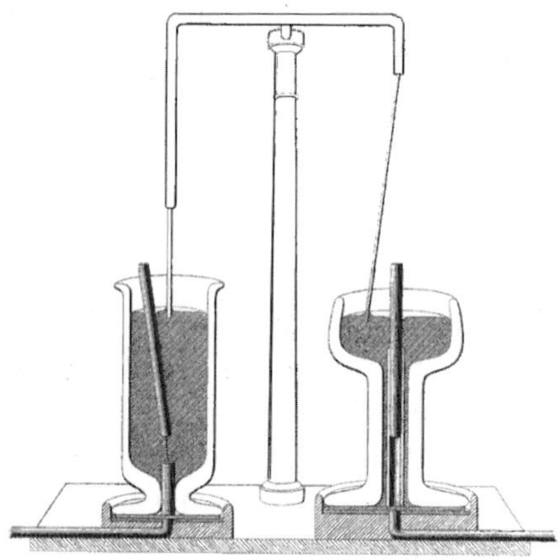

Fig. 3.4. Faraday's apparatus for demonstrating electro-magnetic rotation (Faraday, 1822).

In February 1853 Tyndall was invited to give a Friday evening discourse at the Royal Institution, the topic he chose was *On the Influence of Material Aggregation upon the Manifestation of Force*. At the end of the lecture Tyndall paid tribute to Faraday:[92]

> '[T]he discovery which led to the results brought before you to-night was of this character. That magnet [pointing to the large electro-magnet at the Institution] was the physical birthplace of these results; and if they possess any value they are to be regarded as the returning crumbs of that bread which, in 1846, was cast so liberally upon the waters. I rejoice, ladies and gentlemen, in the opportunity here afforded me of offering my tribute to the greatest worker of the age, and of laying some of the blossoms of that prolific tree which he planted at the feet of the great discoverer of diamagnetism.'

Although this lecture was abstruse it was enthusiastically received by his audience because Tyndall, like Davy and Faraday, could deliver a lecture in a style that appealed to popular audiences. On the basis of this lecture Tyndall was unanimously appointed to the chair of natural philosophy in May 1853.

Not surprisingly for such a long-lived institution, the Royal Institution changed its colours a number of times. After initially being conceived as a practical institution applying science to industry it quickly changed into one whose prime purpose was to deliver lectures that were popular with its members who provided the funding. Davy and Faraday were not content to be simply popular lecturers, though both were excellent in this role, but wanted to pursue original research. The early lecture burden was heavy with Young giving up to sixty lectures a year;[93] when Tyndall was appointed he had to deliver only nineteen lectures a year. Tyndall chose the Royal Institution over other positions because of its laboratory and the chance for research; Faraday was also a big attraction to him. In 1853 a byelaw of the Institution still allowed the Management to dictate an individual's research but Tyndall demanded and received the freedom to pursue whatever research he desired after fulfilling his obligation to deliver nineteen lectures. In 1858 Tyndall's load was reduced to twelve lectures without a reduction in salary on the recommendation of Faraday. Because he was a very conscientious lecturer, Tyndall found himself spending half a year in their preparation which cut into the time he had available for research. However, the reason may in part have been Tyndall's need for extra income. Tyndall's salary at the Institution, which was only £200 a year, was already supplemented by his lectures at the London Institution, his position as an examiner for Military Education at the War Office and from 1859 he was also a professor at the Royal School of Mines. Thus Tyndall's stated reason of wishing to free up time for research seems questionable.

Tyndall's first lecture course in 1853 was *Air and Water*. Although he gave lecture courses on many topics, there were some like his lectures on heat, sound, light and electricity which he constantly revised and gave many times. The vitality and brilliance of Tyndall's lectures can be seen in their published versions and the effort necessary in designing his experiments to obtain his desired effect can be appreciated. Here one fourteen lecture course delivered in 1862: *Heat considered as a Mode of Motion* is briefly discussed.[94] It is appropriate to consider Tyndall's lectures on heat because perhaps these studies were his greatest contribution to physics. At the end of the eighteenth century the caloric theory of heat, in which heat was believed to be a weightless substance held within the pores of matter, ultimately derived from the Greek belief that fire was one of the four elements, was still believed. Rumford and Davy were instrumental in disproving the caloric theory as Tyndall explained in his first lecture:

> 'While engaged in the boring of a cannon at Munich, [in 1798, Rumford] was so forcibly struck by the large amount of heat developed ... that he was induced to devise an apparatus for the special examination of the heat generated by friction. He had constructed a hollow cylinder of iron, into which he fitted a solid plunger, which was caused to press against the bottom of the cylinder. A box which surrounded the cylinder contained 18 lbs [8.2 kg] of water... The cylinder was turned by horse-labour, and after ... two hours thirty minutes ... the water actually boiled.'

Rumford showed that the chips cut from a cannon had the same specific heat as the cannon. Tyndall reported that Rumford held up the chips 'before his opponents [and] demanded whether they believed that the vast amount of heat which he had generated had been all squeezed out of that modicum of crushed metal.' Tyndall obviously could not repeat this experiment in his lecture but substituted a spectacular one. He had a small brass tube sealed at one end which he filled with water nearly to the top and sealed with a cork. The tube was clamped in wooden tongs and rotated fast and after about two and a half minutes the water boiled producing steam of sufficient pressure to project the cork twenty feet (6 m) into the air.

Spectacular experiments at the Royal Institution were established early after its foundation by Davy and Faraday; Tyndall carried on the tradition which has marked the lectures up to the present day. William Lawrence Bragg (1890–1971), the Nobel Prize winning X-ray crystallographer. who was Professor of Natural Philosophy and later a Fullerian Professor of Chemistry at the Institution between 1938 and 1966, summed up the philosophy of the Institutions lectures when he said to his demonstrator Bill Coates (1919–1993) 'Always, if possible Coates, never talk about science, *show* it to them,' which is exactly what Tyndall did.[95] Plate 4 shows Lawrence Bragg delivering a Christmas lecture in 1961 assisted by Bill Coates who is demonstrating a Wimshurst static electricity generator, Lord Brabazon (1884–1964), the President of the Royal Institution, is in the chair. As in Plate 3, the table is full of other apparatus ready for demonstration. No blackboard or screen is present; the demonstrations are simply shown and explained to the audience.

The lecture courses at the Institution, though intended for the general public, explained the latest research. The aspects of scientific research in the nineteenth century differ somewhat from those of the present day. There was, of course, fundamental research aimed at understanding nature and research aimed at solving problems in emerging technology, but there was also significant research into scientific curiosities which neither advanced fundamental knowledge nor solved practical problems. Curiosity is one of the strong motivations of scientists and still finds its way into scientific publications but it no longer occupies a prominent position in an era when funding depends on meeting priorities set by bureaucrats. In the nineteenth century one finds a significant number of scientific papers devoted to curiosities. Tyndall published many such papers including one on the notes produced by the contact of bodies at different temperatures and the sounds produced by the combustion of gases in tubes.[96,97] The experiments on such curiosities are often very interesting and well suited to popular lectures; Tyndall used many of them in his Institution lectures. Such experiments in popular scientific lectures where there is a strong educational element should be distinguished from popular lectures based on science but where the aim was simply entertainment such as were given at the Royal Polytechnic Institution by John Pepper (1821–1900), who was predominantly a showman, and gave his *Strange Lecture* where he used optics to produce eerie phantoms among other similar activities.

Tyndall resigned from the Royal Institution in 1887 and the eminent physicist John William Strutt (1842–1919), who is better known as Lord Rayleigh, the title he inherited on the death of his father in 1873, was appointed to the chair of natural philosophy; a position he held until 1905. Making an address to the Institution after Tyndall's death on his scientific work, Rayleigh remarked on 'the brilliant and often poetic exposition by which his vivid imagination illumined the dry facts of science.' Those Royal Institution lectures of Rayleigh that have been published show that he too had the gift of presenting science to a popular audience in a clear captivating manner, though Arthur Schuster (1851–1934) wrote in his obituary for the Royal Society that Rayleigh was 'not by nature a ready speaker.' Only some of Rayleigh's lectures on individual topics have been published in full such as that on the *Diffraction of Sound.* Although the Michelson–Morley experiments of 1887 put the first nail in the coffin of aether being the medium through which light propagated, there was considerable reluctance, especially among English physicists, to abandon the concept. Both Tyndall and Rayleigh held on to a belief in aether. In one of his first lectures to the Royal Institution, while not mentioning the role of aether in the propagation of light, Rayleigh implied the close connection between the propagation of light and sound in a lecture entitled *Diffraction of Sound.* By using sound with a frequency higher than can be heard by the human ear (greater than about 20,000 hertz) he was able to demonstrate that sound could behave very similarly to light. The French mathematician Siméon Poisson (1781–1840), was not very impressed by Fresnel's wave theory of light and to discredit it he predicted with Fresnel's wave theory the apparently paradoxical effect that that a bright spot should be seen at the centre of a shadow of a circular disc. Fresnel performed the experiment and found that a bright spot appeared at the centre of a shadow as predicted by his theory. Rayleigh showed that the same effect occurred with sound. Using a flame that is very sensitive to sound, which was favoured by Tyndall, Rayleigh showed that not only was there a peak in sound at the centre of the sound shadow of a disc of about 400 mm diameter but also there were lesser peaks in intensity as one moved out from the centre of the shadow as predicted by Eugen von Lommel (1837–1899). These experiments of Rayleigh bear all the hall marks of the lectures of Tyndall, they were visually exciting and informative. Thus Rayleigh continued the tradition of Davy, Faraday, and Tyndall.

3.2.2.1. *Christmas lectures at the Royal Institution*

Faraday introduced the Christmas Lectures for young people in 1826 and the first Christmas lecture series was given by Millington on natural philosophy. Faraday himself gave nineteen Christmas lectures including one every year from 1851 to 1860. The most famous of Faraday's Christmas lectures was *The Chemical History of a Candle*, which he first gave in 1848–1849 and repeated in his last Christmas lecture of 1860–1861.[98] The topic of this series would not seem exciting to today's young people but in 1860 candles were still the most usual means of lighting in the home despite gas being used to light the streets in most towns.[99] Using the familiar household candle[100] Faraday explained combustion and its chemistry by the aid of numerous simple experiments. Starting from the capillary action by which the melted wax is drawn up the wick and the flame is prevented burning down to the wax because the liquid wax extinguishes the flame, Faraday discussed the flame's structure. In the next lecture he went on to discuss the necessity for a supply of air and the products of the combustion of the wax. He showed that one of the products of combustion is water and went on to discuss the different forms of water ice and steam and their properties. Steam was decomposed into oxygen and hydrogen by passing it through a very hot tube filled with iron fillings that oxidize leaving hydrogen which burns. Faraday also decomposed water into oxygen and hydrogen by electrolysis whose laws he had established in 1832. One lecture was given over to a discussion of the atmosphere. In his last lecture Faraday gave much attention to carbon dioxide, or carbonic acid as it was then called, and its importance in the animal and vegetable kingdoms.

Although Faraday's Christmas lectures on a candle was not strictly on physics many were. Tyndall gave twelve Christmas lecture series also mostly on physics. When the great crystallographer William Henry Bragg (1862–1942), Lawrence Bragg's father, gave his first Christmas lecture series *World of Sound* in 1919–1920 he repeated some of Tyndall's experiments with his apparatus stored in the Institution. Since 1966 the Christmas Lectures have been broadcast on television.

3.2.3. *The British Association for the Advancement of Science*

David Brewster (1781–1868), a Scottish natural philosopher and a believer in the decline in science, was prominent in working for the formation of the British

Association and it was William Vernon Harcourt (1789–1871), Canon of York Minster and son of the Archbishop of York, who was midwife at the birth of the BAAS at York in 1831. The original purpose of the BAAS was:

> 'to give a stronger impulse and a more systematic direction to scientific inquiry; to promote the intercourse of those who cultivate Science in different parts of the British Empire with one another and with foreign philosophers; to obtain more general attention for the objects of Science and the removal of any disadvantages of a public kind that may impede its progress.'

However it was also recognized that there two audiences 'visitors coming to meet their fellows in science and the intelligent persons of our own neighbourhoods who hope to be gratified and instructed.'[101]

Brewster chose York for the inaugural meeting of the British Association because of its central position and because of its flourishing local philosophical society headed by Vernon Harcourt. The British Association was to remain an itinerant body with no fixed abode. Membership of the BAAS was open to all; as Vernon Harcourt declared: 'a public testimonial of reputable character and zeal for science is the only passport into our camp.'[102] The initial subscription for the BAAS was £1 per annum or £5 for life, which was increased to £2 per annum or £10 for life in 1845 thus, though membership was open, it was intended for the gentlemen of science and the clergy and aristocrats dominated its early membership.[103] The president of the BAAS was elected each year to hold office until the conclusion of the next meeting. At the second meeting in 1832 held in Oxford the sections of the association started to develop, the first section was mathematics and general physics, which developed into Section A and reports from this section dominated the early years of the BAAS. The mechanical sciences, which were originally a sub-section of mathematics and general physics, became Section G in 1836. The renamed British Science Association now has sixteen sections.

Most of the sectional meetings of the BAAS were specifically for professional scientists but some of the presidential addresses and the evening lectures[104], given from the first meeting in York and formalized from 1842 onwards, were on topics of more general interest. From 1867 one lecture a year was also given to what was called 'the Operative Classes.' The first such lecture, *Matter and Force* [*Energy*], was given by Tyndall in 1867 before the working men of Dundee. Tyndall, a quintessential lecturer, powerfully wove the lecture through consideration of the interaction of matter and demanded complete freedom in such discussions. However as he said: [105]

> 'while I consider science to be alike powerful as an instrument of intellectual culture and as a ministrant to the material wants of man; if you ask me whether it has solved or is likely in our day to solve, the problem of the universe, I must shake my head in doubt... The phenomena of matter and force lie within our intellectual range... But ... the real mystery of the universe remains unsolved.'

This man was not a complete materialist though he provoked much condemnation for his Presidential Address on *Scientific Materialism* seven years later, which was discussed in Chapter 2. The most well known, but poorly documented, comments at a British Association meeting occurred in 1860, during a lecture by John William Draper (1811–1882) of the New York University on social and political aspects of Darwin's theory, when during the ensuing discussion Bishop Samuel Wilberforce (1805–1873) famously turned to Huxley to ask 'whether it was through his grandfather or his grandmother that he claimed descent from a venerable ape.'

3.2.4. *Popular books*

Some writers of popular books in the nineteenth century were professional scientists, like David Brewster who was educated to pursue a career in the Church of Scotland and did become a licentiate but found public speaking too stressful; his fear of public speaking was so great that on one occasion he fainted at a dinner party when invited to say grace and it all but ruled out a normal university career.[106] Although he did pursue original research, mostly in optics, and he published over three hundred original research papers at a time when even most eminent researchers' publications were numbered in the tens rather than hundreds, his main source of income, until he became Principal of the United Colleges of St Leonard and St Salvator at the University of St Andrews in 1838, was from writing and editing popular science.[107] Dionysius Lardner (1793–1859) was another popular science writer who was a professional. Although he was appointed the first Professor of Natural Philosophy and Astronomy at the new London University in 1827 and made a Fellow of the Royal Society a year later, as Brewster had been in 1815, Lardner's research was negligible – in the Transactions of the Royal Society he only published two papers. However, by the 1830s he was well established in London and, when he resigned his professorship in 1831 partly over his salary, he was able to support himself through popular scientific writing and lecturing.[108] Lardner wrote books for the Society for the Diffusion of Useful Knowledge, articles for the Edinburgh Review, and he also edited the *Cabinet Cyclopedia* for the Longman publishing group. He was in fact a remarkably popular member of London's scientific community and when he gave a Friday evening lecture at the Royal Institution in 1834 the audience was one of the largest of the decade, larger even than that attracted by Faraday three years earlier.[108] However, Lardner was not universally liked, Charles Dickens called him 'the prince of humbugs' and his career in Britain came to an end in 1840 when he ran away with the wife of a captain of Dragoons, first to Paris and then to America.[108] The *Cabinet Cyclopedia* was very successful with 129 volumes published before Lardner decamped and at a price of six shilling it was affordable by the middle-class and the more prosperous artisans. Some of the volumes of the *Cabinet Cyclopedia* were devoted to natural philosophy: Lardner himself wrote volumes on mechanics, hydrostatics and pneumatics, heat, and electricity, John Herschel (1792–1871), son of William, contributed his *Preliminary Discourse on the Study of Natural Philosophy* (which will be discussed in Chapter 4) for Lardner's project, and Brewster wrote *A Treatise on Optics* in 1831.[109] In the epigraph to his *Treatise*, Brewster quotes in his review from Charles Babbage's *Reflections on the Decline of Science in England, and on Some of its Causes*, where Babbage recommended that:

> 'the most obvious remedy is to provide the educated classes with a series of works on popular and practical science ... written in simple and perspicuous language, ... which are level to the capacity of ordinary minds,'

and this is precisely what Brewster's *Treatise* was. In Part I he describes the reflection and refraction of light, Part II is on physical optics with a large section devoted to polarized light, Part III explains a number of natural optical phenomena such as the rainbow and how, through the effect of temperature on the refractive index of air a number of phenomena such as mirages in the desert and multiple images of ships seen when they are below the horizon, as for example those shown in Fig. 3.5 which Samuel Vince (1749–1821) had observed at Ramsgate in Kent in 1798. Brewster also has a chapter on his invention, the kaleidoscope.

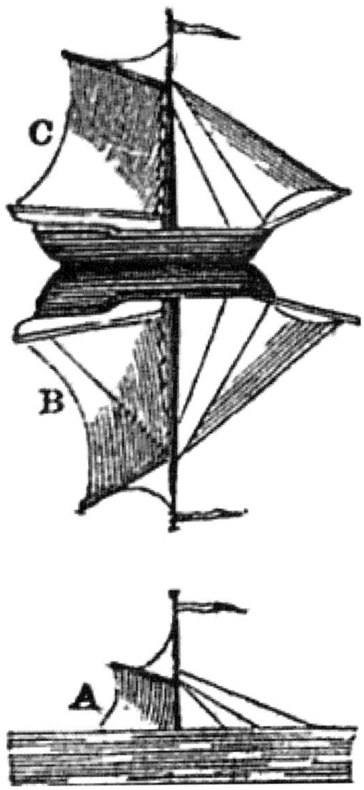

Fig. 3.5. Multiple images of a ship below the horizon. A is the real ship and B, C are images seen by unusual refraction (Brewster, 1853).

In the early nineteenth century it would have been very difficult to rely on science writing for a living. Even Brewster, when asked for advice in 1830 by a young protégé who was considering a career in scientific writing, replied:[72]

> 'I do not object to your making money by your writings, but I am sure that it would be injurious to your happiness to rely on such a source for a permanent portion of your income. The moment you do that you become a professional author, following the worst of professions.'

At about the same time an Edinburgh publisher described Brewster as 'Dr B., who is very clever, but who after all is starving.'[70] Brewster represents the best of popular science writers who used the fruits of their own research in their popular works but there were other writers of popular science who, though they were not actively engaged in scientific research themselves, could write clear and informative popular accounts of science. Some of works of these authors were in part textbooks but the distinction between textbook and popular book was not clear cut. Many of the popular physics writers up to 1830 were teachers such as the Unitarian minister Jeremiah Joyce (1763–1816), Olinthus Gregory, and Thomas Dick.[70]

Joyce published books on natural philosophy, which were mainly for children and young people. His most successful work was the *Scientific Dialogues* initially published in six volumes over 1800 to 1803. The *Dialogues* follow the format developed in the eighteenth century for children's books on natural philosophy and take the form of a conversation between a tutor and his two charges Charles and James, which were the names of the children of the Earl of Stanhope who Joyce had tutored in the early 1790s. The contents of the *Dialogues* covered mechanics, astronomy, hydrostatics, pneumatics, optics, magnetism, electricity and galvanism. What was novel for the time was that it was originally published in six separate volumes at 2s 6d, which was relatively cheap.[110] New editions continued to be published both in Britain and America after Joyce's death, the last edition being published in 1892; it was also translated into French German and Welsh.[110] Joyce's works were published by the notorious Richard Phillips (1767–1840) and were often published, anonymously, under a pseudonym, Phillips' name, or that of another author.[111] He also revised and edited works for publishers such as the 1799 posthumous edition of his fellow Unitarian William Enfield's (1741–1797) *Institutes of Natural Philosophy, Theoretical and Experiments*.[110]

After his early book on astronomy for children, mentioned in §3.1.2.2, was published, Gregory moved to Cambridge where he opened a bookshop and also taught mathematics. In 1802 Gregory published a book on astronomy for adults *A Treatise on Astronomy*, which was strategically dedicated to Charles Hutton

(1737–1823) Professor of Mathematics at the Royal Military Academy in Woolwich.[112,113] The dedication was fruitful and in December of the same year Hutton appointed Gregory the second mathematical master at Woolwich. Gregory also wrote *A Treatise on Mechanics* in two volumes published in 1806 which he justified on the grounds that there had been a decline in natural philosophy and mathematics in Britain quoting from the article on physics in the 1797 edition of the *Encyclopaedia Britannica* written by the Scottish natural philosopher, John Robison (1739–1805):[114]

> 'that while the continent has supplied us with most elaborate and useful treatises on various articles in physical astronomy, practical mechanics, hydraulics, and optics, there has not appeared in Britain half a dozen treatises worth consulting for the last forty years.'

The *Treatise* was in two volumes, the first volume contained the theory and since it was highly mathematical and made use of fluxions (calculus) it cannot be described as a popular physics book, neither can the second volume which was practical and descriptive. There was a third volume of plates. Although not a popular science book, Gregory's *A Treatise on Mechanics* was highly successful.[112] Olinthus Gregory also was one of the editors of a posthumous edition of Jeremiah Joyce's *Scientific Dialogues*.

All of Thomas Dick's many books contained 'a God-praising theology of nature' and some of these have already been discussed in Chapter 2.[115] However in some on natural philosophy and astronomy, these remarks were only in passing and books such as a *Practical Astronomer* first published in 1845 can be classified as popular physics. This book was designed for the general reader and especially for amateur astronomers since it gave information on the construction of telescopes and other astronomical instruments. The reader was spared a discussion on the 'utility of astronomical studies and their moral and religious tendencies,' because that would have increased the length of an already long book.[116] Although Dick was a prolific writer and his books sold well, he did not make very much from his writing because he sold all but one of their copyrights outright for small sums.[115]

During the nineteenth century most of the flow in popular books on physics and astronomy was from Europe to America but there were exceptions. Ormsby Mitchel (1810–1862) graduated from West Point in 1829 but resigned his commission in 1832 to become a professor at Cincinnati College.[117] There was no organized observatory in America in 1842 when Mitchel delivered a series of public lectures on astronomy to see whether there was support for constructing an observatory in Cincinnati. At the end of these lectures Mitchel put his proposal to his audience. The Cincinnati Astronomical Society was formed as a result of this meeting, with Mitchel promising to devote five years to the construction of an observatory. On a tour of Europe to source a suitable telescope, Mitchel decided to commission the construction a refracting telescope using a 300 mm Mertz objective which was available from Munich.[118] This telescope, though of smaller aperture than existing reflecting telescopes, was the second largest refracting telescope in the world.[119] The cornerstone of the observatory was laid in 1843 by John Quincy Adams and the telescope was first used in April 1845.[118] Raising the funds for the observatory was not easy and Mitchel had to fund some of the cost himself. In 1845 the Cincinnati College burnt down and Mitchel had no income. To raise funds, Mitchel launched an astronomy journal the *Sidereal Messenger*, named after Gallileo's *Sidereus Nuncius*, which was intended to be both a professional and popular journal. Every subscriber to the journal could visit the observatory. At the time there were no astronomical journals in America of any kind and no popular ones anywhere. The journal obtained moderate popular support but was not well received by most astronomers and it was discontinued in 1848 for financial reasons.[120] Mitchel also undertook a lecture tour of the Eastern States to raise money out of which came his first popular book on astronomy, *The Planetary and Stellar Worlds: a Popular Exposition of the Great Discoveries and Theories of Modern Astronomy*, was first published in America in 1848 quickly followed by publication in Britain under a different title.[121] *The Planetary and Stellar Worlds* was written in a pleasant lyrical style that would have pleased nineteenth century readers and it is not surprising that it was a success on both sides of the Atlantic. In 1859 Mitchel was appointed as the Director of the Dudley Observatory in Albany, NY. By 1860 Mitchel considered his *Planetary and Stellar Worlds* needed updating and he published *Popular Astronomy*.[122] This book was an excellent introduction to astronomy and went through many editions both in America and Britain. When the American Civil war started in 1861, Mitchel entered the Union Army as a brigadier general and was involved in the abortive attempt to seize a train from the Confederates in Georgia so beloved by movie makers. Mitchel died from yellow fever in 1862. A third book *The Astronomy of the Bible*, complied from lectures Mitchel had given, was published posthumously.

Richard Proctor (1837–1888), an English astronomer, was a truly prolific popular writer, he wrote over fifty books and five hundred essays mostly on astronomy; he also lectured extensively.[123] His early books such as *Handbook of the Stars* published in 1866 did not sell well.[124] His first real success was a small book, *Half-Hours with a Telescope*, first published in 1868, was a straightforward account of the night sky for amateur astronomers and contained the first reasonable map of Mars; it went through twenty editions.[125] Proctor's first major success was, *Other Worlds than our Own*, first published in 1870 and there were twenty-nine subsequent editions by 1909 despite some early critical reviews such as that by Charles Pritchard (1808–1893) who opined that it did not advance any new arguments.[126] In this book Proctor made use of spectroscopic evidence that the Sun is composed of the same elements as the Earth to construct by analogy that the planets are similar in composition to Earth and therefore likely to be capable of supporting life. This analogy was then extended to the stars. The problem was that he then assumed that the planets 'must be inhabited unless we can show very good reason that they cannot be so,' thus throwing the burden of proving a negative on his opponents while he did not attempt to prove his positive contention that they are inhabited, except by purely hypothetical considerations' as Alfred Wallace (1823–1913) stated in his review of the scientific evidence for the plurality of worlds.[127] However he rejected many bizarre speculations such as the peopling of the Sun. During the 1870s Proctor began lecture tours of America and by 1883 he had given nearly four hundred. He also lectured in Australia and Asia.[124] Though he was undoubtedly popular in America he did receive some adverse criticism to the effect that he wrote to sell and deviated from his former high scientific principles.[128] In 1884 he moved to America having remarried an American after his first wife died. Proctor's *magnum opus* was to have been a systematic survey of the whole of the universe. He conceived of this project early in his career as an astronomer and *Saturn and Its System* which he published himself in 1865 was to have been eventually one of its chapters. However, this monograph was did not sell well and when, in 1866 the company in which most of his funds were invested failed, Proctor turned to popular writing to support himself and his family. In 1888, just before his early death from yellow fever, he did publish the first part but much remained unwritten. New editions of Proctor's books were published long after his death. Arthur Cowper Ranyard (1845–1894) completed Proctor's *magnum opus* which was published as *Old and New Astronomy* in 1892. In this book Proctor again showed his perceptive powers in a discussion of the so-called *canals* discovered by Giovanni Schiaparelli (1835–1910) on Mars in 1877, which Proctor thought were due optical effects.[129] Mary Proctor (1862–1957), Proctor's daughter continued writing popular books on astronomy up until the beginning of the Second World War.

3.2.4.1. *Women popular writers*

Women, who did not need to earn a living because they were married, made up another class of popular science writer in the early nineteenth century. Some like

Jane Marcet (1769–1858) used the familiar format of conversations between a teacher and her pupils adopted in the eighteenth century in popular science for women and children. However in the nineteenth century the classification of popular science readers into men, women, and children could be blurred. Her *Conversations on Chemistry*, first published anonymously in 1805 had as a full title *Conversations on Chemistry, Intended More Especially for the Female Sex*. However it was widely read by men and was used by students at mechanics' institutes and medical apprentices though it was not intended as a textbook. Marcet learnt her chemistry from Davy at the Royal Institution. *Conversations on Chemistry* was conceived as a sequel to *Conversations on Natural Philosophy*, which Marcet had written earlier but did not publish until 1818.[130] Initially the reason she did not publish might have been that it was written to help her recover from depression and then, when Joyce's *Scientific Dialogues* was brought to her attention, she lacked the confidence to publish a similar book. Both the natural philosophy and the chemistry book were written in the form of a conversation between a teacher, Mrs Bryan, and her pupils: the flippant Caroline and the serious Emily. Physical chemistry at least comes within the scope of the present book, as Mrs Bryan says in the first conversation, 'chemistry ... is so closely connected with natural philosophy, that the study of one must be incomplete without knowledge of the other.' To the modern reader the conversations seem stilted and lack the sparkle found in the same genre in the eighteenth century, even Caroline is a very proper young lady. The two books were more suited to juveniles than adults[131] and the reason they were popular with students of mechanics' institutes may have been more because their simple language was easy to understand.[132] The often quoted remark of Faraday that *Conversations on Chemistry* 'gave me my foundations in that subject,'[133] is understandable since he was only fourteen or fifteen when he read the book and admitted that his schooling 'was of the most ordinary description, consisting of little more than the rudiments of reading, writing, and arithmetic.'[90] However *Conversations on Chemistry* was very popular, especially in America where there were 23 editions; in England there were 16 editions. Marcet earned some £2000 from *Conversations on Chemistry* between 1805 and 1865.[72]

Mary Somerville (1780–1872) became involved in the sciences in a somewhat similar way to that of Marcet, but she wrote for adults in a formal non-familiar style. She became interested in mathematics in her early teens though her father forbade its study. Her main instruction in higher mathematics came after she married her first husband Samuel Greig (1777–1807), though without his encouragement, by correspondence with the Scottish mathematician William Wallace (1768–1843) with tuition by his brother John.[134] Samuel died in 1807 and in 1812 Mary remarried an army doctor, William Somerville (1771–1860) who did encourage her mathematical and scientific studies.[134] When her husband was stationed in London she mixed with its scientific community. In 1827 Somerville was asked by Henry Brougham to prepare a condensed English version of the first four volumes of Laplace's *Mécanique Céleste* for his Society for the Diffusion of Useful Knowledge (SDUK).[134] Brougham initially proposed two volumes: one giving the popular view while the other was to contain the analysis and the illustrations.[135] It seems curious that Brougham wanted the second volume containing the mathematical analysis for the SDUK which published popular scientific books. However, Somerville saw the analysis was central to Laplace's *Mécanique Céleste*, writing in the introduction to her book *Mechanism of the Heavens* that she did not want to limit her book to giving just Laplace's results 'but rather to explain the methods by which these results are deduced from one general equation of the motion of matter. To accomplish this, without having recourse to the higher branches of mathematics is impossible.'[136] Somerville's *Mechanism of the Heavens* integrated Laplace's analysis and results in one volume and proved unsuitable for the SDUK, but it was accepted by John Murray's publishing house. The book started with a preliminary dissertation by Somerville, which was subsequently published separately as *A Preliminary Dissertation to the Mechanism of the Heavens*. Sales of the *Mechanism of the Heavens* were slow, but when it was adopted as textbook by Cambridge University in 1837 there was a small but steady demand. The *Mechanism of the Heavens* made Somerville's name and she received many honours.[134] In 1834 Somerville's *Connection of the Physical Sciences* was published by John Murray in a cheap format. In this book Somerville attempted to show how all the physical sciences were connected. Some sections of the book do flow from one topic to another but in many there is little connection. However the *Connection of the Physical Sciences* was very successful. French, German, and Italian translations followed and there was a pirated American edition. The work was updated in subsequent editions and it became a review of current physical sciences.[134] Somerville had an understanding of science that was not matched by many popular writers male or female during the nineteenth century.

Agnes Clerke (1842–1907) was a popular writer on astronomy in the style of Mary Somerville. Clerke was educated at home and obtained an interest in astronomy from her father who, though a banker and a lawyer, was an amateur astronomer. She never married and started writing on astronomy shortly before her family moved to London in 1877.[137] *A Popular History of Astronomy during the Nineteenth Century* was her adventurous first book published in 1885.[138] Such a book was needed at the time, because there was no history that gave the most recent developments in astronomy especially the use of the spectroscope, and it soon became a standard work. I have the same opinion of the necessity of history as Clerke did; she wrote in her preface, 'there are many reasons for preferring a history to a formal treatise ... A history tells us ... how we came to know the facts' and, I would add, often gives us a better understanding of them.

Fig. 3.6. The Great Nebula Orion, courtesy NASA, image NASA/JPL-Caltech/STSci.

The late nineteenth century saw an increase in specialization in astronomy, as well as in other sciences. Clerke, outside of the professional ranks, could see the wider implications of new discoveries more clearly and was not afraid to voice her opinions. Some professional astronomers resented a woman non-professional astronomer giving opinions instead of merely describing the work of the male professionals. One such was Richard Gregory (1864–1952) who in 1893 was appointed the assistant editor of *Nature* by the editor Norman Lockyer (1836–1920). The reason for this antagonism may have been the support that Clerke gave to William Huggins over his controversy with Lockyer over the spectra obtained from gaseous nebulae described in later editions of *A Popular History of Astronomy* and in later books such as *Problems in Astrophysics* published in 1903.[124,139] Huggins had started examining the spectra from gaseous nebulae in 1864 and by 1868 he had examined some seventy nebulae one third of which were gaseous.[140] The spectra all of the gaseous nebulae, such as the Great Nebula in Orion shown in Fig. 3.6, consist of a dominant green line at a wavelength of 500.7 nm and two fainter lines also in the green portion of the spectrum.[141] Nitrogen was the only element that produced a line near the dominant line, but on closer inspection of the spectrum of nitrogen, Huggins found that there were in fact two closely spaced lines not one. Huggins speculated that the dominant green line came from a yet undiscovered element which he named *nebulium*. In

1887 Lockyer proposed his *meteoritic hypothesis* that all the bodies in the universe originated in swarms of meteorites.[142] Spectrograms of meteorites showed a series of bands in the green region of the spectrum that Lockyer called fluting and identified them with a compound of magnesium. He found that by increasing the temperature of magnesium the fluted lines became a single line and he identified magnesium with the dominant line in the spectra of gaseous nebulae. If correct the spectra of gaseous nebulae would have added weight to Lockyer's *meteoritic hypothesis*.[141] Huggins disputed Lockyer's claim and he and his wife Margret made new spectrograms in 1889, which when corrected for the Doppler effect caused by the Earth's motion, showed that the dominant line from nebulae differed from the edge of the magnesium band by about 1 Å. James Keeler (1857–1900) at the Link Observatory in California, using what was at the time the largest refracting telescope in the world, made a more accurate analysis of the spectra in 1890 and obtained 500.705 nm for the dominant nebular line and 500.65 nm for the edge of the magnesium spectrum.[141] Clerke discussed the controversy between Huggins and Lockyer in the fourth edition of *A Popular History of Astronomy* and in *Problems in Astrophysics*. Although the proof that that the magnesium fluting edge was not the same as the dominant green line in the spectra of gaseous nebulae did not disprove Lockyer's *meteoritic hypothesis* it was much weakened and it lost currency. However there was also little evidence for Huggins' *nebulium*. As Clerke wrote in her second edition of *The System of the Stars*: [143]

> 'Must we ... admit ... that there is a chemistry transcending text-books? ... To abandon the grand ideal of a universal physical science because of a few outstanding incongruities would be to listen half-heartedly to counsels of impatience.'

She was right but at the beginning of the nineteenth century, as Clerke noted in the preface to her fourth edition of *A Popular History of Astronomy*, there were[138,144]

> 'novel thoughts ... fresh impulses ... "the old order changeth" more and more swiftly ... [and] many of the scientific doctrines ... begin to wear a superannuated aspect.'

In 1927 the American Ira Bowen (1898–1973) showed that the dominant green line in the spectra of gaseous nebulae was emitted by a jump from a metastable state of doubly ionized oxygen to a lower state. Under terrestrial conditions such jumps are very improbable but in low density nebulae the time between atomic impacts are large and atoms can return to a lower energy state with an emission of radiation corresponding to the difference in energy between these metastable states.[145] Thus the riddle of the spectra of gaseous nebulae could not be solved using Newtonian mechanics and had to wait for the new quantum mechanics.

3.2.5. *Popular periodicals*

Low priced periodicals were also important in spreading knowledge especially to the working class who could more readily afford them. In the first half of the nineteenth century the mechanics' magazines provided easily accessible useful knowledge as well as providing some education in physics and astronomy. Popular science periodicals of the first half of the century also assumed that ordinary people had a role to play in the development of science. This assumption changed around the middle of the century as a professional scientist class emerged. Two of the popular science periodicals of the nineteenth century, the *Scientific American* and *Nature* are still in production though they went through a number of reincarnations along the way. The *Scientific American* is the older journal and was first published in 1845. The *Scientific American* started as a four page weekly sold by subscription for $2 a year. Initially it was more like a newspaper than a magazine with very short articles on a variety of subjects but biased towards practical mechanics and reports on patents; it had virtually nothing on natural philosophy or astronomy in the early years.

The *English Mechanic*, which was first published in 1865 as a twelve page weekly magazine costing a penny a copy, was similar in content to the earlier mechanics magazines and it continued to be published until 1926. In 1870 the format of the *English Mechanic* was enlarged and it became more science orientated. The extensive correspondence section of the magazine was very lively. From the early twentieth century the *English Mechanic* had frequent articles on the construction of astronomical telescopes. Those by William Ellison (1864–1836), an Irish clergyman who was an avid amateur telescope maker, were also reprinted in the *Scientific American*.

Norman Lockyer was science editor of the short lived *Reader* and after this journal ceased publication in 1867 Lockyer approached the publisher Alexander Macmillan (1818–1896) with the proposal for a general science journal and *Nature* was the result in 1869. The primarily role seen for *Nature* was 'to place before the general public the grand results of Scientific Work and Scientific Discovery, and to urge the claims of Science to a more general recognition in Education and in Daily Life.'[146] The secondary role of *Nature* was 'to aid Scientific men themselves, by giving early information of all advances made in any branch of Natural Knowledge.' Thus the earlier purpose of popular journals had changed. Some, notably Richard Proctor, still resisted this change. Both

Proctor and Lockyer did original astronomical research, though Lockyer's contribution was far greater than Proctor's, they were also both popularizers and promoters of science. Not surprisingly there was great antagonism between the two men. Lockyer believed in state support for science as initially did Proctor, but coloured by his antagonism with Lockyer and other establishment figures thought that in practice it would stunt it.[124] Two articles by Proctor appeared in the first issue of *Nature* one on a tsunami in the Pacific Ocean in August 1868 and the other on the spectacular Leonid meteor shower seen in November 1866.[147] He also wrote in the *English Mechanic* and carried on a long series of letters with John Beardsley, a Flat Earther who wrote *The Earth Proved not to be a Sphere, nor the Surface of Water upon it Convex*.[148] During the 1870s Proctor continually clashed with Lockyer and in 1881 he set out to challenge Lockyer by publishing a new science weekly *Knowledge* at twopence undercutting the price of *Nature* which had increased its price to sixpence in 1878.[124] Although *Knowledge* sold well, Proctor was forced to increase its price to threepence in 1884. However, this price increase did not solve Proctor's financial problems with *Knowledge* and in 1885 he was forced to turn it into a monthly and raise the price to sixpence.[149] With the transformation to a monthly the readers' role became passive, no longer was it assumed that they would contribute to science.[124] *Knowledge* survived Proctor's sudden death and the editorship was taken over by Ranyard who had previously succeeded Proctor as secretary of the Royal Astronomical Society in 1874 after Proctor had been forced to resign over a quarrel with the president Airy. *Knowledge* continued to be published until 1917.

By the end of the nineteenth century the funding for scientific research was coming more and more from the government. Popular science was important to ensure that there was a scientifically literate and interested public to see that this funding continued. Herbert George Wells (1866–1946), who began his literary career by writing general papers on science, expressed this view in an article published in Nature in 1894.[150]

3.3. Anti-Newtonians

During the eighteenth century most attacks on Newtonian natural philosophy were motivated by religion but by the nineteenth century most Christian sects had come to terms with it and opposition to Newtonian natural philosophy was mostly secular.

Richard Phillips, the radical and notorious publisher, had no scientific education but that did not prevent him publishing his own theory of natural philosophy, which to his great annoyance most saw as nonsense. He started in 1817 with three articles in the *Philosophical Magazine*[151] in which he supposedly disproved Newton's theory of gravity; a book *Proximate Causes of Material Phenomena* published in 1821 gave a fuller account of his theory and extended it to cover all of natural philosophy.[152] According to Phillips:[153]

> 'everyone must feel as an axiom that the orbicular and rotary motions of the earth necessarily give weight to bodies, and laws to their fall, because the moving earth and the bodies are in contact, and partaking of those common motions'

As in many anti-Newtonian tracts, Phillips seized on the undoubted difficulty in Newtonian theory of gravity, how does gravity act at a distance? To overcome this difficulty Phillips proposed that all space was filled with a gas[154] and that this was the medium by which force acted at a distance. He stated that: [153]

> '[The] 'power of gas will be evident on slight consideration; for, if a tube ... of ten feet or a million of feet in length, were filled with hydrogen gas, and a plug were driven into one end, so as to require any known power ... to force it out; then, if a piston were forced with that degree of power into the other end, it is notorious that the rarest gas would expel the plug as effectually as though it were propelled by a continuous rod of iron.'

It is almost impossible to understand Phillip's explanation of the motion of the planets. To him 'the phenomena of the universe are the results of a system of motion producing motion; and of motion generated by motion' whatever that means.[153] Augustus de Morgan (1806–1871), who Phillips tried to convert to his mechanics in 1836, wrote satirically in a *Budget of Paradoxes* that: [155]

> '[He] had four valuable qualities; honesty, zeal, ability, and courage. He applied them all to teaching matters about which he knew nothing; and gained himself an uncomfortable life and a ridiculous memory.'

In 1836 Phillips wrote to François Arago (1786–1853), the French mathematician, physicist, astronomer, and briefly a member of the Executive Power Commission during the Second Republic of 1848:[155]

> 'I cannot expect that on a sudden you and your friends will come to my conclusion, that the present philosophy of the Schools and Universities of Europe, based on faith in witchcraft, magic, &c., is a system of execrable nonsense, by which quacks live on the faith of fools.'

Even Faraday wrote an open letter to Phillips in the *Philosophical Magazine* in 1846 explaining his position.[156] Faraday did not believe in the aether and subscribed the action of gravitational, electrostatic and magnetic forces as similar as did others did in the nineteenth century including Ottaviano-Fabrizio Mossotti

(1791–1863) the exiled Italian physicist to whom he referred. In Faraday's case it was the concept of lines of force that was the link between these forces.

In the early nineteenth century electricity caught the imagination of many including those associated with the mechanics' institutes one of whom was Thomas Simmons Mackintosh, a follower of the socialist Robert Owen (1771–1858). During the 1830s Mackintosh lectured on *The Electrical Theory of the Universe* though this description was somewhat of a misnomer according to him and came from the metropolitan press because he did not appeal to electric action alone. Mackintosh himself stated in his book, which he published in 1838, that:[157]

> 'The principles to which we appeal, are ultimate and universal forces of attraction and repulsion, under whatever form'

and his theory covered both the physical and moral world. In the preface to the book Mackintosh professed that he would 'show Newtonian Philosophy is based upon false principles.' However Mackintosh's scientific credentials can be judged from the fact that he also saw himself as somewhat of an authority on aeronautics and in 1835 designed an airship in the form of a ship's hull powered by sixteen eagles harnessed in frames.

The popular belief in the nineteenth century that electricity was the universal cause of everything, which is not that far removed from the modern scientific concept of a unified theory of everything, spawned other books such as the *The Electric Theory of Astronomy* published in 1886 by the American Benjamin Taylor Kavanaugh (1805-1888), who was a minister in the Methodist Episcopal Church in Newport.[158] Although a clergyman, Kavanaugh did not rely on religion for his argument. Like Mackintosh, he did not believe that universal gravitation was an established fact. However, he allowed that there could be a difference in opinion to 'what causes are in action to propel and keep in uniform motion all the heavenly bodies' and simply offered his electric theory as an explanation.

3.4. Concluding Remarks

At the end of the seventeenth century people were aware that Newton had revolutionized natural philosophy. They did not understand exactly what Newton had done and they thirsted for knowledge of the new natural philosophy. Popular lectures on this new natural philosophy started in the coffee-houses of Oxford and London and spread to provincial towns. Especially during the early eighteenth century these lectures were too expensive for all but the affluent. With the growth of industrialization the lower classes by necessity became more aware of the wider world. Operation of new machinery forced the workers to become more educated even if in only practical matters. Along with developing political awareness they saw the need to understand more fully the machines that they were operating and the forces of nature. Enlightened men in education recognized this thirst for knowledge and began to organize classes in mechanics and natural philosophy, which developed into the mechanics' institutes.

The Royal Institution, first formed to aid the introduction of practical improvements and to educate the workers as well as entertain the wealthy, soon became the preserve of the fashionable. However the money that came from the affluent members' subscriptions enabled it to develop laboratories and fund research. The Royal Society at the beginning of the nineteenth century had lost its vigour and the need for scientists to meet together was not being met. There was also the sense that Britain was falling behind the Continent in science and concerned scientists formed the British Association for the advancement of science which was deliberately not based in London but on a provincial circuit.

Books on natural philosophy began to proliferate in the nineteenth century but did not become cheap enough for workers until the middle of the century. Science began to become a profession but some eminent scientists like David Brewster, had to support themselves by writing popular books. However it was still a period when highly talented people with little formal education could join the ranks of scientists and make very significant contributions. Some popular writers made valuable contributions to science.

From the seventeenth to the twentieth centuries the thirst for knowledge of Newtonian physics among all sections of society was almost unquenchable. Nothing approaching this general enthusiasm for physics has occurred in the second half of the twentieth century or at the start of the twenty-first, despite the intense interest in some popular books like Stephen Hawking's *A Brief History of Time.* There are possible contributing causes for the enthusiasm engendered by physics during the Newtonian period such as the comparative lack of entertainment but I do not think they had a major effect. The primary reason in my view was the bursting on the scene of the new science which promised a new rational light on life as a whole. Of course, at the end of the nineteenth century Newtonian physics became inadequate to explain some of the new experimental data and modern physics began to emerge. However, there will always be a need for classical physics and its development has continued at an even greater rate since the nineteenth century. With most people in the western world receiving at least some education in classical physics, the excitement engendered by it has diminished. Modern physics still entertains as well as educates the thoughtful but

I suspect that it does not touch such a large proportion of the educated populace as classical physics did.

3.5. Notes

[1] Jacob (1976).
[2] Here anodyne is used in its original sense of relieving pain.
[3] Chamberlen (1715).
[4] Stewart (1999).
[5] Willmoth (2004).
[6] Snobelen (2004).
[7] Hauksbee (1709).
[8] Hauksbee constructed the first continuous generator of static electricity and performed important experiments on capillary action.
[9] Halley (1705).
[10] Pumfrey (2004).
[11] Whiston and Hauksbee (1714a).
[12] Whiston and Hauksbee (1714b).
[13] Fara (2004b).
[14] Desaguliers (1717).
[15] Stewart (1986).
[16] Desaguliers (1734).
[17] Morton (1990).
[18] Appleby (1990).
[19] King (nd).
[20] Pickard (1981).
[21] Hales (1727).
[22] Floto (1999).
[23] Newton (1718).
[24] Desaguliers (1727).
[25] Baltimore's journey was to represent the English court at the marriage of the Empress Anna's niece to Prince Anton-Ulrich of Brunswick. Also on the same voyage was the Newtonian popular writer Francesco Algarotti and the son of Desaguliers, Thomas, whose tutor was King. Appleby (1990).
[26] Milburn (1983).
[27] Martin (1743).
[28] Martin (1749).
[29] Demainbray charged a half guineas for thirty four lectures in the 1750s. Morton (1990).
[30] Milburn (2004).
[31] Martin (1747).
[32] Ferguson (1773). The preface to this work is a short autobiography.
[33] Milburn (1985).
[34] Ferguson (1773).
[35] Ferguson (1756).
[36] Ferguson (1754).
[37] Ferguson (1760).
[38] Ditton (1705).

[39] Pemberton (1728). Pemberton, a Fellow of the Royal Society, had superintended the editing of the third edition of the Principia.
[40] Johnson (2004).
[41] Desaguliers (1744).
[42] Also known as Joyce.
[43] Voltaire (1733).
[44] Voltaire (1738). The English version was translated from the French by John Hanna. On the title page it is stated that the work has been 'revised and corrected,' but the changes are only minor. Johnson and Chandrasekar (1990b).
[45] Voltaire had some difficulty in the publication of the French versions of these books. *Lettres Écrites de Londres sur les Anglois*, was first published in London in 1734. Under the title, *Lettres Philosophiques*, they were also published in Rouen later in the same year and promptly condemned by the Parlement de Paris to be burnt. The first publication of *Elémens de la Philosophie de la Neuton* was in an unauthorized pirated edition in Amsterdam in 1738 which had numerous errors.
[46] Johnson and Chandrasekar (1990a).
[47] John Dolland (1706–1761) constructed a composite lens of flint and crown glass in 1758 that approximately eliminated the problem of refragibility and produced an achromatic lens (Johnson and Chandrasekar 1990a).
[48] Johnson and Chandrasekar (1990b).
[49] Voltaire (1738).
[50] The child was not Voltaire's. Émilie had began an affair with the poet Jean François de Saint-Lambert in 1748 and became pregnant. Émilie died a week after giving birth to a daughter, who survived about eighteen months.
[51] Martin (1751).
[52] Fontenelle (1686).
[53] Schiebinger (2006).
[54] Harris (1729).
[55] Algaretti (1739). This English translation was by Elizabeth Cater (1717–1806) a member of Elizabeth Montagu's bluestocking circle which contained men as well as women and the term bluestocking is thought to have arisen from the informal nature of the gatherings where the men wore the more casual blue worsted stocking rather than the formal black silk ones. Only later was the term was applied specifically to a woman devoted to intellectual activities.
[56] Schiebinger (2003).
[57] A lyric which resembles a madrigal.
[58] Mazzotti (2004).
[59] The book was advertised under the title *The Philosophy of Tops and Balls* in the *Monthly Review*, Vol. 24, p. 277, 1761.
[60] Boyle (1661).
[61] Mitchell (1760). The speed of seismic waves in the earth's crust is roughly in the range 2,000–8,000 m/s depending upon the rock.
[62] Secord (1985).
[63] Ferguson (1768).
[64] Weston was primarily a botanist but he had an interest in mathematics and published mathematical problems in the *Ladies Diary*, he must have been a very competent mathematician since it appears that Gregory learnt his significant mathematical skills from him.
[65] Gregory (1799).
[66] Jones (1793).
[67] Eclectic Review, Vol. 2, January–June 1806, pp. 473–474.
[68] Murry (1799).

[69] Murry had earlier written a book on more general topics for young ladies *Mentoria or the Young Ladies Instructor* which was more successful than her sequel on astronomy and natural philosophy.
[70] Fyfe (2005b).
[71] Nature ran at a loss for nearly ten years before its price was increased to sixpence and its size increased from 24 to 28 pages.
[72] Topham (2000).
[73] Tylecote (1957).
[74] This was not the first course on science for tradesmen and master mechanics given in Glasgow; John Anderson had given one some years prior to his death in 1796 but the few practical men who attended were mostly employers since the course cost a guinea and was held during the day. Tylecote (1957).
[75] Stephen and Roderick (1972).
[76] Lee (2004).
[77] Also spelt Steel.
[78] Glasgow Mechanics' Magazine, Vol. 3, No. 72, 7th May 1825.
[79] It was the London Institution that defined a mechanic as 'any person who worked with his hands'. (Stephen and Roderick, 1972).
[80] London Mechanics Magazine, Vol. 2, pp. v–viii.
[81] Shapin and Barnes (1977).
[82] Brougham (1825).
[83] London, Manchester, Liverpool, Sheffield, Birmingham, Edinburgh, and Glasgow (both the Mechanics Institution and the Andersonian Institution).
[84] Hudson (1851).
[85] In the 1865 edition of Chambers' Encyclopedia, vol. VII, pp. 521–522, it was stated that: 'Physics in its narrower sense is equivalent to Natural Philosophy, which, until of late years, was the term more commonly used in Great Britain.'
[86] Auerbach (2001).
[87] Berman (1972).
[88] Cantor (1970).
[89] James (2004b).
[90] James (2004a).
[91] Faraday (1822).
[92] Tyndall (1870).
[93] Howard (2004).
[94] Tyndall (1863).
[95] James (2002).
[96] Tyndall (1854).
[97] Tyndall (1857).
[98] Faraday published the lectures of 1860–1861. Faraday (1961).
[99] Oil lamps were used but until the introduction of paraffin in the 1850s were less used than candles. Gas lighting did not come into its own in the home until the invention of the gas mantle in 1891.
[100] The paraffin wax candle was developed during the 1850s and in 1860 was an example of modernity though it remained expensive until massive oil deposits were found in the USA in 1859. However, by 1879 the price of the paraffin wax candle had fallen to a fifth of the cheapest candle manufactured in 1830. History of Price's Candles, www.prices-candles.co.uk.
[101] Withers *et al.* (2008).
[102] First Report of the BAAS, 1831.

[103] Morrell and Thackray (1981).
[104] Derogatively called by William Buckland in 1832 *Albemarle dilettanti meetings* in reference to the Royal Institution in Albemarle Street that held similar evening meetings. Morrell and Thackray (1981).
[105] Tyndall (1896b).
[106] Morrison-Low (2004).
[107] Brougham obtained a pension of £100 for Brewster in 1832, which was increased to £300 in 1836.
[108] Hays (1981).
[109] Hays (2004).
[110] Issitt (2006).
[111] Issitt (2002).
[112] Gordon (2004).
[113] Gregory (1802).
[114] Gregory (1806).
[115] Astore (2004).
[116] Dick (1845).
[117] Hetherington (1983).
[118] Mitchel (1871).
[119] Lord Rosse's *Leviathan of Parsonstown*, completed in 1845 had a 72 inch (1.83 m) reflector but without an equatorial mount it was difficult to use and apart from determining that some nebulae possess a spiral structure it accomplished little. More successful was William Lassell who built a 24 and a 48 inch (0.61 and 1.22 m) reflecting telescope with equatorial mountings. The largest objective in a refracting telescope at the time was the fifteen inch (381 mm) telescope in the Imperial Observatory at Pulkovo near St Petersburg, Russia. Smith (2002).
[120] McCormmach (1966).
[121] Mitchel (1851).
[122] Mitchel (1860).
[123] Hutchins (2004). The problem was that with such a vast output Proctor often published in haste.
[124] Lightman (2007).
[125] Proctor (1868).
[126] Pritchard (1870).
[127] Wallace (1903).
[128] Saum (1999).
[129] *Canals* was a mistranslation of *canali* meaning channels. Proctor was perceptive but he did believe that there were seas and rivers on Mars.
[130] Marcet (1809, 1824).
[131] The full title is *Conversations on Natural Philosophy in which the Elements of that Subject are Familiarly Explained and adapted to Young People.*
[132] Fyfe (2003).
[133] Jones (1868–1869).
[134] Cresse (2004).
[135] Letter from Brougham to Mary Somerville's husband 27th March 1827. Somerville (2009).
[136] Somerville (1831).
[137] Hollis (2004).
[138] Clerke (1902).
[139] Clerke (1903).
[140] Huggins (1868).
[141] Hirsch (1979).

[142] Lockyer (1887).
[143] Clerke (1905).
[144] The quote is from Tennyson's *Morte d'Arthur*: The old order changeth, yielding place to new, / And God fulfils Himself in many ways.
[145] Bowen (1927).
[146] Lockyer (1919).
[147] In 1833 there had been a unprecedented meteor storm when, according to some estimates, more than two hundred thousand meteors an hour were observed. The Leonid meteor showers are associated with the Tempel-Tuttle comet first discovered independently by the German Ernst Tempel (1821–1889) and the American Horace Tuttle (1837–1923) in 1865/1866. The meteor shower occurs when the comet approaches the sun and gases evaporating from the comet throw off small particles. In 1867 the period of the Tempel-Tuttle comet was calculated to be 33.2 years. Not every time the comet returns is the meteor shower so impressive. In 1899 the meteor shower was unimpressive. The last truly spectacular show was in 1966. Do not wait for the next very spectacular show because it is not expected until 2099.
[148] Beardsley (1872).
[149] Mussell (2007).
[150] Wells (1894).
[151] Phillips (1817a–c).
[152] Phillips (1821).
[153] Phillips (1817c).
[154] Others used that overworked mysterious medium aether but it does appear that Phillips believed that some kind of real gas filled space.
[155] de Morgan (1872).
[156] Faraday (1846).
[157] Mackintosh (1838).
[158] Kavanaugh (1886). Mackintosh's book was reprinted in America in 1846 but there is no evidence that it was known to Kavanaugh.

Chapter 4

Philosophy and Politics

Among the Greeks there was no real division between metaphysics and physics. Aristotle himself did not use the word metaphysics, which he called the first philosophy; it is the term introduced by the first century BC Alexandrian editors of Aristotle. The seventeenth century, especially in England, was a period of revolution in all the affairs of man and, in the later stages of the century physics began to separate from philosophy. Part of the success of Newton's *Principia* was that its natural philosophy was completely separate from metaphysics and speculation on causes. In unpublished draft prefaces written after the publication of the second edition of the *Principia* in 1713, Newton gave his own opinion of metaphysics writing:[1]

> 'Natural philosophy should be founded not on metaphysical opinions, but on its own principles; .... To dispute about the objects of ideas except insofar as they are phenomena is dreaming. In all philosophy we must begin from phenomena and admit no principles of things, no causes, no explanations, except those which are established through phenomena.'

The rise in philosophy in the seventeenth century began with Bacon and his development of the inductive method, which he stated was valid in natural philosophy, politics, and law[2] just as well as it was in philosophy.[3] Deduction in natural philosophy was undervalued by Bacon probably because it often entails mathematics whose importance he underestimated. Bacon foreshadowed that important development in Britain of an experimental rather than a speculative natural philosophy.[4] Hobbes was an empiricist, as was Bacon, but unlike Bacon, whose secretary he was for a time, he valued the mathematical method. Unfortunately Hobbes' mathematical ability was relatively poor, though he did give mathematics lessons to the future King Charles II in Paris, where Hobbes had fled because of his politics after the formation of the Long Parliament, in 1646.[5] Hobbes' two main interests were theoretical and political philosophy but he also was interested in natural philosophy. In natural philosophy Hobbes mainly concentrated on the abstract first principles and is continuous with his theoretical philosophy. His first work on natural philosophy was the *Short Tract on First Principles*[6] probably written sometime between 1630 and 1636.[7] His major work on natural philosophy *De Corpore Philosophia* was published in 1655. For Hobbes the cause of all things is motion. Some kind of background is needed to observe motion, Newton believed in absolute space but Hobbes rejected it, his conception of space was an imaginary phantasm of the mind. There was real space in which a body had extension but this space does not depend on our thought. Hobbes was less clear on real space. The distinction between imaginary and real space probably arose out of his dispute with Descartes. Body and extension are identical in Descartes view which, according to Hobbes, made space phantasmal.[8] Apart from the first principles, Hobbes' main interest in natural philosophy was optics. In his *De Corpore Philosophia* Hobbes gave a *proof* of Snell's law of refraction, which was based on Descartes' *proof* in *La Dioptrique*. Descartes' *proof* was based on the analogy between the refraction of light and the path of a stone thrown into water, which was first mentioned in a tenth century manuscript by the Arab mathematician Ibn Sahl (ca. 940–1000). Unfortunately in order to obtain Snell's law from this analogy where the stone is refracted away from the normal to the surface of the water, one has to assume that light travels more easily in denser substances such as water.[9] Hobbes' political philosophy is given in *Elements of Law*, completed in 1640 just before the beginning of the Long Parliament, *De Cive*, first published in 1642, and the *Leviathan*, published in 1651. Without government every man desires to preserve his own liberty but to dominate others, which leads to war. 'Life' Hobbes tells us is 'nasty, brutish and short.'[10] Monarchy is Hobbes' preferred government. The first sovereign can be freely chosen, but once chosen he determines his succession. Clearly such views were not welcome in Cromwell's Britain and Hobbes was wise to flee to France.

Descartes discussed the elements of philosophy in the preface to the French translation of his *Principia Philosophiae*, which he wrote in the form of a letter to the translator Claude Picot (1614–1668), writing:[11]

> 'the whole of philosophy is like a tree. The roots are metaphysics, the trunk is physics, and the branches emerging from the trunk are all the other sciences, which may be reduced to three principal ones, namely medicine, mechanics and morals. By "morals" I understand the highest and most perfect moral system, which presupposes a complete knowledge of the other sciences and is the ultimate level of wisdom.'

The two most important books on epistemology, or the theory of knowledge, of Descartes were *Discours de la Méthode*, published in French in 1637, and *Meditationes de Prima Philosophia*, published in Latin in 1641.[12] In these books Descartes explained the method of *Cartesian doubt* where he doubted, as far as, possible all things to obtain his theory of knowledge. It is from this method of doubt that Descartes comes to his most famous truth '*cogito ergo sum*', I think therefore I am. He obtains this conclusion by arguing that no demon however cunning could deceive him if he did not exist. This argument did not prove that one had a body but only that one was a thinking entity. The existence of the external world could be doubted, only the existence of God removes that doubt. Descartes' proof of the existence of God relied on scholastic philosophy and is not a real proof. *Principia Philosophiae* gives his natural philosophy, which was briefly discussed in Chapter 1. In Part 1, Descartes again started with the principles of knowledge which are the roots of his philosophy before discussing the principles of material things in Part 2, where he began by arguing for the existence of material things which we could doubt. Again he had to fall back on his premise that God would not deceive us to justify the existence of an 'extended thing that we call body or matter.' To Descartes space was a conceptual abstraction from a body's spatial extension and, since empty space could not have extension, a vacuum was impossible. A body was always in contact with other bodies and when it moved it transferred its contact to other bodies. Indivisible atoms did not exist because, no matter how small, an atom was still extended and hence capable of being divided into smaller parts. The visible universe was discussed in Part 3 and it is here he gave his vortex theory to explain the motion of the planets. Part 4 is on the Earth and it is here that Descartes explained weight as the movement of minute invisible particles pushing a body towards the centre of the Earth. Descartes was the founder of modern philosophy and heralded the new science of the seventeenth century providing the first comprehensive alternative to Aristotelian physics. Unlike most philosophers of the period, Descartes wrote very little on politics. Although Newton rejected much of Cartesian natural philosophy based on unsubstantiated hypotheses and laid emphasis on a mathematical natural philosophy based on experiment, he still owed much to Descartes.[1]

Baruch Spinoza was influenced by Descartes' natural philosophy but did not accept all of his metaphysical ideas. The only book of Spinoza published during his lifetime was *Principles of Cartesian Philosophy*, which was published in 1663.[13] The explicit aim of this work was an exposition of Descartes' ideas but it contains indications of divergence from his metaphysics. In a letter addressed to Henry Oldenburg[14] (1619–1677), Spinoza explained that he had written the *Principles* for use in teaching a youth to whom he did not wish to teach his own views openly.[15] Spinoza reframed Descartes' ideas using a rigorous deductive argument, or the geometric method as he calls it, in the style of Euclid with definitions, axioms, propositions, and demonstrations. The main metaphysical ideas of Spinoza are given in the first two parts of his major work *Ethics,* which was also written using the geometric form; as he explained in the introduction to Part III:[16] 'I shall regard human actions and appetites exactly as if I were dealing with lines, planes, and bodies.' In Spinoza's metaphysics there is only one substance 'God or Nature' whereas Descartes recognized three: God, Mind and Matter. God according to Spinoza had an infinite number of attributes including extension. Everything is an expression of God's nature and ruled by his logic; there is no chance occurrence or free will. Spinoza also believed that the geometric method applied to moral philosophy because it was the essence of his philosophical system that everything could be demonstrated by deduction.

4.1. Metaphysics and Epistemology

Epistemology or the theory of knowledge, and metaphysics, or the theory of existence, are closely linked to natural philosophy. Newton himself never wrote a general philosophical treatise like Descartes' *Meditationes* but he did include general philosophical discussions in the *scholia* of the *Principia* and in the queries of *Opticks*, though of course they are not nearly as extensive as those in Descartes' *Principia Philosophiae*. Locke, the founder of empiricism or the doctrine that knowledge comes from experience, was a contemporary of Newton whose natural or experimental philosophy was also empirical. Hence, though Locke's ideas were different to those of Newton he could, in many respects, be described as Newton's philosopher.

4.1.1. *John Locke* (1632–1704)

The *Essay Concerning Human Understanding* published in 1690 just three years after the first edition of the *Principia*, was Locke's most important work on philosophy.[17] Locke was not without knowledge of natural philosophy. At Christ Church, Oxford, Locke held a studentship, which was equivalent to a fellowship at other colleges, from 1652 until 1675. His first experience of philosophy at Oxford was of Scholastic philosophy and not reassuring, he noted to Jean Le Clerc (1657–1736) that: [18]

> '[he] lost a great deal of time, when he first applied himself to study, because the only philosophy then known at Oxford was the peripatetick, perplexed with obscure terms and stuffed with useless questions.'

Descartes' works were the first that caused Locke 'to relish the study of philosophy,' though Locke did not agree with everything, 'he found that he wrote with great clearness.'[18] Locke started to be interested in medicine in the late 1650s and an interest in natural philosophy may have arisen after he made the acquaintance of Robert Boyle in 1660.[19] When Boyle moved permanently to Oxford in 1664, Locke's notebooks show that he was actively exchanging information with him.[20] In 1666 Locke became friends with Anthony Ashley Cooper (1621–1683), later the first Earl of Shaftesbury, the friendship blossomed and in 1667 Locke moved to Exeter House, the London home of Cooper, as tutor to his eldest son.[19] Locke stayed at Exeter House for the next eight years. In London Locke became close to Thomas Sydenham (1642–1689) who was primarily a physician but was also interested in natural philosophy. Locke was elected to the Royal Society in 1668 four years before Newton but he did not contribute much to its work.

There were three earlier drafts of the *Essay*: Drafts A and B were written in 1671, Draft C was written in 1685 and by the end of 1686 the version was substantially the same as the published version which was finally completed in 1689.[21,22] Drafts A and B were written while Locke was staying at Exeter House. The other versions were written in the Netherlands where Locke had fled in 1683 fearing arrest because of his association with the Earl of Shaftesbury.[23] Draft A voiced the growing evolution of experimental over speculative natural philosophy, which Locke could have obtained from Boyle or Sydenham, and a rejection of inquiry into nature's unobservable causes.[4,20] While Locke was in Amsterdam in 1688 he wrote an anonymous review of the *Principia* for the respected *Bibliothèque Universelle*.[24] It is unlikely that Locke met Newton before his return to England in 1689, once having met, they quickly became friends; there is no evidence that either influenced the other prior to this date.[21] However the views expressed in the first editions of the *Essay* and the *Principia* only differ on one major point, the notion of space and time.[21] Newton's concept of absolute space and time is the foundation on which his mechanics rest. Since Locke believed all knowledge is founded on experience, he could not believe in absolute space and time because it is not possible to arrive at such a concept through experience.[25] He also denied Descartes' identification of space with body [since] extension is an idea belonging to body only but space may ... be considered without it.'[17] Although there is infinite duration, since Locke acknowledged God to be an Eternal Being, he was more equivocal about infinite space. To Locke the space of our experience was relative, which he clearly expressed in a discussion of the idea of place using a chessboard as an example. We determine the positions of the chess pieces from their position on the board. If the chessboard is in a cabin of a ship it may be moved about the cabin, the ship can sail and even if the ship is anchored it moves with the Earth. Locke continued:[17]

> 'our Idea of Place, is nothing else, but such a relative Position ... [and] we can have no Idea of the place of the Universe, though we can of all the parts of ... beyond that, we have not the Idea of any fixed, distinct, particular Beings, in reference to which, we can imagine it to have any relation of distance; but all beyond it is one uniform Space or Expansion, wherein the Mind finds no variety, no marks.'

Locke summed up his views stating:[17]

> 'whether any one will take Space to be only a relation resulting from the Existence of other Beings at a distance; or whether they will think the Words of the most knowing King Solomon, the Heaven, and the Heaven of Heavens, cannot contain Thee; or those more emphatical ones of the inspired Philosopher St. Paul, In Him we live, move, and have our Being, are to be understood in a literal sense, I leave every one to consider.'

A little further on in the *Essay* Locke recognized that space and time are related, writing:

> 'One thing seems strange to me, that whilst all Men manifestly measured Time by the motion of the great and visible Bodies of the World, Time yet should be defined to be the measure of Motion; whereas 'tis obvious to every one that reflects ever so little on it, that to measure Motion, Space is as necessary to be considered as Time.'

Locke made some changes in subsequent editions of the *Essay* as the result of reading the *Principia*.[21] The empiricist Locke had little use for maxims, as he wrote in Chapter VII of Book IV:

> 'Maxims are of little use, where we have clear and distinct Ideas, so they are ... of dangerous use, where our Ideas are not clear and distinct.'

In Section 11 of Chapter VII he discussed the use of maxims in the search for knowledge concluding that they were not of much use in discovering new truths. In the second edition of 1694 he substantially added to this section, using the *Principia* as an example, writing in part:

> '[Maxims] are not of use to help Men forward in the Advancement of Sciences, or new Discoveries of yet unknown Truths. Mr .Newton, in his never enough to be admired Book, has demonstrated several Propositions, which are so many new Truths, before unknown to the World, and are further Advances in Mathematical Knowledge.'

Prior to reading the *Principia*, Locke, like Descartes, believed that action at a distance was impossible but Newton's theory of gravity convinced him that it was possible.[21] In a letter written in 1699 to Edward Stillingfleet (1635–1699), Bishop of Worcester, Locke wrote: [26]

> 'The gravitation of matter towards matter, by ways inconceivable to me, is not only a demonstration that God can, if he pleases, put into bodies powers and ways of operation, above what can be derived from our idea of body, or can be explained by what we know of matter, but also an unquestionable and everywhere visible instance, that he has done so. And therefore in the next edition of my book, I shall take care to have that passage rectified,'

which Locke did in the fourth edition of 1700.

There is also evidence that Locke had some influence over changes that Newton made to the second edition of the *Principia*.[21] One of the most significant changes was the addition of the General Scholium. Locke's near pantheist concept that God constituted duration and space may have influenced Newton when he wrote in the General Scholium that: [27]

> '[God] endures always and is present everywhere, and by existing always and everywhere he constitutes duration and space. Since each and every particle of space is always, and each and every indivisible moment of duration is everywhere, certainly the maker and lord of all things will not be never or nowhere... In him all things are contained and move.'

Another addition to the second edition of the *Principia* was the significant third 'Rule of Philosophising' at the beginning of the more philosophical Book III. This rule stated that:[27]

> 'those qualities of bodies that cannot be intended and remitted and that belong to all bodies on which experiments can be made should be taken as qualities of all bodies universally.'

In the third edition of the *Principia*, Newton added a short section at the end of the discussion of this rule namely:[27]

> 'Yet I am by no means affirming that gravity is essential to bodies. By inherent force I mean only the force of inertia. This is immutable. Gravity is diminished as bodies recede from the earth.'

Three drafts of Rule 3 have been discovered. The first draft, written in 1690, the second draft, which cannot be dated accurately, and a third draft, which seems later than the other two, that combines the main ideas of the other two, and is substantially the same as that published in the second 1713 edition of the *Principia*.[28] Edward McGuire suggested that Locke might well have influenced the second draft of the Rule, which is the more philosophical draft.[28] The language used by Locke and Newton differ. Locke wrote about the primary and secondary qualities of a body while Newton in Rule 3 wrote of 'those qualities of bodies that cannot be intended and remitted', or qualities that cannot be increased and diminished, in other places in the *Principia*, he uses the word *essential*.[29] The primary qualities of a body are those of extension, solidity, figure and mobility; the secondary qualities are colour, taste, etc. or 'meer powers depending on its primary qualities' as Locke described them.[17] As McGuire remarked Newton may have had in mind the difference in operations necessary to change the primary and secondary qualities of bodies.[28] As Locke wrote in the *Essay*:[17]

> 'Pound an almond, and the clear white colour will be altered into a dirty one, and the sweet taste into an oily one. What real alteration can the beating of the pestle make on any body, but an alteration of the texture of it?'

4.1.2. *Gottfried Wilhelm Leibniz* (1646–1716)

The philosopher Leibniz was a polymath of extraordinary dimensions who was an important philosopher but he can be abstruse and difficult to understand. *La Monadologie*,[30] one of Leibniz most famous works on metaphysics, appears to be fanciful and led Russell to write about his first impression in his Preface to his book on Leibniz in 1900 that:[31]

> 'I felt – as many others have felt – that the Monadology was a kind of fantastic fairy tale, coherent perhaps, but wholly arbitrary.'

However, Leibniz was not fanciful, he independently developed calculus,[32] he devised the binary system used in all computers,[33] and he also invented the first workable calculating machine illustrated in Fig. 4.1, for which he was elected a Fellow of the Royal Society.

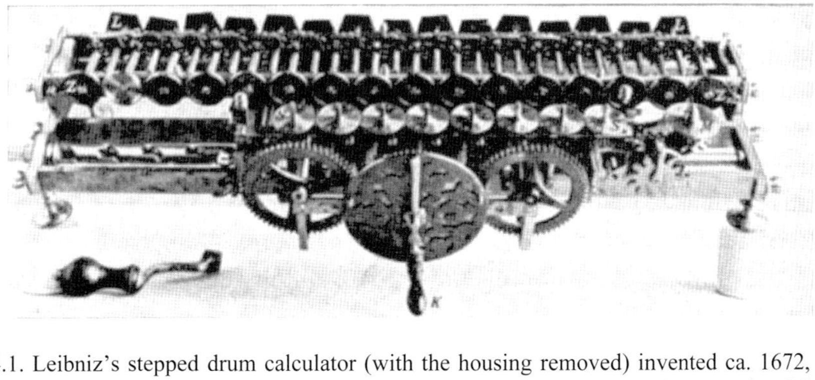

Fig. 4.1. Leibniz's stepped drum calculator (with the housing removed) invented ca. 1672, which performed addition, subtraction, multiplication, and division (Meyer, 1904). Deutsches Museum, München.

One of the starting points in understanding Leibniz's metaphysics is that God and the soul were central and crucial to it. The central role of the soul in Leibniz's metaphysics can be seen in a fragment *Anima quomodo agat in corpus* written in either 1676 or 1677 which begins:[34]

> 'How the soul acts on the body. As God does on the world, that is, not by way of miracle, but through the mechanical laws. And so, if (though impossible) minds were eliminated, leaving the laws of nature, the same things would come about as if there were minds, and even books would be written and read by human machines, understanding nothing. Indeed, it should be known that it is impossible to eliminate minds while preserving the mechanical laws.'

In another fragment written about a year later, Leibniz wrote: 'Substantial form, or soul, is the principle of unity and of duration.' To Leibniz a corporeal substance as well as having extension, is a union of the body and soul that grounds the reality of the physical world.[34] However, there are considerable problems in understanding exactly what was Leibniz's philosophy that have caused an ongoing debate among philosophers since Russell wrote his *A Critical Exposition of the Philosophy of Leibniz* in 1900, which were largely caused by Leibniz's failure to write and publish a *magnum opus*. The vast majority of Leibniz's philosophical writings were unpublished manuscripts and letters and many still remain to be published. It was argued in Chapter 2 that most emphasis should be given to the published works of great men like Newton where there is controversy over their ideas, but Leibniz is an exception because he published so little of his work that his ideas cannot be found in his published works alone. Russell wrote that: 'Suddenly a flood of light was thrown on all the inmost recesses of Leibniz's philosophical edifice' when he first read *Discours de Metaphysique*, written in 1685, and the *Letters to Arnauld*, written 1686–1690 but neither were published until 1846.[31] For most of the twentieth century these two documents were the focus of scholarship on Leibniz but in 1985 Daniel Garber became concerned how the metaphysics could be a foundation for his physics, as he claimed it to be and wrote an iconoclastic essay on Leibniz's middle years[35] and has recently written *Leibniz: Body, Substance, Monad*, which gives an 'account of Leibniz's developing thought on the make-up of the physical world and its metaphysical grounding.'[34] In his book, Garber concentrates on Leibniz's during the years from the later 1670s to the 1690s before he advanced his monadological metaphysical world of spiritual substances and when his metaphysics was grounded in corporeal substances extended unities of matter and form and owed much to the scholastic interpretation of the Aristotelian concept of substance.[35] Leibniz subscribed during this period to a thorough-going mechanical philosophy as he stated in a letter to Thomas Burnett in 1697:[36]

> 'I believe that everything happens mechanically, as Democritus and Descartes desire ... nevertheless everything happens vitally and in accordance with final causes, everything being full of life and perceptions, contrary to the opinions of the Democriteans.'

It is difficult to decide whether Leibniz thought physics to be a science of phenomena or a science of the real world.

Leibniz was a relativist as far as space and motion are concerned, which he expressed most elegantly in an essay written in 1694 while he was in Italy and in which he confronted the issue of the Catholic Church's condemnation of the Copernican system:[37]

> 'With respect to the motions of ... bodies ..., however numerous, moved only by collision with other bodies ... not even an angel could determine with mathematical rigor which ... of that sort is at rest, and which is the centre of motion for others.... It is a wondrous law of nature that no eye, whatever in matter it might be placed, has a sure criterion for telling from phenomena where there is motion, how much motion there is, and what sort it is, or even whether God moves everything around it, or whether He moves that very eye itself.'

The two descriptions of the solar system are equivalent, or in Leibniz's language equipollent. When there are two hypotheses that are equipollent they are empirically adequate but literally inconsistent and in Leibniz's view they can only describe a phenomenon not a reality. However the Newtonian mechanics expressed in the *Principia* had removed the possibility that the Earth was stationary or moving with a uniform velocity. Newtonian mechanics are invariant for any Galilean frame of reference which is moving with a uniform velocity but not for a non-inertial frame of reference.

One of the innovations that Leibniz made in mechanics was the concept of *vis viva*, or living force, which evolved from his metaphysical principle of the equality of cause and effect, its modern term is kinetic energy. Leibniz's use of the term *force* here is equivalent to energy and is not the same as a Newtonian force having different dimensions and is a scalar quantity as was Descartes' quantity of motion. Leibniz described a Newtonian force such as a centrifugal force or the force of gravity as a dead force; he called such forces dead 'since motion does not yet exist in it, but only a solicitation to motion ... [a] living force ... arises from an infinity of continual impressions of dead force.'[37] In his *Brevis demonstratio*[38] written in 1686, Leibniz considered two bodies A and B, with body B four times the mass[39] of body A, both bodies are raised, A through four times the height of B. Leibniz argued that B can be considered as four bodies A so that the work done in raising A and B must be the same. Leibniz now considered the fall of bodies A and B back to their original positions. Using Galileo's theory of free fall, he knew that the velocity of body A would be twice that of body B. The quantity of motion, which Descartes considered would be conserved, of B would be twice that of A, but the living force of both A and B would be the same. By his principle of equality Leibniz argued that the potential to do work must be the same for both bodies and hence his *vis viva* was a measure of that potential and hence conserved.[40] We now know that *vis viva* is not conserved in inelastic impact and only approximately conserved in elastic impact.[41] Leibniz generalized this result and proposed it was the *vis viva* of the universe as a whole that was conserved.[34] The analogy between the free fall of two bodies and the motion of all the bodies in the universe was very loose as Leibniz realised and he attempted other proofs of his concept.[34] Although Leibniz was wrong he was on the right track. *Vis viva* is a measure of kinetic energy and energy, in all its forms, is conserved in an isolated system.

Leibniz knew that Descartes' quantity of motion, which was a scalar quantity, was not conserved but he knew that if the direction of the velocity is taken into account then the vector quantity that Leibniz called quantity of progress (momentum) was conserved as he stated in his *Essay de Dynamique* written in 1691.[42] However Leibniz stated the vectorial nature of the conservation of progress more clearly in his *Explanation of the New System* written in reply to the French philosopher Simon Foucher (1644–1696) in 1696, writing:[43]

> 'There is another law of nature, which I have discovered and demonstrated, and which Descartes did not know. It is that not only is the same quantity of moving force preserved, but also the same quantity of direction towards whatever side in the world is taken. That is to say, drawing any straight line you please, and taking also such and as many bodies as you please, you will find, in considering all these bodies together, without omitting any of those which act upon any one of those you have taken, that there will always be the same quantity of progress in the same direction in all lines parallel to the right line which you have taken, taking care to estimate the sum of the progress by subtracting that of the bodies which move in the direction opposite to that of the bodies which move in the direction taken.'

After 1700 Leibniz almost ceased work on natural philosophy. Leibniz's metaphysics of his later years was based on monads, these were the elements of everything and similar to atoms. The monad had no extension but it had a soul. Although Leibniz's conception of the form of that soul is different to the usual conception and what Leibniz meant by soul was that the monad had perception. Leibniz also wrote that 'the monads have no windows, through which anything could come in or go out,' but here he meant physical substance rather than a non physical perception. Monads could affect one another not through the exchange between them but by a 'pre-established harmony'. If monads were reality what did Leibniz's earlier physics describe? The common view used to be that physics described not the real world but phenomena that we perceived but Garber believes that Leibniz's physics is that of the real world not phenomena and that he was struggling at the end of his life to reconcile his physics and monads but died before he got there.[34] *La Monadologie*, which gives his fullest concept of monads and the real world, was not written until 1714 only two years prior to his death.

4.1.3. *George Berkeley* (1685–1753)

Berkeley, Bishop of Cloyne in County Cork, Ireland, was an Anglo-Irish philosopher who, in contrast to those already discussed, was an idealist who denied the reality of corporeal substance. To him material objects only existed, in a metaphysical sense, through being perceived. At first sight this view would seem to be completely at variance with natural philosophy with its belief in the existence of laws of nature that described how matter behaved. However, Berkeley avoided this problem in a way that gave some credence to the existence of God. Berkeley saw God as overarching everything and, even when no one else was there, God would be there perceiving everything. In the early twentieth century, a chaplain of Trinity College, Oxford, Ronald Knox (1888–1957) wrote a limerick with a reply which neatly sums up Berkeley's position:

There was a young man who said, 'God
Must think it exceedingly odd
 If he find that this tree
 Continues to be
When there's no one about in the Quad.'

 Reply

Dear Sir:
 Your astonishment's odd:
I am always about in the Quad.
 And that's why the tree
 Will continue to be,
Since observed by
 Yours faithfully,
 God.

Berkeley did not object to the attempts of natural philosophers to ascertain the laws of nature, which he believed God had implanted in the universe, but he objected to attempts to account for them. Thus it was not natural philosophy to which Berkeley objected but the metaphysical underpinning of that philosophy. Berkeley's most important philosophical work that described his idealistic immaterialism is mainly found in his *Principles of Human Knowledge*,[44] published in 1710 and the *Three Dialogues Between Hylas and Philonous, in Opposition to Sceptics and Atheists*, [45] published in 1713. In 1721 he published *De Motu*, [46] which was an essay in Latin originally unsuccessfully submitted to the competition of the Paris Academy of Sciences for the best essay on motion. Although Berkeley examined the metaphysical implications of dynamics in this essay, he did not advance his immaterialist views because the French judges of the competition would have been Cartesians and not sympathetic to such a view.[47] After 1721 Berkeley published no other major philosophical work, though he removed some inconsistencies in his *Principles* in its second edition of 1734 Berkeley's final major work *Siris* published in 1743 also contains some metaphysical reflections on natural philosophy but adds nothing new to his idealistic view.[48]

To Berkeley the laws of nature showed:[44]

> 'the uniform working, which so evidently displays the goodness and wisdom of that Governing Spirit whose Will constitutes ... [them], is so far from leading our thoughts to Him, that rather sends them a wandering after second causes.'

Berkeley retained the commonsense view of existence and did not deny 'that the things I see with my eyes and touch with my hands do exist, really exist.'[44] What he denied was the existence of 'that which philosophers call of matter or corporeal substance ... there is no damage done to the rest of mankind who, I dare say will never miss it.'[44] The natural philosopher who was nearest to Berkeley's view was Newton himself since he steadfastly refused to speculate on possible causes for phenomena. Both Newton and Berkeley were, what John Dewey (1859-1952) in the early twentieth century called, instrumentalists that is they believed that the role of natural philosophy was to establish theories, or laws of nature, that described natural phenomena without questioning the reality of the theory. In their view natural philosophy was separate from metaphysics and did not have to explain causes; Newton held this view because causes could only be hypothesized whereas to Berkeley God was the cause of everything. As Berkeley wrote in *De Motu*:[46]

> '*Force, gravity, attraction*, and terms of this sort are useful for reasonings and reckonings about motion and bodies in motion, but not for understanding the simple nature of motion itself or for indicating so many distinct qualities. As for attraction, it was clearly introduced by Newton, not as a true, physical quality, but only as a mathematical hypothesis....

Physically ... a thing is explained not by assigning its truly active and incorporeal cause, but by showing its connection with mechanical principles.'

4.1.4. *Immanuel Kant* (1724–1804)

Immanuel Kant is one of the greatest modern philosophers, he founded German idealism and provided the foundations for much of nineteenth and twentieth century philosophy and he still has an influence today. His most important works concerning epistemology and metaphysics are contained in *The Critique of Pure Reason*, first published in 1781[49] and the *Metaphysical Foundations of Natural Science*, published in 1786.[50] In his early works Kant had attempted to reconcile the science of Leibniz and Newton, but by 1755, when he published his *Universal Natural History and Theory of the Heavens*,[51] he had become fully converted to Newton. When he wrote *The Critique of Pure Reason* he had already decided that the future of metaphysics lay with the Newtonian method.[52]

The most important part of *The Critique* is the theory of space and time. Interpretation of Kant is difficult and much of the literature is still concerned with discerning his exact meaning. Kant wrote the *Prolegomena*, which contains a summary of *The Critique* and is easier to understand, in 1783.[53] Kant wrote that the 'mathematical natural philosophers' (Newtonians) saw space and time as substances or absolute whereas some 'metaphysical natural philosophers' (Leibnizians) regarded space as marked out by objects and time as a succession of events or relative.[54] For Kant space and time are pure intuitions of the mind that are not derived from experience and are not concepts. However, there are also twelve *a priori* concepts that Kant termed categories, after the ten Aristotelian categories, divided into four classes: quantity, quality, relation and modality. What Kant terms antinomies, that is propositions where the thesis and antithesis can both be proved, are obtained if space and time, or any of the categories, are applied to things beyond our experience. Thus nothing can be deduced about whether the world is limited in space and time.

In the *Metaphysical Foundations of Natural Science*, Kant used the metaphysics developed in *The Critique* to underpin Newton's *Principia*.[55] Kant thought that the philosophical principles of proper sciences, such natural philosophy, were capable of being established *a priori* with apodictic, or necessary and absolute certainty. Others, such as chemistry, Kant thought were not true sciences because their principles were only known from experience.[56] At the end of his preface Kant made it clear that mathematical physicists 'should find it not unimportant to treat the metaphysical part, which they cannot leave out in any case, as a special fundamental part of their general physics, and to bring it into union with the mathematical doctrine of motion.'[54]

Metaphysical Foundations, which is concerned with the connection between matter and motion, has four chapters which correspond to the four classes of categories. In the first chapter on phoronomy, or kinematics of motion, matter is defined as that which is movable in space and the magnitude, or quantity of motion, is considered. The second chapter on dynamics defined matter as being movable because it fills a space and considered the quality of motion. The third chapter on mechanics defined matter as the movable insofar as it has moving force, that is, motion can be communicated from one body to another. The final chapter is on phenomenology and considered how moving matter can be an object of experience or how could true motion be distinguished from apparent motion. In the chapter on mechanics, Kant stated three Laws of Mechanics, which were analogous with the three Laws of Motion in Newton's *Principia*. The three laws were:[50]

(1) 'In all changes of corporeal nature the total quantity of matter remains the same, neither increased nor diminished.'
(2) 'Every change in matter has an external cause. (Everybody persists in its state of rest or motion, in the same direction and with the same speed, if it is not compelled by an external cause to leave this state.)'
(3) 'In all communication of motion, action and reaction are always equal to one another.'

Newton's second law: 'A change in motion is proportional to the motive force impressed and takes place along the straight line in which that force is impressed,'[27] could not qualify as one of Kant's three laws because it states a definite relationship between motion and force that can on only be known from experience and cannot be known *a priori*. Kant's second law is Newton's first law and Kant even gave a paraphrase of Newton's law in parenthesis, and his third law is the same as Newton's third law. In the first law Kant expressed a version of the conservation of matter/mass which was never mentioned by Newton. Such a concept had been voiced since the days of the Greek atomists in the fifth century BC and by many subsequent philosophers such as Bacon and Spinoza; Kant in *The Critique* even quotes the line: 'from nothing comes nothing,' which was attributed to Democritus.[57] In the First Analogy in *The Critique*, Kant had given his principle of the permanence of substance and in the *Metaphysical Foundations* he simply stated that: 'From general metaphysics we take as a basis the proposition that in all changes of nature no substance either arises or perishes, and here it is only shown what substance shall be in matter.'[50] In what is almost a slight of hand, Kant changes the statement of his Principle of Permanence in the 1787 second edition of *The Critique* by adding: 'and the quantum of substance is neither increased nor diminished' to make it fit better with his *Metaphysical Foundations*. The necessity of Proposition 2 in Chapter Three becoming Kant's first Law of Mechanics is not apodictic to me especially since his first proposition in that chapter is: 'The quantity of matter in comparison with *every* other matter, can be estimated only by the quantity of motion [momentum] at a given speed.'[50]

4.1.5. *Georg Hegel* (1770–1831)

Hegel is a most difficult philosopher and he has consequently been reinterpreted a number of times. In the nineteenth century the British followers of Hegel came close to denying that his metaphysics was really metaphysics at all but was what, in modern terms, is called category theory – a conceptual interpretation that enables the universal attributes of a family of phenomena to be appreciated and how the different phenomena are interrelated.[58] Until recently it was only Hegel's critics who saw him as a metaphysicist and they simply dismissed his metaphysics. However to Hegel the purpose of philosophy was the rational knowledge of the absolute, which is metaphysics.[59] Hegel developed his metaphysics from Kant although he rejected much of it. Like Kant, Hegel was much influenced by Newton, but whereas for Kant, Newtonian mechanics was a starting point for his metaphysics, Hegel was highly critical of it and attacked not only Newton's philosophy but also some aspects of his physics. There has been some discussion whether Hegel's knowledge of natural philosophy was sufficient to understand properly Newton's *Principia* and the *Opticks*. Hegel's father wished him to enter the church and in 1788 he enrolled Hegel in the Tübinger Stift, a protestant seminary attached to the Eberhard Karls University, in which Kepler had enrolled in 1587. By the end of the eighteenth century, the University had declined considerably and the Stift was the more distinguished of the two institutions.[60] Although Hegel was a dutiful son, it appears that from very early during his time in Tübingen, he never intended to enter the church.[60] In 1781 Christoph Friedrich von Pfleiderer (1736–1821) was appointed a professor of mathematics at the University. Perhaps because of the association of Kepler with the Stift, natural philosophy seems to have been taught to its students and Pfleiderer was Hegel's instructor.[61] However the way natural philosophy was taught in Tübingen, as elsewhere in Germany during the 1790s, gave the impression that Newton had a doubtful claim to have laid the foundations of physics in the *Principia*.[62] In 1801 Hegel went to Jena at the invitation of his friend Schelling from the Stift and in order to teach at the University he had to give a dissertation and the topic he chose was the orbits of the planets.[61] Thus Hegel did have some knowledge of natural philosophy but one probably biased against Newton.

Hegel believed that 'the physical [could not] be proved mathematically,' which is the complete opposite to Newton[63] who in the first sentence of his preface to the *Principia* made it plain that he abandoned the Aristotelian substantial forms and occult qualities and that he was concerned with using mathematics to establish the laws governing natural phenomena. Again at the beginning of Book 3 of the *Principia*, Newton wrote:[27]

> 'I have presented principles of [natural] philosophy that are not, however, philosophical but strictly mathematical – that is, those on which the study of philosophy can be based. These principles are the laws and conditions of motions and of forces, which especially relate to philosophy.'

Newton was careful not to say that he proved the principles but only that he presented them mathematically. Hegel in his *Lectures on Natural Philosophy*, 1823/24, stated that to try to prove physical principles mathematically was 'ridiculous'.[64] Hegel believed the most fundamental laws of nature were phenomenological ones, like Kepler's laws for the motion of the planets; whereas Newton 'set the laws of forces in the place of the laws of phenomena' with laws such as those of gravitation.[65] Hegel objected that:[66]

> 'the Newtonian proof of the proposition that a body governed by the law of gravitation moves in an ellipse around the central body proceeds in general in a conic section, whereas the main point that was to be proven consists precisely in this, that the course of such a body is neither a circle nor any other conic section, but solely the ellipse.'

In a sense Hegel was right since Newton obtained confirmation of his law of gravitation from Kepler's elliptical orbit. One would like to say that Newton's gravitational analysis showed that non-returning comets on hyperbolic trajectories have been observed practically demonstrating that his analysis gave the more general solution. Unfortunately there is no clear evidence that non-returning comets from interstellar space on hyperbolic trajectories have been observed.[67] On the other hand the orbit of the planets is not exactly elliptical because the planets exert a gravitational force on each other. For the smaller planets this effect is negligible but the giant planets Jupiter and Saturn do affect each other's orbit significantly when they are in conjunction. Newton knew of this effect and described it in Book III of the *Principia*. Since Jupiter is more than three times the mass of Saturn, it is the biggest effect is on Saturn's orbit and Newton suggested using the common centre of gravity of the Sun and Jupiter and calculating the force on Saturn towards this common centre, but he did no calculations. It was not until 1785, when Laplace announced to the Paris Academy that the universal gravitation accounted for the anomalies in the motion of Jupiter and Saturn, was Newton fully vindicated.[68] Laplace used an analytical formulation of the theory of gravitation to calculate the orbits of Jupiter and Saturn, using the differential equations of motion independently formulated by Leonhard Euler (1707–1783), Alexis Clairaut (1713–1765), and Jean-Baptiste d'Alembert (1717–1783) in 1747. Hegel knew Laplace's *Exposition du Système du Monde* and quoted from it in his lectures in 1827 and 1830[61] so it is most unlikely that he did not know of his calculation of the orbits of Jupiter and Saturn but he still believed in the superiority of Kepler's elliptical orbits.

A lasting importance of Hegel was his development of the dialectical method which was adapted by Marx to become *Dialectical Materialism*. The dialectical method, in the original sense of seeking knowledge by question and answer, was used by the ancient Greeks but Hegel's immediate influence was Kant. In *The Critique of Pure Reason*, Kant used what he called the *Transcendental Dialectic* to refute that pure reason alone could be used to develop a metaphysical system.[54] Hegel's dialectical method was basically the use of a thesis and its antithesis to produce a synthesis which can then become the thesis and the whole dialectic cycled as many times as needed to come to a satisfactory synthesis. However, Popper believed Hegel used this method 'in an exaggerated and dangerously misleading form,' that led to dogmatism.[69] What Popper saw as the three main elements of the Hegelian dialectic were:[69]

(1) Dialectic opposition to the rejection of realism by Kant.
(2) The incorporation of dialectic into logic.
(3) The application of dialectic to the whole world of natural and spiritual phenomena.

4.1.6. *Philosophers of science*

Kant in his *Metaphysical Foundations* almost completely isolated metaphysics from physics but the gap between the physics and its foundations worried him considerably in his later years and after about 1796 he worked on bridging the gap.[70] This work, which is contained in what is now called the *Opus Postumum* was never properly completed and not published fully until 1936-8[71] and scholars have spent many years discussing what the gap in his critical philosophy was and how he effected the transition from metaphysics to physics. The nineteenth century saw the emergence of the philosophy of science which many believed would replace metaphysics. The philosophy of science is concerned mainly to determine the correct scientific method. Whereas the metaphysicians of the eighteenth century had mainly not been active in natural philosophy, in the nineteenth century many of the philosophers of science were also scientists such as John Herschel and William Whewell, but others such as, Auguste Comte (1798–1857) and John Stuart Mill (1806–1873) were not. Naturally the philosophy of science depends on physics and other sciences and hence is close to being an intimate part of physics but since in the first half of the nineteenth century it formed a continuation of the metaphysics of the eighteenth century it is briefly discussed here. Some physicists do not believe that the philosophy of science is relevant to physics. Richard Feynman had little use for the philosophy of science about which he is attributed as having said: 'is about as useful to scientists as ornithology is to birds.'

John Herschel and William Whewell were close friends at Cambridge University where both excelled in mathematics; Herschel becoming senior wrangler in 1813 and Whewell second wrangler in 1816. However Herschel soon left Cambridge and after a brief flirtation with law joined his father William to work on astronomy but he also did significant research in mathematics, chemistry, and physics, particularly in optics and polarized light. Whewell on the other hand stayed in Cambridge where he became professor of mineralogy in 1828 and Master of Trinity College in 1841. Bacon was the initial source of many of both Herschel's and Whewell's ideas on the philosophy of science. Unlike Herschel, Whewell did not make any significant advances in science but, though he devalued his own scientific achievements, he was nevertheless an eminent man of science. Even Feynman would have had more respect for philosophers of science who also were significant research scientists like Herschel. While undertaking his scientific research, Herschel was contemplating the nature of the scientific method and, when in 1830 he published *The Preliminary Discourse on the Study of Natural Philosophy*,[72] where the second part of the book was on the principles of physical science and the rules by which it should be pursued, he drew on this research experience often using examples from his work on crystal optics and polarized light.[73] Herschel by the use of a bust of Bacon in the frontispiece to *The Preliminary Discourse* proclaimed his work to be Baconian and emphasised its importance in the text. Bacon did not subscribe to induction by simple enumeration but he thought that by using lists of the attributes of objects that he could arrive at a law of the lowest generality. By doing the same with these laws he hoped to obtain laws of greater generality.

Although Newton feigned no hypotheses, it is very difficult to discover causes by induction alone and the usual scientific route is to form a hypothesis of a cause and then to collect data and by induction obtain a law that can be tested. Herschel's view on the use of hypotheses is somewhat unclear, in places he seems to imply that only hypotheses obtained from induction were valid but elsewhere he seems to agree with Feynman that it does not matter how a hypothesis is framed.[74] However, Herschel certainly recognized the danger of using theories not obtained from induction though some of the examples he uses to illustrate the danger are weak. He discussed the theories of Newton and Joseph-Louis Lagrange (1736–1813) for the velocity of sound in which, starting from different models of sound, they arrive at the same equation for its velocity of propagation. Unfortunately there was no essential difference between the two models: Newton in his *Principia* assumed a simple harmonic sound wave but Lagrange in 1759 used a more general model that did not rely on the sound wave being harmonic. To Lagrange's surprise his value for the velocity of sound was the same as Newton's, however this velocity was some 17% less than the best experimental value that was available at the time. Jean-Baptiste Joseph Fourier (1768–1830) had already used what are now known as Fourier series to decompose any heat wave into a series of harmonic waves of increasing frequency in 1807 and more formally in 1822; Herschel should have known in 1822 that because Newton's result was independent of the frequency of the sound the velocity of sound was independent of the wave shape and hence the two models were essentially the same. So the example Herschel used for his argument was poor. It is even more surprising that Herschel compared Lagrange with Newton since Laplace in 1816 had given the *correct* expression for the velocity of sound using a model that differed in one essential feature from that of Newton and Lagrange. Up until Laplace's work it was assumed that sound vibrations in air took place at constant temperature, but Laplace realised that the air movement in sound was too fast for the temperature to remain constant and took place without any loss of heat or, as would be said after 1860, adiabatically. In the above explanation I have used the term *model* in the sense of physical model, such a use did not occur until the twentieth century, and one of the problems is that Herschel's use of the word hypothesis covered theory, analogy, assumption and model. It is not that Herschel freely allowed hypotheses that were not obtained by induction but that he allowed *hypotheses* followed by mathematical deduction to reveal results that could be tested,[74] since he wrote, commenting on the theories of Newton and Lagrange on sound, 'when a theory will bear the test of such extensive comparison [with observed facts], it matters little how it was originally framed.'[72]

Whewell wrote an anonymous review (as was the custom at that time, though the reviewer was easily identified) of Herschel's *Discourse* that overall was enthusiastic.[75] However he did object to the passage:

> 'Such is the tendency of the human mind to speculation, that on the least idea of an analogy between a few phenomena, it leaps forward ... to a cause or law... Almost all our principal inductions must be regarded as a series of ascents and descents, and of conclusions from a few cases, verified by trial on many,'

Whewell thought that Herschel should have reminded the reader of the warning of Bacon on such a course of action; he also did not like some of the 'scientific Gallicism[s]' that Herschel used.[75] Whewell published his own *Philosophy of the Inductive Sciences* in 1840 that, like Herschel's *Discourse*, was much influenced by Bacon, the first edition even started with a list of aphorisms like Bacon.[76,77] Unfortunately Whewell's *Inductive Sciences* was not well received because it was seen to be too close to Kant and his *a priori* principles. Whewell could not have known of Kant's *Opus Postumum* (see above), but he had remarkably similar ideas on the necessity of bridging the gap between *a priori* principles and science.[78] The fourth aphorism of Whewell stated:[76]

> 'The antithesis of *Sense* and *Ideas* is the foundation of the Philosophy of Science. No knowledge can exist without the union, no philosophy without the separation, of these two elements.'

which neatly summed up his position. This aphorism is similar to the second aphorism of Bacon:[79]

> 'Neither the naked hand nor the understanding left to itself can effect much. It is by instruments and helps that the work is done, which are as much wanted for the understanding as for the hand. And as the instruments of the hand either give motion or guide it, so the instruments of the mind supply either suggestions for the understanding or cautions.'

In 1841 Herschel reviewed anonymously both Whewell's *History of the Inductive Sciences*, published in 1837, and his *Philosophy of the Inductive Sciences*.[80] Although Herschel's review of the *History* is generally favourable, he cannot resist introducing his major point of disagreement with Whewell's philosophy writing about the two schools of philosophy:[80]

> 'that which refers all knowledge to *experience*, reserving to the mind only a high degree of activity and excursiveness in collecting, grouping, and systematizing its suggestions – and that which assumes the presence of *innate conceptions* and truths antecedent to experience ... commanding ... universal assent as soon as understood... The author ... belongs to the latter of these Schools ... and ... appears disposed to press its doctrines ... to a very greater extent ... than we have been at all aware of having been before done, except perhaps in the writings of

> some of the later German metaphysicians. We confess in ourselves a leaning, though we trust not a bigoted one, to the other side.'

In 1844 Whewell wrote a letter in reply to Herschel's review keeping the fiction that the reviewer was anonymous.[81] One of the objections that Herschel had of the *Inductive Sciences* was that according to Whewell, induction from experience can only lead to general truths or theories and those ideas are needed to convey universality. Herschel's argument was that universality was conferred on induction from experience 'by the irresistible impulse of the mind to generalize ad infinitum, when nothing in the nature of limitation or opposition offers itself to the imagination,' but as Whewell pointed out there was no essential difference in the two statements.[81] Whewell's letter continued to suggest there was less difference between the two than Herschel believed.

It was not only scientists that criticized Whewell's *Inductive Sciences*. The first major work of John Stuart Mill, the utilitarian philosopher, was *A System of Logic* on the philosophy of science including moral science. In his introduction to Book III *Of Induction*, Mill acknowledged the recent work by Herschel and Whewell but in obvious reference to Whewell stated that their work was 'of unequal but all of very great merit,' and he reserved the most praise for Comte who he called the 'greatest of all.'[82] What Mill most objected to in Whewell's *Inductive Sciences* were the 'Fundamental Ideas' and 'Ideal Conceptions' which owed much to Kant. For Mill induction from experience was the only scientific method; he even argued that induction was used in geometry. Whewell's reply to Mill took the form of a short book, *Of Induction*, which he published in 1849.[83] Whewell made it clear that his comments on Mill's *System of Logic* were confined to the physical sciences but added: 'I think that by attempting at present to include the Moral sciences in the same formulae with the Physical, we open far more controversies than we close.' In regard to induction Whewell argued that on one hand Mill used a too wide interpretation of induction and in some instances, such as Kepler's discovery that the planets revolve in ellipses about the Sun, denied that the discovery was made by induction. Even in his *Philosophy of Science* published over ten years later in 1860 Whewell was still arguing against Mill's *System of Logic* and his support to what in the twentieth century became known as hypothetico-deduction writing:[84]

> 'The notion that the discovery of the laws and causes of phenomena is a loose hap-hazard sort of guessing, which gives "plausible" explanations, accidental coincidences, casual "harmonies," laws, "in some measure analogous " to the true ones, suppositions "tenable" for a time, appears to me to be a misapprehension of the whole nature of science.'

Mill had no first hand experience of physical science and as Whewell wrote:

> 'has introduced cases which are quite foreign to the idea of science ... A discoverer must know the language of science.'

Feynman would not have been so polite. Whewell was even more dismissive of the philosophy of science advocated by Comte who saw all endeavours to determine the causes of phenomena as belonging to an earlier metaphysical phase in the development of the philosophy. However during the nineteenth century Whewell's philosophy of science was rejected by almost everybody and Mill's *System of Logic* became a standard text in Cambridge during his lifetime.[85] Recently there has been a reappraisal of Whewell and he is now seen as one of the most important philosophers of science.[85]

Comte is important for his reaction to the idealism of Kant and Hegel and the development of positivism during the 1830s and early 1840s. Although Comte was primarily what today we would call a social scientist, he began his positivism with a study of the physical sciences. Positivism declares that our senses are the only means we have of knowing nature and that metaphysics has no place in science. Such a view is compatible with classical physics since everything is observable. Positivists are empiricists. Ernst Mach and Wilhelm Ostwald were both much influenced by positivism which made it difficult for them to accept the unobservability of modern physics.

4.1.7. *The North British Group of physicists and engineers*

The North British were a loose group of physicists and engineers with links to Scotland and the north of England, they wrote no works on metaphysics or philosophy of science, but they were major players in trying to unify the new studies of thermodynamics and electromagnetics with Newtonian natural philosophy through the concept of energy and its conservation and as such necessarily had to deal with metaphysical aspects of their work even if some of them thought metaphysics had no role to play in physics.[86] The group was initiated in the 1840s by James Thomson (1822–1892) an engineer and the older brother of the more famous William Thomson, who was also a member of the group. James Joule was co-opted in 1847. Some of the other members of the group were: Macquorn Rankine (1820–1872), professor of engineering, University of Glasgow; James Clerk Maxwell, professor of natural philosophy, University of Aberdeen and later University of Cambridge; Peter Guthrie Tait, professor of natural philosophy in the University of Edinburgh. Hermann von Helmholtz, the famous German physicist and philosopher was also closely linked with the group.[87]

Maxwell and his friend Tait had the extreme opposite views on metaphysics. Maxwell thought that contemporary physics lacked a metaphysical foundation whereas Tait was continually denouncing metaphysics.[88] The first words in Tait's *Thermodynamics* were:[89]

> 'What is Heat? – The metaphysical arguments on this question ... would merely stupefy the reader and tend to prevent him from easily understanding the actual facts'.

In their physics, the methodology of the North British was hypothetico-deductive. They hypothesized a model of the phenomenon they were investigating for which they then formulated the mathematics and compared the predictions with experiment. If the predictions were not falsified by experiment they assumed the model reasonably described the phenomenon. However there is a potential problem with hypothetico-deduction: the mathematical description of different phenomena can be the same or very similar. Thus the fact that the mathematics of a particular model of a phenomenon gives predictions that match experiments does not necessarily mean that the model describes the physics of the phenomenon. During the twentieth century before the advent of high speed computers the fact that two different physical phenomena are described by the same mathematical equations was used to determine the behaviour of one phenomenon from another one where it was easier to measure the variables. For example during the First World War there was the need to know the stresses under torsion in aircraft engine shafts that had keyways. Geoffrey Ingram Taylor (1893–1975), a man in the great British tradition represented by Kelvin and Rayleigh, thought of a soap bubble analogy to the torsion problem which enabled the stresses to be found from the deflection of a pressurised soap bubble formed on a cut-out in a plate of the same shape as the cross-section of the shaft.[90] Now the soap bubble models the torsion stresses accurately but in no way is the soap bubble a physical model of a shaft under torsion.

Although Rankine's name is not now associated with the development of thermodynamics, he was at the forefront of its early development. In 1842 Rankine deduced the fundamental laws of thermodynamics from a hypothesis of molecular vortices but he did not publish this work until 1849 because of lack of experimental data to verify it.[91] Victor Regnault (1810–1878) was commissioned by the French Ministry of Works in 1843 to obtain data on steam for application in steam engines, he also obtained data on air and carbon dioxide, all this data was published in 1847 and was what Rankine needed to verify his molecular vortices model. The papers Rankine published describing the molecular vortices model are not very detailed; he postulated that:[92]

'each atom of matter consists of a nucleus or central point enveloped by an elastic atmosphere, which is retained in its position by attractive forces, and that the elasticity due to heat arises from the centrifugal force of those atmospheres, revolving or oscillating about their nuclei or central points.'

The quantity of heat was the kinetic energy of the radial vortices. From this model, which was even seen to be very artificial at the time, Rankine deduced a number of thermodynamic equations, including one that was later to lead to Rankine identifying a function which became entropy,[93] which were essentially correct and agreed with Regnault's data.[94] Rankine was able to apply the thermodynamics he had developed to the steam engine in more detail than could be done using the thermodynamics of Thomson or Clausius.[95] Here we have an example of a hypothetical model that has no claim to reality except that its physical mathematics closely models reality. Clearly the hypothetico-deduction does not necessarily give any insight to reality even if the quantitative results obtained agree with the phenomenon. From 1850 Clausius and Thomson began the series of papers that formulate thermodynamics as we know it today and Rankine is known for his engineering not his physics. In a review of the second edition of Tait's *Thermodynamics* Maxwell wrote of Rakine's molecular vortices model that 'though the construction and distribution of his vortices may seem to us as complicated and arbitrary as the Cartesian system, his final deductions are simple, necessary, and consistent with facts'.[96] However his general idea of using vortices to model phenomena became fashionable and was used by Maxwell himself and Thomson.

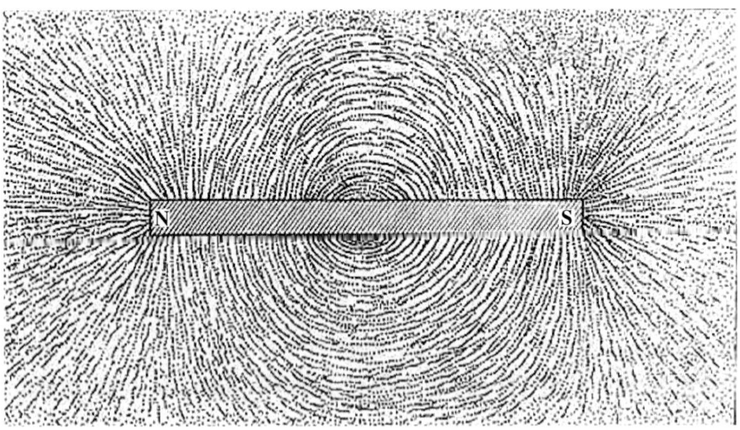

Fig. 4.2. Lines of magnetic force revealed by iron filings (Faraday, 1852).

In 1861 Maxwell published a paper on molecular vortices applied to magnetic phenomena.[97] If iron filings are strewn on a piece of paper supported above a magnet they align themselves in the direction of the lines of magnetic force. A drawing of the result of this simple experiment taken from a paper by Faraday is shown in Fig. 4.2. As Maxwell wrote:[97]

> 'The beautiful illustration ... naturally tends to make us think of the lines of force as something real, ... and we cannot help thinking that in every place where we find these lines of force, some physical state or action must ... to produce the actual phenomena.'

In an earlier paper written just after Maxwell had been elected a fellow of Trinity College, Cambridge, he had considered an analogy of the lines of force to the flow of a fluid 'to assist the imagination, but not to account for the phenomena'.[98] The intention of Maxwell in his 1861 paper was:[97]

> 'to examine magnetic phenomena from a mechanical point of view, and to determine what ... motions of a medium are capable of producing the mechanical phenomena observed.'

Maxwell was seeking a mathematical analogy to magnetism. The lines of magnetic force have a dipolar character and Maxwell noted that a vortex in a medium had a similar dipolar character and proposed that the lines of magnetic force were analogous the molecular vortices. He developed this idea through a series of papers in 1861–1862 which became a focus for the second volume of his *Treatise on Electricity and Magnetism* published in 1873 which was a major scientific work of the nineteenth century. Maxwell did not like the concept of action at a distance which had been developed by the continental physicists and his molecular vortices hypothesis seemed to explain how electromagnetism had an immediate action but it needed a medium or an aether which was acceptable before Michelson and Morley experiments on the propagation of light in 1887. However Maxwell was very much aware that really he had only found an analogy to electromagnetism and not an explanation of its nature.

It is fitting to end this discussion on electromagnetism with the closing remarks of Maxwell to his third paper in his 1861–1862 series:[99]

> 'The facts of electro-magnetism are so complicated and various, that the explanation of any number of them by several different hypotheses must be interesting, not only to physicists, but to all who desire to understand how much evidence the explanation of phenomena lends to the credibility of a theory, or how far we ought to regard a coincidence in the mathematical expression of two sets of phenomena as an indication that these phenomena are of the same kind. We know that partial coincidences of this kind have been discovered and the fact that they are only partial is proved by the divergence of the laws of the two sets of phenomena in other respects. We may chance to find, in the higher parts of physics, instances of more complete coincidence, which may require much investigation to detect their ultimate divergence.'

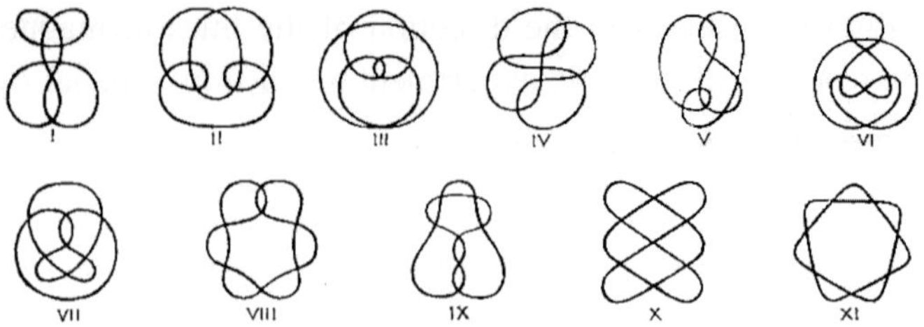

Fig. 4.3. Knots with seven crossings (Tait, 1898).

At the end of his third paper in the series on electromagnetism Maxwell noted he had since seen a paper that Helmholtz had published in 1858 in which Helmholtz had also commented on the 'remarkable analogy between the vortex-motion of fluids and the electro-magnetic action of electric currents'.[100] This paper by Helmholtz, and his witness of the Tait's magnificent display in Edinburgh of smoke rings and their inability to be cut, led Thomson to speculate that atoms might be knotted vortex rings of increasing complexity which in an inviscid medium or aether cannot be created or destroyed giving vortex atoms one of the properties of 'Lucretius's atoms – infinitely perennial specific quality'.[101] Over time Thomson developed the mathematical basis of the vortex atom and many supported his idea. Tait created knot tables classifying them by the number of their crossings and an unknotted vortex ring was thought to be like a hydrogen atom and one with three crossings was thought to represent carbon. Knots with seven crossings are shown in Fig. 4.3. These knotted structures have a worrying similarity to some illustrations of string theory but in postmodern physics we are so far from our experience of reality to be unable to determine whether the images bear any relation to reality. The vortex atom survived until Joseph John Thomson (1856–1940), better known as J.J. Thomson, discovered the electron and proposed the plum pudding model of the atom in 1904 illustrated schematically in Fig. 4.4.

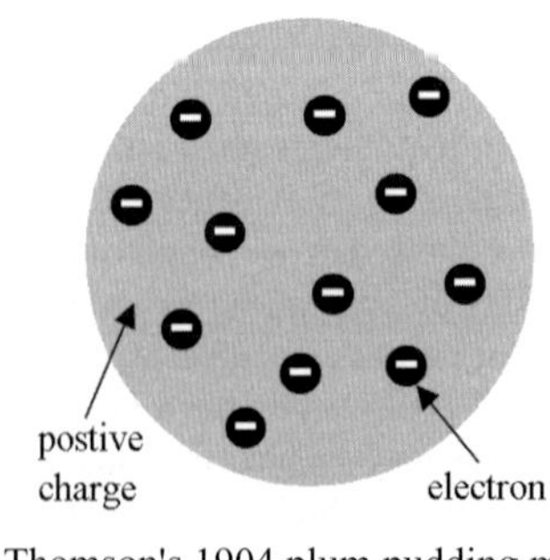

Fig. 4.4. J.J. Thomson's 1904 plum pudding model of an atom.

Maxwell was right, physicists should not ignore metaphysics and physicists should also study the history of science to see the outcome of using unsubstantiated hypotheses like both William and Joseph Thomson's models of the atom.

4.1.8. *Paving the way to the philosophy of modern physics*

From the middle of the nineteenth century developments in physics and epistemology went hand in hand. This linking of physics and philosophy was a necessary beginning for modern physics. One of the first to recognize this necessary development was Helmholtz who initially studied medicine and physiology. Helmholtz was particularly influenced by his teacher Johannes Müller (1801–1858), the foremost German physiologist at that time, and undertook research on the theory of vision. This research led him to reject epistemological realism and adopt the idealistic position that knowledge only comes from the mind. In a lecture *Facts in Perception* given in 1878, Helmholtz outlined his position explaining how his research on vision had led him to adopt Kant's doctrine of *a priori* forms of intuition.[102] However, Helmholtz thought that 'Kant was not critical enough in his critique' and *a priori* forms had to be divided into a general and a narrower form of intuition.[103] The general form of intuition was the true Kantian form whereas the narrower form was exemplified by Euclidian geometry but the general form is non-Euclidian geometry and not limited to three dimensions.[103] Helmholtz had written a series of papers on non-Euclidian geometry in 1868 specifically because of his interest in the perception of space. Bernard Reimann (1826–1866), who formulated non-Euclidian geometry, had lectured on it at the University in Gottingen in 1854 but his famous paper was not published until 1866. While writing his own series of papers on non-Euclidian geometry, Helmholtz became aware of Reimann's work shortly before he published his work in 1868, in which he acknowledged Reimann's prior publication. Surprisingly, Reimann had a similar interest in the perception of space as Helmholtz.[104]

Heinrich Hertz (1857–1894), a student of Helmholtz, inherited his Kantianism but there were crucial differences. In the posthumous publication of Hertz's *Principles of Mechanics* in 1894, Helmholtz wrote a preface in which he highlighted the differences between them; Hertz like Thomson and Maxwell used mechanical explanations derived from *images* whereas Helmholtz used a 'simple representation of physical facts and laws in the most general form, as given in systems of differential equations'.[105] In his introduction Hertz explained exactly what he meant by the Kantian term *Bild* which translates as image or symbol, which represents a phenomenon but is not necessarily an exact mirror of nature but there 'must be a certain conformity between nature and our thought. Experience teaches us that ... [this] requirement can be satisfied'.[106] Hertz explained that the *images* are not necessarily unique and different *images* of the same phenomenon may be permissible and the more appropriate *image* is that which contains the 'essential characteristics' and is the simpler. 'One image may be more suitable for one purpose, another for another; only by gradually testing many images can we finally succeed in obtaining the most appropriate.' Hertz also saw that the correctness of any image can only be judged using our experience which may change over time. The vortex models, described in the previous section that Rankine, Maxwell, and Thomson used, were Hertzian *images*. I am inclined to agree with Helmholtz that a 'simple representation of physical facts' as described by a mathematical model are safer. In modern physics, we will probably never know whether the images used are even remotely similar to reality though the theory may predict the behaviour in question. Boltzmann also subscribed to *Bild* theory but he also saw that science should give an explanation of phenomena as well as being able to predict behaviour.[107] However by the end of the nineteenth century, Boltzmann was aware that mechanical explanations might not be possible.

Ernst Mach though he wrote: 'there is no such thing as the philosophy of Mach,'[108] prepared the way for development of philosophy in modern physics but his own role was somewhat equivocal. '[Kant's] critical idealism was ... the starting point of all my critical thought,' wrote Mach:[108]

> 'but it was impossible for me to retain my allegiance to it.... [I] began to gravitate again towards the views of Berkeley.... [However Berkeley] regards the "elements" as conditioned by an unknown cause external to them (God); accordingly Kant, in order to appear as a sober realist, invents the "thing-in-itself"... [After a while] the superfluity of the role played by "the thing-in-itself" abruptly dawned upon [me, but I believed that] 'the world is our sensation.'

A thing is only known through the sensations it produces. Bertrand Russell coined the term *neutral monism* to describe Mach's philosophy defining it as:[109]

> 'the metaphysical view that the mental and the physical are two ways of organizing or describing the same elements, which are themselves "neutral," that is, neither physical nor mental. This view denies that the mental and the physical are two fundamentally different things. Rather, neutral monism claims the universe consists of only one kind of stuff, in the form of neutral elements that are in themselves neither mental nor physical.'

The importance that Mach stressed on knowledge through sensations posed problems for modern physics since he denied the reality of unobservables such as atoms and molecules. As a young man Mach he accepted the atom but by 1872 in his *History and Root of the Principle of Conservation of Energy* in a discussion of the molecular theory of heat he wrote:[110]

> 'Perhaps one might think that rules for phenomena, which cannot be perceived in the phenomena themselves, can be discovered by means of molecular theory. Only that is not so. In a complete theory, to all details of the phenomenon details of the hypothesis must correspond, and all rules for these hypothetical things must also be directly transferable to the phenomenon. But then molecules are merely a valueless image.'

In his preface to the otherwise unaltered second edition of 1909, Mach stated that 'In this pamphlet ... I made the first attempt to give adequate exposition of my epistemological standpoint.'[110] It was the thing-in-itself to which Mach objected.

Mach's denial of the atom led to an intense debate with Boltzmann. The Austrian physicist Stefan Meyer (1872–1949), an assistant to Boltzmann showed Mach in 1903 the scintillations produced by the impact of alpha particles on the atoms on a zinc sulphide screen and reported that Mach said 'Now I believe in the existence of the atom.'[111] However Mach still denied the existence of atoms until he died. Today almost nobody doubts the existence of atoms. How could such a distinguished physicist even at the end of the nineteenth century deny their existence? The existence of entities such as atoms can be considered at two different levels: the scientific and the epistemological.[112] This was the view of Mach who wrote:[108]

> 'that physics, despite its considerable development, nevertheless constitutes but a portion of a larger collective body of knowledge, and that it is unable, with its limited intellectual implements, created for limited and special purposes, to exhaust all the subject-matter in question.'

A possible explanation for Mach's behaviour is that epistemologically he was led to deny there was any certain knowledge of existence atoms whereas at a scientific level it was 'a good working hypothesis' as Mach, referring to the study of chemistry, explained to the Czech chemist Bohuslav Brauner (1855–1935) just before the beginning of the first World War.[113,114]

What should the assessment of Mach be? Einstein's view was clear writing in Mach's obituary in 1916:[115]

> 'How does it happen that a properly endowed natural scientist comes to concern himself with epistemology? Is there no more valuable work in his specialty? I hear many of my colleagues saying, and I sense it from many more, that they feel this way. I cannot share this sentiment.'

However he thought Mach's view that 'physics is experience, arranged in economical order'[116] neglected 'the free, constructive element in the formation of concepts ... [and] believed that scientific theories arise through discovery and not through invention.'[115] Other physicists have rejected a too rigid approach as Tyndall discussed in a discourse delivered to the British Association in 1870:[117]

> 'Imagination ... left to ramble uncontrolled leads us astray ... but properly controlled by experience and reflection, becomes ... the instrument of discovery in Science,'

4.2. Moral Philosophy

Moral philosophy has been influenced by the methods of natural philosophy and also one of the original aims of natural philosophy was to demonstrate that the ultimate cause of everything was God. Newton himself wrote in query 31 to his second edition of the *Opticks* that:[118]

> 'If Natural Philosophy ... shall at length be perfected, the Bounds of Moral Philosophy will be also enlarged. For so far as we can know by Natural Philosophy what is the first Cause, what Power he has over us, and what Benefits we receive from him, so far our Duty towards him, as well as that towards one another, will appear to us by the Light of Nature.'

Locke believed that 'morality is capable of demonstration, as well as mathematicks,' whereas he suspected that 'natural Philosophy is not capable of being made a Science'.[17] However, Locke in this context is using Science in the Aristotlean sense of *Scientia* that is 'certain and universal knowledge', which cannot be conceived in any other way.[17] Geometric knowledge such as the 'proposition ... that a triangle has three angles equal to two right ones' is such 'certain and universal knowledge' and Locke believed moral knowledge could be known with similar certainty.[17]

The Enlightenment of the eighteenth century grew out of the advances in science and philosophy that had taken place in the previous century and aimed at replacing bigotry and religious enthusiasm with reason and science. Scotland was one of the leading centres of the Enlightenment and the Scottish moral philosophers were an integral part of it clearly seeing the link between moral and natural philosophy. George Turnbull (1698–1748), a regent at Marischal College, Aberdeen was the first to advocate the use of the Newtonian method in moral philosophy. He developed his ideas in the *Principles of Moral Philosophy* published in 1740, which has the quotation from Newton's query 31 in *Opticks*, given above, on its title page; he also stated them in his graduation theses in 1721.[119] In his preface, Turnbull stated:[120]

> 'it was by this important comprehensive hint [that] I was led long ago to apply myself to the study in the same way as to that of the human body, or any other part of Natural Philosophy: that is, to try whether due enquiry into moral nature would not soon enable us to account for moral, as the best Philosophers teach us to explain natural phenomena.'

Turnbull formulated the methods to be used in moral philosophy from the precepts developed by Newton. For example, he rephrased part of Cotes' preface to the second edition of the *Principia* writing:[120]

> 'hypotheses [are not] to be any further admitted, than as questions, about the truth of which it is worth while to enquire; but ... we may proceed in the double method of analysis and synthesis.'

At the beginning of his introduction Turnbull very clearly stated his manifesto that he summarized thus:[120]

> Tho' natural philosophy be distinguished from moral philosophy, yet every enquiry into any part of nature is an enquiry into fact: an enquiry concerning the human mind, its powers, and affections, their operations, is as much an enquiry into fact, as an enquiry concerning the texture of the human body.'

No other Scottish moral philosopher of the Enlightenment was quite as explicit in their application of the Newtonian method but most of them subscribed to it.

Francis Hutcheson (1694–1746) was born in Ulster to a Scottish family. After studying at the University of Glasgow he returned to Ireland where he opened a private academy in Dublin and spent ten years there. It was during this period in Ireland, before he was appointed to the Chair of Moral Philosophy at Glasgow in 1729, that he published anonymously *An Inquiry into the Original of Our Ideas of Beauty and Virtue*.[121] Hutcheson found moral philosophy based on egoistical principles abhorrent and was the first to propose:[121]

> 'that action is best, which accomplishes the greatest happiness for the greatest numbers; and that, worst, which in like manner, occasions misery,'

though he thought that 'the moral importance, or dignity of persons [in the sense of the quantity of public good produced by him] may compensate numbers.' To satisfy his criterion of the best action Hutcheson needed to maximize the quantity of happiness which he attempted to do in the second of the treatises by expressing happiness mathematically in a series of propositions: The moment of public good of a person (M), in the now obsolete sense of measure of good, is given in terms of his benevolence or virtue (B) and his ability to do good (A) by

$$M = B \times A. \tag{4.1}$$

The consequences of our actions on us, or our self interest (I) can be either be good (positive) or bad (negative). Consequently our benevolence is either decreased or increased by our self interest so that

$$B = \frac{M - I}{A}, \tag{4.2}$$

but, if the consequence of our action on us was 'not intended, altho casually, or naturally rebounding to us from the action,' $I = 0$. Since nobody can act above their ability, the maximum moment of public good they can achieve is their ability or $M = A$, and the maximum benevolence of a person is, by definition $B = 1$. Hutcheson then remarks:

> 'And this may shew us the only Foundation for the boasting of the *Stoicks* "That a creature suppos'd innocent, by pursuing *Virtue* with his utmost power, may in *Virtue* equal the Gods." For in their Case if *A* or the *Ability* be Infinite, unless *M* or the *Good* to be produc'd in the whole, be so too, the *Virtue* is not absolutely perfect; and the Quotient [giving the *Virtue* or *Benevolence*] can never surmount Unity.'

Unfortunately Hutcheson did not attempt to quantify benevolence or ability apart from defining that perfect benevolence was unity. Hutcheson dropped the mathematics from his discussion of benevolence from the 1738 fourth and later editions because equations seem to have outraged some readers, as they still seem to do today.[122] However Hutcheson kept his discussions of benevolence to which he believed we are well deposed and likened:[121]

> 'this *universal Benevolence* toward all Men, we may compare to that Principle of *Gravitation*, which perhaps extends to all Bodys in the *Universe*; but, like the *Love* of *Benevolence*, *increases* as the Distance is diminish'd, and is *strongest* when Bodys come to touch each other. Now this *increase* of *Attraction* upon nearer Approach is as necessary to the *Frame* of the *Universe*, as that there should be any *Attraction* at all: For a *general Attraction*, equal in all Distances, would by the Contrariety of such multitudes of equal Forces, put an end to all Regularity of Motion, and perhaps stop it altogether.'

Thomas Reid (1710–1796), another of the Scottish moral philosophers, was prompted by reading Hutcheson's attempt at quantifying benevolence to publish an essay in 1748, which pointed out that sensations and mental states such as pain or pleasure and moral values, though they exist to different degrees, are incapable of being quantified and therefore cannot be expressed mathematically.[123]

David Hume (1711–1776) was an important philosopher whose influence went far beyond the Scottish Enlightenment and who still has a journal *Hume*

Studies devoted to his writings. A child prodigy, Hume probably entered Edinburgh University when he was only ten years old and, along with Latin and Greek, he would have studied scholastic philosophy and maybe moral philosophy; he certainly was taught mechanics, astronomy and optics from the works of Newton and his Scottish commentators: John Keill and David Gregory; Boyle would also have been used particularly for the experimental part of the course.[124] Hume left university in 1725 without graduating and he quickly found that he had 'an insurmountable aversion to every thing but the pursuits of philosophy.'[125] After an abortive attempt by his family to get Hume to study law, a bout of ill health, and a foray into business, he left for France where, staying first briefly in Rheims, but later at La Flèche, in Anjou, he wrote his great work on moral philosophy *A Treatise on Human Nature*[126] before he was twenty-five years old.[124] Hume knew that his *Treatise* would not be well received because of its controversial sceptical nature but instead of provoking the vehement attacks he expected 'it fell dead-born from the press, without reaching such distinction, as even to excite a murmur among the zealots.'[125] This lack of response was galling to Hume, who was never to obtain a university position and had to support himself by his philosophical and historical writing, but as he wrote: 'being naturally of a cheerful and sanguine temper, I very soon recovered the blow.'[125] Hume attempted to make his treatise more saleable by shortening it considerably by cutting out most of the reasons for his conclusions. This work was first published 1748 as *Philosophical Essays Concerning Human Understanding*, the title was changed to *An Enquiry Concerning Human Understanding* in 1758, which initially became better known than his *Treatise*. Kant described reading Hume's *Enquiry* as breaking his 'dogmatic slumber' and giving 'his researches in the field of speculative philosophy a quite different direction'; [127] Kant probably was unaware of Hume's *Treatise*.[128]

Hume, as he stated in the introduction to his *Treatise*, saw the science of man as the premier science on which all other sciences 'even Mathematics, Natural Philosophy, and Natural Religion, are in some measure dependent.'[126] However, he saw that the Newtonian experimental method applied equally to moral philosophy and aspired to be the Newton of moral philosophy. To emphasize this connection Hume subtitled his *Treatise*: *An Attempt to Introduce the Experimental Method of Reasoning into Moral Subjects*, and in the introduction stated that 'the only solid foundation we can give to [the science of man] itself must be laid on experience and observation.'[129] In a discussion of the association of ideas Hume almost paraphrased part of Definition 8 in Newton's *Principia* writing:[129]

> 'Here is a kind of ATTRACTION, which in the mental world will be found to have as extraordinary effects as in the natural, and to shew itself in as many and as various forms. Its effects are every where conspicuous; but as to its causes, they are mostly unknown, and must be resolv'd into *original* qualities of human nature, which I pretend not to explain. Nothing is more requisite for a true philosopher, than to restrain the intemperate desire of searching into causes, and having establish'd any doctrine upon a sufficient number of experiments, rest contented with that, when he sees a farther examination would lead him into obscure and uncertain speculations.'

There is also a reference to the Newtonian method in Hume's *An Enquiry Concerning the Principles of Morals*, published in 1751, where in the first section, *Of the General Principles of Morals*, he wrote:[130]

> 'Men are now cured of their passion for hypotheses and systems in natural philosophy, and will hearken to no arguments but those which are derived from experience. It is full time they should attempt a like reformation in all moral disquisitions; and reject every system of ethics, however subtle or ingenious, which is not founded on fact and observation.'

Hume's view of the certainty of knowledge was similar to that of Locke though he thought that only algebra and arithmetic 'preserve a perfect exactness and certainty ... [and] geometry can scarce be esteem'd a perfect and infallible science' but he used the word science in the modern sense of a body of knowledge that could claim a high probability of proof not absolute certainty.[129] Hume listed seven kinds of the objects of human reason or enquiry: 'resemblance, identity, relations of time and place, proportion in quantity or number, degrees in any quality, contrariety, and causation,' which fell into two categories: those that depend only on ideas and those that also depend upon perception.[129] Only those relationships that only depend upon ideas, such as algebra, can be known with certainty. The others, relations of time and place and especially causation, cannot be known with absolute certainty but only with a degree of probability. Hume even denied:[129]

> '[The] general maxim in philosophy, that *whatever begins to exist, must have a cause of existence*. This is commonly taken for granted in all reasonings, without any proof given or demanded. 'Tis supos'd to be founded on intuition … But … we shall discover in it no mark of any such intuitive certainty.'

In developing the 'rules by which to judge of causes and effects', Hume was influenced by Newton's four 'rules for the study of natural philosophy' given in the *Principia* both in realizing that rules were necessary for moral as well as natural philosophy and he made use of some of Newton's rules. The first of Newton's rules states the principle of simplicity that only the minimum number of causes should be used to explain a phenomenon, the second and third rules are concerned with the principle of uniformity. The second rule states that causes of effects of the same kind should be the same and the third rule that, qualities not dependent on their degree and belong to all bodies on which experiments can be made, should be universal to all such bodies. The fourth rule, added in the third edition of the *Principia*, states that propositions deduced from experiments by induction should be considered probable unless other phenomena show them either liable to exceptions or more certain. Hume had eight rules by which to judge causes and effects, none of which expressed the principle of simplicity but Rules 4–6 express different versions of the principle of uniformity; Rule 4 stated that: 'The same cause always produces the same effect, and the same effect never arises but from the same cause.' Rule 5, which is really a corollary of Rule 4 stated that where different objects produce the same effect, it must be some quality that is common to them, and Rule 6 follows from Rule 5: 'The difference in the effects of two resembling objects must proceed from that particular, in which they differ.' Rule 7: 'when any object encreases or diminishes with encrease or diminution of its cause, 'tis to be regarded as a compounded effect, deriv'd from the union of several different effects, which arise from the union of the several different parts of the cause,' is qualified by the requirement that conclusions should not be drawn from only a few experiments, which is a warning against induction and as such the reverse of the intent of Newton's fourth rule which was to emphasize the that hypotheses should not negate induction without further experimental evidence. Nowhere in the section on rules in the treatise does Hume specifically refer to Newton, but in *An Enquiry Concerning the Principles of Morals*, published in 1751, he refers to *Newton's chief rule*, in the section *Of Justice*, describing it as 'any principle [that] has been found to have a great force and energy in one instance ... [should be ascribed to] like energy in all similar instances.'[130] Here it would seem that Hume was referring to Newton's second rule though it has also been suggested that the reference is to Newton's fourth rule.[131]

Hume was a sceptic, he rejected the principle of induction and he took the empirical philosophy of Locke and Berkeley to its logical conclusion but, as Russell stated, without this principle all science is impossible including the science of man.[128] Thomas Reid was the antithesis of Hume, he believed that:[132]

> 'the wisdom of *philosophy* [was] set in opposition to the *common sense* of mankind. The first pretends to demonstrate a priori that there can be no such thing as a material world, ... and can be nothing else, but sensations in the mind, or images of those sensations in the memory and imagination'

Not surprisingly Reid's first major work, published while he was a Regent at King's College, Aberdeen, was entitled: *An Inquiry into the Human Mind on the Principles of Common Sense.*[132] Reid used the Newtonian method in his refutation of scepticism. Starting from the same Newtonian philosophy as Hume, Reid stated in his introduction that there is 'but one way to knowledge of nature's works, the way of observation and experiment' but he immediately deviated from Hume in praising induction:[132]

> 'By our constitution, we have a strong propensity to trace particular facts and observations to general rules ... This procedure of understanding is familiar to every human creature in the common affairs of life, and it is the only one by which any real discovery in philosophy can be made...[It is] the same method by which Newton discovered the laws of gravitation and the properties of light. His *regulae philosophandi* are maxims of common sense, and are practiced everyday in common life.'

Two of Newton's Rules concerned cause and effect, played down the importance of seeking causes, famously so in the case of gravitation. According to Reid, in ascribing causes we are making hypotheses which Newton condemned.[133] There is a problem with the whole concept of cause. In 1912 Bertrand Russell was to write:[134]

> 'the word "cause" is so inextricably bound up with misleading associations as to make its complete extrusion from the philosophical vocabulary desirable.'

Towards the end of his life Reid, when he was Professor of Moral Philosophy at Glasgow University, was to write similar thoughts from a Christian perspective:[135]

> 'That active power is an attribute, which cannot exist but in some being possessed by that power, I take for granted as a self-evident truth. Whether there can be active power in a subject which has no thought, no understanding, no will, is not so evident The ambiguity of the words, power, cause, agent, ... tends to perplex this question.'

For Reid, as for Newton, there was only the first cause, God, and he went on:[135]

> 'With regard to the operations of nature, it is sufficient for us to know, that, whatever the agents may be, whatever the manner of their operation, or the extent of their power, they depend upon the first cause, and are under his control; and this is all that we know; beyond that we are left in darkness.'

Reid recognized only observable phenomena and empirically verifiable laws and rejected inquiry into all causes other than those that governed methodology for science. It was to be one in which, as Adam Ferguson put it, 'the Regulae

Philosophica [of Newton's *Principia*] . . . must be sustained'[136] and was therefore one of the first proto positivists as Newton himself probably was.

Adam Ferguson (1723–1816) a loyal Whig, born in Perthshire on the boundary between the Scottish highlands and lowlands, was as a young man appointed deputy chaplain to the Scottish Black Watch Regiment during the Jacobite rising of 1745 becoming its principal chaplain in 1747.[137] After nine years with the Black Watch, Ferguson left active service with both the army and the church. In 1759 Ferguson with the help of Hume became Professor of Natural Philosophy at the University of Edinburgh. However natural philosophy was not his bent and in 1764 Ferguson obtained the Chair of Pneumatics (philosophy of the mind) and Moral Philosophy at the University of Edinburgh which was much more to his liking. In 1769 Ferguson published *The Institutes of Moral Philosophy*,[138] this book based on his lecture notes, written almost in point style, was very successful in Britain, Germany and Russia.[137] In retirement Ferguson expanded his *Institutes of Moral Philosophy* into a more readable two volume work *The Principles of Moral and Political Science*.[138] Ferguson took pains to point out the similarities between physical and moral laws; he saw that the laws of nature had two aspects:[138]

> '[T]he term law ... sometimes signifies the fact as it is regularly and invariably established in the course of nature ... however [it] has a farther signification and means some rule of choice, or expression of what is good.'

Laws of the first sort are usually associated with physical laws of the material system, while those of the second sort are usually associated with moral laws of the intellectual system. Ferguson also tried to argue less satisfactorily that some laws of the second sort applied equally well to the material system, such as 'expressions of ... elegance, beauty or utility, in natural subjects,' and that some laws of the first sort also applied to the intellectual system, such as 'facts relating to the mind which are fixed and invariable,' and paraphrased the first two rules of Newton for the study of natural philosophy:[136]

(1) 'We are not to assign, as the cause of any appearance, what is not itself known as a fact in nature.'
(2) 'We are not to deduce effects from causes, which, though real, are unfit to produce the effect.'

Ferguson suggested these rules 'as principles of moral discernment.'

It has been generally asserted that Adam Smith (1723–1790), the moral philosopher and political economist, who wrote the famous economics treatise,

An Inquiry into the Nature and Causes of the Wealth of Nations, was greatly influenced by Newtonianism but direct evidence is very scant. Smith made it all the harder for those who have wished to discover influences on his thought by destroying all his manuscripts before his death, and left behind only his *Wealth of Nations*, *The Theory of Moral Sentiments*, and *Essays on Philosophical Subjects*, which included *The Principles Which Lead and Direct Philosophical Enquiries, Illustrated by the History of Astronomy.*[139] It is this last work, usually referred to, rather misleadingly, as *History of Astronomy* since it is only the last section of four that is concerned with astronomy, which contains some clues to the influence of Newtonianism on his method of reasoning.[140] The first three sections of the *History of Astronomy* are: *Of the Effect of Unexpectedness, Or of Surprise*; *Of Wonder, or of the Effects of Novelty*; and *Of the Origin of Philosophy*, where philosophy is defined as 'the science of the connecting principles of nature' and included natural philosophy. Smith saw that in the Newtonian system the 'parts are all more strictly connected together, than those of any other philosophical hypothesis.' Smith asserted that principles should be formed 'as if they were the real chains which Nature makes use of to bind together her several operations.' Thus there is evidence that Smith used the Newtonian method in his writings.[140]

4.3. Politics

Politics in the seventeenth and eighteenth centuries were inseparable from the new science, philosophy and religion. Modern British political parties had their beginnings during the Exclusion Crisis of 1678–1680 when attempts were made to exclude the Duke of York, later King James II, who was a Catholic, from the succession to the throne. In 1679 the Earl of Shaftesbury (1621–1683) and his supporters, who became known as Whigs, attempted to pass the Exclusion Bill, which was foiled by Charles II dissolving Parliament. It was not until 1680 that a new Parliament met where the Exclusion Bill, though passed by the House of Commons, was rejected by the House of Lords. The supporters of Charles II who opposed the Exclusion Bill became known as Tories. James II came to the throne on the death of Charles II in 1685 and quickly created unrest by trying to achieve religious toleration for Roman Catholics. When a son was born to James II in 1688 bringing about the possibility of another Catholic successor, the crisis came to head. Some Tories combined with the Whigs to invite William of Orange and Mary his wife jointly to the throne. Mary, though the daughter of James II born when he was the Duke of York, was a Protestant. The Glorious Revolution of 1688 finally established the victory of Parliament over the Monarchy, James fled to France, and William and Mary governed with the support of both Tories and

Whigs.[141] Mary, who died in 1694, had no children and, when her sister Anne's surviving child died in 1700, William saw that it was necessary to prevent a Catholic successor to the throne. Robert Hartley (1661–1724) a country Whig, though with close associations with the Tories, saw the Act of Settlement through Parliament in 1701, which limited the succession to Protestants who had not married Catholics.

The Whigs and Tories were not political parties in the sense of modern day parties. In the seventeenth century Whigs generally believed in a limited monarchy and were usually Low Church while some Tories believed in an absolute monarchy and were usually High Church. However there were Country and Court Parties within both Whigs and Tories. Country members thought the influence of the court was corrupting and were more radical than the Court Parties who put more emphasis on authority.

Newton himself was a Whig but was not prominent in politics despite being elected for Parliament for Cambridge University in 1689 and again in 1701; he never spoke in the House of Commons. Newton benefited from being a Whig and became Warden of the Mint in 1696 through the influence of Charles Montagu (1661–1715), later the Earl of Halifax, a prominent Whig politician. At Halifax's urging Newton stood again for Parliament in 1705. Just before the election Newton was knighted by Queen Anne in Cambridge, the first scientist to be so honoured. Halifax was involved in obtaining the knighthood and at least part of the reason for his support was to advance Newton's chances of election since Halifax needed more Whig members in the new Parliament. Despite the knighthood, Newton came last in the election and two Tories were returned for Cambridge University: the sitting member Arthur Annesley and Dixie Windsor.[142]

Prior to Newton, natural philosophy was often associated with enthusiasm, which in the seventeenth century meant not eagerness but supernatural inspiration, and the new experimental philosophy only cast off this association with difficulty. Thus when the Tory Aphra Behn drafted her play *The Emperor of the Moon* in 1684[143] she satirized Dr Baliardo the enthusiastic astronomer, who cavorted about the stage with a very phallic long telescope, aiming to both amuse the audience and pour scorn on the Whigs who showed symptoms of enthusiasm in their support of natural philosophy and in their behaviour during the Exclusion Crisis.[144] The Royal Society had to work hard to remove the taint of enthusiasm from natural philosophy.

Newton's *Principia* had a political impact since it showed that all the bodies of the solar system were equally subject to a single law of nature *gravity* and could be used to argue against an absolute monarchy since all were subject to the same laws.[145] Locke wrote the *Two Treatises of Government*,[146] published anonymously in 1690 primarily to refute Robert Filmer's (1588-1653) *Patriarcha* written in support of the divine right of kings in 1632 but was only published posthumously in 1680.[147] Although Locke, Newton himself, and many other Newtonians were Whigs, some Newtonian natural philosophers, and those interested in it, were Tories. In Edinburgh David Gregory and the physician Archibald Pitcairne (1652–1713), who became an equally convinced Newtonian, were both Tories as were other members of their circle, such as John Keill.[148] Two political newspapers, which commenced publication after the Licensing Act lapsed in 1695, the Tory *Post Boy* and the Whig *Postman* both carried roughly equal advertisements for Newtonian books and lectures, which suggests that there was almost equal support for Newtonianism among Tories though probably the majority of the practitioners were Whig.[149] Even in 1956 it was still asserted by Charles Percy Snow (1905–1980), better known as C.P. Snow, that 'among physicists and biologists ... one finds the Left in strength.'[150]

Although it is frequently asserted that during the eighteenth century Newtonian natural philosophy was used to support political views, specific examples are few and, even in such examples, the reference to Newtonian natural philosophy is often of a very general nature. Desaguliers, a Newtonian Whig, took the advantage of the accession of George II in 1727 to publish in 1728 an ingratiating and rather bad first poem *The Newtonian System of the World, The Best Model of Government*, noting in the dedication to Lord Ilay:[151]

> 'I have consider'd Government as a Phaenomenon, and look'd upon that Form of it most perfect, which did most nearly resemble the Natural Government of our System.... The *limited Monarchy* whereby our Liberties, Rights, and Privileges are so well secured to us, as to make us happier than all the Nations round about us, seems to be a lively Image of our System; ... [and] makes us sensible that *ATTRACTION* is now universal in the Political, as the Philosophical World.'

James Thomson (1700-1748), a Whig Scottish poet, who wrote the words of *Rule Britannia*[152] incorporated Newtonian natural philosophy into the poem *The Seasons*, ostensively a devotional pastoral but also supportive of the Whig ascendancy and the Hanoverians.[153] It has also been suggested that the Leibniz/Clarke exchange on natural philosophy and God, described in Chapter 2, also had overtly political motives.[154]

One of the effects of the success of Newtonian natural philosophy in the eighteenth century was that systematic knowledge of any kind sought to be considered a science and the moral philosophers of the Scottish Enlightenment also promoted politics as a science. Hume was one of the first to describe politics as a science in the modern sense of the word in his *Philosophical Essays concerning Human Understanding*, Hume wrote that:[155]

> 'Moral reasonings are either concerning particular or general facts.... The sciences, which treat of general facts, are politics, natural philosophy, physic, chemistry, &c. where the qualities, causes and effects of a whole species of objects are enquired into.'

Hume's politics are difficult to classify. He saw himself as a moderate Whig but to radical Whigs he was a Tory who defended the royal prerogative. In his *History of Great Britain* Hume tried to be impartial and in Volume 1 published in 1754, which covered the reigns of James I and Charles I, he was assailed by almost everybody Whig and Tory alike for having 'shed generous tear for the fate of Charles I and the Earl of Strafford.'[124] Ferguson, like Hume, saw that:[136]

> 'the affairs of society require the light of science, as well as the direction of a virtuous conduct; insomuch that the recluse by investigating the laws of nature, which relate to the concerns of men, is no less employed for his country ... than those who are most occupied in discharging the functions of state.'

4.3.1. *The Declaration of Independence and the Constitution of the United States*

Many of the influential men in the American Colonies at the time of the Declaration of Independence in 1776 were educated either in Scotland by Newtonians of the Scottish Enlightenment or in the Colonies by Americans who had been so educated themselves. James Wilson (1742–1798), a signatory of both the Declaration of Independence, and the Constitution, was born in Fifeshire and after graduating from the University of St Andrews spent two years studying with Francis Hutcheson, David Hume and Adam Smith at Edinburgh and Glasgow Universities before immigrating to Philadelphia in 1766. Thomas Jefferson (1743–1826), who wrote the first draft of the Declaration of Independence, was taught natural philosophy, metaphysics, and mathematics at the College of William and Mary in Williamsburg, Virginia by William Small (1734–1775) a Scotsman, who received his MA from Marischal College, Aberdeen in 1755, and taught at William and Mary from 1758 to 1764 when he left for London. James Madison (1751–1836), who drafted the Constitution, was educated at the College of New Jersey, which later became Princeton University, whose President was John Witherspoon (1723–1794) a Scottish Presbyterian minister who was educated at the Universities of St Andrews and Edinburgh and influenced by the philosophy of Thomas Reid.

In America after a brief period as an instructor at the College of Philadelphia, Wilson became apprenticed to a prominent Philadelphia lawyer, John Dickinson, and subsequently became a successful lawyer himself. By 1774 Wilson had also become involved in Pennsylvanian politics. His politics were somewhat conservative and at first he refused to support independence openly though, of course, Wilson did finally sign the Declaration of Independence. One of Wilson's ambitions was to be compared to William Blackstone the great English jurist of the eighteenth century. When in 1790 Wilson was appointed the first professor of law at the College of Philadelphia, he immediately set about writing an extensive series of lectures. The ostensive audience for these lectures was the handful of students taking the course, and he did deliver many of his lectures to them, but Wilson planned to publish the lectures as a definitive treatise on American Law.[156] Wilson, and also probably Benjamin Franklin, believed that an important feature of the Constitution was the provision for amendments to it and based their arguments on natural philosophy.[157] In Chapter 10 of his law lectures, Wilson wrote:[158]

> 'There is a *vis inertiae* in publick bodies as well as in matter; and, if left to their natural propensities, they will not be moved without a proportioned propelling cause. If, indeed, the prevailing opinion should be, that the society had not the regular power of altering, on every proper opportunity, its political institutions.'

The first clause in this quotation echoes Definition 3 of the *Principia*:[27, 159]

> 'Inherent force of matter is the power of resisting by which every body, so far as it is able, perseveres in its state either of resting or of moving uniformly straight forward,'

to which Newton added in explanation:

> '[The] inherent force may also be called by the very significant name of *vis inertiae.*'

The last clause of the first sentence, of the Wilson quote, also echoes Newton's second law of motion. Wilson did not reference these clauses to Newton's *Principia* because he assumed that his most important readers would be well aware from whence they came and identification of the reference was not a developed practice. Franklin like Wilson was a signatory to both the Declaration of Independence and the Constitution, However, Franklin did not wholly approve of all the provisions of the Constitution and in his final speech to the Constitutional Convention he said:

> 'There are several parts of this constitution which I do not at present approve, but I am not sure I shall never approve them: For having lived long, I have experienced many instances of

> being obliged by better information, or fuller consideration, to change opinions even on important subjects, which I once thought right, but found to be otherwise.'

This expression of his scientific empiricism makes it likely that he highly valued the provision in Article 5 of the Constitution for amendments.[157] Further on in the speech Franklin voiced the opinion that: 'There is no form of Government but what may be a blessing to the people if well administered.' This opinion of government was common during the Enlightenment and captured in Alexander Pope's *Essay on Man*:

> For forms of government let fools contest;
> Whate'er is best administered is best.

Franklin was one of the leading natural philosophers of the eighteenth century and, like Faraday in the nineteenth century, he was self educated and lacked a thorough mathematical knowledge and may not have well appreciated all the mathematical sections of the *Principia*. Jefferson on the other hand had an excellent university education especially in natural philosophy and mathematics from the hands of William Small. Although Jefferson never was active in scientific research like Franklin, he retained a life long interest in natural philosophy and mathematics. When Jefferson retired from politics in 1812 he was happy in exchanging newspapers 'for Tacitus and Thucydides, [and] for Newton and Euclid'.[160] If anyone was going to introduce Newtonian notions into an official document, Jefferson was most likely to be the one. Bernard Cohen believed that in Jefferson's preamble to the Declaration of Independence: [157]

> 'When in the Course of human events, it becomes necessary for one people to dissolve the political bands which have connected them with another, and to assume among the powers of the earth, the separate and equal station to which the Laws of Nature and of Nature's God entitle them, a decent respect to the opinions of mankind requires that they should declare the causes which impel them to the separation.'

the words "Laws of Nature" were an indirect reference to Newton's Laws of Motion. Certainly John Keill had used this phrase in the title of Chapter XI in his *Introduction to Natural Philosophy*, both in the original Latin edition published in 1702 as well as in his 1720 first English edition; Jefferson had both versions in his library.[157] However, the phrase *Laws of Nature* was used much earlier in a more general sense. Francis Bacon quoted James I as saying that 'Kings ruled by their laws as God did by the laws of nature.'[161] Thus, though it is perfectly possible that Jefferson was thinking about Newton's Laws of Motion when he wrote those words, it is equally possible that the phrase, a part of the English language, was being used in a more general sense as God's laws especially as he follows that phrase by 'and of Nature's God'. Just as nowadays when we refer to a quantum jump or leap, with the mistaken meaning of large rather than small, we are certainly not thinking about quantum mechanics since the misused term entered the English language as a general expression in the late 1940s.

There is no discernable reference to natural philosophy in the United States Constitution but that did not stop Woodrow Wilson stating in a series of lectures given in 1908, when he was President of Princeton University, that the Constitution 'was a sort of unconscious copy of the Newtonian theory of the universe.'[157] At first sight then this section is somewhat negative; natural philosophy had little demonstrable effect on the Declaration of Independence or the Constitution but in the context of the intent of this book to demonstrate the wide influence of physics on our lives, the result is surely positive since so many have believed that it had an influence including Woodrow Wilson, who became President of the USA and was the only statesman of vision at the end of the First World War.

4.3.2. *Marxism and dialectic materialism*

Karl Marx (1818–1883), who called himself a dialectical materialist, was a disciple of Hegel and but he was dismissive of Hegel's philosophy of nature.[162] Marx first came into contact with Hegel when, as an undergraduate, he attended his lectures at the Frederick William University now the Humboldt University of Berlin. Hegel was a supporter of the absolutist Prussian State. Marx at first admired Hegel but his politics were completely opposite. However, Marx saw that Hegel's dialectical theory could be used to support revolutionary as well as reactionary politics. Thus dialectical materialism became the theoretical basis of *Scientific Marxism* and turned Marxism into a dogmatic system that prevented its scientific development.[69]

In the main Marx concentrated on politics and economics leaving most of the philosophy to Friedrich Engels (1820–1895). Engels main writing on dialectics and the development of the natural sciences is contained in the *Dialectics of Nature*, written between 1872 and 1889 but never finished and it was not published until 1925. An English translation of *Dialectics of Nature* with a preface by the socialist geneticist, John Haldane (1892–1964), was published in 1939.[163] Haldane in his preface to the *Dialectics of Nature* wrote:[163]

> 'Marx and Engels were not content to analyse the changes in society. In dialectics they saw the science of the general laws of change, not only in society and in human thought, but in the external world which is mirrored by human thought.'

That is dialectics could be applied to all aspects of human endeavour and fitting to become Marxist dogma. All science had to be forced into this mould whether it fitted or not. Haldane, an apologist for Marxism, believed:

> 'that the sections of the book which deal with biology are the most immediately valuable to scientists to-day. This may of course be because as a biologist I can detect subtleties of Engels' thought which I have missed in the physical sections.'

As someone whose training has been in the physical sciences, I find Engels forced physics into his own mould. Here, of course only the sections of Marx and Engels dialectics that pertain to physics and astronomy are discussed.

Like Hegel, Engels was very dismissive of Newton[164] and at one point in the *Dialectics of Nature* he wrote that compared to Leibniz 'the inductive ass Newton appears as a plagiarist and corrupter.'[163] Hegel, Marx, and Engels believed that Newton's law of gravitation added nothing to Kepler's laws. Possibly one of the difficulties that those believing in the method of dialectics had with gravity is that gravity is always attractive in contrast with electromagnetic forces which can be attractive or repulsive without opposites dialectics falters. In the *Dialectics of Nature* Engels showed a lack of understanding of physics. For example he confused the tangential inertia of a planet orbiting a sun with tangential force writing:

> 'The ordinary school textbook of astronomy follows Newton in explaining the ellipse described as the result of the joint action of two forces, the attraction of the central body and a tangential force driving the planet along the normal to the direction of this attraction.'

Then a few lines on he tried to introduce the opposites of attraction and repulsion, writing:

> 'If the one element of planetary motion, the directly central one, is represented by gravitation, the attraction between the planet and the central body, then the other tangential element appears as a relic, in a derivative or altered form, of the original repulsion of the individual particles of the gaseous sphere.[165] Then the life process of a solar system presents itself as an interplay of attraction and repulsion, in which attraction gradually more and more gets the upper hand owing to repulsion being radiated into space in the form of heat and thus more and more becoming lost to the system.'

Newton was not the only physicist to be disparaged and insulted in the *Dialectics of Nature*. The widely regarded *Treatise on Natural Philosophy* of Thomson and Tait is described as the book where 'thinking is forbidden, only calculation is permitted' and, because Engels did not understand the difference between momentum (mv) and kinetic energy ($mv^2/2$), he went on to write:

'The two contradictory measures of motion are put side by side in this very glaring form. Not so much as the slightest attempt is made to explain the contradiction, or even to disguise it.'

I am surprised the Engels did not seize on this apparent contradiction as an example of dialectics.

There are numerous other examples where Engels showed his lack of understanding of physics and his overbearing desired to force all philosophy into the dialectic mould. Engels was convinced of his own brilliance and anyone who criticized him or Marxism felt his wroth; a powerless nineteenth century Donald Trump. Eugen Dühring (1833–1921), German socialist idealist, enraged Engels by his criticism of Marxism, and he attacked him in *Herr Eugen Dühring's Revolution in Science*, published in German in 1878; an English translation of *Anti-Dühring*, as it is more usually called, was published in 1907.[166] Dühring was just as quarrelsome as Engels but he knew more about physics, though he never had any formal training, and started his career in law, which he had to abandon when he began to go blind.[167] After abandoning law, Dühring returned to the Frederick William University in Berlin and completed a doctorate *On Time, Space, Causality, and the Infinitesimal Analysis of Logic*, in 1861 and this thesis formed the focus for Engels attack. The desire to pour scorn on Dühring's writings outweighed all other considerations as can be judged from Engels' preface to *Anti-Dühring* where he wrote:

'Just as I was completing this preface I received a publishers' notice, composed by Herr Dühring, of a new "authoritative" work of Herr Dühring's: Neue Grundgesetze zur rationellen Physik und Chemie. Conscious as I am of the inadequacy of my knowledge of physics and chemistry, I nevertheless believe that I know my Herr Dühring, and therefore, without having seen the work itself, think that I am entitled to say in advance that the laws of physics and chemistry put forward in it will be worthy to take their place, by their erroneousness or platitudinousness, among the laws of economics, world schematism, etc., which were discovered earlier by Herr Dühring and are examined in this book of mine; and also that the rhigometer, or instrument constructed by Herr Dühring for measuring extremely low temperatures, will serve as a measure not of temperatures either high or low, but simply and solely of the ignorant arrogance of Herr Dühring.'

I finish this section with the closing remarks in Popper's paper on dialectics:[69]

'[D]ialectic is used by Marxists, following the example of Engels' Anti-Duehring, mainly to defend the Marxist system against criticism, and not to criticise it or to develop it.... Thanks to dialectic the anti-dogmatic tendencies [of Marx] have disappeared, and Marxism has established itself as a dogmatism which is elastic enough, by using its dialectic method, to evade any further attack – in short, as a reinforced dogmatism....

The whole development of dialectic should be a warning against speculative philosophy. It should remind us that philosophy must not be made a basis for any sort of scientific

> system and that philosophers should be much more modest in their claims. For their task, which they can fulfil quite usefully, is the study of the methods of science.'

4.4. Concluding Remarks

Newton made physics a subject founded on experiment and mathematical principles independent of philosophy, but it immediately started to influence metaphysics and epistemology, because these branches of philosophy are closely related to physics, and a need for philosophy to be reconciled with physics was soon apparent. Experience and experiment became paramount for philosophy as well as physics and philosophy became largely mechanical until the end of the nineteenth century. The influence has been both ways though resisted by some. Idealists in particular have disputed the reality of matter.

Berkeley was a metaphysical idealist. Kant was an epistemological idealist, who marked a turning point in philosophy and introduced *a priori* concepts, which were pure intuitions of the mind independent of experience, and he exerted a strong influence on the nineteenth century both in philosophy and physics. During the nineteenth century practicing physicists were as important in metaphysics and epistemology as non scientists and the philosophy of science developed out of metaphysics. The problems with the hypothetico-deductive method where the mathematical theory may give good predictions, even if the model has no relation to nature, highlight the need for metaphysics. However not all physicists during the nineteenth century were convinced that metaphysics had much relevance to physics and some still do not today. In the second half of the nineteenth century epistemology became important for the transition to modern physics but, if too rigidly applied, did not foster that transition. Serendipity does occur and speculation should not be discouraged by a rigid epistemology. Karl Popper wrote: [168]

> 'I am inclined to think that scientific discovery is impossible without faith in ideas which are of purely speculative kind, and sometimes even quite hazy: a faith which is completely unwarranted from the point of view of science, and which, to that extent is metaphysical.'

Newtonian physics had a general influence on moral and political philosophy but it was less explicit. Moral philosophy was more influenced by the scientific method than any particular developments in physics. However, Newton's *System of the World* had a large effect on the political philosophy of the eighteenth century since, if all bodies in the universe were subjected to the same laws, should not all men be similarly subject? This analogy was a strong argument against an absolute monarchy. Many of the more important founding fathers of the United States were well versed in natural philosophy and it almost certainly had a general effect on the Declaration of Independence and the Constitution but the text of these documents does not contain any obvious direct references. Physics did not influence the dialectical materialism of Marxism rather Marxists dogmatically attempted to force physics to conform to their dialectics.

I have, of course, not exhausted the professions that have been influenced by Newtonian physics. Law, which has only been mentioned in passing, was certainly influenced by physics especially by its method in the seventeenth century. Matthew Hale (1609–1676), the greatest common lawyer of the seventeenth century, like Bacon was very interested in natural philosophy and saw the laws of nature as Divine commands and used its method in the analysis of law.

Philosophy and politics had a rival influence to physics after the publication of Darwin's *Origin of the Species* in 1859. However, though the life sciences certainly became very much more important after this date, physics, the queen of the sciences, has continued to exert a major influence.

4.5. Notes

[1] Cohen's Guide, Newton (1726).

[2] Bacon was also a lawyer, becoming finally Lord Chancellor, and applied natural philosophy to law. Shapiro (1969).

[3] Bacon (1620).

[4] Anstey (2005).

[5] Malcolm (2004).

[6] An anonymous, untitled, and undated manuscript: British Library Harley MS 6796.

[7] Raylor (2001) and others believe the manuscript is not Hobbes' but the work of Robert Payne (1596–1651), chaplain to the Cavendish family and based only on some of the ideas of Hobbes, who was tutor and companion to William Cavendish.

[8] Herbert (1987).

[9] Light actually travels fastest in a vacuum and is slower in all other transparent media. The problem was that Snell's law, first established empirically in modern times by Willebrord Snel (1580–1626) in 1621, is a simple geometric relationship which can be demonstrated using a variety of assumptions that are not necessarily true. Pierre de Fermat (1601–1665) derived Snell's law in 1657 from his principle that light travels along the shortest path.

[10] Hobbes (1651).

[11] Descartes (1985).

[12] Descartes (1637, 1644).

[13] Spinoza (1663).

[14] Spinoza had been corresponding with Oldenburg since they met in Rijsburg near Leiden in 1661. Oldenburg became a secretary of the Royal Society in 1662.

[15] In the Preface to *Principles of Cartesian Philosophy* written by Ludwig Meyer it is also stated that 'Our author [has] put into geometric form, for a pupil whom he was teaching Descartes' philosophy, the entire second part of the Principles and part of the third'.

[16] Spinoza (1677). The first two parts of Ethics were completed by 1665 and the fifth and final part by 1675. Ethics was first published in Latin in 1677 shortly after Spinoza's death.
[17] Locke (1690a).
[18] Le Clerc (1706).
[19] Milton (2004).
[20] Walmsley (2004).
[21] Rogers (1978).
[22] An abridged French version of the Essay was published in the journal *Bibliothèque Universelle et Historique* in 1688. Milton (2004).
[23] Milton (2004). In 1681 Shaftesbury was charged with treason. A Whig-nominated grand jury found insufficient grounds for his prosecution and he was released. In 1682 two Tories were appointed as sheriffs in London and Shaftesbury, again in danger of arrest, fled to the Amsterdam were he died in 1683.
[24] Axtell (1965).
[25] It has been suggested that Locke's account of space and time was influenced by Newton and the *Principia* in the second and subsequent editions of the Essay. Ducheyne (2009). I do not find that this suggestion is substantiated. Ducheyne seems to confuse infinite and pure space with absolute space.
[26] Locke (1699).
[27] Newton (1726).
[28] McGuire (1968).
[29] Newton did not use a specific word for unessential qualities but other writers of the period used the term accidental in the less frequent modern sense of not essential.
[30] La Monadologie, written in French in 1714 was first published after his death in a German translation in 1721. The original French version was not published until 1840.
[31] Russell (1900).
[32] Leibniz (1684, 1686a, 1693).
[33] Leibniz (1703).
[34] Garber (2009).
[35] Garber (1985).
[36] Leibniz (1697).
[37] Leibniz (1689).
[38] Leibniz (1686b).
[39] Leibniz did not distinguish between mass and weight but that does not affect the argument.
[40] Iltis (1971). Leibniz did not use the term *vis viva* until he wrote the unpublished Essay de Dynamique in 1691 his earlier term was moving force.
[41] In elastic impact between two bodies some energy is left trapped within the bodies and the click of billiard balls tells us that some energy is converted into sound. In both elastic and inelastic impact both the momentum and the total energy (not kinetic energy) is conserved.
[42] Iltis (1971).
[43] Leibniz (1696).
[44] Berkeley (1710).
[45] Berkeley (1713).
[46] Berkeley (1721).
[47] Downing (2005).
[48] The main purpose of Siris was to extol the medicinal virtues of tar water, which he had discovered while in America from 1728 to 1731. What metaphysics there is in Siris is used to reinforce the benefits of tar water.
[49] Substantially revised in 1787. Kant (1787).

[50] Kant (1786).
[51] Kant (1755).
[52] Massimi (2011).
[53] Kant (1783).
[54] Kant (1787).
[55] Friedman (2012).
[56] With developments in chemistry, Kant saw in the 1790s that chemistry might be capable of being demonstrated a true science but died before he could finish establishing its credentials.
[57] The first scientific evidence for the conservation of mass were Lavoisier's experiments on the oxidation of mercury in 1789 three years after the publication of the *Metaphysical Foundations*.
[58] Stern (1994).
[59] Stern (2009).
[60] Pinkard (2008).
[61] Westphal (2008).
[62] Posch (2004). Not until the middle of the nineteenth century did Newton become recognized as the founder of physics in Germany.
[63] Posch (2004). Hegel's Lectures on Natural Philosophy, Berlin, 1823/24.
[64] Posch (2004).
[65] Hegel (1805–1830).
[66] Hegel (1817).
[67] The outgasing tail of a comet when it is near the sun causes a non-gravitational force and non-gravitational accelerations, which makes it impossible to determine whether away from the Sun the trajectory is hyperbolic. Another factor is the giant gas Jupiter and Saturn can perturb the comets trajectory. If true gravitational hyperbolic comets do exist they are very rare. Kresak (1992).
[68] Wilson (1985).
[69] Popper (1940).
[70] Hall (2009).
[71] Selections from his drafts were first published in 1882-4.
[72] Herschel (1830).
[73] Good (1987).
[74] Cobb (2012).
[75] Whewell (1831).
[76] Whewell (1840).
[77] In subsequent editions the aphorisms were moved to the back of the book. Bacon's Novum Organum was just a list of aphorisms. On reflection Whewell saw it was not politic to start his philosophy with a series of unsubstantiated aphorisms.
[78] Ducheyne (2011).
[79] Bacon (1620). The original Latin is similarly awkward in its expression, but the intent seems similar to that of Whewell.
[80] Herschel (1841).
[81] Whewell (1847).
[82] Mill (1846).
[83] Whewell (1849).
[84] Whewell (1860).
[85] Wettersten (1994).
[86] Smith (2004).
[87] Helmholtz's relationship with the North British Group was based on his physics.
[88] Cat (2012).
[89] Tait (1868).

[90] Taylor was unaware that Ludwig Prandtl (1875–1953), the famous early German aerodynamicist, had already suggested the analogy in 1903.
[91] Rankine (1851).
[92] Rankine (1851).These atoms or molecules were not atoms or molecules in the modern sense but just small particles. Whether the nucleus was real, or a portion of the atmosphere in a highly condensed state, or merely a centre of condensation of the atmosphere was left indefinite.
[93] Entropy was originally defined by Clausius in 1865.
[94] Rankine could not determine *a priori* many of the constants in his equations and had to use Regnault's data to determine their value.
[95] Hutchison (1981).
[96] Maxwell (1878).
[97] Maxwell (1861a).
[98] Maxwell (1855, 1856).
[99] Maxwell (1861b)
[100] Helmholtz (1867). A translation by Tait of this paper was published in 1867.
[101] Thomson (1867).
[102] Helmholtz (1878).
[103] D'Agostino (2004).
[104] Richards (1977).
[105] Hertz (1894).Helmholtz's Preface.
[106] Hertz (1894).
[107] De Regt (1999).
[108] Mach (1886).
[109] Russell (1914).
[110] Mach (1872).
[111] Blackmore (1972).
[112] Buchdahl (1951).
[113] Bachtold (2010).
[114] Brauner (1924).
[115] Quoted in Howard (2014).
[116] Mach (1885).1882 address to the Imperial Academy of Sciences, Vienna.
[117] Quotation taken by Tyndall from an address by the President, Benjamin Brodie (1783–1862) to the Royal Society in 1859 and used as the keynote for his discourse on the Scientific Use of the Imagination delivered to the British Association in 1870.
[118] Newton (1718).
[119] Turnbull (1723, 1726, 1740).
[120] Turnbull (1740).
[121] Hutcheson (1725).
[122] Wood (2003).
[123] Reid (1748).
[124] Robertson (2004).
[125] Hume (1777).
[126] Hume (1739, 1740).
[127] Introduction to the Prolegomena, Kant (1783).
[128] Russell (1961).
[129] Hume (1739).
[130] Hume (1751).
[131] Force (1987).
[132] Reid (1764).

[133] Ducheyne (2006).
[134] Russell (1912).
[135] Reid (1788).
[136] Ferguson (1792).
[137] Oz-Salzberger (2004).
[138] Ferguson (1769).
[139] Winch (2004).
[140] Montes (2008).
[141] James made an attempt to recover his throne when he landed in Ireland in 1689. In 1690 his forces were defeated at the Battle of the Boyne. The defeat of the Catholic Jacobites is still celebrated each year by the Northern Irish.
[142] Cambridge University had two seats in the House of Commons.
[143] Probably for Charles II, who died before it could be performed. It was not produced until 1687.
[144] Coppola (2008).
[145] Campion (2009).
[146] Locke (1690b).
[147] The Licensing Act was still in force in 1632 and surprisingly Filmer was denied permission to publish by the King; The reason for the denial is unknown.
[148] Guerrini (1986).
[149] Wigelsworth (2008).
[150] Snow (1956).
[151] Desaguliers (1728).
[152] Originally a song in the masque *Alfred*.
[153] Connell (2009).
[154] Shapin (1981).
[155] Hume (1748).
[156] Conrad (2004).
[157] Cohen (1997).
[158] Wilson (2007).
[159] Cohen (1997) suggested that Wilson was echoing Definition 4 in the first clause, either interpretation is possible but I think that Definition 3 is more likely.
[160] Capon (1959).
[161] Milton (2012).
[162] Stanley (1997).
[163] Engels (1940).
[164] Marx wrote very little on physics but he may not have been so dismissive of Newton. In his notes on his copy of the Leibniz-Clarke he wrote the comments: 'Well said, old Isaak Newton', and 'Bravo, old Newton!' Weston (2012).
[165] Here Engels is referring to the nebular hypothesis of Laplace and Herschel.
[166] Engels (2001).
[167] Dühring's Rule is still used to simplify the calculation of the boiling point of a substance as a function of the pressure. Wisniak (2001).
[168] Popper (1959a).

Chapter 5

Imaginative Arts

The nineteenth century saw the preliminary move in education to dethrone classical literature and to anoint the sciences in its place. The Mason College, Birmingham, which was to become the University of Birmingham, was endowed by Sir Josiah Mason (1795–1881), who started life as a street hawker, and opened in 1880. One of the conditions he placed on the college was that it should not make provision for 'mere literary instruction and education.' Thomas Huxley, that tireless promoter of science, gave the opening address in which he enlarged on this provision in the endowment stating 'that neither the discipline nor the subject-matter of classical education is of such direct value to the student of physical science as to justify the expenditure of valuable time upon either;' and also stating that 'for the purpose of attaining real culture, an exclusively scientific education is at least as effectual as an exclusively literary education.'[1] Huxley noted that such a view was an anathema to most educated men and mentioned Matthew Arnold (1822–1888), the poet and literary critic, who was son of the famous headmaster of Rugby School, to whom the meaning of culture was 'to know the best that has been thought and said in the world,' and that the best was classical literature. Arnold responded to this challenge in his Rede Lecture delivered in Cambridge in 1882 arguing that he who has only studied physical science 'will be unsatisfied, or at any rate incomplete, and even more incomplete than the student of humane letters.' This debate about culture continued during the twentieth century. The most well known debate was in the late 1950s and early 60s between C.P. Snow and Frank Raymond Leavis (1895–1978), better known as F.R. Leavis. Some of their biographical details need to be explained in order to understand the debate properly.[2]

Both men had Cambridge backgrounds. Snow first took an external BSc at London University followed by an MSc, then obtaining a studentship he went to Cambridge where he obtained a PhD in 1930 for research on the infra-red spectra of simple diatomic molecules. On the basis of his research, he became a fellow of Christ's College where up to 1935 he published some twenty papers mainly on spectroscopic studies. In 1932 Snow began publishing novels and popular scientific articles. Snow stopped scientific research in about 1934 because he was disappointed with his achievements and in 1937 he became editor of popular science journal *Discovery*. During World War II Snow, like many other Cambridge Fellows, helped with the war effort, working initially with the Royal Society to deploy scientists for the Ministry of Labour and in 1942 he became the Director of Technical Personnel. After the war, Snow entered the Civil Service as a commissioner in charge of recruiting scientists for the government and he also continued writing novels. In 1964 he was made a life peer by Harold Wilson and entered his government as Parliamentary Secretary of the Ministry of Technology. Snow's forte was in scientific administration, he himself was not satisfied with his scientific research and his novels have not stood the test of time.

Leavis went to Emmanuel College, Cambridge, where he read history and English literature. However his studies were interrupted by World War I during which he served with the Friends' Ambulance Unit; he was a conscientious objector but an agnostic not a Quaker. After the war he returned to Emmanuel College and graduated in 1921. Leavis continued at Emmanuel College as a postgraduate and obtained his PhD in 1924. Although Leavis remained in Cambridge until 1962 he had a rather hard time. In 1927 he was appointed a probationary university lecturer in English literature but that appointment was terminated in 1931. He then moved to Downing College and in 1932 was appointed director of studies in English but Leavis was not elected a fellow until 1936 when he was also appointed a part time university lecturer. He was not to be appointed a full time lecturer until 1947 when he was 52 years old. Leavis rose to reader in 1959 and retired from Cambridge in 1962 but he did not retire from academic life and was active in Oxford and America. Leavis never became a professor at Cambridge but it must be remembered that even in the 1960s there were few professorships available and many very eminent academics retired as readers. Certainly during the second half of the 1950s, when I was at Cambridge, Leavis had a god-like stature and was either worshiped or feared by the English literature students. Leavis left his stamp upon English literature with his journal *Scrutiny*, which he founded in 1932 and remained editor of it until the final issue in 1953. However Leavis was a curmudgeonly figure with few intimates in Cambridge. Snow seems to have benefited from the Cambridge and Oxford collegiate system where it is difficult not to be affected by all aspects of culture by meeting and making friends with people from completely different backgrounds to one's own both socially and academically – unfortunately Leavis seems to have overcome this difficulty.

The debate between Huxley and Arnold was a relatively gentlemanly affair that was necessary at the time to assert the emergence of science as an important part of our culture. The war-like debate between Snow and Leavis was anything but gentlemanly. Snow fired the first shots with his article the *Two Cultures* in the *New Statesman* in 1956.[3] In this article Snow noted how science and the arts had separated so far that neither party interacted to any extent. He also reflected on the esteem that the arts still had compared with science and also the fact that literary figures were apt to hold very right wing views.[4] Music, which is closest to science since its scales and harmony are governed by physics and mathematics, was according to Snow the art which scientists understood the most. Snow's article was characterized by sweeping generalizations for which he gave very little justification. Richard Feynman, who had very original thoughts on physics and life in general, saw the two cultures rather differently to Snow seeing them as: [5]

> 'those two cultures separate people who have and people who have not had ... [the] experience of understanding mathematics well enough to appreciate nature once.'

And I think he might have been right with respect to physics, whose richness is obscured by a lack of understanding of mathematics:[5]

> 'To those who do not know mathematics it is difficult to get across a real feeling as to the beauty, the deepest beauty, of nature.'

Snow's initial salvo did not draw much return fire. However, in 1959 Snow expanded his article in a Rede Lecture at Cambridge. The title for this lecture was *The Two Cultures and the Scientific Revolution*.[6] In this lecture Snow broadened his topic to discuss the need for a scientific revolution, which to my mind did not fit well with a discussion of the two cultures. What Snow said about the necessary scientific revolution needed saying but really it was really more about the necessity for a technological than a scientific revolution. In the 1950s Britain was still recovering from the effects of the Second World War and we were in the depths of the Cold War. More funding and effort was definitely needed if Britain was not to slip into insignificance but that had little to do with the two cultures issue. The lecture was not well written and showed Snow's civil service background.

It took some three years before Leavis returned fire when he marked his retirement in 1962 by delivering his rejoinder *Two Cultures*? *The Significance of C. P. Snow* as a Richmond Lecture in Downing College.[7] His choice of title was very apt because Leavis' lecture was not a discussion of the two cultures but a vitriolic attack on Snow. There is hence very little to say about Leavis' lecture, some of his remarks, stripped of the vitriol, are true but some of the faults ascribed to Snow by Leavis apply equally well to him. For example Snow certainly pontificated in his Rede Lecture, as Leavis remarked at the start of his Raymond Lecture:[8]

> 'If confidence in oneself as a master-mind, qualified by capacity, insight and knowledge to pronounce authoritatively on the frightening problems of our civilisation, is genius, then there can be no doubt about Sir Charles Snow's.'

However Leavis was just as bad, as he almost admitted, writing:[8]

> 'we who ran Scrutiny for twenty years ... were, and knew we were, Cambridge – the essential Cambridge.'

As Aldous Huxley (1894–1963), grandson of Thomas, wrote: the choice between Snow or Leavis was between 'the bland scientism of the *Two Cultures* ... [and the] violent and ill-mannered, one-track, moralistic literarism of the Richmond Lecture'.[9]

When Snow gave his Rede Lecture he noted that the gap between the creative arts and science had widened since his time at Cambridge saying:[10] 'the separation between the scientists and non-scientists is much less bridgeable among the young than it was even thirty years ago'. Indirectly I can partly vouch for Snow's statement myself. When I was writing *Mechanics of Pre-Industrial Technology* with the archaeologist Johan Kamminga, during the late 1980s, I found that archaeologists in the 1920s and 30s had often used mechanics in their papers to explain the technology they were describing whereas those writing since the Second World War usually only made unsubstantiated statements about the mechanical ability of artefacts to perform their task. This observation is for archaeologists, whose discipline lies partly in the sciences as well as the humanities; they are forced to obtain some knowledge of modern physics, such as radiocarbon dating, but largely ignore classical physics. The division of culture is a modern phenomenon. A physicist Norman Robert Campbell (1880–1949) who in 1921 published a small book *What is Science?*[11] 'written in the hope of encouraging the study of science in the classes of the Workers' Educational Association,' and which warned of the danger that 'the W.E.A., which stands for new ideas will become the last stronghold of the reactionary doctrine that science and culture are antagonistic.' If he had been alive during the Two Cultures Debate, he too would have protested that there was only one culture.

It is certain that classical physics had an influence on the creative arts but how much? There are influences that can be demonstrated with certainty through explicit references to physics but other possible influences are not so direct. Some influences are only analogous to developments in physics and these must be carefully sifted to separate the true analogy from the false. It is possible to get carried away with hidden meanings, for example it has been suggested that Shakespeare's Hamlet was really about astronomy and represented the conflict between the Hamlet who personified the Copernican and Claudius who personified the Ptolemaic systems[12] with Rosencranz and Guildenstein personifying the composite system of Tycho Brahe.[13] For the last assertion there is the coincidence that the Latinate names of Rosencranz and Guildenstein appear on the frontispiece to Brahe's *Astronomie Instauratae Mechanica* but Brahe was Danish like Hamlet as were Rosencranz and Guildenstein who had common Danish names. Similar false influences need to be sifted out but not all of the false analogies are as obvious as that quoted.

There is another general assertion that not only did the new science of the sixteenth and seventeenth centuries influence literature but that also literature influenced the new science.[14] This symmetry of influence is very difficult to prove but it is quite possible that it did exist in the time before knowledge started to become compartmentalized. The arts can also aid the dissemination of science.

5.1. The Cultural Revolution of the Seventeenth Century

Although, naturally there is no definite date when the new science started to emerge, the publication of Copernicus' *De Revolutionibus Orbium Coelestium* in 1543 was a watershed event. Because Copernicus was careful to state that his theory was only a hypothesis, it was not initially banned by the Catholic Church. *De Revolutionibus* was freely disseminated until 1616 when the Church suspended it until it was corrected to remove the implication that the Earth really orbited the Sun in contradiction to Holy Scripture. After nine sentences had been modified or removed, *De Revolutionibus* was taken off the index in 1620. The Copernican system was slow to be accepted, may be because of Copernicus' caution. Even Thomas Digges (1546–1595), who was the first to write about the Copernican system in English and whose writing show him to be a committed Copernican, never explicitly stated that the Earth and the planets revolved around the Sun.[15] In the end it was the telescope that showed the truth of heliocentricity. The earliest known telescope was constructed in the Netherlands in 1608. Galileo heard of the invention in 1609 and constructed one himself and published his first astronomical discoveries with it in the *Sidereus Nuncius* (The Sidereal

Messenger) in 1610.[16] With his telescope Galileo was able to resolve more than ten times the number of stars that are visible to the eye alone and the nebulous appearance of the Milky Way was resolved into individual stars. He observed the mountainous appearance of the Moon and estimated that the highest mountains were more than four Italian miles high.[17] He also observed the four major satellites of Jupiter and determined that the satellite furthest from Jupiter whose modern name is *Calisto* had a period of about half a month. A year after Galileo published *Sidereus Nuncius*, he gave the periods of all four major satellites.[18] In addition Galileo, from his observation of the phases of Venus similar to those of the Moon, gave in a letter to Kepler written in December 1610 the first direct evidence of a planet orbiting the Sun. Galileo did not include these observations in *Sidereus Nuncius* because they also occurred after its publication and Kepler published his letter in his preface to *Dioptrics* in 1611. It was *Sidereus Nuncius* more than *De Revolutionibus* that promoted the spread of heliocentricity. The appearance and decay of two new stars, which we now call supernovae, which were visible in 1752 and 1604, showed that the firmament of the stars was not immutable and the appearance of the comet in 1607 that Halley saw again in 1682, and which was named after him, also caused considerable general attention.

Prior to Kepler's *Astronomia Nova* published in 1609, in which he showed that the orbit of Mars was elliptical, everybody, Copernicus included, believed that that all the celestial bodies moved in circles or epicycles which are compounds of circles. *Astronomia Nova* was not very widely accepted; both Galileo and Descartes ignored it. In 1619 Kepler generalized his observations on Mars to include all the planets in the *Harmonices Mundi Epitome* or *Harmony of the World.* However, it was through the *Epitome Astronomia Copernicanae* published in three parts from 1618 to 1621 that Kepler's work became more widely known.

5.1.1. *Literature*

Thus at the end of the Elizabethan Age the Earth was still firmly the centre of the universe in the eyes of most despite Copernicus. John Davies (1569–1626), one of a group of poets who Samuel Johnson despisingly called the *metaphysical poets*, wrote the poem *Orchestra or a Poem of Dancing* in 1595 which was based upon the story in Homer's *Odysse*y, where Penelope Ulysses' queen is waiting for his return and being pestered by suitors. Antinous, one of the prominent suitors, asks her to dance and is refused. A debate occurs in which Antinous claims that the whole universe is organized in a dance[19] and

Only the earth doth stand forever still:
Her rocks remove not, nor her mountains meet,
Although some wits enriched with learning's skill
Say heaven stands firm and that the earth doth fleet,
And swiftly turneth underneath their feet;
Yet, though the earth is ever steadfast seen,
On her broad breast hath dancing ever been

Davies was well educated having gone to Winchester College and then Queen's College, Oxford, but while the poem shows that he knew about the Copernican System, he still believed in the Ptolemaic one. In *Troilus and Cressida*, written probably in 1602, and also loosely based on the *Odyssey*, Shakespeare (1564–1616), who was not so well educated, had Ulysses declaim:

The heavens themselves, the planets and this centre
Observe degree, priority and place,
Insisture, course, proportion, season, form,
Office and custom, in all line of order;
And therefore is the glorious planet Sol
In noble eminence enthroned and sphered
Amidst the other; whose med'cinable eye
Corrects the ill aspects of planets evil,
And posts, like the commandment of a king,
Sans check to good and bad: but when the planets
In evil mixture to disorder wander,
What plagues and what portents! what mutiny!

before the Grecian army camp in Act I scene III, linking the order of the army to that of the universe. The order of the universe is the traditional Greek one where the Earth is the centre surrounded by the celestial crystal spheres of the Moon, Mercury, Venus, the planet Sol (Sun), Mars, Jupiter, Saturn, and the firmament containing the fixed stars. Pietro d'Abano (1257–1316) added two spheres that carried no heavenly bodies: the crystalline sphere that accounted for the precession in the equinoxes and the *Primum Mobile* that moved all the other spheres.

John Donne (1572–1631) was the most significant of the metaphysical poets and is now recognized as one of the great English writers, though his poetry went through a long period where it was not appreciated.[20] The circle and the sphere were seen as perfect forms by Aristotle and they remained powerful literary and

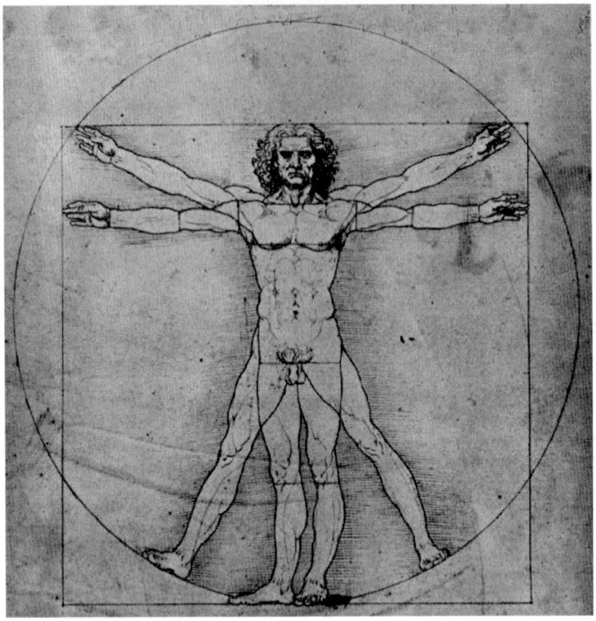

Fig. 5.1. Proportions of the human figure (Leonardo da Vinci, 1492).

visual images. Leonardo da Vinci (1452–1519) saw the circle in the proportions of the human figure and in his famous drawing shown in Fig. 5.1 placed man within a perfect circle. Donne used the circle image widely in his poetry but not more beautifully than many other metaphysical poets. What Marjorie Nicolson (1894–1981), an American literary scholar, saw that set Donne apart from the other metaphysical poets was the 'profusion and intricacy of his circle-images, and because he was the first ... to realise that ... the circle was breaking.'[21] Donne went to Hart Hall at Oxford University in 1584 at the tender age of twelve, probably to avoid subscribing to the Thirty-Nine Articles since he came from a Catholic family.[22] According to his first biographer Izaak Walton (1593–1683), best known as the author of *The Compleat Angler*, Donne was a distinguished student but he left Oxford without taking a degree. In 1587 he fought in the Protestant Army of Prince Maurice in the Low Countries and even if he was not a committed Protestant he was already at least losing his Catholic belief. Donne was appointed secretary to the Keeper of the Great Seal in 1597 and he was elected to parliament in 1601 both of which would have been impossible if he had still been a Catholic. By 1610 King James I was, according to Walton, urging Donne to take holy orders in the Church of England, which he did in 1615 and shortly afterwards became King James' chaplain; in 1621 he was appointed Dean of St Paul's Cathedral. During a visit to Cambridge with King James in 1615, Donne was awarded an honorary doctorate of divinity. Thus Jack Donne became

Dr Donne the divine. The English natural philosophers and astronomers, Thomas Harriot (1560–1621) and William Lower (1570–1615) were friends of Donne.[23] Harriot, who was a leading natural philosopher of the day, made the first astronomical observations with a telescope in England and corresponded with Kepler at least as early as 1606.[24] The first news of Galileo's telescope was sent to England on the day that *Sidereus Nuncius* was published by Henry Wotton (1568–1639) a great friend of Donne's having been at Oxford with him. We know that Donne read Galileo's *Sidereus Nuncius* because he referred to it in 1611 and it had an immediate effect on his writing.[25] Donne also probably read *Astronomia Nova* because Harriot received an early copy of it from Kepler.[26] It is certain that Donne read *De Stella Nova*, published in 1606, which describes Kepler's observations of the supernova of 1604 and of a variable star in the Cygnus Constellation, which brightened to the third magnitude in 1600 before again fading,[27] and was described in *De Stella in Cygno* which formed part of *De Stella Nova* and was mentioned by Donne in 1611.[25] Donne very probably also read Copernicus' *De Revolutionibus*. The knowledge did not flow all one way. Kepler read Donne's *Conclave Ignatii*, or *Ignatius his Conclave*, which satirizes the Jesuits and was published anonymously in 1611. In *Ignatius his Conclave* Donne made Copernicus appear at the gates of hell and say: 'Are these shut against me, to whom all the Heavens were ever open, who was a Soule to the Earth, and gave it motion?' and Donne ridiculed the Catholics writing that 'I had never heard ill of [Copernicus'] life, and therefore might wonder to find him there; yet [then] I remembered, that the Papists have extended the name, and the punishment of Heresie, almost to everything.' Kepler met Donne in 1619 while Donne was chaplain to a diplomatic mission which passed through Linz where Kepler was living. However Kepler knew nothing of Donne, despite having previously read *Conclave Ignatii* because Donne was not identified as its author until after Kepler's death, and he simply wanted his aid in getting permission to dedicate his *Harmony of the World* to King James.[28]

As well as being knowledgeable on the new astronomy, Donne was aware of the developments in natural philosophy through Harriot who was the leader of the *Three Magi* of the *Wizard Earl*, Henry Percy, the ninth Earl of Northumberland (1564–1632), and the mathematicians Walter Warner (?–ca.1644) and Robert Hues (1553–1632). It was through Harriot that Donne obtained his knowledge of the new atomism which abandoned Aristotle and basically went back to Democritus and Epicurus.[29] It is also clear that Donne had read *De Magnete Magneticisque Corporibus et de Magno Magnete Tellure Physiologia Nova*[30] which William Gilbert (1540–1603) published in 1600 because he quoted from Book VI Chapter 3 in his Essays in Divinity which were written in about 1615 but only published posthumously in 1651.[31]

Donne demonstrated his knowledge of the new science in many of his poems and prose works especially in those written in 1610 and 1611. The satire *Ignatius his Conclave* has already been mentioned. In 1610 Donne wrote the *Funerall Elegie* for Elizabeth Drury the only child of Robert Drury (1575–1615) and then *An Anatomie of the World* for the first anniversary of her death the following year.[32] When Donne first became acquainted with Drury is unknown, and he had never met Elizabeth who was only fifteen years old when she died. Donne's financial position was desperate in 1610 and he needed a wealthy patron. Elizabeth's death was devastating for the Drury family and Donne obviously moved them by his action because Drury asked him in 1611 to accompany his family on a European tour. While on the tour Donne also wrote *The Progres of the Soule* for the second anniversary of Elizabeth's death. In the *Funerall Elegie* Donne commented on the new stars or supernovae probably those of 1600 and 1604 described by Kepler:

> But, as when heaven lookes on us with new eyes,
> Those new starres every Artist[33] exercise,
> What place they should assigne to them they doubt,
> Argue, and agree not, till those starres goe out:

In his *Funerall Elegie* Donne had practically exhausted what he could write about a young girl he had never met and when he wrote *An Anatomie of the World* the following year his poem was also about 'the frailty and decay of this whole world,' as the sub-title tells us. Also a far greater Elizabeth than Mistress Drury had died in 1603 – Queen Elizabeth – and Nicolson has made a strong case for reading the poem as being more addressed to her death. The linking of Elizabeth Drury with her must have been flattering to the family.[21] Donne took advantage of the non-standard spelling of the time to differentiate between Elizabeth Drury and Queen Elizabeth by referring to *she* in the former case and *shee* in the second.[34] The line:

> Shee, shee is dead; shee's dead: when thou knowest this,

is a refrain that runs through the poem. Although the 'new philosophy' is by no means a main theme of the poem, the passage given below powerfully enhanced the theme of the decay of the world. The source of Donne's vivid imagery was obtained from his knowledge of the day's astronomy and natural philosophy.

And new Philosophy cals all in doubt;
The Element of fire is quite put out;
The Sunne is lost, and th'earth, and no mans wit
Can well direct him where to looke for it.
And freely men confesse, that this world's spent,
When in the Planets, and the Firmament
They seeke so many new; they see that this
Is crumbled out againe to his Atomis.
'Tis all in pieces, all cohærence gone,
All just supply, and all Relation:
Prince, Subject, Father, Sonne, are things forgot,
For every man alone thinkes he hath got
To be a Phœnix, and that then can bee
None of that kinde, of which he is, but hee.
This is the world's condition now, and now
She that should all parts to reunion bow;
She that had all Magnetique force alone,
To draw, and fasten sundred parts in one;
She whom wise nature had invented then,
When she observ'd that every sort of men
Did in their voyage in this worlds sea stray,
And needed a new compasse for their way;

A third poem, *A Valediction Forbidding Mourning*,[35] was also written in 1611 just prior to Donne taking his leave of his wife Ann to travel with Drury. In the third stanza of this poem Donne commented on the Copernican System of the World:

Moving of th'earth brings harmes and feares,
 Men reckon what it did, and meant,
But trepidation of the spheares,
 Though greater farre, is innocent.

In the first line Donne is referring to the first two motions of the earth: its yearly orbit around the Sun and the daily rotation about its axis despite Nicolson's belief that 'the theories of Copernicus ... as they were concerned with the relative positions of the earth and sun, interested more than they troubled.'[24,36] In third line Donne is using the word *trepidation* in the first recorded astronomical sense of precession of the Earth's axis.[37] Prior to the eighteenth century scientific works were written in Latin; the English the word *trepidation* comes directly from the Latin word *trepidatio* and Copernicus used *trepidatio* in his *De Revolutionibus* with reference to the precession of the equinoxes. Trepidation, according to Copernicus, was the third motion of the earth. The trepidation of the spheres is 'innocent' because it occurs so slowly; Copernicus estimated that it took 25,816 years[38] for the earth's axis to complete a full precession and I suppose that it is 'greater farre' because it apparently shifts the stars.

Although Donne's perfect circle was crumbling in 1611 he did not abandon the image completely. On Christmas Day 1624 Donne preached a sermon in which he said: 'One of the most convenient hieroglyphics of God is a circle ... [and] his sun, and moon, and stars move circularly,' ignoring Kepler. However the circle image, a favourite in the Elizabethan and Jacobean ages, began to disappear and by the eighteenth century was seldom used.[21] By the time that John Milton (1608–1674) began to recite his epic poem *Paradise Lost*[39] to his copyists in 1658 (he was completely blind by 1652) the circle was broken. However it suited Milton's literary purpose when foretelling the events after the crucifixion, to use the perfect circle cosmological image of the Middle Ages. Thus in Book III the narrator describing the attempt of pilgrims who have visited Golgotha (Calvary) after the crucifixion attempted to be assured of ascending to heaven, by disguising themselves as Dominican and Franciscan friars who Milton, hating Catholics, was sure could not enter, said:

> They pass the Planets seven, and pass the fixt,
> And that Crystalline Sphear whose ballance weighs
> The Trepidation talkt, and that first mov'd;
> And now Saint Peter at Heav'ns Wicket seems
> To wait them with his Keys, and now at foot
> Of Heav'ns ascent they lift thir Feet, when loe
> A violent cross wind from either Coast
> Blows them transverse ten thousand Leagues awry ...
> Into a Limbo large and broad, since calld
> The Paradise of Fools, ...

Here the 'planets seven' are the Moon, Mercury, Venus, the Sun, Mars, Jupiter, and Saturn and the 'Crystalline Sphear' causes the trepidation and the precession of the equinox. And in Book VII when the Angel Raphael is describing the Creation, which in *Paradise Lost* is performed at God's command by his Son and the Angels:

He took the golden Compasses, prepar'd
In Gods Eternal store, to circumscribe
This Universe, and all created things:
One foot he center'd, and the other turn'd
Round through the vast profunditie obscure,
And said, thus farr extend, thus farr thy bounds,
This be thy just Circumference, O World.

The 'golden Compasses' are used to construct perfect circles. However in Book VIII, Adam before his Fall, questions the Angel Raphael about the structure of the universe which seems to require the fixed stars to travel such vast distances in their daily rotation around the stationary Earth rather than have the Earth move. Raphael answers, implying that God is amused by both Ptolemy's solution of a stationary Earth with epicycles to account for the observed motion of the planets and also with Copernicus' solution of a stationary Sun with Earth, with its three motions, just another planet:

... whether Heaven move or Earth,
Imports not, if thou reckon right; the rest
From Man or Angel the great Architect
Did wisely to conceal, and not divulge
His secrets to be scanned by them who ought
Rather admire; or, if they list to try
Conjecture, he his fabric of the Heavens
Hath left to their disputes, perhaps to move
His laughter at their quaint opinions wide
Hereafter; when they come to model Heaven
And calculate the stars, how they will wield
The mighty frame; how build, unbuild, contrive
To save appearances; how gird the sphere
With centric and eccentric scribbled o'er,
Cycle and epicycle, orb in orb: ...
And no advantage gain. What if the sun
Be center to the world; and other stars,
By his attractive virtue and their own
Incited, dance about him various rounds?
Their wandering course now high, now low, then hid,
Progressive, retrograde, or standing still,
In six thou seest; and what if seventh to these

> The planet earth, so steadfast though she seem,
> Insensibly three different motions move?

In *Paradise Lost* Milton displays his detailed knowledge of astronomy. Milton even met Galileo. The same Henry Wotton, who was a friend of Donne, gave him advice on his trip to Italy in 1638 and recommended that he visit Florence where Milton met Galileo through Vincenzo Gamba (1606–1649) Galileo's illegitimate son.[40] What was discussed at that meeting is not known, but there are two references to the 'Tuscan Artist' in *Paradise Lost* and his description of the mountains on the Moon and the spots on the Sun. These references are different to the quotations given above in that they are introduced to give topicality to the poem rather than being an essential element of it; Milton also refers to the spice trade in Book II in a similar topical fashion. These elements helped to make the poem lively and would have been appreciated by his early readers, but the contrast between the Ptolemaic and the Copernican Systems, especially in Book VIII, are an integral part of the poem. Also in Book X, the description of the period after the fall of Adam and Eve when 'Decrepit Winter' and 'Solstitial summers heat', become consequences of the Fall, is beautifully achieved by an astronomical reference where again Milton gives both the Ptolemaic and Copernican explanations:

> Some say he bid his Angels turne ascanse
> The Poles of Earth twice ten degrees and more
> From the Suns Axle; they with labour push'd
> Oblique the Centric Globe: Som say the Sun
> Was bid turn Reines from th' Equinoctial Rode
> Like distant breadth to Taurus with the Seav'n
> Atlantick Sisters, and the Spartan Twins
> Up to the Tropic Crab; thence down amaine
> By Leo and the Virgin and the Scales,
> As deep as Capricorne, to bring in change
> Of Seasons to each Clime; else had the Spring
> Perpetual smil'd on Earth with vernant Flours,
> Equal in Days and Nights, ...

Paradise Lost is one of the greatest epic poems in any language and the astronomical imagery, and there are more that are not quoted, is an important feature without which it would not be so great. These elements would have been appreciated by the educated readers since there was an undivided culture in the seventeenth century.

Margaret Cavendish (ca. 1623–1673) neither wrote great literature nor natural philosophy but she did make use of poetry to convey her natural philosophy. Margaret was a friend of Charles I and his queen Henrietta Maria and she left England during the Civil War for Paris with the Queen in 1644.[41] It was in Paris that she met William Cavendish (1592–1676), the Marquess of Newcastle, and married him. Margaret had no formal education in natural philosophy and seems to have learnt hers from her polymath husband. Her philosophy owed little to the scientific revolution; she believed in a vitalist theory of matter and had an interest in Epicurean atomism. In 1651 Margaret returned to England to try to obtain the release of her husband's sequestered estates and while there she published *Poems and Fancies* and *Philosophical Fancies* in 1653.[42] The poems in *Poems and Fancies* mainly expounded Margaret's atomism and I have chosen a shortish one, *A World made by Atomes*, to illustrate her style:

Small Atomes of themselves a World may make,
As being subtle, and of every shape:
And as they dance about, fit places finde,
Such Formes as best agree, make every kinde.
For when we build a house of Bricke, and Stone,
We lay them even, every one by one:
And when we finde a gap that's big, or small,
We seeke out Stones, to fit that place withall.
For when not fit, too big, or little be,
They fall away, and cannot stay we see.
So Atomes, as they dance, finde places fit,
They there remaine, lye close, and fast will sticke.
Those that unfit, the rest that rove about,
Do never leave, untill they thrust them out.
Thus by their severall Motions, and their Formes,
As severall work-men serve each others turnes.
And thus, by chance, may a New World create:
Or else predestinated to worke my Fate.

As Margaret wrote in her introduction she chose poetry 'because [she] thought Errours might better passe there, then in Prose.' Margaret Cavendish was an eccentric for the time both in dress and manners – a seventeenth century hippie. Dorothy Osborne (1627–1695) wrote to her husband to be, William

Temple, in early 1653 about her *Poems and Fancies*: 'the poore woman is a little distracted, shee could never bee soe rediculous else as to venture at writeing book's and in verse too.' Other contemporaries were equally dismissive of her efforts, as was Virginia Woolf last century,[43] but Margaret Cavendish did show how poetry could be used to disseminate natural philosophy and more recently she has been seen in a better light.

5.1.2. *Music*

The influence of physics on music has been on both its theory and composition. Many physicists and astronomers, including Galileo, Kepler, Descartes, Huyghens, Wallis, William Herschel, Helmholtz, Einstein, and Heisenberg were accomplished musicians. Newton, though he wrote on musical theory, never played or diverted himself with music according to John Conduitt (1688–1737). Also Newton told William Stukeley that he did go to the opera once, saying: 'the first act ... I heard with pleasure, the 2nd stretched my patience, at the 3rd I ran away.'[44] Music formed one of the four subjects of the *quadrivium* taught in universities during the Renaissance, the others being arithmetic, geometry and astronomy. From ancient Greece until the seventeenth century, music was a microcosm of the harmony of the spheres. The Pythagoreans used the monochord to study music and knew that the most harmonious musical intervals were obtained when notes obtained from strings whose lengths were based on the fractions formed by the first four natural numbers. Thus the most harmonious intervals are those obtained from strings whose lengths relative to the fundamental string are: the octave (1/2), the fifth (2/3), and the fourth (4/3).

5.1.2.1. *Harmony of the world*

The Pythagoreans believed that the harmony of the planets and musical harmony were linked and just one facet of universal harmony as did Kepler, who in 1619 published *Harmonices Mundi*, which also first gave his third law of planetary motion.[45] Bruce Stephenson has written an excellent book on *Harmonices Mundi* which those readers wishing to understand fully Kepler's musical structure of the planetary system, should consult.[46] Kepler attempted to determine the harmonies that God embodied in the world assuming that the pitch of a note was proportional to the daily arc about the Sun that each planet traversed. Since the planets describe elliptical paths that cause a variation in the speed of the planet being fastest at the perihelion when the planet is closest to the sun and slowest when at the aphelion when the planet is furthest from the Sun, Kepler thought that the interval between the notes of the planets depended upon the eccentricity of their elliptical orbits. Of course Kepler's determination of the harmonies of the planets only gives the intervals not the absolute pitch and in Chapter 6 he used different tunings for each planet because, as Stephenson points out, 'Kepler listens to the planets singly.'[46] With this in mind then, Kepler constructed music of the planets shown in Fig. 5.2. As Stephenson emphasized, this figure does not summarize Kepler's theory but showed 'that the variety of those harmonies includes all eight of the traditional ecclesiastical modes.'[46] The clefts in Fig. 5.2 are those in use in Kepler's time: the cleft for Saturn is the same as our usual bass or F cleft, Jupiter also has a bass cleft but shifted to the top of the staff, Mars uses a tenor cleft with C at the fourth line and is 'almost' because its interval is approximately a fourth, Earth has uses a G cleft placed at the middle of the staff, Venus has a treble cleft in its usual position and because Venus has an almost circular orbit the interval between its notes is much less than a semitone, Mercury with its eccentric orbit has the biggest interval and uses a tenor or C cleft with the C at the second line, and the Moon has a treble clef at the middle line. 'The melodies of the different planets are not intended to be comparable in pitch.'[46]

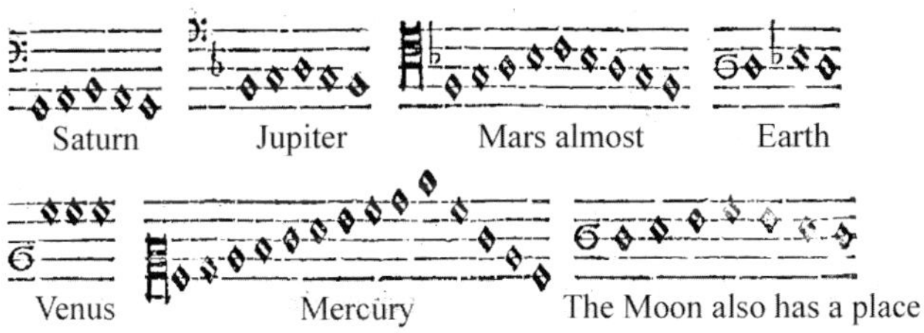

Fig. 5.2. Musical intervals of the planets (Kepler, 1619).

5.1.2.2. *Compositions inspired by Kepler and other Giants of Newton*

Not surprisingly many composers have been fascinated by Kepler and written music and operas inspired by *Harmonices Mundi*. Paul Hindemith (1895–1963), a German born composer, who immigrated to the United States in 1940 and taught music primarily at Yale University until 1953, wrote an opera after he returned to Europe *The Harmony of the World*, which was first produced in 1957. *The Harmony of the World* told the story of Kepler's search for universal harmony in the motion of the planets. At Yale Hindemith taught the jazz musician Willie Ruff in 1952 and introduced him to Kepler's harmony of the spheres. Obviously the harmony of the spheres fascinated Ruff because, in collaboration with John Rodgers a jazz pianist also on the faculty of Yale, he produced a CD *The Harmony of the World: a realization for the ear of Johannes Kepler's astronomical data from Harmonices Mundi 1619* in 1971, which contained a computer-synthesizer realisation of the musical intervals shown in Fig. 5.2.[47]

In 1980 Günter Bergman (1910–1998), a German mathematician and composer, wrote an organ cycle *The Harmony of the World of Jupiter* commissioned by the German Astronomical Society to mark the 350th anniversary of the death of Kepler. Bergman was also fascinated by Kepler's harmony of the spheres but he based his music on the periods of the Galilean moons of Jupiter, Io, Europa, Ganymede, and Callisto known to Kepler.

David Bedford (1937–2011) was commissioned in 1993 by the Royal Observatory in Greenwich to compose a piece to celebrate the renovation of the Great Equatorial Building that contains the 28 inch refracting Great Equatorial Telescope of 1893.[48] Bedford wrote that he 'tried to imagine what it would be like to travel through the cosmos revealed by the first large telescopes'; and the result was a 10 minute composition for electronic keyboard which used some of Kepler's harmony of the planets and which was played in the Great Equatorial Building as background music during the reopening celebrations. The result was interesting but his subsequent expansion of the piece into a CD length suite was repetitious.

In a completely different class was *Kepler*: *An Opera* written by that prolific American postmodernist composer Philip Glass, which premiered in 2009 in the Austrian city of Linz. This work was commissioned by the Linz Landestheater and Linz09 a programme to celebrate the city being named the European Capital of Culture in 2009. The librettist chosen by Linz09 was the Austrian Martina Winkel and the libretto is in German and Latin. Kepler was the Provincial Mathematician of Linz from 1612–1626 and he is very much part of the culture of Linz. The opera starts and ends with Kepler's epitaph 'Once I measured the heavens/Now I measure earthly shadows'. The Thirty Years War started in 1618 and the life and work of Kepler are contrasted with the poetry of Andreas Gryphius (1616–1664) describing the horror of that war.

It is not surprising that Kepler has received the most appreciation of all of Newton's *Giants* by musicians. Three other *Giants* that have inspired composers are Copernicus, Tycho Brahe, and Galileo. The Polish composer Henryk Górecki (1933–2010) celebrated the 500th anniversary of the birth of Copernicus in his Symphony No. 2, *The Copernican* first performed in 1973, which is a choral symphony that contains an excerpt from *De Revolutionibus Orbium Coelestium*. Hakon Børresen (1876–1954), one of the foremost Danish composers of the 20th century, celebrated Tycho Brahe in his 1924 ballet *At Uranienborg: Tycho*

Brahe's Dream. The scene for the *Dream* is his observatory and in it dancers represent the supernova of 1572, the Great Comet of 1577, and the stars. Philip Glass' opera *Galileo Galilei*, which premiered in 2002, is a backwards look at Galileo's life starting with him as an old man and blind after his conviction for heresy in 1633[49] and ending with Galileo as a child listening to an opera of his father Vincenzo Galilei (ca. 1520–1591).[50] There are ten scenes in Glass' one act opera and five of them are directly concerned with Galileo's experimental physics and his astronomy with the remaining half dealing mainly with Galileo's problems with the Catholic Church.

5.1.2.3. *The musical intervals in Western music*

The notion that musical pitch is determined by the frequency of vibration became established in the seventeenth century and in 1636 Marin Mersenne (1588–1648), a Minim friar, published *L'Harmonie Universelle*, in which he gave the relationship between the frequency of a vibrating string and its length, tension, and mass per unit length; Galileo published the same result two years later.[51] It was appreciated in ancient Greece that musical notes contained more than one pitch. In problem 8, book 19 of the Aristotelian *Mechanical Problems*, it was asked: 'Why does the low note contain the sound of the high note?' showing that it was recognized that musical instruments produce a complex note that contained higher frequencies or partials.[52] John Wallis reported independent experiments by William Noble (1650–1681) and Thomas Pigot (1657–1681) in 1677, which showed that a bowed string could make another string, tuned to a pitch an integer number greater than the first string, vibrate in equal aliquot parts[53] at the same pitch with nodes that could be found with paper riders.[54] Joseph Sauveur (1653–1716), a French mathematician, made some of the most significant advances in understanding overtones at the end of the century. In a manuscript, *Traité de la Théorie de la Musique*, written in 1697 Sauveur observed that one could hear the octave, the fifth, and the third etc. in musical notes but gave no explanation.[55] However in a Mémoires of the Académie Royale des Sciences for 1701, after pointing out that a vibrating string has nodal points, being unaware of the work that Wallis had reported, Sauveur made the crucial connection that the nodes corresponded to the overtones.

Whether chords are consonant or dissonant is partly subjective though the simple chords formed by the octave, fifth, and fourth are universally recognized as consonant. Musical scales are constructed so that many chords are consonant. Notes that differ by a small fraction are dissonant because they interfere with each other and produce a rough beating sound. Thus a semitone chord is almost

Table 5.1. Comparison of Newton's just temperament with Western just temperament and Equal temperament.

Interval	Ratio to fundamental		
	Newton's just temperament	Western just temperament	Equal temperament
Unison	1	1	1
Minor second	16/15 = 1.067	25/24 = 1.042	1.059
Major second	9/8 = 1.125	9/8 = 1.125	1.122
Minor third	6/5 = 1.2	6/5 = 1.2	1.189
Major third	5/4 = 1.25	5/4 = 1.25	1.260
Fourth	4/3 = 1.333	4/3 = 1.333	1.335
Diminished fifth	45/32 = 1.406	45/32 = 1.406	1.406
Fifth	3/2 = 1.5	3/2 = 1.5	1.498
Minor sixth	8/5 = 1.6	8/5 = 1.6	1.587
Major sixth	5/3 = 1.667	5/3 = 1.667	1.682
Minor seventh	16/9 = 1.778	9/5 = 1.8	1.782
Major seventh	15/8 = 1.875	15/8 = 1.875	1.888
Octave	2	2	2

universally perceived to be dissonant. The Western just intonation (shown in Table 5.1) is a modification of the Greek scales made in about 1560 by Gioseffo Zarlino (1517–1590), Maitre de Chapelle at St. Mark's, Venice that is widely used in music. The just temperament scale gives the most consonant music because musical instruments do not sound a pure note and the higher partials can produce dissonance if the musical scales are not well chosen as will be discussed later. In a violin the just temperament scales can be played perfectly in any key because the notes are fingered but fixed note instruments like the harpsichord cannot play perfectly in any key if tuned to a just temperament scale. The solution to this problem was first devised not in Europe but in China by Zha Zai-Yu in 1584.[56] The problem with the just temperament scale is that the semitones have different intervals which makes it impossible for the scales in all keys to be in tune on a fixed note instrument. Since there are twelve semitones to an octave, in equal temperament each semitone must have an interval of $2^{1/12}$ and these intervals are given in Table 5.1. While today it is very easy to compute a twelfth root with a pocket calculator, it was a difficult task before 1614 when John Napier (1550–1617) published his tables of natural logarithms to make calculations such as $2^{1/12}$ easier. Chinese algebraists could handle roots of high powers in the thirteenth century but in the first decade of the seventeenth century Zha was unaware of their work and had to resort to a method involving only square and cube roots. Simon Steven (1548–1620) gave the intervals for an equal temperament scale in the first decade of the seventeenth century, quite possibly after he learnt of the work of Zha since he also introduced another Chinese invention, the land yacht, to the West.[56] Newton was not a musician but he was interested in the theory of music. A logarithmic scale is more suitable for comparing scales and Newton devised a method, given in his notes for 1665, in which the equal-tempered semitone is assigned the value of unity and represents an interval of $2^{1/12}$.[57] This system is similar to the modern system where an octave is divided into 1200 logarithmic units called *cents* because each semitone is equal to 100 cents. Newton also devised his own just temperament scale shown in Table 5.1, which is very similar to the standard just temperament scale but is more symmetrical.[57] However it was Sauveur who published the first significant application of the logarithmic method to musical intervals in his Mémoire of 1701 to the Académie Royale des Sciences where, since on a logarithmic scale an interval of an octave is equivalent to $\log_{10}(2) = 0.301$, he divided the octave into 43 intervals, which he called *merides* and each *meride* into 7 *eptamerides* thus making the *eptameride* a tiny interval of $\log_{10}(2)/(7 \times 43) = 1/1000$. Even the *meride* is a small interval, and musicians complained that such small intervals could not be distinguished, but even so the *meride* cannot be used to calculate equal temperament exactly.

5.1.3. *Visual arts*

In art the circle image did not reign supreme for as long as it did in poetry, which is not surprising since a circle not in a plane perpendicular to the line of sight appears in perspective as an ellipse. The conic sections, one of which is the ellipse, were known to the ancient Greeks. The simplest way to draw an ellipse is to use a pencil in a loop of string fixed by two drawing pins at the foci of the ellipse. This method was known in Byzantium in the sixth century but not known in the west until the late sixteenth or early seventeenth century.[58] Leonardo da Vinci was drawing ellipses in the first decade of the sixteenth century and devised specialized compasses for drawing them (see Fig. 5.3); the pencil holding leg to the right of the drawing swivels around and slides up and down the inclined leg describing a cone which, intersecting with the plane of the paper, draws an ellipse. Some writers distinguish between ovals that appear to the eye to be ellipses, but are in fact only

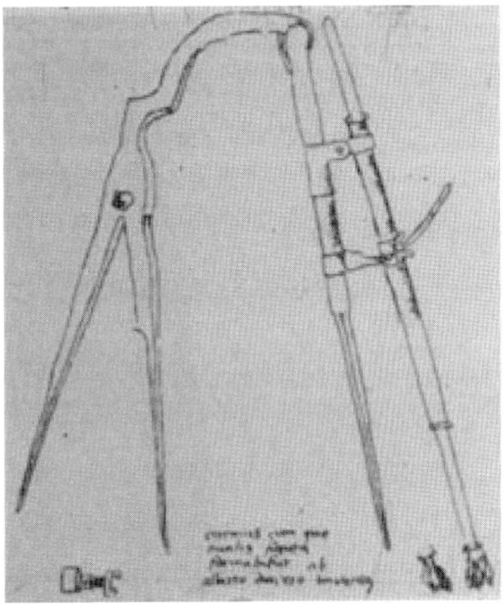

Fig. 5.3. Leonardo da Vinci's compasses for constructing ellipses.

approximations to ellipses, and true ellipses. I believe this distinction to be very artificial, what is important is the use of ellipse-like oval images where previously only the circle image was used. The first use of ellipse-like images in painting according to Erwin Panofsky (1892–1968) was by Antonio da Correggio (1489–1534) in the *Madonna with St Francis* painted in 1514 for the altarpiece in his local Franciscan monastery in Correggio shown in Plate 5.[59] In architecture ellipse-like ovals have been used since classical times – the plan of the Colosseum in Rome has an ellipse-like oval floor plan but its use has nothing to do with physics or astronomy.[60]

Galileo grew up in an artistic environment and he had a lifetime friendship with Lodovico Cardi (1559–1613) the Florentine painter known as Cigoli (1559–1613) because of his birthplace Castelvecchio di Cigoli.[59] The two friends delighted in each others activities; Galileo drew and Cigoli made astronomical observations. both Galileo and Cigoli made drawings of sunspots seen with the aid of the telescope. In Cigoli's fresco on the dome of Santa Maria Maggiore in the Lateran, Rome completed in 1612, Cigoli depicted the Madonna as the Immaculate Queen of Heaven with her foot resting on the Moon which shows the craters seen by both Galileo and Cigoli (see Fig. 5.4).[59]

Fig. 5.4. Ludovico Cigoli, The Virgin of the Immaculate Conception, Fresco on the Dome of the Pauline Chapel of the Santa Maria Maggiore, Rome, 1612. Panofsky (1956) Fig. 1, with permission University of Chicago Press.

Newton identified colour with musical harmony as Aristotle had. He obtained a spectrum from sunlight shining through a small hole onto a prism and thence onto paper and asked 'an assistant whose eyes for distinguishing colours were more critical than mine' to draw lines demarking the seven colours of the spectrum.[61] Likening the range of visible light to an octave, Newton found that the colour divisions corresponded to a scale very similar to that which he had proposed for music, shown in Table 5.1. However the spectrum obtained by Newton was not a pure spectrum since he used a large aperture (approximately 8.5 mm diameter) to admit the sunlight and did not focus the light on the paper screen. Thus any apparent agreement between the colours and his musical harmony was fortuitous. In the seventeenth century, though some such as Huygens and Hooke believed in the wave theory of light, Newton's theory was corpuscular and by linking the spectrum of light to music he was not proposing that light consisted of waves but that simply there was a harmony in numbers similar to that which occurs in music. The frequency of the light spectrum was not obtained until 1802 and Young's experiments.[62] It is surprising that the frequency of visible light at the violet end of the spectrum is roughly twice that at the red end and hence visible light does occupy almost the same interval as an octave,[63] though Newton placed the fundamental note at the violet end of the spectrum and this correspondence has no physical significance. In query 14 to the *Opticks*, Newton also speculated that:[61]

> 'the harmony and discord of Colours arise from proportions of the vibrations propagated through fibres of the optick Nerves into the Brain, as the harmony and discord of Sounds arises fro proportions of the vibrations of the Air. For some Colours are agreeable, as those of Gold and Indigo, and others disagree.'

5.2. The Enlightenment

Knowledge until the sixteenth century was largely based on the synthesis of dogma, law, and custom and controlled by the Catholic Church. Scholasticism dominated the universities. The Reformation broke the monopoly of the Catholic Church but many of the protestant churches were just as dogmatic as it was. New science in the sixteenth and seventeenth centuries challenged scholasticism but at first natural philosophy remained speculative. The experimental method, which Bacon foreshadowed, was an important weapon against scholasticism and dogma, became an element in the Enlightenment. Newton showed the full power of the experimental method and the revolution in natural philosophy was extended to all areas of culture.

5.2.1. *Literature*

The poet Richard Glover (1712–1785) was right when, in the poem eulogizing Newton, he wrote 'Newton demands the Muse,'[64] because Newton had a great influence on poetry, especially during the first half of the eighteenth century. Henry Pemberton (1694–1771) used Glover's poem as the frontispiece to his *A View of Sir Isaac Newton's Philosophy* published in 1728.

It was in the coffee houses of London in the eighteenth century that gentlemen of all professions met. Button's Coffee House[65] in Russell St, Covent Garden was frequented by literary figures including Richard Steele and his great friend, Joseph Addison (1672–1719), who used it as their editorial office for the *Guardian* which they began to publish in 1713. Button's still serves coffee today but in paper cups for it now bears the name of Starbucks and the white marble head of a lion with wide-open jaws, designed by William Hogarth (1697–1764) to receive letters and papers, and installed in 1713 by Addison for the *Guardian*, has been replaced by a trash bin.[66,67] The Latin inscription beneath the lion was taken from two of the epigrams of Marcus Valerius Martialis (ca. 40–104) and read:

> 'Those talons are kept for mighty necks; it is not only on the beast he has chosen that he feeds.'

Writers and poets flocked to Button's. Jonathan Swift (1667–1745), yet to write *Gulliver's Travels*, but already an established writer and poet, first met Addison at Button's; the poets John Hughes (1677–1720), John Gay (1685–1732), and Alexander Pope (1688–1744) together with the satirist John Arbuthnot (1667–1735), also a Fellow of the Royal Society, regularly met here.[68] It was in Button's Coffee House that Whiston gave his first astronomical lectures and Pope, Gay, and may be Hughes, attended his lectures.[69] In *The Ecstasy* written probably in 1720 but only published posthumously, Hughes certainly accurately described the solar system and the stars of the Milky Way and he did liken Newton's soul to a meteor 'that daily travels [through the solar system] in search of knowledge for mankind below,' but in fact his cosmology described Descartes' vortex theory for the motion of the planets not Newtonian:

> New solid Globes their weight, self-balanc'd, bear,
>   Unpropp'd admidst the fluid Air,
> And all, around the Central Sun, in circling Eddies roll'd.
>   Unequal in their Courses, see they advance,
> And form the Planetary Dance!

However, later in the poem Hughes did show that he had knowledge of the speed of light and 'how sun-born colours rise in rays distinct.'

Pope certainly made use of Whiston's astronomical lectures in his *Essay on Man*, published between 1733 and 1734, and Nicolson and Roussau opined that its 'magnificent first Epistle ... would never have been written had Pope not heard Whiston's coffee-house lectures' and Newtonianism shines throughout his work.[69] In the *Essay on Man*, which is written in four epistles, the theme of the first Epistle, which overarched the whole poem, was the relationship of man to the Universe and God. He announced this theme at the beginning of the first section of the first epistle:

> Say first, of God above, or man below
> What can we reason, but from what we know?
> Of man, what see we but his station here,
> From which to reason, or to which refer?
> Through worlds unnumbered though the God be known,
> 'Tis ours to trace Him only in our own.
> He, who through vast immensity can pierce,
> See worlds on worlds compose one universe,
> Observe how system into system runs,
> What other planets circle other suns,
> What varied being peoples every star,
> May tell why Heaven has made us as we are.

Pope saw the order of the Universe as:

> The general order, since the whole began,
> Is kept in nature, and is kept in man.

He did not voice Newton's doubts that the Universe was stable and in need of God's occasional intervention to restore order that Newton mentioned in query 31 of the *Opticks*,[70] but saw Newton's laws alone as maintaining that order:

> And, if each system in gradation roll
> Alike essential to the amazing whole,
> The least confusion but in one, not all
> That system only, but the whole must fall.
> Let earth unbalanced from her orbit fly,
> Planets and suns run lawless through the sky;

Let ruling angels from their spheres be hurled,
Being on being wrecked, and world on world;
Heaven's whole foundations to their centre nod,
And nature tremble to the throne of God.
All this dread order break – for whom? for thee?
Vile worm! – Oh, madness! pride! impiety!

Pope saw 'the proper study of mankind is man./ Know, then, thyself, presume not God to scan.'[71] However, he gloried in man's scientific achievements:

Go, wondrous creature! mount where science guides,
Go, measure earth, weigh air, and state the tides;
Instruct the planets in what orbs to run,
Correct old time, and regulate the sun;

But Pope also gave a caveat to his own epitaph of Newton: 'Nature and Nature's Laws lay hid in Night./ God said, Let Newton be! and All was Light', and wrote:

Superior beings, when of late they saw
A mortal man unfold all Nature's law,
Admired such wisdom in an earthly shape
And showed a Newton as we show an ape.
Could he, whose rules the rapid comet bind,
Describe or fix one movement of his mind?

Pope also introduced Newton's natural philosophy as well as his astronomy:

See plastic Nature working to this end,
The single atoms each to other tend,
Attract, attracted to, the next in place
Formed and impelled its neighbour to embrace.

Pope was entranced by Newtonianism and the new science, which for him required little justification. However such a view was by no means universal. There was a growing feeling during the seventeenth century that, since much of the new science had no practical effect on life, it had little value.[72] This view was held mainly laymen but there were notable natural philosophers such as Bacon and Boyle who thought that science should have a utilitarian aspect as well as advancing pure knowledge.[72] Such a view, that science should useful, has had many adherents since the seventeenth century. Pope's friend Swift was certainly of the view that science should be practical and that science for science's sake was immoral and he brought to bear his great satirical power to ridicule what he saw as a false taste in learning. Many of the habitués of Button's Coffee House, including Pope, had similar dislike of all forms of falsity and pomposity in learning and in 1714 they formed the Scriblerus Club and to write satires ridiculing such people in the form of memoirs of a fictitious Martinus Scriblerus. Swift, one of the greatest satirist of them all, had started writing satires much earlier. In the seventeenth century a quarrel began between those who believed the Ancients, that is, the ancient Greeks, represented the best in literature and those who believed that they were surpassed by the Moderns. The Quarrel began in France under the name *Querelle des Anciens et des Modernes*. Charles Perrault (1628–1703), who was the Secretary of the Académie des Belles Lettres until he was dismissed in 1695 and started writing fairy stories for children, opened the Quarrel for the Moderns with his *Le Siècle de Louis le Grand* in 1687 and Fontenelle quickly supported him with *Digression sur les Anciens et les Modernes* in 1688. The Quarrel, though it was not between the arts and science, had many of the characteristics of the two cultures clash in the nineteenth and twentieth centuries particularly in the animosity that it engendered. In England the statesman William Temple (1628–1699), who had retired to Moor Park near Farnham, Surrey, argued against the Moderns in his essay *On Ancient and Modern Learning*, published in 1690, and attacked fellows of the Royal Society. Temple was unwise in this action because his knowledge of the literary scene was limited and he came under attack for his views, some of which were plainly wrong. Swift had joined Temple at Moor Park as his friend and secretary in 1688 and felt compelled to defend him, he did so through his satire *The Battle of the Books*, which he wrote in 1697 but which was not published until 1704. In *The Battle of the Books*, Swift had the books in St James Library take up the controversy and come to a 'decisive battle'.[73] The battle is not limited to literary books, though to Temple these contained the most important areas of learning. Fontenelle on the other hand had stressed the excellence of the new science as well as suggesting that the Moderns may have surpassed the Ancients in arts and letters. Thus Swift had Descartes, Gassendi, and Hobbes the leaders of the bowmen for the Moderns 'whose strength was such, that they could shoot their arrows beyond the atmosphere' and in the army of the Ancients, 'Homer led the horse ... Euclid was chief engineer ... [and] Aristotle commanded the bowmen.' Although Swift defended Temple, his attitude to the controversy is ambiguous and he pretended that the discovered manuscript was damaged and no conclusion to the battle is given.

The earnest Whiston was a prime candidate for satire and a god given gift to the Scriblerus Group. In the *Memoirs of Martinus Scriblerus*, Whiston was satirized for his comet theory of the Great Deluge[74] and also for his proposal for his screwball method of determining longitude at sea. Whiston combined forces with the mathematician Humphrey Ditton (1675–1715) in 1714 to propose having vessels, stationed at fixed locations, fire star shells at fixed times; the longitude could then be calculated from the time difference in seeing the flash of the shell and hearing it. Arbuthnot declared that the proposal anticipated a burlesque proposal of his own intended for the *Memoirs*. The proposal also gave rise to the scatological ditty *On the Longitude*:

The longitude miss'd on
By wicked Will Whiston;
And not better hit on
By good master Ditton.

So Ditton and Whiston
May both be bepist on;
And Whiston and Ditton
May both be beshit on.

Sing Ditton,
Beshit on
And Whiston,
Bepist on.

Sing Ditton and Whiston.
And Whiston and Ditton,
Beshit and bepist on,
Bepist and beshit on.

The authorship of this ditty has been variously attributed to Jonathan Swift, John Gay or Thomas Parnell (1679–1718) another member of the Scriblerus Group. Whiston's scheme for determining longitude was satirized by Hogarth in the last Plate of *The Rake's Progress* which shows Tom Rakewell in Bedlam (see Fig. 5.5). Among the lunatics there is an artist behind the door drawing on the wall. One drawing on the wall near the doorway to the right shows a mortar firing a star whose trajectory arches over a drawing of a globe to explode just above of the word longitude.[75] Underneath the mortar there is a ship which may be determining its longitude by Whiston's proposed method, which obviously Hogarth rightly thought was crazy.

Fig. 5.5. William Hogarth, The Rake's Progress, Plate 8, Print, 35.5 × 41 cm. Tom is in the foreground being examined by a doctor. The graffiti on the wall of the mortar firing a shell has been enhanced so that it is easily visible, courtesy British Museum.

Jonathan Swift was an Ancient at heart and did not quite have Pope's enthusiasm for Newton and the new science. His greatest satire on science, including physics and astronomy, is contained in the *Voyage to Laputa*, which is the third part of what is now usually called *Gulliver's Travels*, first published in 1726.[76] Laputa is a 'flying or floating island' circular in shape with a diameter of 7837 yards (a microcosm of the earth, which Newton had estimated had a diameter of 7846 miles),[72] the four circular lakes symmetrically placed at the centre of the island leave no doubt that the island was designed like a button and probably as a reference to Swift's favourite coffee house.[77] The Island was supported and propelled by a 'loadstone of a prodigious size.' The minds of the inhabitants of Laputa 'are so taken up with intense speculations, that they neither can speak, nor attend to the discourses of others, without being rouzed' by an attendant with a flapper, a bladder filled with dried peas. The caricature of the absent minded professor has a long history. The Monarch ruled not only Laputa but the continent of Balnibarbi whose capital was Lagado where the lord was Munodi. A delegation from Lagado went to Laputa some forty years earlier and 'came back with very little smattering in mathematics;' a reference to Newton's remark to William Derham about 'little Smatterers in Mathematicks'. Like the Moderns, this delegation wanted to change every thing on their return. and obtained a royal patent for an Academy of Projectors (a reference to the Royal Society) that could change the rules for everything so that 'one man shall do the work of ten; a palace may be built in a week, of materials so durable as to last for ever, without repairing; all the fruits of the earth shall come to maturity at whatever season we think fit to chuse, and increase an hundred fold more than they do at present.' The inventions proposed by the professors of the Academy of Projectors included that of 'a most ingenious architect who had contrived a new method for building houses by beginning at the roof and working down to the foundations, which he justified to me by the practice of those two prudent insects the bee and the spider.' Swift indulged his love of scatology with the most ancient of projectors who intended to recycle human excrement into food. Deanne Swift (1707–1783) thought that his great uncle Jonathan used his satires 'not against the whole tribe of chymists, projectors, and mathematicians in general; but against those, and those only, who despise the useful branches of science, and waste their lives in the pursuit of aerial vanities and extravagances.'[78] One of Samuel Johnson's essays in the *Idler* series published in 1759 was entitled *What have ye done*, and begins:[79]

> 'When the philosophers of the last age were first congregated into the Royal Society, great expectations were raised of the sudden progress of useful arts; the time was supposed to be near when engines should turn by a perpetual motion, and health be secured by the universal medicine; when learning would be facilitated by a real character, and commerce extended by ships which could reach their ports in defiance of the tempest. But improvement is naturally slow. The society met and parted without any visible diminution of the miseries of life. The gout and stone were still painful, the ground that was not plowed brought no harvest, and neither oranges nor grapes would grow upon the hawthorne. At last, those who were disappointed began to be angry; those likewise who hated innovation were glad to gain an opportunity of ridiculing men who had depreciated, perhaps with too much arrogance, the knowledge of antiquity. And it appears from some of their earliest apologies, that the philosophers felt with great sensibility the unwelcome importunities of these who were daily asking, "What have ye done?"'

Table 5.2. Summary of papers and letters published in Vols. 16 and 31 of the Philosophical Transactions. The percentage of each category is shown in brackets.

Subject matter	Vol. 16 (1686–1692) 90 papers and letters		Vol. 31 (1720–1721) 55 papers and letters	
	Pure science	Applied science	Pure science	Applied science
Physics and Mathematics (including Astronomy, Music, Technology, Meteorology, Tides, and Oceanography)	21 (23)	27 (30)	16 (29)	11 (20)
Miscellaneous	10 (11)	0(0)	3 (5)	1 (2)
Anatomy (including Physiology of animals and plants)	2 (2)		5 (9)	
Medicine (including Surgery, Pharmacy, and Bills of Mortality)	13 (14)		12 (22)	
Natural History	8 (9)		7 (12)	
Travel and Geography	6 (7)		0 (0)	
Antiquities	3 (3)		0 (0)	

Where did Swift obtain his view that much of science was useless? In the early 1720s when Swift was writing *Gulliver's Travels* the *Philosophical Transactions* of the Royal Society were widely read, as was the *Miscellanea Curiosa*, a collection of such papers from mainly the 1690s published in three volumes in 1705–7. In the nineteenth century Charles Babbage thought that science declined in England after the death of Newton but examination of the *Philosophical Transactions* do not seem to confirm his opinion.[80] A simple survey of the papers published from 1720 to 1779 show little change in the type of papers published.[80] It is difficult to judge the eighteenth century perception of the quality of the papers but I thought that there may have been a change in type of paper from the late seventeenth century to the 1720s when Swift was writing *Gulliver's Travels* and I examined two volumes of the *Philosophical Transactions*: Vol. 16 (1686–1692) and Vol. 31 (1720–1721) summarized in Table 5.2. With such a small sample the percentage difference in the number of papers in each category is not significant, nor is the difference between pure and applied science in the physics and mathematics section, where about half the papers fell into the applied category, significant. The standard of the papers is somewhat subjective but some, such as those published by Edmund Halley, are excellent, and many had practical applications, at the other end of the scale there were a few silly papers. The silliest in the two volumes occurred in Vol. 16 and came from Ireland and it won this distinction by a wide margin. The title of the paper was *An Extraordinary Effect of the Power of Imagination* and it is so silly that it warrants quoting in part (fortunately it is very short):

> 'One Elizabeth Dooly ... aged 13 years ... Her mother being with child of her was frightened by a cow as she milked it, thrown down and hit on her temple ... by the cow's teat. This child has exactly in that place, a piece of flesh resembling a cow's teat ... 'tis perforated and she weeps through it... and it grows in proportion to the rest of her Body.... This is lookt upon to be as strange an instance of the strength of imagination as can be produced.'

I must emphasize that none of the other papers even remotely approached the silliness of this one. However the projects of the Academy of Projectors, described by Swift, are no sillier and it is even possible that it influenced his description of the Struldbruggs or Immortals who were born with a red circular spot in the forehead which grew larger with age from the size of a silver three-penny piece to the size of a shilling. I remember that in my university department we had a member of staff who delighted in reading out aloud excerpts of silly items in journals to our tea room, some of which he probably made up because another member of staff told me later that he could not find them when he looked in the journal himself. Could it not be possible that in Button's Coffee House one of the Scriblerus Club members' did the same? Certainly the projects of the Academy were impractical but what characterizes them even more than their impracticality is their silliness.

It has been suggested that Fontenelle, who Swift had earlier attacked in the *Battle of the Books*, may have been the catalyst for his attack on the frivolousness of science with the immediate cause being the republication of his defence of the Moderns in a 1699 Preface in the *Memoirs of the Royal Academy at Paris* as a Preface in *Miscellanea Curiosa*.[72] However, I do not believe that Fontenelle's preface could have influenced Swift in this way. Fontenelle's preface starts by recognizing that many laymen look upon mathematics and natural philosophy as useless, 'because they are crabbed and not easily learnt' but goes on to argue what usefulness comes from pure science. The first example he gave was observation of the four major satellites of Jupiter which can be used to determine longitude. Fontenelle went on to give examples of the usefulness of a whole range of sciences in a well argued discourse. Such arguments about the usefulness of science have been necessary many times since the time of Swift and no doubt will be needed many times in the future. The arguments given by Fontenelle are equal to the best and his preface could not possibly have had any influence on Swift's satires. Swift did not despise science nor was he ignorant of it as is shown in the *Voyage to Laputa* where the Laputans' telescopes 'far excelling ours in goodness' had discovered two satellites revolving around Mars 'the innermost three Martian diameters from the centre of the planet ... and the outermost, five... The former has a period of ten hours, the latter 21½ hours.' The existence of these satellites, which were not discovered until 1877, was pure speculation on Swift's part but the ratio of the periods of their orbits compared with their orbital distances was not.[81] Kepler's third law gives that the square of the period of a satellite is proportional to the cube of its distance from the planet. Using the distances and periods given by Swift for the satellites, Kepler's law is followed if:

$$\left(\frac{10}{21.5}\right)^2 = \left(\frac{3}{5}\right)^3,$$

the first ratio is 0.2163 and the second 0.2160. These two ratios are so close that either Swift made the calculation from Kepler's law himself or he asked one of his natural philosopher friends to do it for him, in either case it shows a regard for the laws of nature discovered by natural philosophers. Some scientists, as well as some literary figures, are pompous and lack common sense. It was the singers not their song that Swift satirized. Marjorie Nicolson and Nora Mohler (1897–1984) have tentatively identified some of those who were satirized in *Voyage to Laputa* by Swift.[72] The first of the Projectors who was working on a scheme to extract sun-beams from cucumbers to be stored in 'vials hermetically sealed, and let out to warm the air in raw inclement summers' is believed to be modelled on Stephen Hales who made many biophysical experiments during the 1720s which he published in the Philosophical Transactions and later in *Vegetable Staticks*.[82] In experiment 20 of *Vegetable Staticks*, Hales explained: 'The impulse of the Sun-beams giving the moisture of the earth a brisk undulating motive, which watery particles, when separated or rarefied by heat, do descend in the form of vapour.' The man born blind, who identified colours by touch, is probably based on John Vermaasen, who said he identified colours by their asperity; this story was related to Boyle by John Finch (1626–1682). Swift even took a shot at Newton. In Glubbdubdrift, the Island of Sorcerers, they could call up people from the past and among those who Gulliver caused to be called were Descartes and Gassendi whom he asks to explain their systems to Aristotle. While Aristotle 'freely acknowledged is own mistakes ... he found that Gassendi, who had made the doctrine of Epicurus [atomism] as palatable as he could, and the Vortices of Descartes were equally exploded. He predicted the same fate to *Attraction* ... He said the new systems of nature were but new fashions, which would vary in every age; and even those who pretend to demonstrate them from mathematical principles, would flourish but a short period of time, and be out of vogue when that was determined.' Here Swift was most perceptive seeing that even Newton's gravity would not go unchallenged but he did not know that it would take more than two hundred years. What an irony that *Gulliver's Travels* was for long seen as a book for children. I disagree with my late engineering colleague, Bill Johnson (1922–2010), Swift was a friend of science.[81]

The Scottish poet James Thomson (1700–1748) was also a friend of science. He entered the University of Edinburgh in 1715 where he studied natural philosophy as part of his course until 1719 but did not take a degree. James Thomson then entered Divinity Hall where he studied to become a Presbyterian minister but he must have lost his calling because, when he left for London in 1725, ordination was to be a last resort; a literary career was his desire.[83] In London he met Pope, Arbuthnot, and Gay but was not a member of their circle. James Thomson's most famous poem was *The Seasons*, which he began with *Winter* published in 1726 followed by *Summer* in 1727. *Winter* is the season of the Ancients, whereas *Summer* extols the Moderns. In the original text of *Summer* James Thomson extolled many Moderns including Bacon, Isaac Barrow (1630–1677) the first Lucasian Professor at Cambridge who resigned his professorship in favour of Newton in 1669, Boyle, Locke, and Newton himself. James Thomson lengthened his text in subsequent editions but removed some of the worthies he extolled including Barrow. In the completed *The Seasons* published in 1730, James Thomson wrote in *Summer*:[84]

> Why need I name thy Boyle, whose pious search
> Amid the dark recesses of his works,
> The great Creator sought? And why thy Locke,
> Who made the whole internal world his own?
> Let Newton, pure intelligence, whom God
> To mortals lent, to trace His boundless works
> From laws sublimely simple, speak thy fame
> In all philosophy....

Voltaire met Swift in London soon after he had completed his *Gulliver's Travels*. Swift's satire obviously impressed Voltaire because he went on to write what he called *contes philosophiques* (fictional stories that criticized the foibles of notable professionals and members of society) of which the most famous is his *Candide*. One of Voltaire's first *contes philosophiques*, was Le Voyage du Baron de Gangen written about 1734. Little is known of this *conte*, which has become lost but another *conte*, *Micromégas*,[85] is thought to be based on it. *Micromégas*, completed in 1739 just after Voltaire's *Elements of Sir Isaac Newton's Philosophy* is much influenced by natural philosophy.[86] In my comments on *Micromégas* that follow, I have taken the liberty of extensively using Roger

Pearson's wonderfully illuminating notes to *Micromégas*, contained in the Oxford Edition of *Candide and other Stories*.

Voltaire was ambivalent on the question of the plurality of the worlds but *Micromégas* is the story of Micromégas an inhabitant of a planet of the sky's brightest star, Sirius, who travels first to Saturn and then to Earth via Jupiter. 'Monsieur Micromégas, an inhabitant of the land of Sirius, measures ... [eight leagues (39 km)] from tip to toe, and since ... our globe is nine thousand leagues in circumference, [one] will find ... that the globe which fostered him must be exactly twenty-one million, six hundred thousand times greater in circumference than our little Earth.' In his calculation of the height of a stellar inhabitant Voltaire is satirizing the German philosopher Christian Wolff (1679–1754) who calculated the height of a Jovian inhabitant on the basis that the size of the pupils in their eyes assuming that they would need to receive as much light as we do on Earth.[87] In explaining how cultivated Micromégas was Voltaire satirized Blaise Pascal (1623–1662): 'while not yet two hundred and fifty years of age and still a pupil at the Jesuit College of his planet ... he independently solved more than fifty of Euclid's Problems. Which is eighteen more than Blaise Pascal, who solved thirty-two of them for his own amusement [when a child] or so his sister says.' The religious Pascal was frequently a target for Voltaire's wit. Voltaire goes on to satirize Fontenelle and Moreau de Maupertuis (1698–1759) whose ship Micromégas saw, during a visit to Earth,[88] in the Baltic Sea where Maupertius had sailed to visit the Arctic Circle to measure a length of a degree of arc of the meridian. Where Micromégas saw that the Earth 'flattens out at the poles' Newton had calculated that according to the theory of gravity the Earth was not a true sphere but was flattened at the poles and a oblate ellipsoid. Jacques Cassini (1677–1756) and his father Giovanni Domenico (1625–1712) believed that from both experiments on the period of pendulums and geodesic measurements that on the contrary the earth was a prolate ellipsoid or lemon shaped. In 1735 another expedition to what then was Peru but is now Ecuador headed by Charles Marie de la Condamine (1701–1774), had measured the length of a degree of arc of the meridian at the Equator. The result of these measurements proved that Newton was right the Earth is an oblate ellipsoid. Micromégas gently picked up the ship and the geometers climbed down on to the Sirian's fingers with two Lapp girls. Maupertius had met two Lapp sisters in Tornio on the Gulf of Bothnia near the Arctic Circle, who eventually made their way to Paris much to his embarrassment because one was pregnant. '"Aha!" [Micromégas said], "I have caught nature in the act."'[89] Micromégas wrote a nice book on philosophy for the humans so that 'they would discover what was what.' The humans took the book 'to Paris to the Academy of Sciences. But when the secretary opened it, he found nothing but blank pages. "Aha", he said, "just as I thought."' Swift met his match in Voltaire.

5.2.2. *Music*

Until the sixteenth century there were twelve modes in music that is there were twelve different ways an octave was divided into seven intervals. Renaissance music was based on all these modes and derived its variety from them. By the beginning of the seventeenth century the number of modes had reduced to two, our present major and minor keys, and variety was obtained by change of key rather than mode. Baroque music of the seventeenth and first half of the eighteenth century saw the complete adoption of tonality in music over modality. The composers of the Baroque period that are regularly heard in today's concerts: Arcangelo Corelli (1653–1713), Antonio Vivaldi (1678–1741), Jean-Phillippe Rameau (1683–1764), George Frideric Handel (1685–1759), Domenico Scarlatti (1685–1757), and the great Johann Sebastian Bach (1685–1750), were the products of the final transition to tonality. Rameau, who has been called the Newton of music, made advances in forming a theory of harmony, which were revolutionary.[55] D'Alembert praised Rameau in his *Preliminary Discourse* to the *Encyclopedia* for 'finding the principle of harmony and of melody in the chord root.'[90]

In his *Traité de l'Harmonie* published in 1722, and succeeding treatises up to 1761, Rameau established a system of harmonic principles for the music of the eighteenth century.[91] However he was continually altering his argument in a dozen publications spread over a period of forty years and it is difficult to get to grips with core of his propositions and the difficulty is compounded by errors in both empirical observations and in theoretical arguments.[55] The great mathematician d'Alembert did publish an account of Rameau's theory, *Élémens de Musique Théorique et Pratique Suivant les Principes de M. Rameau*, in 1752 and this publication helped the spread of Rameau's work, but Rameau found fault in some of his interpretations. The note of a musical instrument contains higher partials and it is the partials of the notes of a chord that determine whether the chord is harmonious. However Rameau only became aware of the higher partials of musical notes through a review of his *Traité* by Louis Bertrand Castel (1688–1757), a Jesuit French mathematician in which he referred to the work of Sauveur and remarked that nature gave the same system that Rameau had discovered from aliquot divisions of a monochord.[55] Because Rameau used a stringed instrument in his studies, his notes contained all the partials but all wind instruments do not sound every harmonic partial and percussion instruments have non-harmonic partials. When he wrote his *Traité* Rameau gave only a mathematical theory of harmony but by 1726 he was beginning to realize that a physical justification was desirable and, in *Génération Harmonique* published in 1737, he gave a detailed discussion of the relationship of his mathematical rules and the physics of music.[92] When two notes are close together they are heard as a single tone of the average frequency that varies in loudness at a frequency equal to half of the difference between the two frequencies of the notes, this phenomenon is known as beating and is a cause of dissonance. Beats can also be used as an aid in tuning and have been known to organists and organ builders since the Renaissance. Although Sauveur had some understanding of partials and their effect on dissonant beating, it was not until the nineteenth century that they were fully understood. In 1733 the Swiss mathematician and physicist Daniel Bernoulli (1700–1782) had realized that a vibrating string could have an infinite number of modes of vibration and by 1750 had seen that these could be represented by an infinite series of sines of the form $\sin(n\pi x/l)$, where l is the length of the string, x is the distance from one end, and n is an integer but this result was not proved until Fourier's analysis of the early nineteenth century. The sound waves of musical notes were not successfully visualized until the American physicist, Dayton Miller (1866–1941), invented his *Phonodeik* in 1908 which enabled a photographic record of a sound to be made. Figure 5.6 gives the wave forms of various wind instruments all playing middle C in scientific pitch (256 Hz) taken from recordings made by Miller on his *Phonodeik* in about 1916 compared with a pure note. The quality of a musical note depends upon the higher partials that are present and their strength. The flute's clear and simple tone arises because only the fundamental and its octave are strong; the double octave is also present. The note of the clarinet contains reasonably strong higher partials up to about the eighth. In the oboe the fundamental is weak and most energy is in the fourth and fifth partials. Bernoulli attacked Rameau arguing that his *corps sonore* was a fiction because he assumed all instruments were like the monochord and contained all the harmonic partials.[55] Nevertheless Rameau had a great influence on musicians during the eighteenth century especially through his conception of the fundamental bass, the lowest note that defines and codifies the structure of the chords. Even Mozart used the concept of the fundamental bass when teaching harmony and composition to pupils during the 1780s.[93]

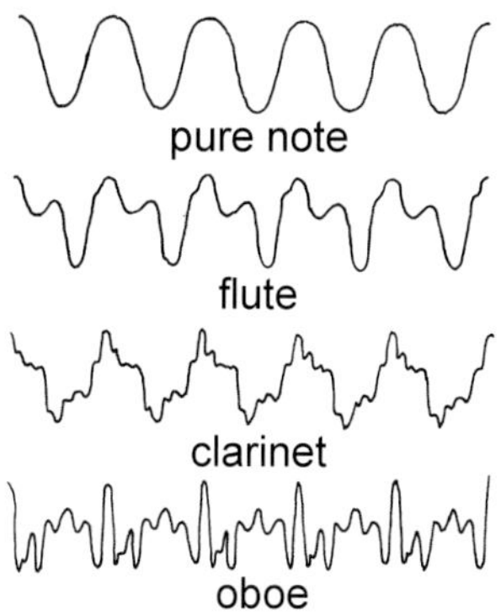

Fig. 5.6. Wave forms of notes of a flute, clarinet, and oboe compared with a pure note (after Miller 1937).

In the classical period from about the mid eighteenth century, harmony became subordinate to melody with the best known composers being Joseph Haydn (1732–1809), and Wolfgang Amadeus Mozart (1756–1791). The symphony and the sonata with well-balanced melodies were developed during this period. Although there is no direct link between the classical period and natural philosophy there is a similar structural framework and order.

It is somewhat surprising that there are few direct links between natural philosophy or astronomy and music, even opera, during the Age of Enlightenment, with the almost deification of Newton, one would have imagined that he would be the subject of at least one. Two of the few exceptions are Haydn's comic opera *Il Mondo della Luna* (The World of the Moon), in which the libretto by Carlo Goldoni (1707–1793) made fun of the interest in the plurality of the worlds, and the oratorio that he wrote at the end of his life *The Seasons* where the libretto by Gottfried van Swieten (1733–1803) was based on James Thomson's poem *The Seasons* but is almost unrecognizable and has very little to do with natural philosophy.

5.2.3. *Visual arts*

During the Age of Enlightenment the style of art changed from late Baroque or Rococo to Neoclassical. The Baroque saw art in the grand style dominated by religious themes. By the beginning of the eighteenth century artists had begun to reject the grandeur and symmetry of the high Baroque and a simpler more naturalistic style, though still ornate, developed that is either referred to as late Baroque or Rococo. The simplification continued in Neoclassicism that developed around the mid eighteenth century. Subject matter changed too, with the decline in influence of the Catholic Church, religious subjects were fewer and art often reflected the new discoveries of man. Genre paintings reflecting scenes representing every day life emerged. Baroque art was largely commissioned by the aristocratic few but by the mid eighteenth century more could afford art. Arbiters of taste began to be established, two of the first British art critics were Jonathan Richardson (1665–1745) and Joshua Reynolds (1723–1792), the first President of the Royal Academy of the Arts. Natural philosophy and astronomy became subject matter for genre works.

Astronomy created intense interest during the eighteenth century and featured in many genre paintings. In Italy there was an intimate relationship between art and science. In Bologna, one of the most important centres for astronomy, an aristocratic amateur astronomer, Luigi Ferdinando Marsili (1658–1730) built-up the best collection of astronomical instruments in Italy and wanted to donate his collection and library to the city, provided suitable accommodation could be found. Marsili and Eustachio Manfredi (1674–1739), the dean of the astronomy school, sought financial assistance from Pope Clement XI, who was sympathetic to science, to form what was to be known as the Accademia Clementina. To encourage Clement's interest, Marsili in 1711 commissioned a series of eight landscape paintings from the prominent Baroque painter Donato Creti (1671–1749) that depicted the seven known planets, the Sun, the Moon, and a comet, as well as displaying Marsili's astronomical instruments to good advantage.[94] In these paintings the images of the planets were painted as if they were viewed through a telescope by a Bolognese miniature painter Raimondo Manzini (1668–1744) to make sure that their details were clearly seen.

In the eighteenth century art became a marketable commodity. Owen MacSwinny (1676–1754), who was an entrepreneur with an interest in the arts, conceived of marketing a book of engravings in the 1720s and commissioned a series of twenty-four allegorical paintings to commemorate recent great British men.[95] Each painting was dominated by a large urn in which it was to be supposed that the hero's remains were deposited. One of this series of paintings, shown in Plate 6 featured Newton and was painted by Giovanni Battista Pittoni (1687–1767), his brother Domenico (d. 1771) and Giuseppe Valeriani (1708–1762).[96] A representation of Newton's famous prism experiment is a feature of the painting.[97]

Joseph Wright of Derby (1734–1797) came from a professional family and, while not a scientist himself, he had friends who were, including Erasmus Darwin (1708–1802), the grandfather of Charles, and John Whitehouse (1713–1788), a clockmaker and geologist and other members of the Lunar Society of Birmingham, who met monthly on the Monday nearest the full Moon to discuss the practical applications of scientific knowledge. Wright earned his living by painting portraits of industrialists and other wealthy clients but he also painted a number of candlelight genre paintings, which had science or industrial subjects, after the chiaroscuro style of the Dutch. The best known of these paintings are *A Philosopher Giving a Lecture on the Orrery* (1766) and *An Experiment on a Bird in the Air Pump* (1768). This painting, illustrated in Plate 7, contains a vacuum pump which the lecturer is using to extract the air from a glass bowl on a tall stand containing a white cockatoo, an exotic bird for the eighteenth century. Similar experiments were commonly performed at popular lectures on natural philosophy. There have been many interpretations of this painting. Is Wright's painting about science or about human emotions? One can interpret the painting as personifying science in the person of the rather dishevelled lecturer in a red gown who is looking straight at the observer of the painting not at the cockatoo, which has collapsed, and could be playing God with the power of life or death. Although the lecturer's right hand is not on the handle of the pump and his left hand appears to be about to lift the cap on the vessel of the vacuum pump to resuscitate the cockatoo, which will be difficult if there is much of a vacuum in the vessel. However, with such an interpretation, would not all the lecturer's audience be awe struck? Whereas only the two young girls seem affected by the fate of the cockatoo and the father appears to be gesticulating with his right hand that the cockatoo will be alright. Only the man on the left in the foreground is rapt in the experiment. The elderly man on the right in the foreground is staring intently on the skull immersed in liquid at the centre of the table. These different reactions are in stark contrast with *A Philosopher Giving a Lecture on the Orrery* where all the audience is intent on the lecturer. Whatever the message that Wright was trying to convey, the painting is powerful and always attracts attention at the National Gallery, London where it has been hung since 1863.[98]

Fig. 5.7. Robert Guilliaume Dardel, *Descartes Piercing the Clouds of Ignorance*, Terracotta, 43.2 cm, 1782, with permission The Wallace Collection, London.

I will finish this section with the classical terracotta statuette by Robert Guillaume Dardel (1749–1821) *Descartes Piercing the Clouds of Ignorance*, first exhibited in 1782, shown in Fig. 5.7. In this statuette Descartes is shown as almost succeeding in letting the sunlight of science break free from the clouds of ignorance. Dardel's patron at the time was the aristocrat the Prince of Condé[99] and, at the same exhibition that the *Descartes* statuette was shown, he also exhibited a sculpture of his patron.[100] In 1793 when Dardell began a sculpture of Newton, after the French Revolution and the abolishment of the monarchy in 1792, his patron was Jacques-Louis David (1748–1825) the pre-eminent artist of the period, who was a supporter of the French Revolution and a friend of Robespierre. The terracotta model of Newton has been lost but the its title *Newton discovers and displays Truth* is very informative, Newton holds a prism in one hand to indicate the theory of colours, and in the other a magnetic circle to designate his system of attraction.[101] In the French Revolution, Newton was often used to symbolize the god of reason and featured prominently in political tracts.[100]

5.3. The Romantic Age

In the Age of Enlightenment reason was supreme and science was its backbone. God was sidelined and subject to his own laws of nature. As in science, the imaginative arts could be reduced to principles. All was ordered. It was social change that essentially bought the Age of Enlightenment to an end. At the beginning of the eighteenth century there were the lower orders which became the lower class by the end of the century, no longer preordained or immutable and above them the middle and upper classes. The Industrial Revolution was largely the cause of the change. In practice the ordering of society had never been fixed but the change in term coincided with a recognition that there was some fluidity in social position and that movement from one class up or down to another could be effected by the individual. Reason was stressed during the Age of Enlightenment but on its own it was not enough and in the Romantic Age the individual and their emotions became as important as reason. The Romantic Age was at its height in literature from the beginning of the French Revolution in 1789 to the Parliamentary Reforms in Britain of 1832. In music the Romantic Age lasted much longer there but there do not seem to be any significant links with physics.

Young Englishmen at the end of the Age of Enlightenment, before the Reign of Terror began in 1793, embraced the French Revolution. William Wordsworth (1770–1850) went to Revolutionary France briefly in 1790 and returned again the next year after he graduated from Cambridge when he was twenty-one. In France he became enraptured with the Republicans and with Annette Vallon who bore his child in 1792. Afterwards he became disenchanted in the Republicans because of the Reign of Terror, and Wordsworth left Vallon and returned to England. However those years in Revolutionary France had lasting impact and later in 1805 he composed *The French Revolution as It Appeared to Enthusiasts at Its Commencement*, a poem that is almost a hymn to the Romantic Age that begins:

> Oh! pleasant exercise of hope and joy!
> For mighty were the auxiliars which then stood
> Upon our side, we who were strong in love!
> Bliss was it in that dawn to be alive,

5.3.1. *Poets and their arts*

The British Romantic Movement was born out of the intense politics surrounding the French Revolution. Wordsworth had entered St John's College, Cambridge two years before the start of the Revolution and the slightly younger Samuel Taylor Coleridge (1772–1834) entered Jesus College two years after its start. The older William Blake (1757–1827) did not have any formal schooling and entered into an apprenticeship in etching and engraving prints when he was fourteen years old, which he completed in 1779 and enrolled as a student in the Royal Academy of Arts. Blake was interested in poetry from an early age, his first poems being written when he was twelve years old. With Blake it is impossible to separate his visual art from his poetry and both will be discussed in this section. The three best known second generation Romantic Poets, George Gordon Byron (1788–1824) who became Lord Byron on the death of his great-uncle in 1798, Percy Busshe Shelley (1792–1822), and John Keats (1795–1821); all died young before the end of the Romantic Age. Byron and Shelley went to Cambridge and Oxford University respectively whereas the younger Keats trained as surgeon at Guy's Hospital, London. The American Edgar Allan Poe (1809–1849), better known as a writer of tales of mystery and the macabre, thought there was beauty in science, and wrote romantic poems. The position on science among these seven poets varied from Shelley who had a passion for it, especially chemistry,[102] to Keats and Blake who were anti-science.

The interest in science by the Romantic Poets was complimented by an interest in poetry by men of science such as Erasmus Darwin who used poetry in *The Botanic Garden* to popularize his science, Humphry Davy who wrote his thoughts on life in poems and Wordsworth's friend, William Rowan Hamilton (1805–1865) the mathematician and physicist. Looking back on his early life in 1847, Hamilton wrote to John Herschel that 'it would really seem to have been at one time a toss-up, whether I should turn out a rhymer or an analyst.'[103] In 1829 Wordsworth, who Hamilton met in 1827, visited him in Ireland and the two spoke of the conflicts between poetry and science, which they had corresponded about earlier. As a result of this visit Hamilton wrote *On Poetry*[104] renouncing poetry in the last verse but confirming the unity of human culture:

> Spirit of Beauty! though my life be now
> Bound to thy sister Truth by solemn vow,
> Though I must leave thy sacred hill,
> Yet be thine inward influence with me still:
>   And with a constant hope inspire,
>   And with a never-quenched desire,
> To see the glory of your joint abode,
> The home and birth-place, by the throne of God!

5.3.1.1. *Poets receptive to science*

During the 1790s Coleridge thoroughly approved of Newton. In *Religious Musings* written between 1794 and 1796, he linked Newton with Milton.

> ... To Milton's trump
> The high groves of the renovated Earth
> Unbosom their glad echoes: inly hushed,
> Adoring Newton his serener eye
> Raises to heaven: and he of mortal kind
> Wisest, he[105] first who marked the ideal tribes
> Up the fine fibres through the sentient brain.
> Lo ! Priestley there, patriot, and saint, and sage,
> Him, full of years, from his loved native land
> Statesmen blood-stained and priests idolatrous
> By dark lies maddening the blind multitude
> Drove with vain hate....

Joseph Priestley, the dissenting minister and natural philosophy, was another of Coleridge's heroes who supported the French Revolution and who went with his family to America in 1794 to escape the reaction against people like him. Coleridge and Wordsworth were millenarians, that is, they believed that the second coming of Christ would be soon, as were all those mentioned in *Religious Musings*. However a decade later Coleridge became disenchanted by mechanical philosophy believing that it left no room for spirituality and was antagonistic towards Newtonianism.[106]

In the second edition to *Lyrical Ballads*, containing poems by both Wordsworth and Coleridge that was published in 1800, Wordsworth included a preface in which he compares the poet and the man of science and wrote about the proper subjects for poetry in a way that agreed with Ezra Pound's (1885–1972) ideas in the twentieth century:[107]

> 'Poetry is the first and last of all knowledge – it is as immortal as the heart of man. If the labours of Men of science should ever create any material revolution, direct or indirect, in our condition, and in the impressions which we habitually receive, the Poet ... will be ready to follow the steps of the Man of science, not only in those general indirect effects, but he will be at his side, carrying sensation into the midst of the objects of the science itself. The remotest discoveries of the Chemist, the Botanist, or Mineralogist, will be as proper objects of the Poet's art as any upon which it can be employed, if the time should ever come when these things shall be familiar to us, and the relations under which they are contemplated by the followers of these respective sciences shall be manifestly and palpably material to us as enjoying and suffering beings. If the time should ever come when what is now called science, thus familiarized to men, shall be ready to put on, as it were, a form of flesh and blood, the Poet will lend his divine spirit to aid the transfiguration, and will welcome the Being thus produced, as a dear and genuine inmate of the household of man.'

However Wordsworth in his poetry also often criticized men of science. In Book IV of his long poem *The Excursion* first published in 1814 but written more than five years earlier, he condemned such men for analyzing and dissecting nature until all the grandeur was eliminated:[108]

> Now, shall our great Discoverers, "he exclaimed,
> Raising his voice triumphantly," obtain
> From sense and reason, less than these obtained,
> Though far misled? Shall men for whom our age
> Unbaffled powers of vision hath prepared,
> To explore the world without and world within,
> Be joyless as the blind? Ambitious spirits –
> Whom earth, at this late season, hath produced
> To regulate the moving spheres, and weigh
> The planets in the hollow of their hand;
> And they who rather dive than soar, whose pains
> Have solved the elements, or analysed
> The thinking principle – shall they in fact
> Prove a degraded Race? and what avails
> Renown, if their presumption make them such?
> Inquire of ancient Wisdom; go, demand
> Of mighty Nature, if ' twas ever meant

That we should pry far off yet be unraised;
That we should pore, and dwindle as we pore,
Viewing all objects unremittingly
In disconnection dead and spiritless;
And still dividing, and dividing still,
Break down all grandeur, still unsatisfied
With the perverse attempt, while littleness
May yet become more little; waging thus
An impious warfare with the very life
Of our own souls!
And if indeed there be
An all-pervading Spirit, upon whom
Our dark foundations rest, could he design
That this magnificent effect of power,
The earth we tread, the sky that we behold
By day, and all the pomp which night reveals;
That these – and that superior mystery
Our vital frame, so fearfully devised,
And the dread soul within it – should exist
Only to be examined, pondered, searched,
Probed, vexed, and criticised?

In 1829 when Hamilton questioned Wordsworth about this criticism he replied that it was not those who elevated the mind to God to whom he objected but those who collected facts for their own sake or material ends.[109] And some forty years, after he started writing the autobiographical *Prelude* in 1798,[110] Wordsworth added two last lines in his description of Newton's statue in the antechapel of Trinity College in Book III, which describes his time in Cambridge, changing what was a bleak description of the statue to a glorification of Newton's achievements and science:

Near me hung Trinity's loquacious clock,
Who never let the quarters, night or day,
Slip by him unproclaimed, and told the hours
Twice over with a male and female voice.
Her pealing organ was my neighbour too;
And from my pillow, looking forth by light
Of moon or favouring stars, I could behold
The antechapel where the statue stood

Of Newton with his prism and silent face,
The marble index of a mind for ever
Voyaging through strange seas of Thought, alone.

Byron showed little interest in science though he did eulogize Newton beautifully in the Tenth Canto of Don Juan written in 1823

When Newton saw an apple fall, he found
  In that slight startle from his contemplation –
'T is said (for I 'll not answer above ground
  For any sage's creed or calculation) –
A mode of proving that the earth turn'd round
  In a most natural whirl, called 'gravitation';
And this is the sole mortal who could grapple,
Since Adam, with a fall or with an apple.

Shelley became interested in science at a young age. When he was ten years old, Shelley attended Syon House, a boarding school in Islington, London. At the time even public schools did not have full time science masters and at Syon House during the time Shelley was there between 1802 and 1804, Adam Walker[111] (ca. 1730–1821) a successful itinerant science lecturer gave lectures probably based on his own successful book *A System of Familiar Philosophy*[112] published in 1799.[113] Walker's lectures, especially those on chemistry and electricity, fascinated Shelley.[114] In 1804 Shelley moved to Eton where Walker also gave lectures. Walker was an amateur scientist and, though he was reasonably good, he had some outdated and peculiar notions of his own that he emphasized, such as, that fire, light, the caloric or phlogiston, and electricity were all manifestations of the same principle, which particularly appealed to Shelley.[113] Walker's favourite hypothesis was used by Shelley in the Fourth Act of *Prometheus Unbound* as a vehicle for science though it does not add to the story.[102] Desmond King-Hele, a physicist, has given some interesting interpretations of the Fourth Act, for example when Ione describes the white 'winged infant' driving the moon's chariot:

Its limbs gleam white, through the wind-flowing folds
Of its white robe, woof of etherial pearl.
Its hair is white – the brightness of white light
Scattered in strings; ...

King-Hele suggests that:[102]

> '"The wind-flowing folds" are the straggling lunar mountain ranges; the white hair "scattered in strings" probably refers to the bright ray systems seen around some craters at full moon. The "eyes" of the winged infant pour out "fire that is not brightness", presumably the infrared radiation discovered by Herschel in 1800.'

A little later Panthea, describing an opening in the forest, responds:

> And from the other opening in the wood
> Rushes, with loud and whirlwind harmony,
> A sphere, which is as many thousand spheres,
> Solid as crystal, yet through all its mass
> Flow, as through empty space, music and light:
> Ten thousand orbs involving and involved,
> Purple and azure, white, and green, and golden,
> Sphere within sphere; and every space between
> Peopled with unimaginable shapes,
> Such as ghosts dream dwell in the lampless deep,
> Yet each inter-transpicuous, and they whirl
> Over each other with a thousand motions,
> Upon a thousand sightless axles spinning,
> And with the force of self-destroying swiftness,
> Intensely, slowly, solemnly, roll on,
> Kindling with mingled sounds, and many tones,
> Intelligible words and music wild.

Here King-Hele suggests that Shelley used the image of whirling molecules that produce words and music. King-Hele may be right in his detailed interpretations in Act 4 of *Prometheus Unbound* but what it is certain is that Shelley's love of science permeated his poetry. Alfred Whitehead was prompted to write: 'If Shelley had been born a hundred years later, the twentieth century would have seen a Newton among chemists,'[115]

The American Poe in the Preface to his long prose poem *Eureka: An Essay on the Material and Spiritual Universe* published in 1848 wrote:[116]

> 'I offer this Book of Truths, not in its character of Truth-Teller, but for the Beauty that abounds in its Truth; constituting it true.'

In the prose poem itself, Poe made a good attempt at solving what has become known as *Olbers' Paradox*:, which was first posed by Kepler in 1610: Why is the night sky dark if there are an infinite number of stars? In 1823 Heinrich Olbers (1758–1840) suggested that light was absorbed by interstellar space.[117] In *Eureka* Poe wrote:

> 'Were the succession of stars endless, then the background of the sky would present us an uniform luminosity, like that displayed by the Galaxy – *since there could be absolutely no point, in all that background, at which would not exist a star*. The only mode, therefore, in which, under such a state of affairs, we could comprehend the *voids* which our telescopes find in innumerable directions, would be by supposing the distance of the invisible background so immense that no ray from it has yet been able to reach us at all.'

Olbers' Paradox was not fully resolved until the 1980s and it is more appropriate to leave further discussion of it until Chapter 7.

Although Poe believed that science was beneficial to the arts, thought that science could dull the poet as he expressed in his sonnet *To Science* written in 1829:[118]

> Science! true daughter of Old Time thou art!
> Who alterest all things with thy peering eyes.
> Why preyest thou thus upon the poet's heart,
> Vulture, whose wings are dull realities?
> How should he love thee? or how deem thee wise,
> Who wouldst not leave him in his wandering
> To seek for treasure in the jewelled skies,
> Albeit he soared with an undaunted wing?
> Hast thou not dragged Diana from her car?
> And driven the Hamadryad from the wood
> To seek a shelter in some happier star?
> Hast thou not torn the Naiad from her flood,
> The Elfin from the green grass, and from me
> The summer dream beneath the tamarind tree?

5.3.1.2. *Poets antagonistic to science*

As Aldous Huxley noted in his *Literature and Science* many poets have detested Newton because he cut the 'connections between the stars and the heavenly host'.[9] Keats objected to the demystification of nature by science and in *Lamia* written in 1819 wrote:[119]

> ... Do not all charms fly
> At the mere touch of cold philosophy?
> There was an awful rainbow once in heaven:
> We know her woof, her texture; she is given
> In the dull catalogue of common things.
> Philosophy will clip an Angel's wings,
> Conquer all mysteries by rule and line,
> Empty the haunted air, and gnomed mine –
> Unweave a rainbow, as it erewhile made
> The tender-person'd Lamia melt into a shade.

It is interesting to speculate that Poe was influenced by Keats' poem when he wrote his sonnet *To Science* the two quotations seem to express the same thought. What is certain is that the influence of *Lamia*, and especially the passage quoted, continues in the twenty-first century.

Of all the Romantic Poets it was Blake who most consistently rejected science. Blake was a visionary, literally as well as metaphorically, and saw science as the destroyer of visions and imagination. Although Newton decried hypotheses, significant scientific advances need an initial vision which can then be tested by experiment or predictive deductions. Poets need to be visionaries but testing their poetry is more problematic and one has to rely on subjective criteria such as the test of the poet Alfred Housman (1859–1936): does it make the hairs on the back of your neck stand on end?

From 1795 to about 1805, Blake produced a series of twelve large colour prints, which were probably in six paired sets based mostly on the Bible.[120] One pair of this series was *Newton* and *Nebuchadnezzar* whose composition is very similar:[121] Newton is sitting but crouched down and Nebuchadnezzar is on his knees and hands. Blake depicted Newton, in his print shown in Plate 8, as a naked young man seated on seaweed covered rock, with squids and sea anemones or polypuses attached to it, leaning down drawing with a compass.[122] In Blake's mind Newton, like the Chaldean wise men called before Nebuchadnezzar, was denied divine wisdom.[120] It was not that Blake denied the genius of Newton but he attacked what he saw as Newton's materialism in leaving out God and all the immaterial aspects of life.[123] Blake also identified Satan as Newton's ruler writing 'O Satan ... Art thou not Newton's Pantocrator' in his epic poem *Milton*.[124] In Blake's *Europe a Prophecy* 'A mighty Spirit leap'd from the land of Albion,/Nam'd Newton,' who for Blake was the modern representation of Urizen, the personification of reason and science. Blake imagined that Newton used his compass to encircle and capture nature. Thus it is curious that Eduardo Paolozzi (1924–2005), the Scottish sculptor of Italian immigrants, chose Blake's *Newton* as the model for his 1995 bronze statue to commemorate Newton that stands in the courtyard of the British Library London.

Blake saw Newton as one of the unholy trinity of Bacon, Newton, and Locke and he attacked them especially in his epic poem *Jerusalem*:[125]

> ... [T]he Spectre like a hoar frost & a Mildew rose over Albion,
> Saying, I am God, O Sons of Men! I am your Rational Power!
> Am I not Bacon & Newton & Locke who teach Humility to Man?
> Who teach Doubt & Experiment ...

In Blake's language 'the Spectre of Man [is] the Holy Reasoning Power, And in its Holiness is closed the Abomination of Desolation,' and Humility 'is the teaching of the Spectre'.[125] In the unfinished poem *The Everlasting Gospel*, which Blake left scattered through his *Note-Book*, he aimed to prove that Jesus was not meek and mild and that humility should not be shown even to God; humility meant forsaking man's godlike potential.[123] Later in the poem Blake showed that, while he might think that Bacon, Newton, and Locke were destroyers of visions, he recognized their greatness by pairing them with his heroes Milton, Shakespeare, and Chaucer:[125]

> The Druid Spectre was Annihilate, loud thund'ring, rejoicing terrific, vanishing,
> Fourfold Annihilation, & at the clangor of the Arrows of Intellect
> The innumerable Chariots of the Almighty appear'd in Heaven,
> And Bacon & Newton & Locke, & Milton & Shakspear & Chaucer.

5.3.2. *Visual arts*

By the eighteenth century colourmen were providing artists with ready-made paint. Thomas and Richard Rowney, originally wigmakers, started producing artists' colours in 1783 and two of the Romantic painters, Joseph Mallord William Turner (1775–1851) and John Constable (1776–1837), who excelled in the use of colour, were among their customers.[126] Obviously colour has always been extremely important to painters and many in the past such, as Leonardo da Vinci, have left their thoughts on it. Modern colour theory starts with Newton, who was the first to demonstrate with the prism that white light was composed of colours. He identified seven colours to reflect the seven notes of music and produced the first colour wheel with the colour segments in proportion to the musical notes.[127] Newton also briefly discussed the difference in behaviour of coloured light and pigments in Part II of Book I of the *Opticks*. Coloured lights are combined in an additive process but, since coloured pigments absorb colours other than the colour of the pigment, when pigments are combined, as in paint, the process is a subtractive one. Thus when red, green, and blue lights are combined a reasonable white light is produced, as shown in Plate 9a. The secondary colours are produced by addition: yellow is produced by adding red and green, green and blue give cyan, and magenta is produced by adding blue and red. Television and computer monitor screens use red green, and blue (RGB system). In mixing pigments because light is being subtracted the primary and secondary colours are reversed. If we have three primary pigments cyan, yellow, and magenta that are used in three colour printers,[128] the three secondary colours are blue, green, and red, and the combination of the three primary colours tend towards black as all light becomes absorbed as shown in Plate 9b. Why only three primary colours are needed to produce all the colour hues is because in reasonably bright light colour is registered on our retina by three different cone cells which are most sensitive to blue, green, and red light.[129] Turner was appointed Professor of Perspective at the Royal Academy of the Arts in 1807 and over a period of some seventeen years he delivered a number of lectures on colour.[130] Turner was one of the few artists of his time who understood the difference between combinations of coloured lights and pigments and observed that red, yellow and blue lights combined to give a reasonable white light while with pigments the result is a muddy colour. The distinction between the additive mixture of light and the subtractive mixture of pigments was not clarified until 1867 and the publication of Helmholtz's *Handbuch der physiologischen Optik*.[131]

Newton's theory of colours did not go unchallenged by natural philosophers or artists during the eighteenth and the first half of the nineteenth century. Many of these challenges were due to confusion between physical colours, that is light of different wave lengths, and physiological colours that we perceive with our eyes. Thus in 1831 we find Brewster stating that all the colours of the spectrum of sunlight can be formed from red, yellow, and blue and there are not seven independent colours as defined by Newton.[132] Young had appreciated the difference between physical and physiological colours and in 1801; he originally gave red, yellow, and blue as the colours we perceived but, as a result of experiments by Wollaston, changed these to red, green, and violet in 1802.[62] A triad of primary colours had been proposed by Moses Harris (1730–ca. 1788), a naturalist and artist, in his *Natural System of Colours* published in 1770, his primary, or as he described them grand or principal, colours were red yellow and blue. From these three primary colours, Harris obtained three secondary colours to complete what he called his prismatic colour wheel.[130] In his lectures on colour, Turner used Harris' prismatic wheel. Turner understood that, since mixing pigments is a subtractive process, it does not produce a brilliant hue and in the 1820s he experimented with a version of what was to become known as *pointillism* in parts of some of his watercolours. If two colours are laid side by side in the form of lines or dots they become perceived by the eye as the additive hue, if they are viewed so that the individual lines or dots cannot be resolved by the eye, which is brighter than the hue obtained by mixing the pigments. Turner, in some of his watercolour River Scenery series, commissioned for a series of etchings and hence quite small, used very fine lines or dots that are difficult to see in parts of the painting to increase their brilliance. In *Shields, on the River Tyne*, shown in Plate 10, the area of shade on the wharf in the lower right hand corner has tiny dots of red and blue-green; the smoke rising from the fire is a light green wash covered with fine lines of vermillion.[130] Unfortunately these details are too small to be seen in the reproduction in Plate 10.

By the time of the Neo-Impressionists, colour theory as developed by Young, Maxwell, and Helmholtz, was available from a number of sources. In 1881 Georges-Pierre Seurat (1859–1891) first became aware of the American physicist and amateur artist, Ogden Rood (1831–1902) and his book *Modern Chromatics* through a newly published French translation.[133] Rood gave a practical account of the development of colour theory that exactly described *divisionism*, that pointillism was originally called, which was to be used by Seurat in his ground breaking painting *Sunday Afternoon on the Island of La Grande Jatte*, shown in Plate 11, which he painted between 1884 and 1886.[134] In his book Rood, wrote:

> 'when different colours are placed side by side in lines or dots, and then viewed at ... a distance ... the blending is more or less accomplished by the eye of the beholder... But before this distance is reached there is a stage in which the colours are blended, though somewhat imperfectly, so that the surface seems to flicker or glimmer.... This communicates a soft and peculiar brilliancy to the surface, and gives it a certain appearance of transparency.'

Unfortunately *La Grande Jatte* is not as brilliant now as it was when it was first exhibited because Seurat used unstable paints. Even as early as 1892 Seurat's friend and art critic Félix Fénéon[135] (1861–1944) wrote: 'Because of the colours which Seurat used toward the end of 1885 and in 1886, this painting of historical importance has lost its luminous charm: while the reds and blues are preserved, the Veronese greens are now olive greenish, and the orange tones which represented light now represent nothing but holes.'[136]

Toward the end of his life Turner was influenced by Goethe's theory of colours in *Zur Farbenlehre*, originally published in 1810 but which he read and annotated in the English translation by his friend and fellow painter Charles

Eastlake (1793–1865) in 1840.[137] Goethe thought that he had disproved Newton's theory of colour. Whereas Newton viewed the image of a slit illuminated by the Sun and projected though a prism on a screen, Goethe looked through a prism. When he first tried this experiment in 1790 he looked at a white wall and to his surprise saw no spectrum at all, he then turned and looked through a window again he saw no spectrum but instead coloured fringes around the horizontal lattices of the window.[138] Goethe reasoned wrongly that no colours were seen when he viewed the white wall because there was no contrasting darkness and conceived the notion that colours were produced by the interplay of darkness and light that occurred when he viewed the lattice window. Essentially it was this effect that led Goethe to believe that Newton was wrong. Goethe thought that white light was homogeneous and not the result of combination of the colours of the spectrum. Aristotle also thought that colours were produced by the interplay of light and darkness. Even though Turner annotated his copy of Eastlake's translation it is not clear to what extent he accepted Goethe's theory or rejected Newton's. However, Eastlake clearly believed that Goethe was more relevant to painting than Newton.[139]

Goethe saw no colours when he viewed a plain white wall because light entered the prism from all directions and so, though each ray of white light was split into a spectrum of colours, these colours combined with all the spectra from rays entering the prism from different directions to produce white light. When Goethe looked at the latticed window, he did not receive light from all directions near the lattice and so he saw only a partial spectrum. The effect of looking at a lattice through a prism is similar to viewing uniformly illuminated white rectangular panels on a black background as Goethe described in Part 2 of *Zur Farbenlehre*. The white panels have coloured bands along their horizontal edges, as shown in Plate 12, if the prism is horizontal with an apex bottommost, bands of violet and blue are seen along the top edges of the two broadest white panels whereas along the bottom edges there are red and yellow bands between these coloured bands the light is white. On the third and narrowest band the coloured bands overlap and the blue mixes with the yellow and forms green. Newton described a similar effect in his *Opticks* where he allowed sunlight to pass through a wide slit positioned near to a prism, his illustration of this experiment is shown in Fig. 5.8, where Newton indicated the spectra obtained from rays of light coming from the extreme top and bottom of the slit. Newton described the resulting colour effects seen on a piece of paper placed at MN or mn:[140]

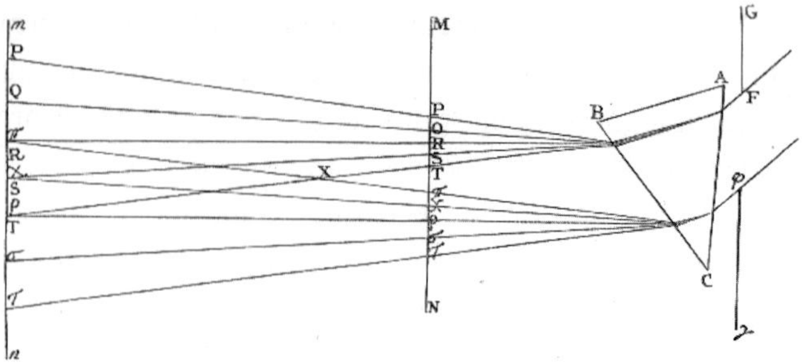

Fig. 5.8. Newton's figure from *Opticks* illustrating the colours seen when sunlight from a wide slit falls on a prism: P and π are rays of deepest violet; Q and χ are rays of middle sort between indigo and blue; R and ρ are rays of middle sort of green; S and σ are rays of middle sort between yellow and orange; T and τ are rays of deepest red. *Opticks*, Book I, Part II, Fig. 12, Newton (1718).

> 'If the paper MN be so near the prism that PT and πτ do not interfere with one another, the distance between them Tπ will be illuminated by all sorts of rays in that proportion to one another which they have at their very first coming out of the prism, and consequently be white. But the spaces PT and πτ hand either hand will not be illuminated by them all, and therefore appear coloured. And particularly at P, where the outmost violet-making rays fall alone, the colour must be the deepest violet. At Q where the violet-making and indigo-making rays are mixed, it must be a violet inclining much to indigo. At R where the violet-making, indigo-making, blue-making, and one half of the green-making rays are mixed, their colours must [Newton here referred to the use of his colour wheel to determine the compound colour from the quality and quantity of the primary colours] compound a middle colour between indigo and blue. At S where all the rays are mixed except the red-making and orange making, their colours ... compound a faint blue verging more to green than indigo. And in the progress from S to T, this blue will grow more and more faint and dilute, till at T ... it ends in whiteness.'

Newton then explained how the colour at τ is the deepest red graduating to yellow at χ becoming fainter again until it is white at π. If the paper is moved to mn where the rays from the edges of the slit merge over the centre part colours fill the whole space between P and τ and the centre is 'a fuller green than before.' As Newton noted the effect is the same if a white rectangle on a black background is viewed through a prism and also the colours that are visible when there is a black rectangle on a white background. Goethe discussed the identical problem in *Zur Farbenlehre* and gave a very similar diagram (see Plate 13) but whereas Newton gave a simple explanation of the phenomena, Goethe only wrote vaguely of 'the movement of the dark against the light,' he also remarked that 'an objective experiment with a dark object has hitherto scarcely been thought of.' Unless Goethe ignored much of what Newton wrote in *Opticks*, he could not have read it very thoroughly.

The two paintings of Turner that were most influenced by Goethe are the two he exhibited at the Royal Academy in 1843: *Shade and Darkness – the Evening of the Deluge*, and *Light and Colour* (*Goethe's Theory*) – *the Morning after the Deluge – Moses writing the Book of Genesis*, which is shown in Plate 14. Turner was somewhat of a poet and he had a habit of adding a verse to his paintings when he exhibited them: *Light and Colour* had a verse from his *Fallacies of Hope* which was much influenced by the minor poet and clergyman, John Langhorne (1735–1779):[141]

> The ark stood firm on Ararat; th' returning sun
> Exhaled earth's humid bubbles, and emulous of light,
> Reflected her lost forms, each in prismatic guise
> Hope's harbinger, ephemeral as the summer fly
> Which rises, flits, expands and dies.

Turner seldom discussed the meaning of his paintings and there is much debate about these two paintings. Clearly Turner was thinking about Goethe's theory of colour because he told us so in the full title. Although the small scale reproduction shown here at first sight looks like an abstract painting it is an impression of the morning after the deluge. On close inspection the figure of Moses can be seen, about a quarter of the painting down from top centre, seated on a cloud clad mountain presumably writing the *Book of Genesis*; below him at the centre is a bronze serpent coiled around a pole, which God instructed Moses to make as an antidote to the serpents who were biting the Israelites because they had spoken against God and Moses for leading them out of Egypt into the wilderness. In the bottom left hand quarter of the painting there are many human bodies floating in the retreating waters of the flood. All is contained in a giant iridescent bubble – Goethe referred to the shimmering colours seen on bubbles and the octagonal frame in which *Light and Colour* was originally exhibited would have emphasized the bubble even more. Even though Turner was very restrained with his use of blue in *Light and Colour* it is yellow and blue, the two colours that Goethe thought were the fundamental colours, which dominate the painting. For Goethe, yellow was allied to light and blue with darkness.

Although Turner's *Shade and Darkness* and *Light and Colour* were not understood when first exhibited they are now well appreciated; in the display caption in the Tate Gallery *Light and Colour* is described as: 'This triumphant explosion of light brilliantly exploits the warm side of the spectrum. It celebrates God's Covenant with Man after the Flood.' Before leaving Turner and his use of natural philosophy, the reader is recommended to read the French philosopher, Michel Serres', stylistic essay *Turner Translates Carnot* in which he argues that

Turner consciously or unconsciously was influenced by thermodynamics. Whether Serres' thesis is right or not, the essay is interesting reading.[142]

Constable never experimented with dots or fine lines of adjacent colour to intensify colour brilliancy but he did become known for adding tiny dabs of gold, silver, ivory, and especially white, highlights to his oil paintings so much so that they became known as Constable's snow. He was just as interested in colour theory as Turner and Constable and painted rainbows with more sympathy than anyone else during the Romantic Period but it was not until the about 1830 that he really studied rainbows scientifically. Newton explained most of the phenomena of rainbows in his *Opticks*. It is significant that Goethe did not attempt to explain the optics of a rainbow. Small raindrops less an about half a millimetre are spherical, because surface tension dominates the forces acting on them, but larger drops are flattened by the reduction in air pressure around their mid-horizontal plane as they fall. A 4 mm diameter raindrop is on average about 20% fatter than its height and their shape oscillates because of vortex shedding; very few raindrops are this large so that Newton's assumption of spherical shape was correct. When Constable was studying rainbows there was little known about the shape of raindrops; some sheets of stationery paper exist with Constable's sketches of the mechanism of the formation of rainbows and the possible shapes of raindrops.

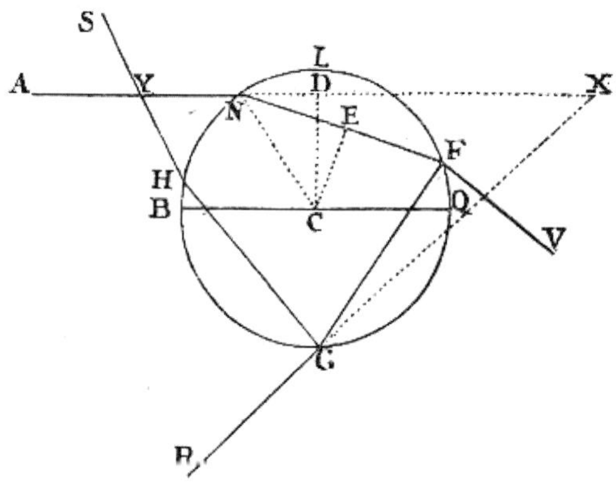

Fig. 5.9. Newton's illustration of the refraction and reflection of light in a raindrop. *Opticks*, Book I, Part II, Fig.14, Newton (1718).

Just as a pure spectrum of colours can only be seen from a prism if a narrow slit admits the sunlight, so too a pure spectrum of colours is not visible in a rainbow. A primary, or inner, rainbow is formed by sunlight being refracted as it enters a raindrop, reflected, and then refracted again as it leaves the drop. Newton's illustration of this phenomenon is shown in Fig. 5.9. A ray of sunlight AN enters the drop and is refracted along NF where part of the light refracts out

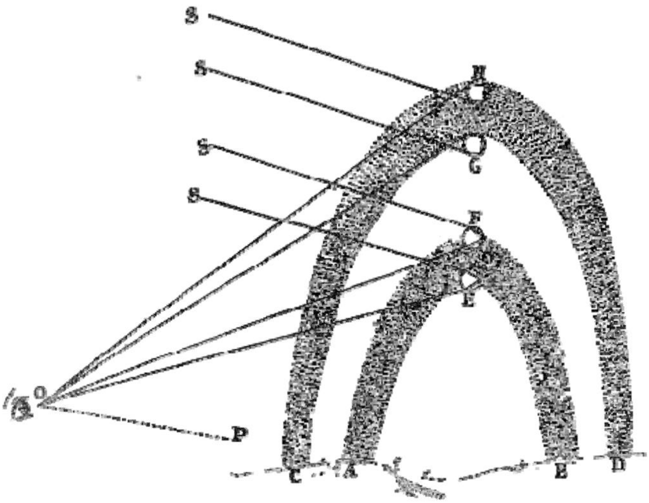

Fig. 5.10. Newton's illustration of primary and secondary rainbows. *Opticks*, Book I, Part II, Fig. 15, Newton (1718).

of the drop along FV and part is reflected along FG; at G part of the ray is refracted out of the raindrop along GR to form the primary rainbow and part is reflected along GH and refracted out along HS to form the secondary rainbow. The angle AXR for the primary rainbow depends on the position of N on the drop. As the point N moves from B to L the angle AXR varies slowly and for the extreme red end of the spectrum Newton calculated that it reaches a maximum at 42° 2′, for the extreme violet end it is 40° 17′. These are the angles at which the spectrum is brightest and they define the half angles of the cone of the spectrum for the primary rainbow that we see. Newton's illustration of the inner primary and outer secondary rainbow is shown in Fig. 5.10 where OP is parallel to the sunlight falling on the raindrops. In the primary rainbow the inner most portion is violet with the angle POE equal to 40° 17′ and the outer portion is red with the angle POF equal to 42° 2′ Because in the secondary rainbow the ray emerging from the raindrop crosses over the ray entering the drop, the innermost rays of the secondary rainbow are red and the outermost rays are violet with the angles POG and POH equal to 50° 51′ and 54° 7′ respectively. In 1812 when Constable painted *Landscape with Double Rainbow* he got the order of the colours in the secondary rainbow wrong. However by 1831, when he is assumed to have painted the watercolour *View from Hampstead with a Double Rainbow*, shown in Plate 15, both rainbows were scientifically correct.[143] Constable sketched an illustration of the optics of primary and secondary rainbows very similar to Fig. 5.9. The secondary rainbow is fainter than the primary because some light is lost each time light is reflected; it is also wider. Higher order rainbows are possible but these are so faint that they can be rarely seen. In the letterpress to the engraving of his *Stoke by Neyland*, made in 1830, Constable included a short passage on the correct features of rainbows.[144]

Scientific accuracy in a painting does not necessarily make great art. In 1831, Constable painted the famous *Salisbury Cathedral from the Meadows*, shown in Plate 16, which has a dramatic rendering of a rainbow arching over and protecting the cathedral but since the sunlight comes from the right hand side of the painting a rainbow would not be visible because the Sun must be directly behind the observer for a rainbow to be seen.[145] Constable selected some lines from James Thomson's *The Seasons* to be printed alongside the painting's title in the Royal Academy catalogue.

> As from the face of heaven the scatter'd clouds
> Tumultuous rove, the interminable sky
> Sublimer swells, and o'er the world expands
> A purer azure. Through the lightened air
> A higher lustre and a clearer calm
> Diffusive tremble; while, as if in sign
> Of danger past, a glittering robe of joy,
> Set off abundant by the yellow ray,
> Invests the fields; and nature smiles revived.

5.4. Victorian Age and Afterwards

The two scientific topics that most influenced the literature during the Victorian Age were evolution and thermodynamics and frequently both are significant influences on a piece of literature but only the thermodynamic influences are considered here.[115] Thermodynamics had little influence on the other two imaginative arts discussed.

Although the first law of thermodynamics is optimistic – the universe will not run down like a clock – the second law is pessimistic. William Thomson wrote in 1852:[146]

> 'within a finite period of time to come the earth must ... be, unfit for the habitation of man as at present constituted, unless operations have been, or are to be performed, which are impossible under the laws to which the known operations going on at present in the material world are subject.'

In 1930 James Jeans, coined the term *heat death* writing in the second edition of *The Universe Around Us* that 'The second law of thermodynamics predicts that there can be but one end to the universe – a "heat-death" in which the total energy of the universe is uniformly distributed, and all the substance of the universe is at the same temperature.'[147] The morbid fascination of the Victorians with death may have made the acceptance of the implications of the second law of thermodynamics easier. Tait and others, such as Helmholtz, also discussed the application of the thermodynamic laws to the cycle of life.[148] The plants converting the carbon dioxide exhaled by animals back to oxygen and carbon and Tait stated: [148]

> 'that if plants were not assisted by some external supply of energy, here would be something equivalent to the perpetual motion on the grandest conceivable scale.... But it is quite obvious that a process of that kind cannot go on without a supply of energy from without. The raising of energy from the lower form to the higher always requires external application of some fresh energy, which is itself degraded in the process.'

Many in the Victorian Age saw analogies between the physical and social world, Balfour Stewart (1828–1887), a Scottish physicist, and Norman Lockyer wrote an article for a popular magazine in 1868 on the analogies:[149]

> 'In the social world a man may degrade his energy, so also in the physical world may energy be degraded; in both worlds, when degradation is once accomplished, a complete recovery would appear to be impossible, unless energy of a superior form [mechanical work or a high temperature heat source] be communicated from without.... Long before any of these [thermodynamic] laws were known the superiority of certain kinds of energy were instinctively recognised.'

Helmholtz was a force in music and painting in the nineteenth century. In 1863 Helmholtz published (in German) his ground breaking work: *On the Sensations of Tone as a Physiological Basis for the Theory of Music*, which was first translated into English in 1870.[150] Also between 1871 and 1873 he delivered a series of lectures on the relation of optics to painting, which were translated into English in 1881.[151] The former of these works is by far the most important. Helmholtz's aim was to connect the boundaries of physics and physiology to both the theory of music and painting.

5.4.1. *Literature*

In seeking the influence of thermodynamics on the literature, as with all fields, there is a whole spectrum from the speculative with no real evidence to the well documented evidence of influence. Even to assert the possibility that a consciously thermodynamic influence exists it is necessary to first demonstrate that the influence is significant, not just a phrase in one line of a poem, and that the author had some definite connection with thermodynamics. The problem of separating real from imagined influences is compounded because thermos-dynamics is about *energy*, this word originally entered the English language at the end of the sixteenth century to express vigour of expression and developed in the seventeenth century to include the exercise of any kind of power, and was only in 1807 adopted by Thomas Young to describe the capacity for doing work of some kind. Previously physical *energy* had been covered by that overworked fourteenth century word *force*; in its current usage in physics its meaning is that which causes motion.[152] When science adopts a word in general use as the description of a particular phenomenon it is never clear when that word is used by non-scientists whether they are using it in its general sense or in its particular scientific meaning. Thus the use of the word energy in literature, even reasonably frequent use, does not necessarily imply thermodymanic influence. Entropy does not have this problem because it has no common usage but it is interesting that when it is used by non-scientists it is frequently treated as being almost synonymous with energy. Because entropy is not a common word, it was not often used in literature, even if the writer was really invoking the second law of thermodynamics, until after World War II.

There have even been attempts to infer that some literary figures have understood the laws of thermodynamics before they have been completely realized by physicists themselves.[153] One example of this is Alfred Tennyson (1809–1892). *In Memoriam* was written for his close friend Arthur Henry Hallam (1811–1833) who died suddenly at the age of twenty-two. Tennyson started this long poem, really a series of poems, a few months after the death of Hallam and completed it in 1849 well before the physicists themselves had developed the second law of thermodynamics; even the first law of thermodynamics was not formulated until the 1840s. It is true that Thomson realized that Carnot had effectively made use of both laws in 1824, but that was in hindsight, neither Carnot nor anybody else at the time, realized the full import of his work. Tennyson was not without interests in astronomy and geology but he demonstrated no interest in what in 1854 was to become known as thermodynamics.[154] Certainly there are phrases and verses such as

> ‘The stars,’ she whispers, ‘blindly run;
> A web is wov’n across the sky;
> From out waste places comes a cry,
> And murmurs from the dying sun:

that could be construed as a reference to the second law but there are other simpler interpretations. Here the 'priestess in the vaults of Death' is whispering with her 'lying lip' the 'waste places' are the 'vaults of Death' and the 'dying sun' is simply the setting of the Sun, that is, the dying Hallam.

Another problem is with the statistical mechanics interpretation of the second law of thermodynamics and the idea that disorder flows from order and not the reverse is probably as old as man himself and does not necessarily have anything to do with thermodynamics. Eddington quoted the children's rhyme Humpty Dumpty:

> Humpty Dumpty sat on a wall
> Humpty Dumpty had a great fall.
> All the king's horses and all the king's men
> Couldn't put Humpty together again.

as an unconscious form of the second law.[155] There are many examples in literature after the middle of the nineteenth century where the influence may be due to the time honoured concept that disorder flows from order rather than to the second law. Thus in many cases it is impossible to say to a high degree of certainty that thermodynamics has been an influence. One of the first candidates for the real influence of thermodynamics on a literary figure is the poet Gerard Manley Hopkins (1844–1889).[156,157] However, though there is hard evidence that Hopkins was interested in astronomy and optics, it is only conjectured that he had an interest in thermodynamics. Hopkins went up to Oxford in 1863 and eventually read *Literae Humaniores* or *Greats* which then included elements of the mathematical sciences including physics. In 1864 his interest in astronomy, which continued for the rest of his life, is evident in this poetic fragment:

> ___ I am like a slip of comet,
> Scarce worth discovery, in some corner seen
> Bridging the slender difference of two stars,
> Come out of space, or suddenly engender'd
> By heady elements, for no man knows:
> But when she sights the sun she grows and sizes
> And spins her skirts out, while her central star
> Shakes its cocooning mists; and so she comes
> To fields of light; millions of travelling rays
> Pierce her; she hangs upon the flame-cased sun,
> And sucks the light as full as Gideon's fleece:

> But then her tether calls her; she falls off,
> And as she dwindles shreds her smock of gold
> Amidst the sistering planets, till she comes
> To single Saturn, last and solitary;
> And then goes out into the cavernous dark.
> So I go out: my little sweet is done:
> I have drawn heat from this contagious sun:
> To not ungentle death now forth I run.

And during his final year at Oxford he showed his interest in optics and astronomy in an essay he wrote *The Tests of Progressive Science*, which contained references to the use of spectroscopy in determining the elements present in astronomical bodies.[158] Hopkins was brought up an Anglican but converted to Catholicism after graduating in 1867 and one year later entered a Jesuit seminary to train for the priesthood. In the late nineteenth century the Jesuits were favourably inclined to astronomy and physics. Hopkins spent three short periods at Stonyhurst College in Lancashire, from 1870 to 1873, for three months in 1878, and for a year and four months from late 1882.[159] Science had a strong presence at Stonyhurst, which had its own observatory. From 1868 to his death Stephen Perry (1833–1889), who was elected to a fellowship of the Royal Society in 1874, carried out significant astronomical and meteorological research at Stonyhurst and very likely he influenced Hopkins.[160] Hopkins general interest in science at Stonyhurst is demonstrated by his having had two letters published in *Nature* on an atmospheric phenomenon during his third period there.

The nearest we come to evidence that Hopkins was interested in thermodynamics is his interest in Tyndall and his writings. In 1868 Hopkins took an alpine walking holiday in Switzerland and met Tyndall at the foot of the Matterhorn. It seems that Hopkins maintained an interest in Tyndall who translated many of the works of Helmholtz and wrote various popular articles on thermodynamics and other topics. Hopkins certainly read Tyndall's infamous 1874 Belfast address to the British Association on scientific materialism[161] and wrote to his mother that he found 'it interesting and eloquent, though it made me most mad'.[162] In his address Tyndall discussed the first but not the second law of thermodynamics. So there is the possibility that Hopkins was informed about at least some aspects of thermodynamics. The Christian religion offers both a problem and a solution to an acceptance of the second law which Hopkins' poetry seems to express. The problem is that the laws of nature detract from 'God's Grandeur', as Hopkins expressed in his poem written in 1877. The second law of thermodynamics is insistent on an end to the universe. I think that

Hopkins' solution, which he seems to articulate in his poetry, is 'That Nature is a Heraclitean Fire' and, before the final whimper and the fire goes out, the 'Resurrection' takes charge and 'this Jack is immortal diamond'.[163] At least this is the gist of those critics who find that Hopkins was significantly influenced by thermodynamics. Although there is possibly some truth in what they write, I cannot subscribe to all their particular examples and still all may be in their eyes not Hopkins'.[164] I end this short consideration of Hopkins as something of a second Lord Kelvin with one of his last and best poems written in his sprung rhythm that the respected literary critic Gillian Beer certainly believes was influenced by thermodynamics:[165]

***That Nature is a Heraclitean Fire and of the comfort of the Resurrection*:**

Cloud-puffball, torn tufts, tossed pillows | flaunt forth, then chevy on an air-
Built thoroughfare: heaven-roysterers, in gay-gangs | they throng; they glitter in
marches.
Down roughcast, down dazzling whitewash, | wherever an elm arches,
Shivelights and shadowtackle in long | lashes lace, lance, and pair.
Delightfully the bright wind boisterous | ropes, wrestles, beats earth bare
Of yestertempest's creases; | in pool and rut peel parches
Squandering ooze to squeezed | dough, crust, dust; stanches, starches
Squadroned masks and manmarks | treadmire toil there
Footfretted in it. Million-fueled, | nature's bonfire burns on.
But quench her bonniest, dearest | to her, her clearest-selved spark
Man, how fast his firedint, | his mark on mind, is gone!
Both are in an unfathomable, all is in an enormous dark
Drowned. O pity and indig | nation! Manshape, that shone
Sheer off, disseveral, a star, | death blots
black out; nor mark
Is any of him at all so stark
But vastness blurs and time | beats level. Enough! the Resurrection,
A heart's-clarion! Away grief's gasping, | joyless days, dejection.
Across my foundering deck shone
A beacon, an eternal beam, | Flesh fade, and mortal trash
Fall to the residuary worm; | world's wildfire, leave but ash:
In a flash, at a trumpet crash,
I am all at once what Christ is, | since he was what I am, and
This Jack, joke, poor potsherd, | patch, matchwood, immortal diamond,
Is immortal diamond.

The second example of the possible influence thermodynamics on a Victorian English literary figure presented here is George Eliot (1819–1880) one of the greatest Victorian novelists. She had little formal education but with the early encouragement of Charles Bray (1811–1884) she educated herself and met many of the Victorian intelligentsia. Eliot knew many who were interested in thermodynamics and read their works.[166] She met Herbert Spencer the popular Victorian philosopher in 1851 and, though Eliot was at first romantically drawn to him, he became just a lifetime friend. Spencer believed that all evolution, both of the universe and biological, could be explained by the first law of thermodynamics but he did not really understand the second law.[167] On a visit to Germany in 1854–1855 with George Lewes (1817–1878), Eliot met Helmholtz. In London both Eliot and Spencer were on visiting terms with Tyndall. In the late 1860s and early 1870s she also had regular contact with Maxwell and attended some of his lectures. She was familiar with the kinetic theory of gases[168] and she read thermodynamic works by Helmholtz, William Grove (1811–1896) and Tyndall's *Fragments of Science*, which contained the elements of thermodynamics.[166] The mature Eliot certainly had knowledge of thermos-dynamics by the time that she wrote *Middlemarch* which was published in eight instalments, as was common in the nineteenth century, during 1871–72. Thermodynamics covers living as well as inert things and as Tait said in the introduction to his 1874 Edinburgh lectures on the physical sciences: [148]

> 'It was a grand step in science which showed that just as the consumption of fuel is necessary to the working of a steam-engine, or to the steady light of a candle, so the living engine requires food to supply its expenditure in the forms of muscular work and animal heat...Now it may be startling to some of you, especially if you have not particularly considered the matter, to hear it surmised that possibly we may, by the help of physical principles, especially that of the Dissipation of Energy, some time attain to a notion of what constitutes Life – mere vitality, nothing higher.'

Eliot thought that thermodynamic concepts could be extended further to cover social life. There is thread running through *Middlemarch* of dissipated or 'wasted energy' that is unproductively expended in an analogous manner to the loss of mechanical work in heat engines.[166] Dissipation of energy cannot be avoided in the real irreversible world and in *Middlemarch* it occurs to both those characters, who have major faults such as Casaubon, Featherstone, Rosamond, and Bulstrode as well as to the heroes of the work, Dorothea and Lydgate. In her last book, *Daniel Deronda* (1876), Eliot made direct reference to the science of thermodynamics writing that:

> 'Heat is a great agent and a useful word, but considered as a means of explaining the universe it requires an extensive knowledge of differences; and as a means of explaining character "sensitiveness" is in much the same predicament.'

However *Deronda* does not have the same thread of wasted energy throughout the book as *Middlemarch* does.

In the second half of the nineteenth century the science fiction genre blossomed. There were two types of science fiction: the scientific novel which had the didactic aims of popularizing science and the true science friction which had no such aims. Examples writers of the first type are Jules Verne (1828–1905) and, especially in the context of this section, Camille Flammarion (1842–1925) who wrote *La Fin du Monde* (1893), such writers will not be considered here because their novels are examples of direct application of science and technology for its own sake. In a similar differentiation, and to limit an already burgeoning chapter, only writers who used science fiction to discuss more general topics such as social and political issues will be mentioned. The prime example of this type of science fiction writer is H. G. Wells. Although Wells never wrote a story where thermodynamics was a central theme, he did use thermodynamics to express the pessimism of the *fin de siècle* in both *The Time Machine* (1895) and *The War of the Worlds* (1898).[169] The Time Traveller built a time machine and travelled to the future, his first stop was 802,701 years into the future. Eventually the Time Traveller was to go thirty million years into the future so that at eight hundred thousand odd years the world still had some way to go but there was evidence of dissipation of energy and an approach to uniformity at even this distance into the future:[170]

> 'It seemed to me that I had happened upon humanity upon the wane. The ruddy sunset set me thinking of mankind.... No doubt the exquisite beauty of the buildings I saw was the outcome of the now purposeless energy of mankind before it settled down into perfect harmony[,] ... the fate of energy in security: it takes to art and eroticism and then come languor and decay.'

Humanity had devolved into the androgynous Eloian elite and the brutish Morlockian workers. Escaping the Morlocks the Time Traveller went forward thirty million years and he saw the Sun grow large and red, as the Earth approached nearer the Sun, the Moon vanished, and the alternations of day and night grew slower, as:[170]

> 'the work of the tidal drag was done' by the Sun brought the Earth 'to rest with one face to the sun.... [An] abominable desolation ... hung over the world.... [The Time Traveller goes further into the future] watching with a strange fascination the sun grow larger and duller.'

An eclipse of the Sun, possibly by Mars, caused a 'great darkness'; the cold smote him to his marrow and he returned to the Victorian Age. Wells tells us himself in the preface of a 1931 edition that he wrote the *Time Machine* under the influence of thermodynamics and the pessimism of the *fin de siècle* brought about by the approach of the Sun's death in some millions of years and apologizes for believing in: [170]

> 'the dreadful lies about the about the "inevitable" freezing up of the world ... [and that] life would be over in a million years or less.'

These "lies" were words of William Thomson who in 1862 wrote in a popular article in Macmillan's Magazine[171] that the 'inhabitants of the earth cannot continue to enjoy the light and heat essential to their life, for many million years longer.'[172] It seems Wells' change in heart was brought about by him reading the optimistic popular version of cosmology given by Jeans in his *The Universe Around Us*, who in 1928 was the first to propose, the now discredited, Steady State cosmology. In Jeans' book, Wells read that Sun would last millions of millions of years. Certainly the predictions of heat death that Wells used in the *Time Machine* were too low but eventually the second law will prevail and everything in the universe become cold and uniform so that no energy can be utilized.[173] Thermodynamics does not play a central part in Wells' *War of the Worlds* but he did write that the probable reason that the Martians invaded the Earth was that Mars is much smaller than the Earth and therefore cooled more quickly and 'that [the] stage of exhaustion, which to us is still incredibly remote, has become a present-day problem for the inhabitants of Mars' and cause them to invade the Earth. Wells was a journalist and liked to add real topical facts to his fiction. Although it has nothing to do with thermodynamics, it is interesting that Wells made use of the observation of a strange light on Mars at the Nice Observatory in 1894 where Henri Perrotin (1845–1904) was the Director and he even gave the correct reference of its publication as a letter in *Nature* on 2nd August; it was even speculated that it could have been Martians trying to signal the Earth![174] In the *War of Worlds* Wells wrote: 'I am inclined to think that this blaze [which in the letter to *Nature* was also speculated to be perhaps a forest fire] may have been the casting of the huge gun' which the Martians used to fire a spaceship to Earth.

Joseph Conrad (1857–1924) admired the work of Wells and the admiration was mutual. The heat death of the Sun and Wells' interpretation in *The Time Machine* was in Conrad's mind when he wrote *Heart of Darkness* (1899). Earlier in 1897 he had written to his new friend Robert Cunninghame Graham

(1852–1936), the socialist author, discussing the old sailor Singleton in his novella *The Nigger of the Narcissus*: [175]

> 'Nothing can touch him but the curse of decay – the sun, the stars one by one, and in another instant shall spread a frozen darkness over the whole universe.'

The horror of the great darkness remained with Conrad and he repeated similar thoughts in an appreciation of the American writer Henry James (1843–1916) for the North American Review in 1905.[176] Such pessimism was common at the *fin de siècle*; in 1902 Bertrand Russell wrote: [177]

> 'All the noonday brightness of human genius, are destined to extinction in the vast death of the solar system, and that the whole temple of Man's achievement must inevitably be buried beneath the debris of a universe in ruins.'

Conrad right from the beginning of *Heart of Darkness*, when Marlow, Conrad's *alter ego* who appeared in several of his novels, is on the cruising yawl *Nellie* at anchor on the sea-reach of the Thames with his friends and about to tell his tale, darkness is mentioned menacingly:

> 'The air was dark above Gravesend, and farther back still seemed condensed into a mournful gloom, brooding motionless over the biggest and greatest, town on earth.'

Heart of Darkness is a dark story, and darkness is mentioned nearly fifty times, but behind the darkness there is the much bigger menace of heat death – the setting sun on the *Nellie* is a 'dull red without rays and without heat, as if about to go out suddenly, stricken to death.' The idea of wasted energy was also used by Conrad. Before Marlow starts his story he discusses the difference between the nineteenth century colonist and the Roman invaders of Britain saying 'What saves us is efficiency – a devotion to efficiency,' during his description of the voyage out to the African country,[178] where he was to captain the steamer trading in ivory and they come across a French warship shelling the jungle in vain:

> 'a tiny projectile would give a feeble screech – and nothing happened. Nothing could happen,'

When speaking of the traders, who Marlow called pilgrims because they invariably carried long staves, and who attempted to shoot a troublesome hippopotamus he said:

> 'The pilgrims used to turn out in a body and empty every rifle they could lay their hands on at him.... All this energy was wasted.' By the end of the Heart of Darkness, Marlow's dark story has infected the narrator with his *fin de siècle* pessimism and the death of the sun

> appears almost imminent: the way of the *Nellie* 'was barred by a black bank of clouds, and the tranquil waterway ... seemed to lead into the heart of an immense darkness.'

There is also darkness in *The Secret Agent* but the key adjective used is gloomy, which is used to describe frankness, haughtiness, fury, and sympathy as well the more usual London weather. In the 1920 *Author's Note*, Conrad described the London of the story as 'indifferent to heaven's frowns and smiles; a cruel devourer of the world's light.' Sunny is not a way one would describe London but in *The Secret Agent* the sun is 'bloodshot' or 'rusty' and ever declining. Even the old terrorist, Karl Yundt, had 'no visions of blood-red flags waving, or metaphorical lurid suns of vengeance rising above the horizon of a doomed society.' As Winnie, the sister of the doomed Greenwich bomb thrower, poor half-witted stumbling Stevie, reflected, 'everything decays, wears out, in this world ... [all that she excepted] from the rule of decay affecting all things human and some things divine' was sisterly devotion. Decay, wasted energy point towards the lesson to be learnt from thermodynamics – even the Chief Inspector, as 'the principal expert in anarchist procedure' is named, Heat. Verloc would have advanced 'the disintegration of the old morality' by the 'violent disintegration' of Stevie, if the boy had not been identified by his address sewn inside his overcoat by Winnie, Verloc's wife. In the final paragraph 'the incorruptible Professor', an anarchist who specialized in explosives and was once an assistant demonstrator in chemistry at a technical institute, is used to illustrate the impossibility of the aim of terrorists in a world where the social, as well as the physical, world is governed by the second law of thermodynamics.

> 'And the incorruptible Professor walked too, averting his eyes from the odious multitude of mankind. He had no future. He disdained it. He was a force. His thoughts caressed the images of ruin and destruction. He walked frail, insignificant, shabby, miserable – and terrible in the simplicity of his idea calling madness and despair to the regeneration of the world. Nobody looked at him.'

The last novel of Aldous Huxley *Island* published in 1962, was in some ways an echo of Conrad's *Heart of Darkness*.[179] Both featured cannibalism. In Heart of Darkness there is actual cannibalism, either indulged in or tolerated by Kurtz, as well as the cannibalistic exploitation of the Congo by Europe; in *Island* visions of cannibalism are conjured up by Will Farnaby during a moksha-medicine trance in the form of praying mantises and there is a similar cannibalistic exploitation of the island Pala by Western oil companies. Kurtz is affected by 'The horror! The horror!' and Farnaby by 'the Essential Horror' neither of which are fully explained. Both stories are pessimistic and both have some thermodynamic influences. In *Heart of Darkness* thermodynamics is not explicitly mentioned though the second law of thermodynamics permeates the story; in *Island* both entropy and the second law of thermodynamics are explicitly mentioned but thermodynamics is not central to the story and the pessimism is not due to the second law. Huxley came from a scientific family; science was not a foreign language, so it is not surprising that he would use scientific words in his vocabulary without necessarily carrying scientific implications beyond the immediate metaphor.

In the above discussion I am reasonably sure that the literary figures were influenced by thermodynamics. I end this section with a short discussion of the modernist Thomas Stearns Eliot (1888–1965), better known as T.S. Eliot, and his 1925 poem *The Hollow Men*. Katherine Edbury and other critics believe that in this poem Eliot was influenced by the second law of thermodynamics, the evidence is inconclusive and I am doubtful but they may be right.[180] The apparent evidence for Eliot being influenced by thermodynamics starts with the epigram to the poem: 'Mistah Kurtz – he dead,' which is a quotation from Joseph Conrad's *Heart of Darkness*. Conrad was genuinely influenced by the second law of thermodynamics but Eliot's use of this quote for an epigram almost certainly had nothing to do with the second law but was used simply because Kurtz was a hollow man.[181] However the references to fading and dying star in twilight kingdoms and especially the last verse:

> This is the way the world ends
> This is the way the world ends
> This is the way the world ends
> Not with a bang but a whimper.

seems awfully like a poetic rendering of the second law. Eliot probably had a better grasp of the laws of thermodynamics than most literary figures of the time. He went to Harvard in 1906 and, after an eclectic choice of subjects grouped around comparative literature, he obtained a BA and MA. In the autumn of 1910, Eliot went to the Sorbonne in Paris for a year's postgraduate study where he attended lectures by Henri Bergson (1859–1941) and others.[182] Bergson, a major French philosopher, had an early interest in physics but eventually decided to study philosophy, whose doctoral thesis was on time and free will. In his thesis Bergson examined what elements of physics, including thermodynamics, could be applied to human action. Eliot returned to Harvard in 1911 where he commenced a doctoral thesis on Bergson's British idealist critic Francis Herbert Bradley (1846–1924). Bertrand Russell, a leader in the realists' reaction to nineteenth century idealism, was a visiting professor at Harvard in 1914 and Eliot was one of his students. The scientific background of Russell and the Lowell lectures that he delivered, published later as a book with the title *Our Knowledge of the External world as a field for Scientific Method in Philosophy*, could not have failed to have influenced Eliot.[183] In his earlier years Eliot had been receptive of science[184] but after the early 1920s he became disenchanted with science and in his letters he mentioned at least twice that 'unless you know a good deal about science you don't know enough to AVOID it.'[185] This change in Eliot's view of science may have been due to his approaching conversion to Christianity and Anglo-Catholicism in 1927.[184] Some of the attributions of the influence of thermodynamics on literature in the nineteenth century have less evidence to support them than the speculation about *The Hollow Men*.

5.4.2. *Helmholtz and music*

Helmholtz saw that music was closely connected with pure sensation and in his *Sensations of Tone* he answered why there was the connection of harmony with the ratios of the first few prime numbers that Pythagoras had first observed '[150] This observation could be explained because Helmholtz had knowledge of Fourier's analysis of periodic variables and the work of the anatomist Alfonso Corti (1822–1876) on the ear, especially the structure of the cochlea and its nerve fibres, which earlier writers on the theory of music did not have.[186]

The law by which the ear analyzes a musical note, based on Fourier's analysis, was first stated by Georg Ohm (1787–1854) in 1843 and Helmholtz developed this law and 'prove[d] that the human ear really does analyse musical tones according to the law of simple vibrations' in his *Sensations of Tone*.[187] Helmholtz also calculated, in his appendix, the strength of the partials of a string plucked or struck at particular position from the dynamics of a string. Thus Helmholtz gave the definitive theory of partials. But that is physics. We do not analyze the actual sound vibrations when we listen to music but perceive them through our hearing system. Along with the partials of a note we perceive other tones because of this system. Two combinational tones, other than the overtones, we perceive are a difference and a summation tone. If two pure notes of frequency f_1 and f_2 ($f_2 > f_1$) are sounded we also perceive notes of $f_2 - f_1$ and $f_1 + f_2$. The difference tone had been discovered a century earlier by the German organist Georg Sorge (1703–1778) and became more widely known through the Italian Baroque composer Giuseppe Tartini (1692–1770) and were commonly called *Tartini tones* but even Helmholtz did not perceive the summation tone, which is very weak, until after he had demonstrated its existence analytically. In all the previous work on vibration it had been assumed that the vibrations were

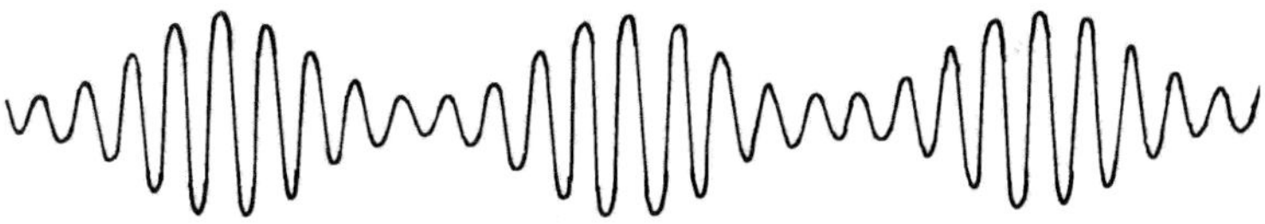

Fig. 5.11. Beats formed by two pure notes of similar frequencies.

infinitely small and linearly proportional to the disturbing force. Helmholtz analyzed the vibration of a system where the restoring force, F, is non-linear and given by

$$F = ax + bx^2, \tag{5.1}$$

where x is the displacement. Such as system excited by two frequencies, f_1 and f_2, contains in addition to sympathetic frequencies that would be would present if the non-linear constant b was zero, a difference tone $f_2 - f_1$, and a summation tone $f_1 + f_2$. We perceive sound through the vibration of the tympanic membrane which if it had a non-linear response would explain these extra terms. Almost as a footnote to his appendix on the mathematical analysis, Helmholtz also makes the observation that the form of Eq. (5.1) is unsymmetrical as well as non-linear. The ear's

> 'drum skin is especially distinguished by its want of symmetry, because it is forcibly bent inwards by the handle of the hammer, and ... that this peculiar form of the tympanic membrane conditions the generation of combinational tones.'

Because the combinational tones are weak they are usually not detected when the two notes are sounded by musical instruments and do not have any significant effect on dissonance. Dissonance is caused by beats. When two pure tones of different but similar frequencies f_1 and f_2 are sounded together the combined sound is a note of frequency $(f_2 + f_1)/2$ that varies in amplitude at a frequency $(f_2 - f_1)/2$ as shown in Fig. 5.11. This variation in amplitude causes a throbbing of the sound called beating. If the beat frequency is low the effect is pleasant rather than disturbing and similar to vibrato but as the beat frequency increases it becomes harsh or rough. The perceived roughness increases to a maximum and then decreases again until the roughness is imperceptible as the beat frequency becomes large. The beat frequency which produces the maximum roughness increases with the frequency of the notes as does the frequency at which roughness become imperceptible. Helmholtz gave no explicit results of his own but in his fourth edition of 1877 he included the results of the American physicist Alfred Mayer (1836–1897) who connected the perception of roughness to the phenomenon of aural persistence and showed that, when beats blended into a

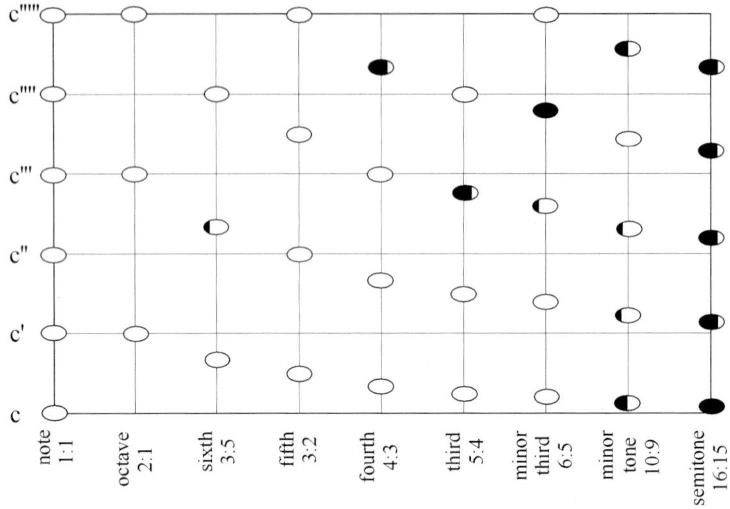

Fig. 5.12. Roughness of partials of various intervals based on Helmholtz (1895), ⬭ zero roughness, ◐ quarter roughness, ◐ half roughness, ◑ three quarter roughness, ⬬ complete roughness.

smooth continuous sound because of aural persistence, they were no longer objectionable.[188] The beat frequency that causes the maximum roughness Mayer perceived as one third that that caused the beats to become imperceptible but he gave, and Helmholtz used, a factor of one quarter, which was the perception of Emma Seiler (1821–1886) the singer and voice teacher who in earlier years had worked with Helmholtz. Mayer found that the beat frequency for maximum roughness varied linearly with the frequency of the lower note. Helmholtz chose Mayer's results to construct curves for the roughness caused by the fundamental and the higher partial notes when a middle c is played on one violin and an increasingly higher note is played on another from c′ to c‴; these curves are reproduced in numerous books on music. I have chosen to give a simpler presentation, based on Helmholtz's conception of dissonance, that I think more clearly shows why the ratios of the first few numbers produce consonance and ratios of higher numbers dissonance. In Fig. 5.12 the note c with all the harmonic partials up to c′′′′′ is shown on the left; to the right are the consonant intervals of an octave (1:1), a sixth (2:1), a fifth (3:2), a fourth (4:3), a major and minor third (5:4 and 6:5), which only contain the first six numbers, and the dissonant intervals of a minor tone (10:9) and a semitone (16:15), which contain intervals formed from numbers higher than six. The partials of these intervals show whether or not they form rough beats with the partials of first note (the code is indicated in the figure caption); the more black that is seen in the partials of intervals, especially in the lower partials, the rougher the interval; The interval of an octave is of course perfectly consonant and so is a fifth, the sixth, the fourth, and both the major and minor thirds are all considered consonant but they contain a few beating partials, which give maximum roughness in the highest partials but only little roughness in the lower partials. However the minor tone and especially the semitone contain black in the lower partials and are dissonant. Thus Helmholtz fully explained why intervals formed from the first six numbers are the most consonant, the lower partials do not interfere and produce no significant roughness.

5.4.3. *Helmholtz and painting*

Although Helmholtz stated that music very closely depended on the nature of our perceptions, other arts such as painting are also affected by our senses. [151] In the four lectures on painting that Helmholtz gave in 1873, he covered form, shade, colour, and harmony of colour.[151]

The effect that photography had on the first two topics in Helmholtz's lectures is seen from his many references to it. Stereoscopic photographs had been made since the 1840s and in discussing perspective Helmholtz remarked upon 'the wonderful vividness which the solid form of objects acquires when good stereoscopic images are viewed in the stereoscope.' Although two dimensional paintings cannot perfectly reproduce a three-dimensional view, Helmholtz discussed the ways that an artist can give the impression of three-dimensionality. He also mentioned that since we are very familiar with some forms like houses that we are preconditioned to perceive some elements in a painting as three-dimensional. Shadows too are used to convey that impression. Rather surprisingly at first sight Helmholtz discussed why the sky is blue in his section on form. Perhaps the reason why it appeared in that section and not in the section of colour was because it was not known until Rayleigh's final paper on the transmission of light through the atmosphere that the strong scattering of the shorter blue wavelengths which causes the blue of the sky is by the molecules in the air and not by suspended small particles.[189] Helmholtz also explained why sunsets are red; the light is transmitted through much more air at sunset and sunrise and since the scattering of light is inversely proportional to the fourth power of the wavelength of light the blue end of the spectrum is more quickly lost and the sun appears red. Unlike the blue of the sky, sunsets are significantly affected by atmospheric pollution, the more the pollution the redder the sun.

In his section on colour Helmholtz not surprisingly discussed how we perceive colour through nerves sensitive to red, green, and blue as supposed by Thomas Young. He also noted that our eye is much more sensitive to slight variations in the intensity of blue than red. Helmholtz discussed the harmony in colours caused by the physiology of the eye and did not compare colours to music as Newton did. To Helmholtz the harmony or discord in colours is related to the way we perceive colour.

5.5. Concluding Remarks

The influence of classical physics on the imaginative arts did not of course end with the twentieth century but, since art is largely led by the young who eagerly seek after the new, it was the emerging modern physics that has most captured the imagination in the twentieth century. Although individual artists continued to be influenced by classical physics after the nineteenth century there was no longer movements whose inspiration could be so identified as a whole.

The chapter began with a discussion of the two cultures debate which dominated the interaction between the sciences and art in the middle of the last century. As I have tried to show that debate was in many ways an artificial one. There were curmudgeonly figures in both the sciences and arts, and still are, who see human endeavour as being separable into isolated houses with one being grander and more important than the other. I too can be seen as artificially separating the physics from science. I have done this not because I believe that physics is better, though I do believe that physics is the more fundamental science since the life sciences were necessarily built upon physics, but because the book would have been too unwieldy if the life sciences were included and also to do so would have been beyond my capability. I hope that this chapter adequately demonstrates that the imaginative arts have never been isolated from classical physics.

Both the sciences and the arts rely on imagination and some of the most influential figures, like Leonardo da Vinci, are impossible to classify as either a scientist or an artist because they saw no distinction between the two. Most of us are not polymaths and till a smaller field but the start of every endeavour is imagination. What distinguishes the sciences from the arts is that imagination has to be formally tested before it becomes science. This step in science is absolutely crucial. I have not considered the influence of the arts on science because it is difficult to prove. On occasions I am certain the imaginative first step in science has been inspired by the arts but that is almost impossible to prove. Thus I have not included any speculations as to science originating in the arts and I have also tried to eliminate those influences of science on the arts that cannot be demonstrated but may exist.

5.6. Notes

[1] Huxley (1964).
[2] The biographic details of Snow and Leavis come mainly from the Dictionary of National Biography. Weintraub (2004); MacKillop (2004).
[3] Snow (1956).
[4] I personally have not found that right wing views are the preserve of the humanities. The most right wing person I have ever known was a Russian scientist who lived in the Soviet Republic until the fall of the Berlin Wall and was one of their trusted people. When I visited him in Moscow in the 1960s he had a servant who slept on a pullout pallet in the kitchen whom he treated like a serf.
[5] Feynman (1965).
[6] Published in book form Snow (1959).
[7] Published in book form Leavis (1962).
[8] Leavis (1962).
[9] Huxley (1963).
[10] Snow (1959).
[11] Campbell (1921).
[12] Because Ptolemy's first name was also Claudius.
[13] Usher (2006).
[14] Marchitello (2011).
[15] Pumfrey (2011).
[16] Galilei (1610).
[17] More than 5 kilometres high. The highest mountain on the Moon is 5.5 kilometres high and there are two others that are more than 4 kilometres high.
[18] In *Discorso intorno i Galleggianti*, Galileo gave the periods of the four major satellites (using their modern names), which are compared with modern values in brackets: *Io* 1.77 (1.769) days, *Europa* 3.56 (3.551) days, *Ganymede* 7.17 (7.154) days, *Calisto* 16.75 (16.689) days.
[19] One of the characteristics of the metaphysical poets was that they used metaphysical conceits where metaphors are treated as if they were real. Thus in Donne's poem *The Sun Rising* the Sun who comes through the window and discovers the lovers is treated as an actual person.
[20] Samuel Johnson attacked Donne's poetry particularly vigorously in his Life of Cowley published in 1781.
[21] Nicolson (1950).
[22] Donne's biographical details are taken from Colclough (2004).
[23] Oliver (2002). Harroit was the most important astronomer of the two and tutored Lower. Roche (2004).
[24] Nicolson (1935).
[25] Donne (1611a).
[26] Russell (1964).
[27] The modern designation of this star is P Cygni. By 1626 it has decreased in magnitude so much that it was not visible to the naked eye since then it has varied up and down in brightness and is on average a star of the fifth magnitude.
[28] Bernstein (1997). What is not clear is why Kepler, a protestant, found Ignatius his Conclave an 'insolent satire' as he stated in a footnote to his science fiction novella *Somnium, Sive Astronomia Lunaris*.
[29] Kargon (1966).
[30] Gilbert (1600).
[31] As was usual at the time, Donne did not explicitly mention Gilbert but, from the text, it is obviously him. Donne (1651).

[32] The *Funerall Elegie* was published along with *The first Anniversary.* Donne (1611b).
[33] *Artist* here is used in the late sixteenth century sense of a learned man.
[34] In other poems Donne used the similar device of spelling he, *hee* when referring to God. The text of Milton's *Paradise Lost* also exhibits non-standard spelling sometimes doubling the letter e, sometimes omitting the letter as in *thir* instead of *their*. Milton was completely blind by 1652 and recited his poem to a copyist so one would expect that the spelling used after this date was due to the copyist or the printer. It has been suggested that the double ee was used for emphasis but such as use does not seem to be consistent and hence no significance can be attached to the differences in spelling. Adams (1954).
[35] Donne (1635).
[36] It has been suggested that the first line refers to earthquakes. There were earthquakes felt in London in the 1580 and 1601, the 1580 Dover Straits earthquake caused consternation and some half dozen chimney stacks to fall in London. The largest European earthquake immediately before Donne wrote *A Valediction Forbidding Mourning*, was the Lucerne earthquake of 1601. However, it is highly unlikely that the Metaphor King would have mixed his metaphors so that the first two lines of the stanza referred to earthquakes and the third and fourth lines referred to astronomy.
[37] The first recorded use of the word trepidation in its more usual sense was by Francis Bacon in 1605.
[38] The modern figure for a complete precession of the earth's axis is 25,770 years.
[39] The quotations are taken from the 1674 edition where Milton revised *Paradise Lost* into twelve books. Milton (1674). The original spelling in the poem has been retained.
[40] Campbell (2004).
[41] Margaret Cavendish's biographical details are taken from Fitzmaurice (2004).
[42] Cavendish (1653a,b). Margaret had intended to publish both in the same book but she was late in getting *Philosophical Fancies* to the printers and so it was published separately. Sarasohn (2010).
[43] Woolf (1927).
[44] Westfall (1980).
[45] Kepler (1619). See §1.1.2.
[46] Stephenson (1994).
[47] Ruff and Rodgers (1979).
[48] The largest of its kind in the UK and the seventh largest in the world.
[49] Galileo did not go blind until four years after his heresy conviction and in the meantime wrote his equally famous treatise *Discourse Concerning the Two New Sciences.* Blindness did not stop Galileo working and he was dictating ideas on the theory of impact to Vincenzo Viviani and Evangelista Torricelli when he succumbed to a fever which caused his death in 1642.
[50] Vincenzo was a renowned musician but the first real opera was not written until the very end of the sixteenth century after he had died.
[51] Galilei (1638).
[52] The partials of a note are the individual Fourier components each with a frequency of a whole multiple of the fundamental.
[53] An aliquot part of something is a part which goes into the whole an integer number of times.
[54] Wallis (1677).
[55] Christensen (1993).
[56] Needham and Wang (1962).
[57] Gouk (1986).
[58] West (1978).
[59] Panofsky (1956).
[60] Huerta (2007).
[61] Newton (1704).

[62] Young (1802).
[63] The ratio of the frequencies of the extreme violet and red ends of the visible spectrum is about 1.97.
[64] Pemberton (1728). Marjorie Hope Nicolson used this quotation for her very interesting book on the eighteenth century poets and Newton's *Opticks.* Nicolson (1963).
[65] At the time it was the custom to name a coffee house not after the owner but after the head waiter. Button had been a servant of Lady Warwick whom Addison married; at first he was a waiter and only subsequently became the owner.
[66] Green (2012).
[67] Ukers (1922).
[68] Addison and Steele were Whigs and Button's Coffee House was predominantly Tory and Swift, Gay, Pope, and Arbuthnot sided more with the Tories. This difference in politics created tension between them especially between Swift and Steele. By 1715 Pope and Gay were frequenting Will's Coffee House on the opposite side of Russell St.
[69] Nicolson and Rousseau (1968).
[70] See §2.1.
[71] I have reversed the order of these lines.
[72] Nicolson and Mohler (1937).
[73] Swift (1710).
[74] See §2.3.
[75] Quennell (1956).
[76] Swift (1726).
[77] Worth (1991).
[78] Swift (1755).
[79] Quoted as a frontispiece in Patey (1991).
[80] Sorrenson (1996).
[81] Johnson (1995).
[82] See §3.1.1.
[83] Sambrook (2004).
[84] Robertson (1908).
[85] Voltaire (1752).
[86] Wade (1950).
[87] Crowe (1986).
[88] During the journey by comet Micromégas, passing close by Mars, observed its two moons described by Swift yet 'which have escaped the attention of our astronomers.'
[89] 'I have caught nature in the act' is a phrase from one of Fontenelle's works, which was quoted by Lord Bolingbroke when Fontenelle was himself discovered in a compromising situation with the Marquise de Tencin (1682–1749), a Parisian salon hostess. Voltaire was supportive to Maupertuis and helped him obtain a royal pension for the sisters and wrote a poem for their cause.
[90] d'Alembert (1751).
[91] Rameau (1722).
[92] Philip Gossett's introduction to *Traité de l'Harmonie.* Rameau (1722).
[93] Lester (2002).
[94] Johns (1992).
[95] Charles Lennox (1701–1750), the grandson of Charles II and the second Duke of Richmond, was an important patron for the paintings of which only nine eventually appeared in his *Tombeaux des Princes, Grands Capitaines et Autres Hommes Illustrés, Qui ont Fleuri dans la Grande-Bretagne vers la Fin du XVII et le Commencement du XVIII Siècle* published in 1741.

[96] Cohen (1979).
[97] A later version of this painting exists where the urn has been replaced by a representation of Newton's tomb in Westminster Abbey and the prism experiment has been eliminated.
[98] Helmers (2001).
[99] Louis Joseph de Bourbon (1736–1818), Prince of Condé from 1740. He was known as the Grand Condé.
[100] Fara (2003).
[101] Fara (2003). The French title was 'Newton découvre et montre la Vérité qui d'une main tient un prisme pour marquer la théorie des couleurs, et de l'autre un cercle aimanté pour designer son système de l'attraction.'
[102] King-Hele (1992).
[103] Brown (2012).
[104] Graves (1859).
[105] Coleridge's notes indicate that this 'he' was David Hartley (1705–1757), a philosopher who was one of Coleridge's hero's; he named his first son David Harley.
[106] Sysak (1993).
[107] Wordsworth (1963).
[108] Wordsworth (2007).
[109] Thomas and Ober (1989).
[110] The *Prelude* was first published posthumously by his wife Mary in 1850.
[111] Walker was a friend of Joseph Priestley and other members of the Lunar Society.
[112] Walker (1799).
[113] King-Hele (1984).
[114] King-Hele suggests that James Lind (1736–1812) a physician, who was also an astronomer and meteorologist, was an even greater scientific influence on Shelley, while he was at Eton, than Walker. King-Hele (1992).
[115] Whitehead (1948).
[116] Poe (1848).
[117] Harrison (1987).
[118] Poe (1929).
[119] Keats (1820).
[120] Lindsay (1989).
[121] Bindman (1977) thought that Nebuchadnezzar and Newton were representatives of science and the senses.
[122] It is interesting is that another pair of the twelve large plate series: *Elohim Creating Adam* and *Satan Exulting over Eve* have been suggested as a companion pair to *Newton* and *Nebuchadnezzar*. Lindsay (1989). Newton is also very similar in appearance to Satan. Although this seems to suggest that Blake was identifying Newton with Satan, it unlikely since Blake used an image of the same young man to represent many different people including Christ as a child in his watercolour *Christ in the Carpenter's Shop*, where Christ is represented as a miniature adult.
[123] Damon (2013).
[124] Blake (1907).
[125] Blake (1904).
[126] Barnett *et al*. (2006).
[127] Newton (1718), Fig. 11, Book I, Part II.
[128] Colour printers also use black in what is known as the four-colour CMYK system.
[129] The concept of the retina having three receptors for colours is usually attributed to Thomas Young and his 1801 Bakerian Lecture at the Royal Society but George Palmer (1740–1795) had similar ideas in 1770. Mollon (1997).

[130] Finley (1967).
[131] Schweizer (1982).
[132] Brewster (1834).
[133] First published in America in 1879. Rood (1879).
[134] There is a huge difference in the size of Turner's Shields and La Grande Jatte, which is some thirteen times bigger at 206 × 305 cm. Pissarro stated that a picture should be viewed at a distance 'which allows its colours to blend,' which should be about three times the diagonal of the canvas (Homer 1964). Thus the size of the dots or lines should be about the same relative to the size of the picture. However, Turner's dots and lines are also much finer relative to the picture size.
[135] Who coined the term Neo-Impressionist for Seurat and his group.
[136] Fiedler (1989).
[137] Goethe (1810).
[138] Duck (1987).
[139] Finley (1991).
[140] Newton (1718).
[141] Ziff (1964). Langhorne is now mainly remembered for the translation, with his brother of *Plutarch's Lives*.
[142] Serres (1982).
[143] The paper of this painting has an 1832 watermark.
[144] Beckett (1970).
[145] Schweizer (1982).
[146] Thomson (1852).
[147] Jeans (1930). It is frequently stated that Clausius introduced the German word *Wärmetod* in his lecture delivered before the 41st Meeting of the German Scientific Association, at Frankfort on the Maine in 1867, which was translated and published in English in 1868. Clausius (1868), but in the paragraph where he is supposed to have used *Wärmetod* the word is translated as 'unchanging death'. I have not been able to locate the German text of this lecture but I very much doubt that *Wärmetod* could have been translated as 'unchanging death'. The first use of the word *Wärmetod* in relation to the maximizing of entropy that I have been able to find is by Svante Arrenhius in his book *Das Werden der Welten* published in 1907.
[148] Tait (1876b).
[149] Stewart and Lockyer (1868).
[150] Helmholtz (1895).
[151] Helmholtz (1893).
[152] First used in its physical sense by Thomas Salusbury (died 1665 or 1666). Salusbury (1665) Also subsequently first used by Newton in 1686.
[153] Gold (2002).
[154] Meadows (1992).
[155] Eddington (1928).
[156] Nixon (2002).
[157] Goss (2011).
[158] Daw (2011).
[159] White(2004).
[160] Clerke (2004).
[161] See §2.6.
[162] Banfield (2007).

[163] When Hopkins sent this poem to his friend and posthumous editor, Robert Bridges, he wrote that the sonnet on the Heraclitean fire 'distilled a great deal of early Greek philosophical thought but the liquor of the distillation did not taste very Greek'.

[164] The most detailed of Hopkins thermodynamic critics is Nixon (2002).

[165] Beer (1991).

[166] Tondre (2012).

[167] Bertrand Russell wrote to Beatrice Webb in 1923 that 'I don't know whether he [Spencer] was ever made to realize the implications of the second law of thermodynamics; if so, he might well be upset. The law says that everything tends to uniformity and a dead level, diminishing (not increasing) heterogeneity.'

[168] Rosenthal (2010).

[169] Wells (1989, 1931).

[170] Wells (1931).

[171] Thomson (1862).

[172] Fortunately Thomson hedged his statement by adding 'unless sources now unknown to us are prepared in the great storehouse of creation.' Rutherford gave an account of a lecture he gave at the Royal Institution in 1904 on the influence of radioactive heating on estimates of the Earth's age where Thomson, or Lord Kelvin as he was then, was present in the audience, 'Kelvin fell asleep, but as I came to the important point, I saw the old bird sit up ... [A] sudden inspiration came, and I said Lord Kelvin had limited the age of the Earth provided no new source [of heat] was discovered.... Behold! The old boy beamed on me.' Eve (1939).

[173] In 1925 when Wells wrote his *Outline of History* he still believed what he had written in the time machine about the cooling of the Sun writing at the end of Chapter 1: 'There will be a time when the day will be as long as a year is now, and the cooling sun, shorn of its beams will hang motionless in the heavens.'

[174] Anon. (1894). It is ironic that Orson Welles (1915–1985) was to do something of the same kind to the *War of the Worlds*, when in 1938 he caused consternation in the USA by broadcasting it as if it was a news bulletin.

[175] Watts (1969).

[176] Conrad (1921).

[177] Russell (1910).

[178] Undoubtly the Congo where Conrad had been the captain of a steamer on the Congo River.

[179] Meckier (2003).

[180] Edbury (2012).

[181] If additional evidence is needed originally Eliot used a different quotation from the *Heart of Darkness* as an epigram to his earlier poem *Waste Land* but was persuaded to drop it by his friend Ezra Pound.

[182] Bush (2004).

[183] Russell (1975).

[184] Whitworth (2011).

[185] Letter to Richard Addington 24 February 1927. Eliot (2012).

[186] Even in the third edition (1870) of *Sensations of Tone*, Helmholtz was able to improve parts of his work by using more recent physiological and anatomical researches on the ear.

[187] Ohm (1843).

[188] Mayer (1875).

[189] Rayleigh published a series of papers from 1871 and it was not until the final paper Rayleigh (1889) that he was certain that the blue of the sky was caused by the molecules of the air.

Part III

Modern and Postmodern Period

Chapter 6

Philosophy, Politics, and Religion

Despite the hiatus of the late nineteenth and twentieth centuries that saw the fragmentation of knowledge, the development of modern physics has been so profound that it has become entwined again with philosophy, politics, and religion. The influence of classical physics on philosophy and religion could be dealt with in separate chapters, though even then their boundaries were not clear cut, but a discussion on modern physics is best handled by treating them together.

In his 1956 Tarner Lecture at Trinity College, Cambridge, Schrödinger made the point of emphasizing how it is important, when discussing the effect of physics on philosophy or religion, not to fall into Kant's trap who:[1]

> 'accepted physical science in the form it had reached during his lifetime as some thing more or less final and he busied himself to account for its statements philosophically. This happening to a great genius ought to be a warning to philosophers ever after.'

Stephen Hawking and Leonard Mlodinow come close to Kant in *The Grand Design* where they pronounce the death of philosophy and the potential obsolescence of God.[2] To be charitable when Hawking and Mlodinow wrote 'philosophy is dead' they probably meant that philosophy and physics are inseparable but they are arrogant enough to believe that now only physicists are competent to comment on philosophy. The Finish philosopher Georg Henrik von Wright (1916–2003) who succeeded Ludwig Wittgenstein (1889–1951) as professor at Cambridge was equally as scathing about physicists who 'express their non-professional views on the "big questions" on the nature of mind and matter' meaning in particular Eddington, whose non-professional views are now remembered while von Wright's professional views have largely been forgotten.[3]

The effect of modern physics on philosophy and religion has been as revolutionary as Newtonian natural philosophy was but it does not seem to have created such a shock for the general public; there has been no event that moved the public as Tyndall's Presidential Address to the British Association at Belfast in 1874 did.[4] Excellent summaries of its effect on religion and philosophy are given in the final chapters in two popular books on modern physics: *The Nature of the Physical World* by Arthur Eddington and *The Revolution in Physics* by Ernst Zimmer.[5] A possible reason for the difference in the impact is that classical physics is about concepts that correspond closely with everyday experience; it deals with everyday real objects and has a categorical framework of four dependent components:[6]

(1) The objects are located in our everyday three-dimensional Euclidean space and one moment in time is succeeded by another.
(2) The usual rules of logic that were symbolized by George Boole (1815–1864) apply.
(3) Every event has a prior cause. A sufficient reason and logical determinism makes propositions either true or false.
(4) An observer does not in principle affect events, though measurement of them extracts some energy.

The same categorical framework was mostly adopted in philosophy and theology prior to the twentieth century and, even when classical physics was used in philosophical and theological works to support opposite points of view, there was little dissent with their categorical framework. Modern physics is outside of our everyday experience and understanding and needs a different categorical framework. The theory of relativity is the easiest of the modern theories to understand, though even here space and time are entwined and non-Euclidean. With quantum mechanics the position is more difficult; the orthodox view believed by most physicists is that much is indeterminate, though there are still some who believe in a deterministic version. The position of string theory is even more uncertain. Thus in modern physics there is no clear cut categorical framework on which to hang a discussion and many of the attempts at philosophical and theological discussion descend into a discussion of categorical frameworks where one opinion is pitted against another with little commonality in language. Thomas Filk and Albrecht von Müller argue plausibly that quantum mechanics applies 'to the transition from possibilities to facts,' or in other words:[6]

> 'it is a theory about … how facts "are born", i.e. the "status nascendi" of facts. In contrast, classical physics deals with facts, not the coming into being of facts.'

String theory has a similar status. Even after some ninety years of the development of the orthodox version of quantum mechanics, there are few who really understand it. As Richard Feynman said during his 1964 Messenger Lectures at Cornell University:[7]

> 'There was a time when the newspapers said that only twelve men understood the theory of relativity. I do not believe there ever was such a time. There might have been a time when only one man did, because he was the only guy who caught on, before he wrote his paper. But after people read the paper a lot of people understood the theory of relativity in some way or other, certainly more than twelve. On the other hand, I think I can safely say that nobody understands quantum mechanics.'

Newton's natural philosophy led to an almost self evident belief in the design argument for theism, which is not there in modern physics. It is far from self evident in the present world that there is a cosmic purpose let alone something that is undeniably and everlastingly good. Some like Stephen Weinberg, the Nobel Prize recipient in 1979, find this lack of purpose pessimistic writing that:[8]

> 'The more the universe seems comprehensible, the more it also seems pointless… The effort to understand the universe is one of the very few things that lifts human life above the level of farce, and gives some of the grace of tragedy.'

I personally do not find this lack of purpose pessimistic on the contrary the thought of being a mere cog in a wheel of a purposefully built machine is pessimistic to me. The modern physics' indeterminism promises a freedom that was not there in classical physics. Arguments making use of classical physics in religion seem much more plausible to me than the apologetic arguments for religious belief based on modern physics. Another difference between the proponents using physics to argue for theism is that during the classical period there were many more physicists who were theists than atheists; in the era of modern physics that has been reversed and, apart from a few notable exceptions, by far the most physicists are atheists.[9]

6.1. Spiritualism, Aether, Continuity, and Discontinuity[10]

Psychic phenomena and spiritualism were fairly widely believed at the fin de siècle and there were a surprising number of eminent physicists and chemists who dabbled in the occult: J.J. Thomson, Rayleigh, William Crookes (1832–1919) and Oliver Lodge (1851–1940) were members of the Society for Psychical Research, which was founded in 1882. In 1900 Ernst Mach wrote an article in which he tried to show why physicists of the time were prone to believe in psychic phenomena.[11] Mach pointed out that physicists were as prone to fetishism as savages were, stating that:[12]

> 'So long as we consider heat, electricity, and magnetism as mysterious and impalpable entities residing in bodies and imparting to them their known wonderful properties, we still stand on the level of fetishism ... When a chemist [Crookes] who has achieved fame by his

beautiful discoveries in his science espouses spiritualism; when a noted physicist [Lodge] does the same … it is certain that the intellectual malady of which I have here been speaking is very deeply seated, and that not alone in the minds of the non scientific public.'

The discovery of the then mysterious X-rays, and alpha, beta and gamma rays at the turn of the century aided the belief in psychic phenomena. Classical physics had dealt with continuous entities but in the early twentieth century it was coming to terms with the idea that it could be discontinuous at the scale of the atom which was difficult for those physicists raised on continuum classical physics to accept. Thomson and Lodge though accepting sub-atomic particles like electrons[13] clung to the idea of continuity by retaining the mysterious all pervading aether. In his Presidential Address to the 1909 meeting of the British Association in Winnipeg, Canada, Thomson said:[14]

'... matter occupies ... but an insignificant fraction of the universe, it forms but minute islands in the great ocean of the ether the substance with which the whole universe is filled.... [aether] proves itself a physical agent; not indeed tangible, but yet concretely real.'

Lodge was an equally enthusiastic believer in the aether despite having performed an experiment with rotating discs in 1891 to investigate whether a moving body could drag the aether along with it that returned a negative result.[15] Aether has had a long history in physics but the interpretation of this mysterious intangible medium has varied considerably. For Aristotle the Earth was the centre of the universe and, in the sublunary sphere, matter was composed of the four elements but superlunary space was permeated by a fifth element: his aether.[16] Newton had an ambivalent attitude to aether. He was not completely satisfied by gravity's action at a distance or with light propagating in a vacuum. Newton mentioned aether in the *Principia* but refers to it more frequently in the queries to the second edition of his *Opticks*. In query 21, Newton, assumed that light propagates like sound and calculated the ratio of aether's elasticity to density compared with air from his *Principia* calculation[17] of the propagation of sound in air:[18]

'Sounds move about 1140 *English* Feet in a second Minute of Time, … [and l]ight moves from the Sun to us in about seven or eight Minutes of Time, which is a distance about 70000000 *English* Miles greater …[but] I do not know what this Æther is'

With the development of a wave theory of light by Thomas Young and others at the beginning of the nineteenth century, interest in aether burgeoned. One would have imagined that after, the Michelson and Morley experiments on the velocity of light which proved that there was no aether wind relative to the Earth, interest in aether would have diminished. However the converse was true and even

Einstein wrote about the aether into the 1930s, howbeit he identified his aether with the metric tensor of general relativity.[19] What happened at the end of the nineteenth century was that the existence of aether, along with the new invisible rays, was used by physicists who believed in spiritualism to explain the psychic phenomena. Lodge, who contributed to research on atoms and X-rays in the 1890s, was the President of the Society for Psychical Research from 1901–1903 and a true believer in spiritualism.[20] For Lodge aether was essential for both the physical and spiritual worlds. In his book *My Philosophy*, published in 1933, he wrote:[21]

> '… when our existence continues amid other surroundings we shall not be in ordinary association with matter as we are here, and therefore shall not appear to the senses of our fellows left behind on the earth. We shall have a body or mode of manifestation suited to our new surroundings and shall be fully perceptible to our fellows in like case. My hypothesis is that this body or more refined mode of manifestation will be composed of ether, and may be properly spoken of as an etheric body, or what St. Paul called a spiritual body.'

Twenty years earlier he made an impassioned plea for retaining 'Continuity' and was 'a fervent believer in the Ether of Space' in a long Presidential Address to the meeting of the British Association at Birmingham where Lodge was Principal of the University.[22] Lodge went on to say:

> 'Matter in motion, Ether under strain, constitute the fundamental concrete things we have to do with in physics.'

He even saw that if 'the Principle of Relativity, in an extreme sense establishe[d] itself, it seems even time would become discontinuous.' At the end of his address Lodge briefly discussed the psychic saying:

> 'Science may not be able to reveal human destiny but it certainly should not obscure it…. I am one of those who think that the methods of Science are not so limited in their scope as has been thought … the Psychic region can be studied and brought under law too.'

The day after Presidential Address *The Times* published an article under the title of *The New Agnosticism*, which they believed Lodge was advocating. The new agnosticism was not of religion but of science as a result of its 'limited scope'. The writer of the article contrasted Lodge's Presidential Address with Tyndall's Belfast Address on scientific materialism attacking religion.[4] Lodge was expected to make more explicit remarks about psychic phenomenon in his address, which was eagerly awaited. *The Times* of the 11th September had two other articles on Lodge's Address in addition to the *New Agnosticism* article: a full report of the address and an article on his views expressed on *Continuity and Evolution*. Lodge kept aether alive until 1940.

6.2. Reality, Determinism, and Causality

Although classical physics could not avoid all questions of reality, such as the nature of light, it was not until the nineteenth century that these questions become pressing. The theories of special and general relativity did raise the question of the nature of space and time but did not present unanswerable reality problems. However with quantum mechanics reality became impossible to understand and the determinism and causality of classical physics, according to the orthodox Copenhagen interpretation, disappeared. Quantum mechanics has been so extraordinarily successful in predicting the advanced properties of materials such optical, semiconductors, superconductivity, etc. that one would imagine it gave a clearer insight into reality than classical physics. Nothing could be further from the truth; the reality of quantum mechanics is completely outside everyday reality. In discussing the change in state of an electron Eddington wrote:[23]

> 'a particular change of state of the atom which cannot really be associated with movements in space as macroscopically conceived. *Something unknown is doing we don't know what*— that is what our theory amounts to.... I have read something like it elsewhere'

and then he quoted from the last verse of Lewis Carroll's nonsense poem the *Jabberwocy*:

'Twas brillig and the slithy toves
Did gyre and gimble in the wabe

Eddington gave the quotation scientific sense by continuing: 'by contemplating eight circulating electrons in one atom and seven circulating electrons in another we begin to realize the difference between oxygen and nitrogen.' Or in other words: 'Eight slithy toves gyre and gimble in the oxygen wabe; seven in nitrogen.' Schrödinger in 1935 gave a similar example of the different nature of quantum mechanics in his paradoxical cat:[24]

> 'A cat is penned up in a steel chamber, along with the following diabolical device: ... in a Geiger counter there is a tiny bit of radioactive substance, so small, that perhaps in the course of one hour one of the atoms decays, but also, with equal probability, perhaps none; if it happens, the counter tube discharges and through a relay releases a hammer which shatters a small flask of hydrocyanic acid. If one has left this entire system to itself for an hour, one would say that the cat still lives if meanwhile no atom has decayed. The first atomic decay would have poisoned it. The [wave] function [for the state of the atom and also] of the entire

> system would express this by having in it the living and the dead cat ... mixed or smeared out in equal parts.'

As Schrödinger wrote this situation is ridiculous. A cat obeys the usual logical principle of the extended middle, it is either living or dead, it cannot be in a superimposed state. In quantum mechanics some logical principles are violated and by its nature quantum mechanics never gives a definite answer to the question of the exact state of a physical entity. Heisenberg thought that 'insistence on the postulate of complete logical clarification would make science impossible.'[25] The philosopher who came nearest to the philosophy behind modern physics in the classical period of physics was Kant in his *Critique of Pure Reason* who saw knowledge as part *a priori* as well as being derived from experience.

Mathematical and physical realities have a commonality. Heisenberg believed that perhaps the only experimental situations that can arise in nature are those expressible by mathematical formalism.[25] The idea that mathematics is not just the handmaiden of physical reality but an intimate part of that reality is not confined to modern physics it was part of the seventeenth revolution in physics. Galileo wrote in *The Assayer* published in 1623 that:[26]

> 'Philosophy is written in this grand book, the universe, which stands continually open to our gaze. But the book cannot be understood unless one first learns to comprehend the language and read the letters in which it is composed. It is written in the language of mathematics, and its characters are triangles, circles, and other geometric figures without which it is humanly impossible to understand a single word of it; without these, one wanders about in a dark labyrinth.'

Galileo was right, the language of physics is mathematical. Newton's greatest contribution to physics was his insistence that the principles of physics were mathematical. Physics has marched with mathematics ever since, not necessarily in step and usually with mathematics leading, but in quantum mechanics physics has merged with mathematics and become inseparable as Dirac recognized in his introductory remarks to a paper on magnetic monopoles in 1931 writing:[27]

> 'The most powerful method of advance that can be suggested at present is to employ all the resources of pure mathematics in attempts to perfect and generalise the mathematical formalism that forms the existing basis of theoretical physics, and after each success in this direction, to try to interpret the new mathematical features in terms of physical entities.'

Dirac also took the relation of mathematics to physics as the topic for his James Scott Prize Lecture in Edinburgh in 1931.[28] In the 1960s the Hungarian born physicist and mathematician Eugene Wigner (1902–1995), joint recipient of the

Nobel Prize for Physics in 1963 for his work on the atomic nucleus and its elementary particles through the understanding of symmetry, saw:[29]

> 'The miracle of the appropriateness of the language of mathematics for the formulation of the laws of physics is a wonderful gift which we neither understand nor deserve.'

All physics can be expressed by mathematics and while all mathematical objects do not relate to physical principles, a large proportion of them do, while the rest are pure mathematical abstractions some of which might in the future be seen to apply to aspects of physics. At least 'the apprehension of mathematical truth to the perception of physical objects, and thus of mathematical reality to the physical universe' is the view of Platonists.[30] Mathematical Platonists see mathematical objects as abstractions that are independent of the physical world and thus, if there are other universes completely independent causally from ours, the mathematics in those universes inhabited by intelligent agents, will be no different to ours. Mathematics is the multiversal language. The Swedish-American cosmologist, Max Tegmark, argues in his new popular book *Our Mathematical Universe* 'that the ultimate reality is purely mathematical.'[31] However this statement is not the same as the mathematics is the ultimate reality. That colossus of mathematics, David Hilbert (1862–1943) gave a talk in 1925 *On the Infinite* in which he made the point that:[32]

> 'The infinite is nowhere to be found in reality, no matter what experiences, observations, and knowledge are appealed to.'

Although Platonism has little to do with Plato, the conception of the ultimate mathematical reality had a long gestation starting with Galileo's remarks in *The Assayer*. Just as every elementary particle has its antiparticle so do philosophical notions. Contemporary nominalism, the anti-notion of Platonism, rejects abstract objects like mathematical structures. Before one jumps to the conclusion that nominalism represents a very narrow view, it is worthwhile to consider that Newton was not completely immune to nominalism.[33] Newton was the student of Isaac Barrow (1630–1677), the Lucasian Professor of Mathematics at Cambridge before him.[34] Barrow had a nominalistic objection to analytic mathematical methods, favouring a geometric approach and not surprisingly Newton was influenced by him. However by the time of writing the *Principia*, Newton had developed his own style which involved 'the possibility of working out the mathematical consequences of assumptions that are related to possible physical conditions, without having to discuss the physical reality of those conditions.'[35] What physics describes is not reality but a model of reality that is continually being improved though it always remains just that, a model. Wittgenstein, who had a background in physics and trained as aeronautical engineer at Manchester University, expressed similar thoughts in *Tractatus*, the notes for which he wrote while a soldier in the German army during World War I and completed while a prisoner of war in Casino, Italy during 1918. *Tractatus* was written as a series of numbered propositions and those relevant to the idea that physics only deals with a model of reality are:[36]

2.063 The total reality is the world.
2.1 We make to ourselves pictures of facts.
2.11 The picture presents the facts in logical space, the existence and non-existence of atomic facts.
2.12 The picture is a model of reality.
2.17 What the picture must have in common with reality in order to be able to represent it after its manner – rightly or falsely – is its form of representation.
2.223 In order to discover whether the picture is true or false we must compare it with reality.
2.224 It cannot be discovered from the picture alone whether it is true or false.

The reality of unobservable objects in quantum mechanics exists in their mathematical formalism. In the orthodox Copenhagen interpretation the fundamental particles of quantum mechanics, such as electrons, only possess physical reality while they are being observed. As Heisenberg expressed it:[25]

> 'In the experiments about atomic events we have to do with things and facts, with phenomena that are just as real as any phenomena in daily life. But the atoms or the elementary particles themselves are not as real; they form a world of potentialities or possibilities rather than one of things or facts.'

Once the indeterminist character of the orthodox Copenhagen interpretation of quantum mechanics had been realized there was a flurry of discussions on its implications on the question of whether we lived in a deterministic world, in which Eddington took a leading part. Physicists naturally approach philosophical questions from a different perspective than philosophers and Eddington came into criticism for being rather naïve in his philosophy. We need the rigorous approach of philosophers to ensure that we know as exactly as we can, what we mean by such words as determinism whether we are discussing its ontological nature or its role in the epistemological theory of knowledge. However not being so grounded in physics, philosophers also can be naïve in their use of thought experiments based on physics. There has been considerable recent speculation as to whether classical physics is really as deterministic as it has been assumed. An example of apparent indeterminacy in Newtonian mechanics is the case of a ball resting on the top of a dome, of a particular shape though its shape does not really matter, and the question of when it will roll down has been given.[37] The mathematical analysis is performed correctly and amazingly the solution is not unique: the ball can either roll down the dome or remain as short or as long a time as it wishes before rolling down. Any first year undergraduate in physics or engineering will tell you that this is simply the question of whether equilibrium is stable or unstable.[38] A ball placed at the top of a dome is in equilibrium but it is unstable so that if the ball or the dome is not absolutely perfect or if there is a slight vibration or a breath of wind it will roll down either immediately or after some time. There are two morals to this story: thought experiments need to be approached with caution and do not necessarily illustrate reality, and no discipline is independent of others and this latter moral is the more important. As far as physics and philosophy are considered Eddington's remarks in 1933 are as apt as ever:[39]

> 'in approaching the more fundamental problems of science the physicist finds it necessary to adopt a more philosophical outlook than is habitual to him; and if the philosopher wishes to handle results ascertained scientifically he must enter to some extent into the scientific outlook.'

Fig. 6.1. Time exposure of Double Pendulum with LED attached to its end. Courtesy George Ioannidis.

Although the example given above certainly does not show that classical physics can lead to indeterminism there are phenomena in classical physics where the motion is extremely sensitive to the initial conditions and do appear to generate indeterminate random motion but whose motion is in fact deterministic but chaotic. A simple example of chaotic motion is the double pendulum consisting of two equal rigid arms shown in Fig. 6.1. The importance of chaos to philosophy was realized in the early twentieth century by Duhem and Poincaré and the latter recognized the weather as an example of chaos.[40]

Eddington, a convinced indeterminist, wrote that before the orthodox Copenhagen interpretation of quantum mechanics there were two types of regularities in the natural world:[39]

> 'Primary regularities occurred because they were governed by strict causation; the beginning of the sequence predetermined the end. Secondary regularities occurred, not by strict causation, but because there was a strong probability of their occurrence.'

Note that the second class of regularity pertained to some of classical physics though it was not necessarily initially recognized as such. The gas laws are of the second type of regularity since gas pressure is due to the motion of gas molecules. According to Eddington: [39]

> 'this dual source of regularity is no longer accepted. The changes that have occurred in the fundamental conceptions of physics ... [since 1925] brought it about that all the observed regularities are now attributed to the second source.'

It now appears that at the sub atomic level everything is indeterminate but it depends exactly what is meant by determinism as to whether this behaviour means that the world is indeterminate. But as Eddington commented:[23]

> '[Although] human life is proverbially uncertain; few things are more certain than the solvency of a life-insurance company. The average law is so trustworthy that it may be considered predestined that half the children now born will survive the age of x years. But that does not tell us whether the span of life of young A. McB. is already written in the book of fate.'

In most situations we are dealing with many more atoms than there are children in the world and the individual atoms are not identifiable as A. McB is; we know the behaviour of a very large number of atoms to such an extremely high probability that is by any practical measure the same as certainty. Nevertheless, since the macro-world is built with micro-bricks which seem to be indeterminate, we live in an indeterminate world that on the macroscale is predictable. At a joint meeting of the Aristotelian and Mind Societies in Trinity College, Cambridge in 1931 two philosophers, Charlie Dunbar Broad (1887–1971), better known as simply C.D. Broad, who was a frequent broadcaster on the BBC's Third Programme in the 1950s, Richard Bevan Braithwaite (1900–1990) and Eddington discussed determinacy and indeterminism. In Broad's definition of indeterminism a single indeterminate phenomenon is sufficient to overthrow determinism; Eddington's 'scientific doctrine of indeterminism is not that there exist occasional exceptions to deterministic law, but that every phenomenon is to a greater or lesser extent indeterminate.'[41] Determinism and predictability are not synonymous though determinism implies that in principle all events are predictable and Eddington's remark that:[41]

> 'The entrenchment of determinism in physics was due to the fact that throughout the whole range of macroscopic phenomena the unpredictability is negligibly small, so that for all practical purposes a strictly causal scheme is the most obvious and simplest way of treating them.'

Although the indeterminism of quantum mechanics does not necessarily remove predictability, it does considerably weaken the notion of a strictly materialistic world devoid of free will as was emphasized by Heisenberg in his *Physics and Philosophy*.[25] Bohr expressed a similar view of free will made possible by 'the frustration of attempts to order experience regarding our own consciousness as a causal chain of events, originally suggested by the mechanical conception of nature.'[42] Religion that advances some code of morality can only exist if we have a degree of free will. Eddington in his *Nature of the Physical World* concluded with a discussion of the wider implications of modern physics writing: 'It will perhaps be said that the conclusion to be drawn from these arguments from modern science, is that religion first became possible for a reasonable scientific man about the year 1927.' He then asked: 'Is it merely a well-meaning kind of nonsense for a physicist to affirm this necessity for an outlook beyond physics?' before giving what I believe to be the only answer: 'It is worse nonsense to deny it.' Einstein, the patriarch of cosmic religion, retained a belief in strict determinism and denied the existence of free will until he died.

6.2.1. *Relativity and spacetime*

Einstein's theory of relativity shows that the Newton was wrong in his conception of absolute space and time. The meaning of before, simultaneous with and after applied to two separate events only has meaning relative to the inertial frame of the observer as is easily demonstrated for two dimensional spacetime, where the gravitational field is weak, using the Minkowski diagram[43] shown in Fig. 6.2. In the reference frame Σ event A occurs simultaneously with an event B but to an observer in a frame Σ' travelling in the positive x direction, event A occurs after B and to an observer travelling in frame Σ'' in the negative x direction event A before B. In fact provided B remains outside of the light cone of event A, its past and future relative to A has no meaning since its existence can never be communicated to an observer or an event at A. Only points such as

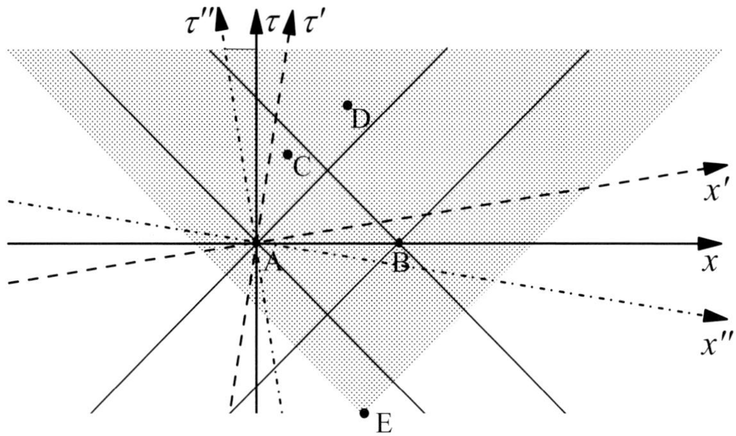

Fig. 6.2. Order of events in Minkowski spacetime.

C and D in the upper light cone emanating from event A can be said to be after A. Thus only events where there is no possibility of a direct casual relationship are in that hazy region of spacetime where there is no absolute past or future to an event. Within the light cones events can still have direct causation and the further back in time the greater is the distance between events that can have causality thus an event at E can cause an event at C either directly or through a causal chain E→A→C and cause an event at D either directly or through the causal chains E→A→C→D or E→B→D. Although the special theory of relativity cannot strictly apply from the Big Bang, a causal chain could exist back to that moment without violating the theory of relativity so that relativity can be deterministic even though the order of events without a possible direct causal relationship is not fixed. Thus Schrödinger was being somewhat over enthusiastic when he said in the Tarner Lectures[1] at Trinity, College, Cambridge in 1956 that relativity:

> 'meant the dethronement of time as a rigid tyrant imposed on us from outside, a liberation from the unbreakable rule of "before and after"... [and that] the thought that the whole "timetable" is probably not quite as serious as it appears at first sight. And this thought is a religious thought, nay I should call it the religious thought.'

Nevertheless time has always been important in theology.

God even in ancient Greek times was usually conceived as being eternal, which causes theological problems. If God knows all because he is eternal, how can man have free will? Boethius, a sixth century Roman philosopher gave a definition of eternity as being 'the complete possession all at once of illimitable life' that enabled him to argue that a timeless God and free will were not incompatible. St Thomas Aquinas adopted Boethius' arguments in his *Summa*

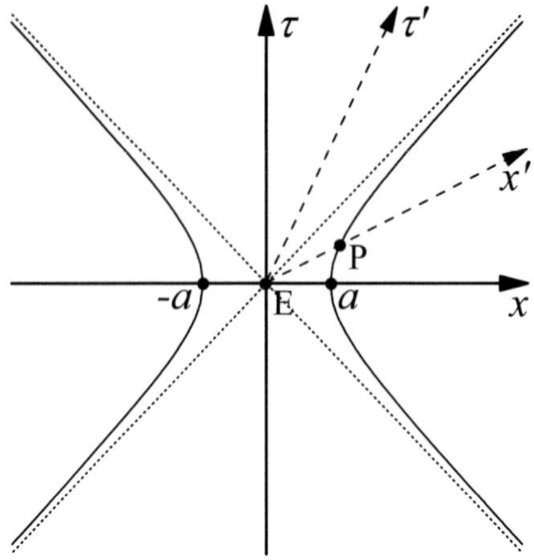

Fig. 6.3. Two-dimensional space-time representation of the simultaneity of eternity, E, and temporal event, P, (after Jammer 1999).

Theologica. Eleonore Stump and Norman Kretzmann (1926–1998) saw that simultaneity in relativity was related to Boethius concept of the simultaneity between eternity and the temporal moment.[44] The renowned philosopher of science, Max Jammer (1915–2010), developed the idea of Stump and Kretzmann using the special theory of relativity, cleverly making use of the fact that, outside of the light cones of an event, past, now, and future have no meaning.[42] Using for simplicity a two-dimensional spacetime Jammer illustrated how it is possible for eternity and temporal events to be simultaneous for one particular set of histories of a temporal event; his argument can be extended to four-dimensional spacetime. Eternity, in Fig. 6.3, is represented by the point E(0, 0) in the reference frame x, τ. Because the world-lines that contain temporal events, P, that are simultaneous with eternity is accelerating relative to the reference frame local inertia frames E'_p for the events P need to be chosen where the time in the local frame for the event and eternity are both zero. Normalized time, ct, in the local frame, τ', is given by

$$\tau' = \gamma[(\tau - \tau_P) - \beta_P(x - x_P)], \tag{6.1}$$

where v_P is the velocity of the event P along its world-line, w, in the reference frame and $\beta_P = v_P/c$. For τ' to be zero at both E and P, $\beta_P = \tau_P/x_P$ and only hyperbolic world lines given by $x^2 = a^2 + \tau^2$, where $\pm a$ is the intersection of a world line with the x-axis, have this property. Two of the set of hyperbolic world-lines are shown in Fig. 6.3. As Jammer acknowledged, there is the problem that God's eternity is located spatially but eternity can only be simultaneous with temporal events if it too is of the physical world and, according to the theory of relativity, time cannot be separated from space, thus God seems to need spatiality. These speculations on God's nature based on the theory of relativity should not be pressed too hard but the argument that God's timelessness allows man free will is also supported since the world-lines of temporal events are outside of eternity's light cone and as such cannot be affected physically. However the significance of Jammer's observations is not in the details that might be deduced from them but the fact that he showed that the seemingly impossibility of the simultaneity between eternity and the temporal moment is made possible by the special theory of relativity.

One would think that, since the special theory of relativity is comparatively simple, that the question whether it implies that everything is predetermined or that the future is not fully determined but to some degree open would have been resolved, but it is not. Einstein was a 'convinced determinist' as he stated in a letter in 1930 to Rabbi Abraham Geller of Brooklyn, New York who had suggested that the probabilistic nature of quantum mechanics undermined determinism.[45] In this letter Einstein is clearly writing about the philosophy of science and not about his religious belief. One would have thought that Einstein would have used the theory of relativity to support his belief in an unrestricted causal determinism, he did not, but he must have believed that relativity supported determinism as Newtonian mechanics had done. Einstein was very much influenced by Spinoza's philosophy and religious thought[42] and it is likely that he obtained his firm adherence to determinism, at least initially, from Spinoza. When asked to write an essay on the ethical significance of Spinoza's philosophy in 1932 Einstein declined because he wrote:[46]

> 'I do not have the professional knowledge to write a scholarly article about Spinoza. But what I think about this man I can express in a few words. Spinoza was the first to apply with strict consistency the idea of an all-pervasive determinism to human thought.'

The question whether we live in a deterministic world, is a religious question because absolute causal determinism negates free will. Einstein found this thought disturbing and in an article on his belief in 1930 he wrote that a saying of Arthur Schopenhauer (1788–1860) 'A man can surely do what he wills to do, but he cannot determine what he wills' was consoling.[47] In an article published in the New York Times Magazine in 1930 Einstein did comment on the religious implication of a belief in determinism writing that:[48]

> 'The man who is thoroughly convinced of the universal operation of the law of causation cannot for a moment entertain the idea of a being who interferes in the course of events - provided, of course, that he takes the hypothesis of causality really seriously.'

This statement by Einstein is an exception to his general rejection of the idea that religious beliefs can be deduced from science. Spinoza believed in a form of cosmic religion as did Einstein. God, in Einstein's view, did not intervene in the affairs of man but he strongly believed that religion promoted science. At the end of a response to a greeting by the Liberal Minister's Club of New York, Einstein wrote in 1948 that:[49]

> 'While it is true that scientific results are entirely independent from religious or moral considerations, those individuals to whom we owe the great creative achievements of science were all of them imbued with the truly religious conviction that this universe of ours is something perfect and susceptible to the rational striving for knowledge. If this conviction had not been a strongly emotional one and if those searching for knowledge had not been inspired by Spinoza's Amor Dei Intellectualis, they would hardly have been capable of that untiring devotion which alone enables man to attain his greatest achievements.'

Einstein found the indeterminacy of quantum mechanics profoundly disturbing.

6.2.2. *The philosophical problems of Bohr's quantum mechanics*

Einstein could not come to terms with the observer created realism of quantum mechanics. In a letter sent by Einstein which was intended to be read at the dinner of the Royal Society in honour of Newton's bicentennial in 1927 he wrote: [50]

> 'But the last word has not yet been said. May the spirit of Newton's method give us power to restore unison between physical reality and the profoundest characteristic of Newton's teaching – strict causality.'

No doubt Einstein was thinking about Heisenberg's uncertainty principle when he wrote that letter. Much later Heisenberg, in *Physics and Philosophy* based on his Gifford Lectures given at St Andrews University, Scotland in the winter of 1955-56, wrote of Einstein:[25]

> 'When [he] has criticized quantum theory he has done so from the basis of dogmatic realism,... [which] claims that there are no statements concerning the material world that cannot be objectivated. This is a very natural attitude. Every scientist who does research work feels that he is looking for something that is objectively true.'

In 1935 Einstein, Boris Podolsky (1896–1966), and Nathan Rosen (1909–1995) all working at the Institute for Advanced Studies in Princeton, NJ attempted to show that the description of physical reality given by quantum mechanics was not complete because of the uncertainty in the values of pairs of variables that do not commute. Einstein and his co-workers stated that:[51]

> 'If, without in any way disturbing a system, we can predict with certainty ... the value of a physical quantity, then there exists an element of physical reality corresponding to this physical quantity.'

In an entangled quantum state consisting of two sub-systems far enough away from each other to prevent interaction, they thought that they could show that correlation in the variables could be used to predict the variable in one system from the other without actually measuring it and so side-step uncertainty. From their thought experiment they deduced 'that either (1) the quantum mechanical description of reality given by the wave function is not complete or (2) when the operators corresponding to two physical quantities do not commute the two quantities cannot have simultaneous reality.' Thus they thought that 'the wave function does not provide a complete description of the physical reality.' Bohr responded to the Einstein-Podolsky-Rosen paper in 1935 by pointing out its ambiguities. In a discussion of the well-known diffraction of electrons by a slit, Bohr wrote:[52]

> 'we are not dealing with an incomplete description characterized by the arbitrary picking out of different elements of physical reality at the cost of sacrificing other such elements, but with a rational discrimination between essentially different experimental arrangements and procedures which are suited either for an unambiguous use of the idea of space location, or for a legitimate application of the conservation theorem of momentum.'

In essence Bohr thought that the important qualifier 'without in any way disturbing a system' in the statement of Einstein-Podolsky-Rosen quoted above could not be met. The orthodox Bohr rejected quantum reality and stated that:[53]

> 'There is no quantum world. There is only an abstract quantum description. It is wrong to think that the task of physics is to find out how nature is.'

There seemed to be another problem for the completion of the description of reality in quantum mechanics; in 1932 the Hungarian born polymath John von Neumann (1903–1957) had given a proof that a hidden variable extension of quantum mechanics was impossible.[54] Einstein was not the only physicist that found the uncertainty in quantum mechanics troubling. The American physicist David Bohm (1917–1992) also believed quantum theory should have physical reality and in 1952 he suggested an interpretation using hidden variables, not contemplated by Neumann that provided: [55]

> 'a consistent alternative to the usual assumption that no objective and precisely definable description of reality ... at the quantum level of accuracy ... [where the] reality at the quantum level [was] at least in principle not fundamentally different from that at the classical level.'

However Einstein did not accept Bohm's theory. Almost immediately, in a letter to Born, Einstein wrote: [56]

> 'Have you noticed that Bohm believes ... that he is able to interpret the quantum theory in deterministic terms? That way [hidden variables] seems too cheap to me.'

In 1953 Einstein gave his reasons for rejecting Bohm's theory based on consideration of a standard example of a one-dimensional particle in an elastic sided box that showed the motion in the macro system did not agree with classical physics.[57] Pauli and Heisenberg had other objections.

The final words on the Einstein-Podolsky-Rosen paper were given by John Bell (1928–1990), a Northern Irish physicist who was on sabbatical leave from CERN in Geneva, in two very famous papers that were published in the reverse order in which they were written.[58] In the first paper[59] Bell showed that Neumann's 1932 proof that hidden variables were not compatible with quantum mechanics was not general and there was another class of hidden variables that were compatible. However for a causal mechanism to exist the model has to be non-local and as Bell wrote:[60]

> 'The Einstein-Podolsky-Rosen is resolved in the way which Einstein would have liked the least.'

Einstein rejected actions at a distance, referring to them in a letter to Born in 1947, as 'spooky'. In his comments on this letter Born wrote:

> 'I too had considered this postulate to be one which could claim absolute validity. But the realities of physical experience had taught me that this postulate is not an a priori principle but a time dependent rule which must be, and can be, replaced by a more general one.'

In his 1966 paper, Bell did not prove that any hidden variable account of quantum mechanics must be non-local as the penultimate sentence in that paper stated; in his second paper[61] he gave the proof. Experimental violation of, what is known as, Bell's inequalities have confirmed that entangled photon pairs, similar to the quantum entanglements envisioned by Einstein-Podolsky-Rosen, must be considered as a non-separable object though separated by a space-like interval but the photons themselves have no known realty.[62] In a recent experiment the lower bound speed of the spooky quantum teleportation between two such quantum entanglements has estimated at four orders of magnitude greater than the speed of light.[63] Richard Feynman described such phenomena as 'impossible *absolutely* impossible to explain in any classical way.'[64]

Einstein was the most famous physicist to disagree with the orthodox interpretation of quantum mechanics; Popper was the most famous philosopher to disagree. Popper first published his famous *Logic der Forschung*[65] in 1934 not long after 'the new gospel of irrationality in physics was first publicly preached by Bohr in Como at the International Congress of Physics of 1927.'[66] Popper is correct, modern physics has much in common with religion, and he was not a believer in Bohr's orthodoxy. Despite the incredible success of orthodox quantum mechanics, Popper believed that physics was in crisis for two basic reasons: subjectivism in physics and the belief that quantum theory had reached the 'complete and final truth.'[67] In Popper's view the physical world is real he believed that Heisenberg's uncertainty relations, which were viewed as the limits to our subjective knowledge, were in fact objective statistical scatter relations. A statement such as: 'the probability of *a* given *b* is equal to *r*' can be interpreted either subjectively or objectively.[68] The subjective interpretations, as used in orthodox quantum mechanics, regard probability as a measure of incompleteness of knowledge. However probability can be interpreted objectively 'as a statement that can in principle be objectively *tested*, by means of statistical tests.'[68] In the first 1934 edition of *Logic der Forschung*, Popper used an objective frequency interpretation of probability *r*, which is the relative frequency that have the outcome *a* in a series of experiments under the conditions *b*, but in the first 1959 English edition, he introduced his "idea that probabilities are physical propensities, comparable to Newtonian forces.'[68] Objective probability is the propensity to realize singular events. Popper's concept of propensity as an objective interpretation of probability was generally well received.[69] Although the concept of probability as a propensity of an event removes subjectivity from quantum mechanics it does not make it deterministic.[69]

6.2.3. *Politics and quantum mechanics*

Quantum mechanics began in a period of intense political upheaval. Lenin (1870–1924)[70] and Leon Trotsky (1879–1940) played leading roles in the Russian October Revolution of 1917. Trotsky, the Marxist theorist, failed in his opposition to Joseph Stalin (1878–1953) and was expelled from the Communist Party in 1927. In Germany, Adolf Hitler (1889–1945) formed the National Socialist German Workers Party, known as the Nazi Party in Britain and the West. After coming to power in 1933 on 43.9% of the votes, Hitler used the excuse of the infamous Reichstag Fire to pass the Enabling Act which transformed Hitler's party into a dictatorship. In the 1920s and 1930s it was impossible not to be affected by politics and physicists working on the new quantum mechanics could not ignore it. After the Second World War things did not much improve, the Cold War polarized political opinions. Joseph McCarthy (1908–1957) was not much better than the other Joseph and during the first half of the 1950s hounded all the professions for real or imagined un-American activities. David Bohm, a Marxist who at the time was an assistant to Einstein at Princeton University, was called before the Un-American Activities Committee in 1949 to testify against a former colleague of his who was at the Berkeley Radiation Laboratory with him during the war and had been suspected of spying. Bohm refused to testify and was indicted for contempt of Congress. Princeton University suspended him. The suspension was removed when Bohm was acquitted but his contract ended three weeks later and his contract was not renewed in what was clearly a political decision.[71] Effectively blacklisted in the USA, Bohm applied for a position at Manchester University, which he did not obtain. However, with support from Einstein he did obtain a professorship at the University of Sâo Paulo, Brazil. Such treatment of innocent people was not limited to America.

Soviet politics were all invasive and affected all aspects of life. Lenin was the architect of the Russian Revolution and Soviet power. The first attempt at revolution in 1905 was largely unorganized. Its immediate cause was a massacre of more than two hundred workers led by a priest who attempted to deliver a petition to the Tsar Nicolas II at the Winter Palace in St Petersburg. The failure of this revolution was a big setback for Lenin and between 1905 and the Revolution of 1917 there was an ideological fight mainly between the Mensheviks and the Bolsheviks led by Lenin. This fight was waged at all levels including science. Some of Lenin's opponents were Russian scientists who were followers of the positivist philosophy of Ernst Mach. Dialectical materialism had to be the universal ideology. To further this aim Lenin wrote *Materialism and Empirio-Criticism*, which was published in Moscow in 1909.[72] Lenin saw Mach as an extreme idealist, though he admitted that Mach did have some materialistic views, but the decisive part of his book was aimed at criticism of his ideas and other physicists such as Helmholtz.[73] Machists' views were seen as reactionary by Lenin.

Bohm's first major task during his exile in Brazil[74] was to complete a textbook *Quantum Theory*, based on his lectures at Princeton University in 1947–1948, which was published in 1951.[75] This book basically followed the orthodox Copenhagen interpretation of quantum mechanics and Bohm's dialectical materialism is only apparent in a few passages.[71] However, Bohm then immediately started work on a deterministic realist version of quantum theory. While this new theory of Bohm seems to correspond to the kind of dialectical materialism advanced by Lenin in *Materialism and Empirio-Criticism*, it was generally only French Marxist physicists who supported Bohm's interpretation. It seems that Bohm was not consciously influenced by dialectical materialism while he was writing his new version of quantum mechanics.[71] In an interview with Maurice Wilkins (1916–2004), joint Nobel Prize recipient with Crick and Watson for their work on DNA, Bohm said in1986:[76]

> 'Before I got to Brazil I was beginning to get interested in philosophy more strongly because for one thing this question of causality interested me. The Bohr interpretation, which I had favored, seemed to say there was no question of objectivity or causality or anything. I sat here with this new interpretation. It came up again so I wanted to understand what was meant by causality. I began going to what people had said about it philosophically.'

In the interview Bohm explicitly stated that the 'new interpretation' that he was talking about was his 1952 paper in the *Physical Review*.[55] The result of Bohm's consideration of the philosophy of causality was his book *Causality and Chance in Modern Physics*, which he began in Brazil and finished when he moved to Israel in 1955.[77] Bohm's use of dialectics in this book was influenced by Mario Schoenberg[76] (1914–1990) a Brazilian physicist who was a communist, in parts of the book there are close similarities with Friedrich Engels' *Dialectics of Nature*.[71] In his interview with Wilkins, Bohm made the curious remark that: 'I never found a communist other than Mario Schoenberg to whom it [Marxist dialectical theory] did get through.'[76] Bohm's deterministic realist version of quantum mechanics was largely ignored. Even behind the Iron Curtain Bohm's interpretation was discussed but neglected despite similar interpretations by Russian physicists.[71]

Bohm's new interpretation of the nature of quantum mechanics brought Léon Rosenfeld (1904–1974) to the defence of Bohr's Copenhagen interpretation. Since 1927 much of the Copenhagen interpretation centred on Bohr's *Complementarity Principle* and this was the focus of Rosenfeld's defence. Bohr was not the easiest of writers and what exactly Bohr meant by complementarity has been a source of debate since 1927. Here only a single sentence explanation of Bohr is given:[78]

> '[E]vidence obtained under different experimental conditions cannot be comprehended within a single picture, but must be regarded as complementary in the sense that only the totality of the phenomena exhausts the possible information about the objects.'

If you do not understand that you are in good company for, during the Bohr-Einstein debates, Einstein wrote:[78]

'Bohr's principle of complementarity, the sharp formulation of which, ... I have been unable to achieve despite much effort which I have expended on it.'

Bohr saw Heisenberg's *Uncertainty Principle* came under the umbrella of the *Complementarity Principle* and when he read Heisenberg's prepublication paper he asked him to add an addendum in proof that in part read:[79]

'Bohr has brought to my attention that I have overlooked essential points in the course of discussions in this paper. Above all the uncertainty in our observation does not arise exclusively from the occurrence of discontinuities, but is tied directly to the demand that we ascribe equal validity to the quite different experiments which show up in the corpuscular theory on one hand, and in the wave theory on the other hand.'

Like Bohm, Rosenfeld was a Marxist and he was a collaborator of Bohr from 1930 until Bohr's death in 1962; one might almost call him Bohr's bulldog, what Wolfgang Pauli did call him in a letter was the $\sqrt{\text{Trotsky} \times \text{Bohr}}$.[80] During the Second World War Rosenfeld was a professor of theoretical physics in the Occupied Netherlands. Politics was thrust on him and his Marxist views were confirmed and he was as keen to refute Bohm's version of the Marxist's dialectic method as he was to refute his interpretation of quantum theory. Rosenfeld believed that the Copenhagen version of quantum mechanics with its concept of complementarity was as an example of the true Marxist dialectical method and he expressed this belief in a paper *L'Évidence de la Complémentarité* for a volume to celebrate de Broglie's 60th birthday published in 1953. A revised English version entitled *Strife about Complementarity* was published in the same year.[81] In this paper Rosenfeld outlines how the transformation of the old deterministic realist interpretation of quantum mechanics into the new interpretation is an example of the pure dialectical method as interpreted by Engels:

'The impossibility of integrating the quantum action into the body of deterministic laws of classical physics corresponds to the phase of "negation"; in fact, as Engels[82] observes, the dialectic negation consists in recognising a limitation to the validity of some concept, in the present instance the concept of classical determinism. This phase, dominated by the contradictions of old quantum theory, ends with the formulation of quantum mechanics and its interpretation in the frame of complementarity. This is the "synthesis" in which those contradictions vanish to make room for a new harmony.'

Rosenfeld rejected Lenin's Marxist philosophy and dismissed Stalin's view as rubbish.[80] In a very quotable passage in his paper, Rosenfeld compares quantum theory with Picasso's art:[81]

'Quantum theory rather calls to mind Picasso's art. It does in the last resort make use of concepts which describe everyday experience, but these concepts no longer occur in their familiar interconnections.'

Max Born 'emphatically den[ied] that the development of physics c[ould] be used as a paradigma for the truth of the Marxistic doctrine,' in an unpublished rejoinder to Rosenfeld's *Strife about Complementarity* in 1955.[83] Born was a liberal with a small l but a man with a large compassion for humanity. The Cold War was in full swing in 1955. That year Bertrand Russell, and Albert Einstein, only just a few days before Einstein's death, had issued a manifesto calling on world leaders to seek a peaceful resolution to international conflict, to which Born was one of the signatories. The Pugwash Conferences on Science and World Affairs, which started in 1957, drew their inspiration from that Manifesto. Although it has nothing and everything to do with physics, it is fit to quote the opening sentence of the Manifesto:

> 'In the tragic situation which confronts humanity, we feel that scientists should assemble in conference to appraise the perils that have arisen as a result of the development of weapons of mass destruction, and to discuss a resolution in the spirit of the appended draft.'

Born sent a copy of his rejoinder to Rosenfeld. He did not publish it probably because he did not want to increase political tensions by publicly criticizing Marxism at the same time as he was actively working for the resolution of tensions. What Born objected to was the concept that the dialectic materialism was the answer to all the philosophical problems of physics:[83]

> 'The claim that these entirely new ideas are in nuce[84] contained in the doctrine of dialectic materialism means a derogation of the efforts of several generations of physicists and in particular of Bohr's genius. There is, of course, also in modern physics a metaphysical component, since there is no logical way from experiment to theory .... But the metaphysical assumptions (about space and time, cause and chance, matter and field) are a continuous development of previous ones and not wild guesses as those of Hegelian and Marxistic teaching.'

This episode in the 1950s when quantum mechanics became embroiled in politics did not end with the decade but broke out again in 1966. The spark for this renewed battle was a preprint of a paper by Klaus Tausk, an Austrian, who had immigrated to Brazil with his Jewish parents in 1938, and was a doctoral student of Schönberg at the University of Sâo Paulo. In 1966 Tausk was working under a scholarship at the International Centre for Theoretical Physics (ICTP) in Trieste, Italy.[85] In the preprint Tausk attacked a paper on measurement problems in quantum mechanics written by Adriana Daneri, Angelo Loinger, and Giovanni Prosperi, supporters of Bohr's Copenhagen interpretation of quantum mechanics. To make things worse, Tausk had without authorization put the unrefereed preprint in an official ICTP cover and distributed it widely. A copy came into the hands of Rosenfeld, who in 1965 had written a paper in support of Daneri,

Loinger, and Prosperi. Rosenfeld and Loinger were enraged by Tausk's paper and attacked both him and ICTP. Abdus Salam (1926–1996) the Director of ICTP wrote an appeasing letter to Rosenfeld which stated in part that:

> 'Mr. Tausk is a special pupil of Mario Schönberg in Brazil. I have not had a chance to see him yet. He is due to leave us at the end of this month to join the Vigier [a Marxist]group in Paris. I would request you that you may consider this episode as part of the old battles and in no case an expression of opinion from the Centre here.'

Tausk was not without his supporters, among who were numbered Bohm and John Bell. Eugene Wigner also wrote a paper with Josef Jauch (1914–1974) and Mutsuo Yanase criticizing the paper of Daneri, Loinger, and Prosperi. The Marxist Bohm had a similar idealistic view to the right-wing Wigner who supported the American arms race. At the time people thought there was a political side to the argument. The Austrian born physicist Otto Frisch (1904–1979) certainly thought there was one, writing to Hugo Tausk, Klaus' father and his cousin:

> '… anyone here in the West who doubts the orthodox interpretation – even for objective reasons – is suspect of communism. All this with the complexities and meaninglessness of a religious war, complete with converts: the greatest defender of the orthodoxy is a communist, and many in the opposition are fully bourgeoisie.'

When Tausk returned to the University of Sao Paulo, Schönberg, who was Rosenfeld's friend, was extremely angry with him and would not attend his defence of his thesis. Tausk's research career was effectively ended by this episode and he did no further significant research.

Rosenfeld would brook no objection to the Copenhagen interpretation or his version of Marxist's dialectic.

6.2.4. *Bohm's causal interpretation of quantum theory*

In 1957 Bohm moved to England first to Bristol University and then in 1961 to Birbeck College, University of London. In England, especially at Birbeck College where Basil Hiley who was to become Bohm's collaborator joined the college as an assistant lecturer on the same day, Bohm found more conducive conditions and he continued to work on the philosophy of quantum theory. Bohm did not believe that quantum theory could be described in a mechanical way which is why he thought that the term quantum mechanics was a misnomer.[86] Orthodox quantum mechanics was unsatisfactory to Bohm; it worked extremely well and gave results that agreed with experiments but it did not give an adequate picture of reality. What Bohm was trying to do from 1951 was to give 'an ontological interpretation of quantum theory,' which was the subtitle to his last book *The Undivided Universe* written with Hiley, on which he was putting the final finishing touches on the day he suddenly died from a heart attack.[87] Bohm considered that what was of prime importance in his approach was not that he was trying to give a deterministic and causal interpretation of quantum theory but that he was attempting to show the nature of quantum theory reality, or its ontology. His theory predicted the same results as orthodox quantum mechanics but contained 'the classical limit within it ... without the need for any special assumptions' and contained the potential for extensions 'that go beyond quantum theory.' There was no attempt at replacing orthodox quantum mechanics; Bohm believed 'that there should be a kind of dialogue between different interpretations rather than a struggle to establish the primacy of any one of them.'

The roots of *The Undivided Universe* go back to Bohm's 1952 paper. The 1964 and 1966 papers of Bell showed that Bohm's hidden variables were admissible provided that one accepted quantum non-locality. Bohm and Hiley showed that non-locality does not infer that information can travel faster than light but it still did not fit easily with relativity. However in a series of papers Bohm and Hiley showed that non-locality and relativity could be accommodated.[88] For Bohm the interconnectivity of things or their wholeness was a significant feature of the new quantum theory. This interconnectivity is not mechanical but an interaction between fundamental entities. Bohm used the analogy with a holographic image formed by the interference between a coherent light from a laser and the light scattered by an object where each part of the image contains implicit information from the whole of the object. The wholeness of quantum theory and its implicit order was explored in Bohm's *Wholeness and the Implicate Order* published in 1980.[89] In classical physics, the order of every part is explicit but in quantum theory the total order is implicit in every part. In the last chapter of *Wholeness and the Implicate Order* Bohm generalized his approach to quantum theory to encompass all existence:

> '[I]n the implicate order of the totality of existence is enfolded within each region of space (and time)... [W]e are suggesting that the implicate order applies both to matter (living and non-living) and to consciousness, and that it can therefore make possible an understanding of the general relationship of these two... One reason why we do not generally notice the primacy of the implicate order is that we have become so habituated to the explicate order, and have emphasized it so much in our thought and language, that we tend strongly to feel that our primary experience is of that which is explicate and manifest.'

Bohm's quantum theory is determinist, the Copenhagen interpretation of quantum mechanics is indeterminist, and yet the mathematical models of both theories give the same predictions. Such apparent contradictions weaken the influence of modern physics on both philosophy and religious belief and suggest that mathematics may be the only reality. Another possibility that was explored by Bohm in *Wholeness and the Implicate Order* was that language controlled our perception of reality, which is not dissimilar to the early ideas of Wittgenstein in *Tractatus*. In particular Bohm thought that 'the subject-verb-object structure of modern languages implies that all action arises in a separate subject, and acts either on a separate object, or else reflexively on itself,' whereas some ancient languages, such as Hebrew, focused on the verb but modern Hebrew usage is now similar to English. Bohm experimented in a flowing form of language, which he called the *rheomode* that would be more appropriate to quantum theory. However, he did not pursue this experiment after *Wholeness and the Implicate Order*; perhaps like Wittgenstein, who later realized that an ideal formal language that accurately pictures the world was impossible, he had second thoughts.

6.2.5. *The Age of Aquarius and quantum mechanics*

Alternative religion has always held an attraction for some. In the nineteenth century it was theosophy that was the rage. The last quarter of the twentieth century saw the development of a New Age Movement which is a smorgasbord of earlier alternative religions such as theosophy and the Eastern religions such as Buddhism, Hinduism, and Taoism.[90] The New Age Movements typically rejected rationalism and hence science but illogically, or perhaps not with their eclectic choice of beliefs, they seized on the new physics, which was seen to legitimize their worldview. They saw classical physics as supporting materialism whereas modern physics was holistic with the observer and the observed becoming blurred. Thus New Age Science is an integral part of the New Age Movement since there are parallels between Eastern mystical religions and modern physics. Even prominent physicists have noted these parallels. In his address to the Physical and Biological Conference at Bologna in 1937, Bohr stated:[91]

> 'For a parallel to the lesson of atomic theory ... we must in fact turn to quite other branches of science, such as psychology, or even to that kind of epistemological problems with which already thinkers like Buddha and Lao Tse have been confronted, when trying to harmonize our position as spectators and actors in the great drama of existence.'

However he went on to make it clear that this did not imply any acceptance of 'mysticism foreign to the true spirit of science.' Bohm went further than Bohr in his discussion of the concept of wholeness writing:[89]

> 'Why ... do we not drop our fragmentary Western approach and adopt these Eastern notions which include not only a self-world view that denies fragmentation but also the techniques of meditation that lead the whole process of mental operation nonverbally to the sort of quiet state of orderly and smooth flow needed to end fragmentation.'

However that did not mean that Bohm was accepting Eastern culture as a whole; 'both in the East and in the West, true insight may have been turned into something false by the procedure of learning mechanically through conformity.'

Two popular writers who exemplify the approach of the New Age movement to modern physics are the American Gary Zukav and the Austrian born American Fritjof Capra. Zukav had no formal training in physics. Capra did a doctorate at the University of Vienna in theoretical physics but has published very few research papers and apparently none since 1979. Both Zukav and Capra have published very successful New Age Science books which have inspired many imitations.[92] The two most successful books are: Zukav's *The Dancing Wu Li Masters* first published in 1979 and Capra's *The Tao of Physics* published in 1975.[93] The reviews of these and similar books by physicists were mostly negative, not because they noted parallels with Eastern religions, but because they implied an identity between physics and Eastern mysticism not simply an analogy.[94]

6.3. Cosmology and the Big Bang

Although astronomy has long been recognized as a science, cosmology, the study of the universe as a whole, has only very recently had the status of a science conferred on it. Earlier cosmology belonged to religion and philosophy and only started to become a science and part of physics in the second half of the nineteenth century. Helge Kragh cites Clausius' claim in 1865 that, the laws of thermodynamics are so general that they extend to the universe as a whole, as the first step of modern cosmology becoming a science.[95] It is not coincidental that the first use of the word astrophysics to describe that branch of astronomy that deals with the physical properties of celestial phenomena and objects was in *Nature* in 1870. Mach and other positivists denied that physical laws could be applied to the universe as a whole. The cosmological theories of the 1930s were seen by Herbert Dingle (1890–1978), the rather maverick English physicist, as giving '[i]nstead of principles from phenomena ... a pseudo-science of invertebrate cosmythology.'[96] By the 1950s cosmology was becoming science and in 1952 Hermann Bondi (1919–2005) published *Cosmology* in which in the first sentence of the preface he stated its main aim was 'to present cosmology as a branch of physics in its own right.'[97] Following a review by the astrophysicist

Gerald Whitrow (1912–2000) of *Cosmology* on the BBC Third Programme in 1952, The British Journal for the Philosophy of Science published a discussion, between Whitrow and Bondi as to on whether cosmology was a science.[98] Whitrow stated his position that:

> 'Physical cosmology has been, still is and, from its very nature, will, I believe, remain a border-line subject between the special sciences and philosophy.'

Bondi believed in Popper's definition of science and argued that cosmology was a science, and hence part of physics, because cosmological theories were capable of being falsified by evidence from observations. However, since cosmological theories were chosen initially to agree with known facts, they could not immediately be part of science. They were in the 1950s really only hypotheses, it was not until the 1960s that new facts, such as the background radiation in the sky, started to be obtained that could be compared with the predictions of the theories that cosmology began to emerge as a fully fledged science.

Most religions have something to say about how the universe and we humans came into being. For Christians the exactly details as described in the Bible have now less importance to most, except fundamentalist sectarians, than they had prior to the twentieth century when classical physics reigned. That the universe was *creatio ex nihilo* was part of the Christian creed by the time of the New Testament, and this was affirmed at the Fourth Lateran Council in 1215. This creed is also part of the Anglican belief though it is not clearly stated in the King James version of the Bible because of textual problems in the translation.[99] The Big Bang was christened by Fred Hoyle (1915–2001) during a 20 minute radio broadcast entitled *Continuous Creation* where he expounded his rival Steady State cosmic theory.[100] Many, including Pope Pius XII, saw the Big Bang as agreeing with the Christian creed. Fred Hoyle was partly so strongly opposed to the Big Bang theory because he was an atheist and thought that religious belief was interfering with the judgement of some of its proponents. However there are atheists who see the Big Bang theory as confirming their atheism.

Einstein believed in a static universe and introduced the cosmological constant, Λ, into his equations of general relativity in 1917 to obtain one However the American astronomer, Vesto Slipher (1875–1969), who began observing the Doppler shift in the spectrum of spiral galaxies in the first decade of the twentieth century, gave Eddington in 1922 a list of the velocities of 41 galaxies, 36 of whose spectra showed a redshift showing that they were receding from us and only 5 gave a blue shift; these results indicated that the universe was not static.[101] Georges Lemaître (1894–1966), A Belgium mathematician and Jesuit priest was the first to propose a version of the Big

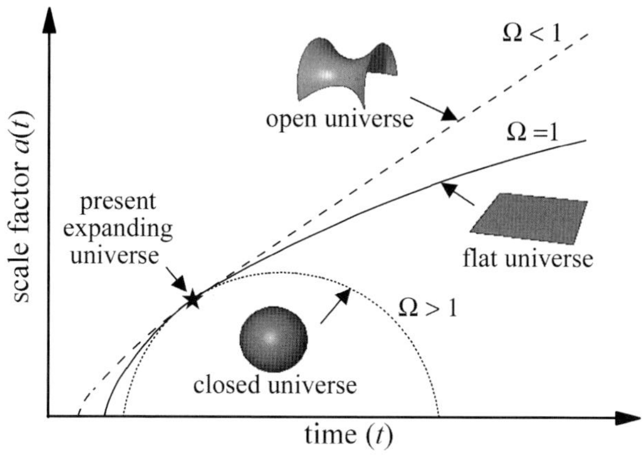

Fig. 6.4. The three classes of Friedmann universes for a zero cosmological constant. The three types of universe are represented by two dimensional model universes.

Bang. His assumption, which became known as the *cosmological principle*, was that the universe is spatially homogeneous and isotropic on a scale that encompasses galaxies clusters and superclusters of galaxies, which is really an extension of the Copernican principle that the Earth is not special and not the centre of the universe. Friedmann had already in 1922 solved Einstein's general relativity equations both with and without a cosmological constant for a homogeneous and isotropic universe and showed that there was a range of solutions and that they included an expansion phase but he was more interested in the mathematics than the physics.[102] The solutions without a cosmological constant modelled three types of universe depending on their energy density as shown in Fig. 6.4. A flat Euclidean universe that expands forever, though at an ever decreasing rate, occurs if at initiation it has precisely a critical density but it is unstable. The type of universe depends upon the density parameter, Ω, which is the ratio of the density of the universe to its critical value. If Ω is ever so slightly greater than one, the universe initially expands but eventually gravity prevails and the universe contracts and the density increases, in this case the universe is always finite and closed. If $\Omega < 1$, the density decreases as the universe continually expands. Willem de Sitter also obtained a solution in 1917 for the limiting case, where the density of matter in the universe is negligible, shortly after Einstein gave his first solution where the universe expands exponentially.

During 1923 Lemaître worked in Cambridge with Eddington and knew of the redshifts in the galaxies. After he had returned to Belgium, Lemaître obtained an exact dynamic solution to Einstein's general theory of relativity assuming a closed universe whose radius of curvature, as estimated by Hubble,[103] was 2.7×10^8 parsecs[104] (about 8.8×10^8 light years) from which he showed that the universe was expanding.[105] In the same paper Lemaître also established that galaxies recede from us at a velocity which is proportional to their distance, a result that is now known as Hubble's law, with a constant of proportionality of 625 km s^{-1}/Mps.[106] In his Presidential address to the Mathematical Association in January 1931 Eddington had discussed the universe, which he believed was closed in spacetime, that is, though the universe is finite there is no edge to space in the same way that there is no edge of to the Earth because it is spherical, there can only be an end in time and he stated:[107]

> 'Following time backwards, we find more and more organization in the world. If we are not stopped earlier, we must come to the time when the matter and energy of the world had the maximum possible organization. To go back further is impossible. We have come to an abrupt end of space-time – only we generally call it the "beginning".'

He then said that 'philosophically, the notion of a beginning of the present order of Nature is repugnant to me' without any explanation. Lemaître, who had been thinking about the implication of an expanding universe since his 1927 paper, was moved by Eddington's remark to write a letter to *Nature* where he explained his speculative view of a quantum beginning of the universe:[108]

> '[T]he present state of quantum theory suggests a beginning of the world very different from the present order of Nature.... If the world has begun with a single quantum, the notions of space and time would fail to have any meaning at the beginning... We could conceive the beginning of the universe in the form of a unique atom, the atomic weight of which is the total mass of the universe. This highly unstable atom would divide in smaller and smaller atoms by a kind of super-radio-active process.'

With this short letter the *primeval atom*[109] and the first version of the Big Bang were born.

Although in his youth Lemaître had toyed with trying to integrate both physics and the Bible,[102] he came to believe that science and religion were different facets of knowledge.[110] However, Lemaître did not believe that science was completely independent from religion or philosophy because he did write that:[111]

> 'Philosophy and theology, when kept in isolation from scientific thought, either change into an outdated self-enclosed system, or become a dangerous ideology.'

Also in his original version of his letter to *Nature* Lemaître had ended it with the sentence:[102]

> 'I think that everyone who believes in a supreme being supporting every being and every acting, believe also that God is essentially hidden and may be glad to see how present physics provides a veil hiding the creation.'

but he removed this sentence before sending the letter to *Nature* probably because he wished to avoid mixing religion with physics. Yet Lemaître was to become a significant influence on the thinking of the Catholic Church.[112] In 1936 Pope Pius XI refounded the Pontifical Academy of the Sciences and Lemaître was one of his first appointments to it. Lemaître was a member of the Academy for the rest of his life becoming president in 1960. A close relationship grew between the next Pope, Pius XII, and Lemaître that enabled this Pontiff to develop an understanding of the new cosmology well before the concept of the Big Bang became widely accepted.

Stephen Hawking and Roger Penrose have demonstrated that the general theory of relativity shows that gravity is unstable and that there is a 'big bang initial phase which is common to most relativistic expanding models'.[113] However the general theory of relativity cannot hold right at the Big Bang singularity and there must be a quantum effect when the curvature is more than 10^{15} m^{-1}. The Hot Big Bang is the standard cosmological model where the initial temperature of the universe, compressed into an unimaginably small space, is unimaginably high. The standard cosmological model is known as the Friedmann–Lemaître–Robertson–Walker model. In the late 1920s and early 1930s the American Howard Robertson (1903–1961) and the English Geoffrey Walker (1909–2001) showed that the Friedmann and Lemaître model was the most general consistent with homogeneity and isotropy of the universe. The Robertson–Walker metric, has only two parameters: the scale factor R, which is roughly the radius of the universe, and the curvature of the spatial sections of the universe. In the standard Big Bang model a singularity is at the beginning of time. Singularities are not physical realities but the mathematical consequence of extending a physical theory, in this case the general theory of relativity, beyond the limits of its validity. Provided the form of the early universe away from the singularity is consistent with what can be deduced about it physically, the presence of the mathematical singularity is not important, though of course nothing can be learnt about the creation of the universe.

The cosmological constant has been measured with increasing accuracy starting with the discovery of remnants of the primeval fireball of the Big Bang by Arno Penzias and Robert Wilson of Bell Laboratories, who were testing a sensitive horn antenna which was designed for detecting low levels of microwave radiation, when they found a low level of radiation in 1964, which they first thought was simply interference noise, but they subsequently realized it to be cosmic microwave background radiation (CMBR) with a temperature of 3.5 K.[114] With the equipment available in 1964 no anisotropies in the radiation could be detected and the CMBR temperature appeared to be uniform in all directions indicating a homogeneous and isotropic universe. However the space probes: Cosmic Background Explorer (COBE) satellite launched in 1989, the Wilkinson Microwave Anisotropy Probe (WMAP) launched in 2001, and the most recently the European Space agency's Planck satellite launched in 2009, have provided increasingly more accurate information. The CMBR temperature obtained from the Planck mission gives a background temperature of 2.726 K with very tiny temperature fluctuations.[115] The CMBR is the oldest radiation in the universe emitted only 380,000 years after the Big Bang when the universe first became transparent to radiation. The temperature fluctuations originated from the regions of the early universe that had slightly different densities. The cosmological constant, Λ, obtained from the Planck collaboration data is 0.693.

6.3.1. *Inflation*

The parameters of the standard cosmological model can be determined from the tiny fluctuations in the CMBR to an extremely high precision. The total density parameter Ω has been calculated to be one to a percentage-level precision and hence the present universe is extremely close to being completely spatially flat with zero curvature.[115] With the standard cosmological model, the total density at one second after the Big Bang would have had to have been $\Omega = 1$ to an accuracy of better than 10^{-16}.[116] Such fine tuning seems improbable. This problem and others in applying the standard cosmological model from the Big Bang are nicely summarized by Alan Guth.[117] There is a horizon problem. We observe microwave radiation from 380,000 years after the Big Bang but the size of the Hubble horizon then, the distance over which radiation can travel, was some hundred times smaller than the size of the universe, thus we receive radiation from parts of the universe which have had no possible causal connection; how then is the radiation so uniform? Another problem is that the spatial and spacetime curvatures of the universe are at present comparable but, in the infant universe at the Planck time ($\sim 10^{-43}$s) after the Big Bang, the spacetime curvature implied by the standard model would be very much larger than the spatial curvature by a factor of something like 10^{30}; such a huge asymmetry in curvature appears to be wrong.[118] There is also a problem with the entropy of the universe. Expansion of the universe in the standard model takes place adiabatically with no change in entropy why then is the entropy of the cosmic radiation in the current universe so high?[119] To solve these problems Guth and

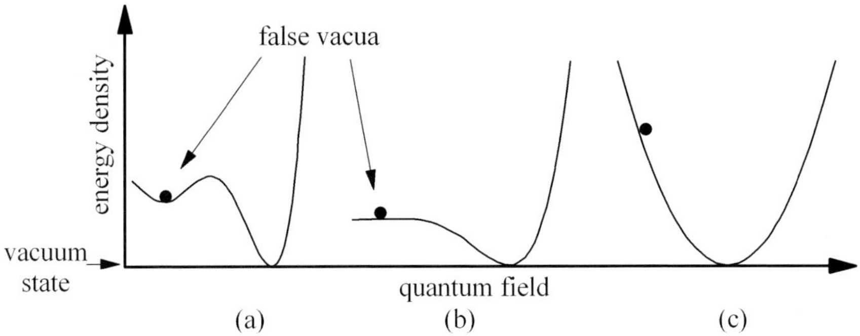

Fig. 6.5. Energy density responsible for inflation (after Guth, 1997, 2007): (a) Locally stable field of original inflation theory. (b) Metastable field of the new inflation theory. (c) Unstable inflation.

others in the early 1980s proposed that the in the embryo universe went through a very short period of intense exponential expansion called *inflation*.[120] Although he did not know it at the time, Guth had rediscovered de Sitter's solution to the equations of general relativity.[116] From 10^{-36} to 10^{-32} seconds after the Big Bang the universe underwent inflation when its radius increased from 10^{-58} m to 1m. Inflation is caused by what is called a *false vacuum*, which is a rather misleading term. In quantum field theory a vacuum is not an empty state but simply the quantum state with the lowest energy. A false vacuum is not the lowest state but a local minimum separated from the true vacuum by barriers from which escape is possible by jumping or tunnelling as shown in Fig. 6.5a and can have a very high energy density. There are more than one inflationary theories but in all a small volume of the infant universe obtained a false vacuum state. It was a high energy false vacuum that caused the initial exponential expansion of the infant universe. Inflation resolved the flatness problem. Instead of the total density being driven away from its critical density it is driven towards its critical value and hence one second after the Big Bang Ω became extremely close to one as is required if the universe was to evolve to its present state of flatness. In one theory of inflation the universe went through supercooling during a phase transformation that resulted in an increase in entropy, which accounts for the large entropy of the present universe and also produced random bubbles that led to a non homogeneous universe.[121]

One inflationary scenario advanced by a Russian physicist, Andrei Linde in 1982 overcame some of the exit problems in Guth's original inflation model. The difference in the inflation predicted by Linde and that of Guth was that the false vacuum was not in a locally stable state as shown in Fig. 6.5a but in a metastable state indicated in Fig. 6.5b. The decay of the false vacuum is then much gentler and 'the whole observable part of the universe is contained inside one bubble, so we see no inhomogeneities caused by the wall collisions of other bubbles.'[122]

This inflationary scenario led to what has become known as *eternal inflation*, first proposed by Paul Steinhardt who, with Andreas Albrecht, had independently proposed a similar inflationary scenario within weeks of Linde.[116] In the new inflationary scenario the energy density drop is not so precipitous as in the original inflation scenario and the rate of loss of energy density decreases exponentially and the false vacuum can actually expand and is not lost when a universe is formed and the process can continue eternally with universe after universe being formed. These universes which are independent of each other are called bubble or pocket universe. The pocket universes are highly unlikely to be identical and develop with different physical constants.

Not even a metastable state is needed for an inflationary scenario and that inflation can start from an unstable state, as shown in Fig. 6.5c, under chaotic inflation. It might be asked that, if eternal inflation continues infinitely into the future can it not have been occurring infinitely in the past, but Guth has shown it not to be possible. These and other topics on inflation are covered by Guth.[123]

Not all have been enthusiastic supporters of the concept of the Big Bang. Although Popper, as a young man, was 'an ardent admirer of Friedmann's suggestion', which simplified Einstein's solution by removing the need for a cosmological constant,' he came to reject the Big Bang, which he thought was an immature theory, because of 'the number of auxiliary hypotheses [was] simply intolerable: according to [his] theory of science' and not science.[124]

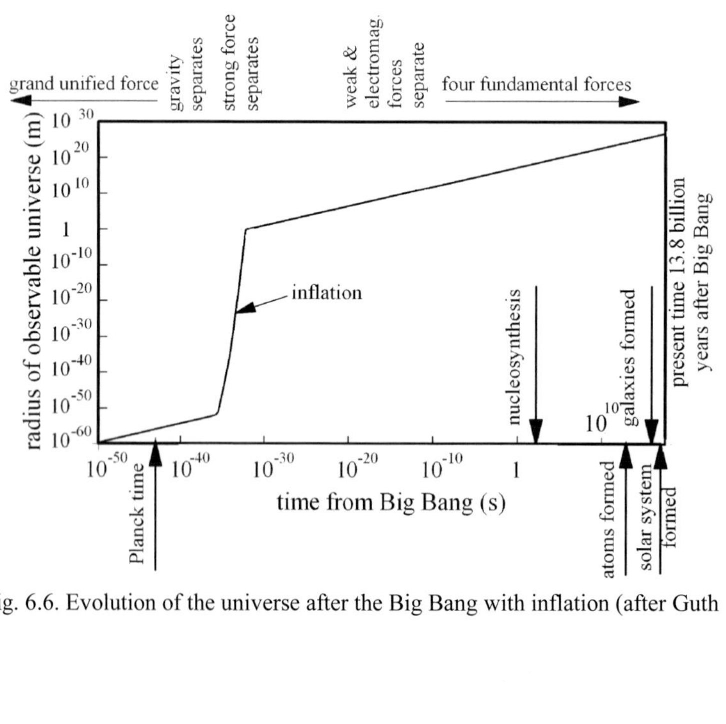

Fig. 6.6. Evolution of the universe after the Big Bang with inflation (after Guth 1997).

A simple timeline for the Big Bang model with inflation is shown in Fig. 6.6. Nucleosynthesis, the formation of the nuclei of atoms began about 3 minutes after the Big Bang and atoms began forming about 380,000 years later. The first stars and galaxies formed about 300 million years after the Big Bang and our solar system formed at about 9 billion years. The present age of the universe is estimated to be 13.8 billion years.

6.3.2. *Dark matter and energy*

Although inflation seems to have solved many of the problems with the standard cosmological model there are still major questions left unanswered. The normal matter of the universe is composed of baryons[125] which in astronomy is the term for all normal matter. The results from the satellites enable the baryonic matter to be estimated from the relative heights of the acoustic peaks in the CMBR and it has proved to be far too small if the total density, Ω, is one. To account for the apparent total matter in the universe a non-baryonic matter, which cannot be detected directly, called dark matter, which neither emits nor interferes with electromagnetic waves, has had to be postulated. Dark cold matter (dunkle kalte Materie) was first mentioned by Fritz Zwicky (1898–1974) in a paper in German which he published in 1933.[126] Zwicky estimated the mass of eight galaxies in the Coma cluster from their luminosity but, when he compared this value to the mass he obtained from the dispersion of their radial velocities, he found it too small by a factor of 400. Although Zwicky overestimated the discrepancy because he used a value of 558 km s^{-1}/Mps for Hubble's parameter which was much too high, using the correct value still gives a discrepancy factor of 50.[126] Still Zwicky's comment:

> 'If this [discrepancy] is confirmed we would arrive at the astonishing conclusion that dark matter is present [in Coma] with a much greater density than luminous matter.'

was essentially correct. From the acoustic peaks in the CMBR recorded during the Planck mission, it is deduced that dark non baryonic matter contributes 26.9% of the equivalent energy of the universe, whereas the baryonic matter, which we can see, only contributes 4.9%. The remaining unaccounted 68.2% of energy is called dark energy. The Planck team stated that:[115]

> 'A major challenge for cosmology is to elucidate the nature of the dark energy driving the accelerated expansion of the Universe. Perhaps the most straightforward explanation is that dark energy is a cosmological constant.'

Thus nothing is directly known about an astounding 95.1% of the energy of the universe. However, dark energy may account for the acceleration of the expansion of the universe first discovered in 1998 from a study of the luminosity of supernovae. The leaders of the research teams: Saul Perimutter, Lawrence Berkeley National Laboratory, Brian Schmidt, Australian National University, and Adam Reiss, John Hopkins University, jointly received the Nobel Prize for Physics in 2011 for their research which had been supposed would reveal not acceleration but a deceleration in the expansion of the universe.

The standard cosmological model assumes that the general theory of relativity is valid on cosmological scales but has only been accurately tested in the solar system. The most accurate assessment of the general theory of relativity was made in 2002 when the Cassini spacecraft, launched in 1997, on a mission to Saturn, passed near the Sun and the frequency shift of radio photons was measured, from which the γ parameter was calculated to be $1.00002 \pm 2 \times 10^{-15}$; in Eddington's parameterized post-Newtonian formalism for the general theory of relativity $\gamma = 1$.[127] Hence within the solar system the general theory of relativity is accurate to about 0.002%. Irwin Shapiro's delay in light from some binary pulsars such as PSWR B1534+12, which are at a distance of about 700 pc from the solar system, is about 1%.[128] Thus at a distance of some hundreds of parsecs the accuracy decreases by two orders magnitude. The scale of the universe is Hubble's length of 5,000 Mpc, seven orders of magnitude more distant than the binary stars. Either dark matter is real or the general theory of relativity does not hold over distances commensurate with the cosmos. Imagine people's reaction if Newton's theory of gravity on Earth had needed the hypothesized existence of dark matter for it to apply to the Solar System.[129] Apart from the precession of the perihelion of Mercury, the only apparent inaccuracy in Newtonian gravitational theory found prior to the twentieth century was the slight irregularity in the orbit of Uranus which was subsequently found to be due to the undiscovered planet Neptune.[130] In 1987, Virginia Trimble stated:[131]

> 'a mainstream astronomer who seriously doubted that we are somehow not readily seeing 90% or more of the stuff in the Universe has found himself in the position of having to justify his discordant views.'

There are some who do question the reality of the unseen stuff and whether the general theory of relativity holds on the scale of the universe. William Vanderburgh discussed two alternatives to the general theory of relativity not because he necessarily thought they were viable but simply to show that there are alternatives.[132] As Vanderburgh remarked, Newton seems to have foreseen the type of philosophical problems that have arisen with the general theory of relativity in his Fourth Rule, added to the third edition of the *Principia*:[133]

> 'In experimental philosophy, propositions gathered from phenomena by induction should be considered either exactly or very nearly true notwithstanding any contrary hypotheses, until yet other phenomena make such propositions either more exact or liable to exceptions.'

Although, as Vanderburgh's pointed out, the interpretation of this rule here is not clear since the validity of the general theory of relativity at the scale of the universe seems to require the hypothesis of the existence of dark matter. Pavel Kroupa, has also questioned the reliance of the standard model of cosmology on the hypothesized existence of dark matter and energy.[134] Rather refreshingly Kroupa begins his paper with a quote from Karl Popper[135]:

> 'For if we are uncritical we shall always find what we want: we shall look for, and find, confirmations, and we shall look away from, and not see, whatever might be dangerous to our pet theories. In this way it is only too easy to obtain what appears to be overwhelming evidence in favor of a theory which, if approached critically, would have been refuted.'

Kroupa believes that the standard model of cosmology is self contradictory and 'is falsified and that dark-matter sub-structures do not exist.'

The cosmological principle is a cornerstone of the standard cosmological model and the evidence for dark energy comes from the assumption of a homogeneous universe. If there was a huge local void in the universe with a radius of some 200 Mpc, the expansion of supernovae could be modelled without dark energy[136] but swapping one assumption for another does not seem a fruitful course of action. However, there is some other evidence that the cosmological principle is not perfect. The CMBR at any frequency would be uniform in all directions according to the cosmological principle but, because the galaxy cluster that includes our Milky Way galaxy and the solar system is moving at a velocity of 370 km/s relative to the reference frame of the CMBR, there appears to be dipole anisotropy. However the magnitude appears to be about four times larger than it should be, implying some anisotropy in the universe which changes with time and challenges the cosmological principle.[137]

6.3.3. *Creation*

The explanation by the standard cosmological model as to how the universe evolved to its present state may be the best there is but it does not answer the question that is crucial in philosophy and religion: was the universe created? This question is metaphysics not physics because any answer that is given cannot be falsified but, unless one believes the question to be a purely religious one independent of any physical reality, the answer must surely be influenced by physics. The problem is that physical cosmology is still in the hypothesis stage, it is based on the general theory of relativity whose accuracy at cosmic distances is unknown, and certainly must be invalid at the Big Bang. Thus all that can be done is to make reasoned speculations that almost certainly will change as our knowledge of physical cosmology develops. But what else is there other than revelation.

The first attempt at giving a materialist explanation of creation can be traced back to an attempt at theoretically determining the mass of the universe from the assumption that its total energy is zero which was made by Arthur Hass (1884–1941) in 1936.[138] It is possible for the total energy to be zero because the gravitational potential is negative and the mass energy is positive. Pascual Jordan wrote in 1947:[139]

> 'The sum of the individual energies of all the elementary particles in the world is within an order of magnitude of their mutual gravitational potential energy; and the possibility suggests itself of tightening this empirical order-of-magnitude equality by hypothesizing that the total energy of the universe is zero.'

In these earlier flirtations with the concept that the energy of the universe is zero, Jordan did speculate that it might be possible for a star to be created out of nothing but the possibility the entire universe might be created out of nothing was not considered probably because the Big Bang theory had not yet emerged. In 1973 Edward Tryon wrote a short paper, published in *Nature* that asked whether the universe was a vacuum fluctuation, which created much interest at the time.[140] However within a decade the consensus was that the case was non-proven: the total energy of the universe depends upon the spacetime geometry, in some cases it is zero but in others it is not.[141] An alternative way of creating the universe out of nothing by quantum tunnelling[142] was proposed by a number of researchers in the early 1980s but their conception of what out of nothing means differs.[143] In his 1982 paper Alexander Vilenkin considered that 'the universe is spontaneously created from literally *nothing*,' but by 1984 he had modified that view writing: 'by "nothing" I mean a state with no classical spacetime.'[144] David Atkatz and Heinz Pagels concluded that quantum tunnelling can only produce a spatially compact universe, that is, one where the initial state is closed space. On the other hand the closed spacetime model of James Hartle and Stephen Hawking has no initial boundary.[145] In contrast Richard Gott and Li-Xin Li believe that asking 'how to make something out of nothing?' is the wrong question, instead one should ask: 'Do the laws of physics prevent the universe from being its own mother?'[146]

In 1951 Pope Pius XII, in a quite extraordinary address to the Plenary Session of a meeting at the Pontifical Academy of the Sciences, affirmed that:

> 'Contrary to rash statements in the past, the more true science advances, the more it discovers God, almost as though He were standing, vigilant and waiting, behind every door which science opens.'

and discussed how the Big Bang model of the cosmos was in accord with the teaching of the Church.[112] Pius XII began his address by examining the proofs of the *quinque viæ* or five ways for a belief in God, summarized in the *Summa Theologica* (1265–74) of the 'Angelic Doctor' Aquinas, 'on the basis of the new scientific discoveries especially those based upon the movement and order of the universe.' That the Pope should consider that the new cosmology could impinge on the five ways of Aquinas is at first sight extraordinary but probably it is logical since some of the five ways come straight from Aristotle, who for long held the position of Chief Scientist to the Catholic Church. It is also clear that in Pius' eyes the modern cosmology would not change any doctrine but that it could be used to strengthen it in the modern world, for he said:

> 'Rather than a revision of the philosophic proof, it is a question of scrutinising the physical bases from which those arguments derive.... But there is no fear of surprises: science itself remains firmly grounded in that world which today, as yesterday, presents itself in those five "modes of being" from which the philosophic demonstration of the existence of God takes its motives and force.'

Pius XII thought, as Fred Hoyle did, that the Big Bang theory agreed with the Christian revelation:

> '[A] mind illuminated and enriched by modern scientific knowledge ... recognises the work of the omnipotent Creator, Whose power, aroused by the mighty "fiat" pronounced billions of years ago by the Creative Spirit, unfolded itself in the universe and, with a gesture of generous love, called into existence matter, fraught with energy. Indeed, it seems that the science of today, by going back in one leap millions of centuries, has succeeded in being a witness to that primordial *Fiat Lux*, when, out of nothing, there burst forth with matter a sea of light and radiation, while the particles of chemical elements split and reunited in millions of galaxies.... [I]t is worth noting that modern exponents of the natural sciences consider the idea of the creation of the universe entirely reconcilable with their scientific conception, and indeed they are spontaneously brought to it by their researches, though only a few decades ago such a "hypothesis" was rejected as absolutely irreconcilable with the present status of science. As late as 1911, the celebrated physicist Svante Arrhenius declared that "the opinion that something can proceed from nothing is in contrast with the present status of science, according to which matter is immutable."'

There was no doubt in the mind of Pius XII that the modern cosmology was true to the Church's doctrine:

> 'Today ... the need is not to find God for the first time, but rather to recognise Him as a Father, to revere Him as Legislator, to fear Him as Judge; it is urgent for the salvation of all peoples that they adore the Son, the loving Redeemer of mankind and they bend the knee to the gentle urgings of the Spirit, fruitful Sanctifier of souls.'

The principle of sufficient reason (PSR), usually associated with Leibniz because it was one of his logical premises but whose original can be traced back to the ancient Greeks and has many interpretations, has been used on both sides of arguments as to whether the Big Bang cosmology supports a theistic or atheistic view of the creation of the universe.[147] The PSR and the related principles such *ex nihilo nihil* (nothing comes from nothing) and the causality principle that every thing has a cause, are seen by many people as self evident but that only really means that the principle has become a dogma through oft repetition and, since such people offer no other reason for their truth, it only requires one intelligent person to say they not self evident to them to negate them much after the style of Popper's principle of falsification. One such intelligent person was Hume.[148]

6.3.4. *Theism, atheism and the Big Bang cosmology*

An industry has developed among both theists and atheists to use the Big Bang cosmology to support their arguments, frequently making use of variants of the PSR. A good example of the arguments[149] is provided by *Theism, Atheism and Big Bang Cosmology* written by two Americans: William Lane Craig, an evangelical Christian theologian and philosopher, and Quentin Smith, an atheist philosopher.[150] In the first part Craig presents a theistic cosmological argument with Smith responding, Smith presents the atheistic argument in the second part with Craig responding, and in the third part Craig and Smith discuss the religious implications of Stephen Hawking's quantum cosmology. Here only a summary of the cosmological arguments can be given and many of their nuances are necessarily omitted.

The Big Bang scenario that there was a beginning to the universe fits evangelical Christianity and is readily accepted by Craig. He now only has to invoke his 'first premiss, that everything that begins to exist has a cause of its existence,' to have a theist's case. The first premise is not defended because not only does Craig not have the space but also he thinks 'it to be somewhat unnecessary as well. For the first premise is so intuitively obvious.' This premise is crucial for the theists' argument and Craig should not have dismissed it so casually. That C.D. Broad, though an atheist said:

> 'I cannot really believe in anything beginning to exist without being caused ... by something else which existed before and up to the moment when the entity in question began to exist.'

is no support for Craig's position. Smith obviously seizes on Craig's causality principle and argues that the quantum nature of the universe at its beginning negates this principle because of the innate indeterminacy of quantum theory. Craig responds 'that if the universe did originate from nothing, then that fact does point to a supernatural cause of its origin.' Craig argues that Smith 'in no way proves that the universe began to exist without a cause, but only that its beginning to exist was unpredictable,' which Craig can accept. The Penrose-Hawking singularity theorems prove 'that an approximately homogeneous and isotropic "open" [or closed] cosmological model must have a singularity in the past.[151] Hawking has also demonstrated the *principle of ignorance*, that is, what comes out of the singularity in the general theory of relativity cannot be fully predicted. However, as Hawking stated: [152]

> '[W]hat the singularity theorems really indicate is that the gravitational field becomes so strong that quantum gravitational effects become important.... [and] it is not necessary to postulate new laws for the singularities.'

The exact structure of the universe obviously depends upon the exact nature of the initial slight inhomogeneity and anisotropy. The development of the structure of the universe from the initial singularity is chaotic but it is in principle predictable but not predictable by us. If there is a God, He could predict the nature and structure of the universe. Craig asks 'whether the whole of being could come out of non-being' and answers 'here a negative answer seems obvious,' rejecting the vacuum fluctuation models mentioned above.

In the next chapter Smith presents the atheists' argument: 'that the classical Big Bang cosmology is inconsistent with theism,' admitting that 'the non-theistic interpretation remains largely undeveloped and unpromulgated.' Smith notes that the classic Big Bang cosmology has been abandoned as has been discussed above. In his argument Smith uses the anthropic principle almost exclusively that God would have ensured an animate universe, which I think is inappropriate. The argument is changed from whether atheism or theism is the more likely given the Big Bang cosmology to comparing atheism with a benevolent Christian-like god whose end purpose is assumed to be to create humans. Hence Smith fails in this chapter to demonstrate that the atheist's position is more tenable than the theist's. Since I do not agree with his arguments about the unpredictability of the universe, I also think he fails to demonstrate that benevolent god is less likely than atheism. In the following chapter Craig makes good use of this lapse in Smith's argument stating:

> 'It does not follow …that necessarily, if God chose to create the universe, then its first state either is animate or leads by physical necessity to an animate state. For there is no inconsistency in His choosing to create an inanimate universe.'

In his last chapter in Part 2 of the book Smith sets out to show that the Big Bang cosmology is inconsistent with theism. However he persists in maintaining that God necessarily must create an animate universe and adds nothing significant to the debate. Craig in his reply also does not add any more significant points. I personally find neither of the arguments given by Craig and Smith convincing.

The two chapters in the third part of the book, discussing Hawking's viewpoint, are quite separate. Although the necessary quantum nature of the early universe is mentioned in the first two parts of the book, the main thrust of the argument is concentrated in the classical Big Bang cosmology. I think the debate would have been better if the current best cosmological model was more the focus of both Craig and Smith.

There have been many more who have debated whether cosmology supports theism or atheism. Some of those supporting the theist cause have been physicists such as Russell Stannard and John Polkinghorne, who became an Anglican priest, but they are very much in a minority. The majority of physicists 'have not only no religious faith but no interest in the issue,' as Stephen Weinberg has stated.[153] The biggest problem in using cosmology to argue whether the universe was created by a Supreme Being, or that it arose out of nothing by some kind of uncaused quantum fluctuation, is that the present cosmological model may be the best available but I doubt that there is a cosmologist who believes that it gives anything like an exact account of the moment of creation. Many will argue that it will be impossible ever to know exactly how it was created. String theories are so new and comparatively untried that they have not yet contributed significantly to the debate. Perhaps the best line to take in this debate is not to try to answer the religious question but to attempt to decide the philosophical question of whether cosmology ever will be able to determine how the universe came into being. Only if this question is answered positively is it worthwhile asking the religious question.

6.4. The Anthropic Principle

Newtonian mechanics of the motion in the solar system was used in the eighteenth century to develop the design argument for the existence of God. In the twentieth century the argument has been reworked to become the *anthropic principle.*[154] In the theists' version of the principle, God set the initial conditions of our universe at the Big Bang so that humans must evolve. It is obvious that since we humans inhabit this universe, any physical theory must result in a universe where our evolution was possible but putting ourselves forward as the be all and end all of the universe is a human centric theological notation. Galaxies, stars, planets, the elements, and sub-atomic particles must also be possible; our only importance in this weak anthropic principle is that we can observe our universe. The *a posteriori* nature of a weak anthropic principle[155] is unobjectionable. The strong anthropic principle, on the other hand, states that the initial conditions of our universe had to be such so that humans must evolve. This *a priori* principle cannot be justified scientifically and is a theological principle that is objectionable to all except religious believers.[156] As Sherrilyn Roush has argued the strong anthropic principle is anti-Copernican.[157] Did Copernicus get rid of geocentricism only to have it replaced by anthropocentricism in the twentieth century? A third anthropic principle was proposed by John Barrow and Frank Tipler: 'Intelligent, information-processing must come into existence in the universe, and, once it comes into existence, will never die out.'[158] Martin Gardner (1914–2010), an American popular mathematics and science writer, referred to this principle in a review of Barrow and Tipler's book as the 'Completely Ridiculous Anthropic Principle,' which has the acronym CRAP and I do not dispute his judgement.[159] In all applications of the anthropic principle it is assumed that the possible alternative universes have the same physical laws as ours, the only difference being the physical parameters. If the universes of the multiverse are spawned from other universes, such an assumption may be correct, since there is no hint whatsoever that laws of physics can mutate like living organisms. In any case such an assumption is necessary because, if the alternative universes have different physical laws of which we have no knowledge, nothing can be said about them. However it is a sweeping assumption that in itself shows the weakness of the anthropic principle; we are trying to make judgements about why our universe is like it is without having anything but a large number of unsubstantiated hypotheses. More of Newton's cautionary, *hypotheses non fingo*, would be welcome in many works discussing the anthropic principle.

The weak anthropic principle does provide one test for any Big Bang theory, it must account for the observable universe. There are twenty six parameters that define the standard model of quantum mechanics and eleven parameters that define the concordance cosmological model. Not all of the thirty six parameters are considered essential in determining whether a universe is capable of developing life, and there only five or six that are really critical but which five does vary somewhat.[160] I have decided to base this section on the six chosen by the eminent cosmologist, Martin Rees, the present Astronomer Royal. Rees lists the six parameters in his popular book *Just Six Numbers*:[161]

- Ratio of electromagnetic and gravitational force *N*.
- Nuclear efficiency in converting mass into energy ε.
- Density parameter Ω.
- Cosmological constant Λ.
- The amplitude of primordial fluctuations *Q*.
- The number of spatial dimensions of our world *D*.

Table 6.1. Some physical constants and their values in our universe.

Constant	Symbol	Value
Total density	Ω	0.9995
Cosmological constant	Λ	0.693.
Fine structure constant	α	7.297×10^{-3}
Gravitational coupling constant	α_G	5.905×10^{-39}
Mass ratio proton to electron	μ	1836
Ratio electromagnetic to gravitational force	N	1.236×10^{36}
amplitude of primordial fluctuations	Q	2×10^{-5}
Number of spatial dimensions	D	3

The density parameter, Ω, and cosmological constant, Λ, have already been discussed above and I will very briefly discuss the rest of these parameters chosen by Rees as needing to be tuned for intelligent life to evolve in a universe. Rees' six parameters are all non-dimensional. The conjectured hypothetical universes, or those a multiverse, are completely independent of each other and we cannot use arbitrary yardsticks from our own universe to measure parameters in another universe. However, if all the universes follow the same physical laws, the natural Planck units can be used. The physical parameters that are used in this discussion and their values in our universe are given in Table 6.1.

6.4.1. *Ratio of the electromagnetic to gravitational force N*

A planet, if it has potential to develop intelligent life, must orbit a star which gives off radiant energy and has a long life so that there is time for such life to evolve. Our Sun has existed for some 4.6 billion years and the Earth formed less than a billion years later. The first life formed on Earth about 3.5 billion years ago which is the order of time necessary for intelligent life to evolve but we cannot use Earth years to measure time in either a hypothetical universe or a universe of the conjectured multiverse. In Planck time 3.5 billion years is of the order of 10^{60}. A normal star is a 'gravitationally bound fusion reactor' for converting hydrogen into helium whose gravitational collapse is balanced by the pressure created in the reactor. A star needs to be of a size sufficient to last this long before it begins to collapse and both a star's mass and lifetime are a consequence of a weak gravitation force. Three physical parameters, which determine the mass and lifetime of a star are:

The fine structure constant:

$$\alpha = \frac{k_e e^2}{\hbar c} = \left(\frac{e}{q_{Pl}}\right)^2, \tag{6.2}$$

where e is the charge of an electron and q_{Pl} is the Planck Charge.
The gravitational coupling constant:

$$\alpha_G = \frac{Gm_p^2}{\hbar c} = \left(\frac{m_p}{m_{Pl}}\right)^2, \tag{6.3}$$

where m_p is the mass of a proton and m_{Pl} is the Planck Mass.
The ratio of the mass of a proton to that of an electron, m_e,

$$\mu = m_p / m_e. \tag{6.4}$$

The values of these three constants in our universe are given in Table 6.1.

The ratio of the electromagnetic to the gravitational forces for two protons, which is independent of the distance between them because both forces follow an inverse square law, is given by

$$N = \frac{F_e}{F_G} = \frac{\alpha}{\alpha_G} = \left[\frac{(e/q_{Pl})}{(m_p/m_{Pl})}\right]^2 \approx 10^{36}. \tag{6.5}$$

This nondimensional number, the first of those discussed by Rees, is very large because gravity is extremely weak compared with electromagnetic forces.

Fig. 6.7. The asteroid Vesta mean diameter 525 km Courtesy NASA/JPL-Caltech.

Gravity is always positive but the electric forces between protons and electrons in molecules cancel out and as the number of molecules increase the gravitational force increases while the electric charge does not. However until the aggregation of matter becomes very large, the self gravity, that is the force holding an aggregation of matter together, is small. Even the self gravitational forces in Vesta (see Fig. 6.7), one of the largest asteroids in the solar system, with a mean diameter of 525 km and mass 2.6×10^{20} kg, is not quite large enough for gravity to pull it into a sphere. The dwarf planet Ceres, which is 959 km in diameter and mass 9.4×10^{20} kg, is large enough and is almost a true sphere.

The reason why N is so large in our universe is because our existence depended upon there being a planet orbiting a normal star, like our Sun, and that star needed to have a lifetime long enough for us to evolve. Here I am using only the weak anthropic principle and I do not imply that this had to happen but just stating that if it had not happened I would not be sitting writing this book. Since the necessary size of the star and its lifetime are closely related, only the lifetime is discussed here. The lifetime of a normal star is related to a characteristic time is given by[162]

$$\frac{t_c}{t_{Pl}} = \frac{\alpha^2}{\alpha_G^{3/2}} \mu^2. \tag{6.6}$$

With the physical parameters of our universe, the characteristic time given by Eq. (6.6) is of the order of 10^{59} which is similar to the estimate of the time taken for us to evolve. Even if the gravitational constant, G, changed by several orders of magnitude it would not greatly affect the interaction of atoms and molecules but it would affect both the magnitude and lifetime of stars.[161] If G was a hundred times larger it would decrease the characteristic lifetime of a star in Planck units a thousand fold, which might not be sufficient for intelligent life to evolve. As

Rees has stated: 'G need not be fine-tuned – merely exceedingly weak so that α_G^{-1} is indeed very large.'[162]

6.4.2. *Nuclear efficiency in converting mass into energy*

Our solar system was formed out of the debris of older stars much larger than our Sun that had exploded in supernovae. These stars, though they start by generating energy from the nuclear fusion of hydrogen into helium are massive enough for the force of gravity to raise the core temperature sufficiently to progress to fusion of higher atomic weight elements. Nuclear fusion can form elements up to iron with the release of energy. Elements above iron in the periodic table can only be formed by fusion by the aid of additional energy. Thus one of the most important reactions that enabled the evolution of our universe is the nuclear fusion of hydrogen into helium under extremely high pressure. Jupiter, the largest planet of the solar system being two and a half times the mass of all the other planets combined, consists mainly of hydrogen plus about 10% helium but it mass is only about one thousandth of that of the Sun and too small to generated a high enough pressure to cause the nuclear fusion of hydrogen into helium.

When hydrogen fuses to form helium there is a loss in mass of 0.7% which is released in the form of energy. The fractional loss in energy of the fusion process is often called the nuclear efficiency, ε, and is related to the strong nuclear force that binds the nucleus of atoms together. The nuclear fusion of hydrogen into helium takes place in three stages. The first stage, the formation of a nucleus of deuterium, a heavy isotope of hydrogen consisting of a proton and a neutron from two protons, is crucial. This nucleus is stable because of the strong binding force, which is reflected in the value of $\varepsilon = 0.007$. If the strong force was much smaller and cause the nuclear efficiency ε to be less than 0.006, this crucial first stage would not be stable and the fusion process would not occur.[161] However if $\varepsilon > 0.008$ the bounding between protons would be so strong that there could be no free protons and no hydrogen. This certainly looks like fine tuning, but only takes into account of one parameter, what is important is the strong force that holds protons together compared with the electromagnetic force, which forces protons apart, or nuclear efficiency, ε, compared with the fine structure constant, α. It is not a question of individual fine tuning of physical constants like ε but looking at the five or six parameter space of the significant constants.

6.4.3. *The amplitude of primordial fluctuations Q*

In his presentation speech at the Nobel Prize Award Ceremony for Physics to John Mather and George Smoot in 2006 for their work on the Cosmic Background Explorer Satellite (COBE), Per Carlson, the Chairman of the Nobel Committee, said that the tiny primordial fluctuations as measured from the CMBR represented the 'seed of the structures in the universe.' If there were no fluctuations then all matter would have been spread evenly in the Big Bang and stars and galaxies would probably not have developed and, if they had been, gravity would have swallowed everything in black holes. Instead we have a Goldilocks universe. The ratio, Q, of the energy released in the formation of the stars, galaxies, and clusters of galaxies to their rest energy Mc^2 is a measure of the texture of our universe.[161] In our universe $Q \approx 2 \times 10^{-5}$. If Q were greater than about 10^{-4} the galaxies would be too dense and stars would have close encounters and the universe would be too violent to have stars with a comparatively stable planetary systems.[163] If the primordial fluctuations were much less so that Q were less than 10^{-6} there would not be texture in the universe and gravity would not be able to form stars.[161]

6.4.4. *Number of spatial dimensions D*

Newton realized that the inverse square law of gravity not only made the orbits of isolated planets elliptical but made their orbits stable so that planets did not either plunge into the Sun or disappear into space if their orbits were perturbed. As discussed in Chapter 1, the inverse square law is a consequence of the geometry of three dimensional space. Rees began his discussion of the number of space dimensions by the stark statement: 'Our space has three dimensions.'[161] Most would agree that our senses seem to perceive that there are but three dimensions but that does not amount to a proof that there are just three dimensions. We can easily imagine a one or two-dimensional space but I doubt that there are many, if any, who can really conceive of space with more than three dimensions. Kant's early discussion of the number of dimensions in space was continued in modern times in a paper published in 1918 by the Austrian born physicist Paul Ehrenfest (1880–1933) a close friend of Einstein and Bohr, who became a Dutch citizen in 1922.[164] In his introduction to that paper Ehrenfest admitted that perhaps the question 'Why space has just three dimensions?' has no sense but purposely did not address this issue in his paper. However, Craig Callender has addressed this question in a thought provoking article.[165] Callender and others dispute the claim that it can be shown that our universe must have three spatial dimensions from the argument that only that number confer stability to planetary orbits or any other singular feature of physics. Victor Stenger (1935–2014) also made the point that spacetime is our invention, a model, and not necessarily a reality and hence nor is its dimensionality.[160] In passing it should be mentioned that the stability of planets argument has been extended to spacetime and relativity.[166] One '*raison d'être* for the observed dimensionality of the physical world' comes from Appendix II of the third edition of Einstein's *The Meaning of Relativity* where he noted that, for four dimensional spacetime, 'the gravitational equations for empty space determine their field just as strongly as do Maxwell's equations.'[167] However, what about superstring theories with ten space dimensions, how are such theories compatible with the notion that there can only be three space dimensions? The claim of those that follow the argument that there can only be three space dimensions has had to be amended so that at the macroscale it has to appear that there are only three by having the extra dimensions curled to form a cylinder whose size is of the order of the Planck length, or some other means of hiding any extra dimensions, whereas at a fundamental level, such as reigned at times of the order of Planck time after the Big Bang, there was no limitation on the number of dimensions of space.

Callender is right to state that in their enthusiasm to ask why there must be three space dimensions, most have side stepped the first crucial point: 'How many spatial dimensions are there?'[165] Leaving this problem with the argument aside, Callender asks why stability is necessary. Stability is not part of any law it is only an anthropic principle and is no proof that there must be three dimensions. Stenger's argument[160] that 'since the spacetime model is a human invention, so must be the dimensionality of spacetime,' seems valid. Modern physics is very much about symmetry, not just geometric symmetry but also symmetries in the laws of nature, increasing the number of dimensions in our model of space in M-theory to ten greatly increases the number of available symmetries and enables the merging of the four fundamental forces at distances of the order of the Planck length. However the number of space dimensions that are not curled up in such a small space that they are unobservable is still three. Kant saw that incongruent counterparts, like left and right hands, showed that space is not a property of objects.[168] The mathematician August Möbius (1790–1868) pointed out in 1827 that the left hand could be flipped into a right hand if there were four space dimensions so that the incongruent became congruent, which is again shows that the number of symmetries increase with the number of space dimensions.[169] Wittgenstein made the same remark in *Tractatus* writing:[36]

> 'The Kantian problem of the right and left hand which cannot be made to cover one another already exists in the plane, and even in one-dimensional space; where the two congruent

> figures ... cannot be made to cover one another without moving them out of this [one-dimensional] space. The right and left hand are in fact completely congruent. And the fact that they cannot be made to cover one another has nothing to do with it. A right-hand glove could be put on a left hand if it could be turned round in four-dimensional space.'

The last of Rees' numbers, the number of dimensions, has not the same status as his other five.

6.4.5. *What can be learned from the anthropic principle?*

The first question is whether our universe is fine-tuned. Rees does not claim that these parameters have to be as fine-tuned as others, especially theist writers who often, without any justification, have claimed incredibly fine-tuning is necessary.[160] However if it is really necessary to fine-tune five parameters to five unique values then the probability of this occurring by chance becomes quite slim even if the tuning of each parameter is not particularly fine. For example, if the probability of obtaining near enough tuning in any one parameter is one in four, then the probability of achieving close enough tuning in all five becomes 0.25^5 which is less than one in a thousand. However, as Stenger pointed out, there are not just five unique values of the parameters that are necessary for life to evolve but a range of all these five values which could make the probability of it occurring by chance substantially greater.[160] To make his point, Stenger wrote a programme to look at model or toy universes determined by four physical constants: the fine structure constant, strong nuclear force coupling constant, and the mass of an electron and proton, which calculated eight conditions necessary for intelligent life using the four physical constants chosen randomly that varied by ten and two orders of magnitude from those of our universe, running his programme ten thousand times in each case. For the purpose of this demonstration neither the particular physical constants chosen as significant nor the eight conditions considered necessary for life are particularly important. With physical constants chosen randomly over a range of ten orders of magnitude from the values in our universe, intelligent life was possible in 13% of his toy universes. When the range of constants was reduced to two orders of magnitude, 37% of the universes could possibly support intelligent life.

An astrophysicist, Fred Adams, has examined the range in the values of three cosmological parameters that would allow normal stars similar to our Sun to evolve.[170] He found that about a quarter of the three parameter space would allow for the existence of normal stars. By reducing the number of physical constants involved and only looking at one aspect of the anthropic principle, Adams' result would necessarily overestimate the occurrence of normal stars, but his result is similar to that of Stenger.

Tegmark and his co-authors, who included Rees, also have recognized the need to examine the effect of more than one physical constant at a time.[163] In an approach limited to examining the values of the constants necessary for a stable planetary system, they chose four constants: the dark energy density, the baryon mass per photon, the cold dark matter mass per photon, and the amplitude of primordial fluctuations. They had the problem faced by all of how to present their results, since a four-dimensional parameter space cannot be visualized, and gave a series of two-dimensional slices through this space. Such an approach is justified since their purpose was to present results to other physicists but if the purpose of the work is to discuss fine-tuning and the anthropic principle for a wider readership, I believe that the approach of Stenger and Adams is more desirable.

Thus it seems that the fine-tuning necessary to satisfy the anthropic principle may not be as fine as many have stated. However let us now turn to examine the implications if the fine-tuning is such that the right parameters are not likely to be obtained on a first throw of cosmological dice. The possible explanations of how such apparent fine-tuning has arisen so that we find ourselves in this universe are discussed in the following subsections.

6.4.5.1. *No fine-tuning was necessary because as yet undiscovered laws of nature determine the constants of particle physics and cosmological constants*

None of these parameters can at present be calculated *a priori* but have been found from observation and experiment. Superstring theory has no adjustable parameters but the problem is there is not just a single set of parameters but an extremely large number of different possible parameters and seems more compatible with the multiverse explanation of the anthropic principle. One can only speculate whether a theory of everything will ever be found that enables these parameters to be calculated, in which case there would be no anthropic principle, and Einstein's musings,[171] remembered by his assistant Ernst Straus (1922–1983) as to whether God could have created the world any differently, would have to be answered in the negative. However cosmological constants seem even less likely to be preordained than those of particle physics since the very slight inhomogeneity in the CMBR suggests that chance set its value and the distribution of stars and galaxies was not predetermined. The only constant that has been calculated, the cosmological constant, Λ, is an embarrassment. One contribution to the cosmological constant is proportional the vacuum energy density, which is the quantum state with the lowest energy. However, using estimates of the vacuum energy, results in a contribution to the cosmological constant that is 120 orders of magnitude too large.[172] Thus this explanation for the value of the constants that determine our universe may prove to be unlikely.

6.4.5.2. *A Supreme Being created the universe so that it would support intelligent life*

This explanation readily explains both the weak and strong anthropic principle but because it is not the only possible explanation does not in itself constitute a strong argument for a Supreme Being.

6.4.5.3. *The physical and cosmological constants have the values that have enabled intelligent life to evolve by chance*

I think that this explanation is too readily dismissed by most. If the quantum theory and cosmological constant can take a large range of values, and those for any universe occur purely by chance, then any set of constants is as likely as any other including those that satisfy the anthropic principle. We can only observe a universe which satisfies the anthropic principle and we have no knowledge whether there have been or will be any other universes. If God was playing dice we do not know whether he rolled the dice only once and came up with a double six. In fact if he obtained any other result and a different universe was created in which intelligent beings could not exist, that would be just as unique result as our own universe. There is no rational reason why a universe in which intelligent beings can evolve is special. It is just that we can only observe a throw of a double six; any other result such as throwing a five and a two, though we would not see it, is just as unlikely. It is humans who have placed mystic importance in a double six as they did when they put the Earth at the centre of the universe.

Rees is unsatisfied by 'the hard-headed response … that we couldn't exist if these numbers weren't adjusted in the appropriate "special" way; we manifestly are here, so there's nothing to be surprised about.'[161] However, I find the hard-headed explanation acceptable.

6.4.5.4. *Our universe is not the only one but there is a whole multiverse of universes*

The universes in a multiverse are causally independent of each other. Although William James (1842–1910), an influential American philosopher and psychologist, coined the term, multiverse was not used by physicists or cosmologists before the 1980s but that does not mean that the concept was not discussed.[173] Boltzmann in considering the thermodynamics of the universe suggested in 1895 that it would consist of many subuniverses whose entropy would vary from low to high while the entropy of the total universe had a maximum.[173] Since the 1980s publications on the multiverse have developed into a torrent which has resulted in almost a race to see who can propose the most outlandish similar to the earlier rush to populate every body in our universe. Much of this interest since the 1990s stems from the anthropic principle; the controversy is whether the multiverse is a legitimate topic of physics.

There is no clear or universal classification of the different kinds of multiverse that have been proposed. Indeed as George Ellis and his co-authors write: 'There is … vagueness about the proposed nature of multiverses'.[174] Tegmark, perhaps the most vocal proponent of multiverses, has proposed four levels:[175]

- *Level I multiverse*

Inflation spins off pocket universes while inflation continues. Since all the universes start with the same Big Bang, it is believed that all the universes will have the same physical laws but their initial conditions are different. It has been questioned whether the *Level I* multiverse should be classified as a multiverse at all but is rather a 'multidomain universe'.[174]

- *Level II multiverse*

Here the multiverse is a true multiverse where inflation combined with the string landscape scenario is assumed to produce universes with different physical constants and different particle physics. Bernard Carr believes that this level warrants four subdivisions.[176]

- *Level III multiverse*

Tegmark based his *Level III* multiverse on the parallel universes in the many-worlds interpretation of quantum physics that Hugh Everett (1930–1982) proposed in his doctoral dissertation at Princeton University in 1957.[177]

- *Level IV multiverse*

This is the most controversial of Tegmark's proposed and it expresses his belief that the ultimate reality is mathematics and that 'all mathematical structures exist physically'.[175]

Rees and some other cosmologists favour the multiverse as an explanation of the anthropic principle. Even the Christian apologist, Polkinghorne, believes that it is an 'even-handed choice between many universes and creation,' but as Polkinghorne emphasized 'both proposals are metaphysical in character'.[178] However Rees firmly believes that the multiverse is part of physics. Ellis takes objection to Rees' position and accuses him shifting his criterion of whether the multiverse forms part of physics from being *observable in principle* to being *real*, arguing that it has at least to be observable in principle to be considered part of physics.[179] Predictions about unobservable entities cannot be tested scientifically but, as Don Page a Canadian physicist and an evangelical Christian notes: [180]

> 'One cannot test scientifically a theory that makes predictions about what is unobservable, but one can test a theory that makes use of unobservable entities to explain and predict the observable ones.'

A multiverse must contain a universe that has the potential to be similar to ours or it is immediately falsified. The problem, as Kragh points out, is that multiverse theories are not specific and cannot predict whether the parameters necessary for our universe are observed.[173]

Many cosmologists believe that philosophers should not lecture them on how physics is done. But is that what most philosophers do? I think that such cosmologists are over sensitive because they have no real justification for their theories. Anything is permissible in the first stage of a hypothesis but an objective criterion is needed for it to become a theory. Popper is still the philosopher of choice for most physicists; the multiverse hypothesis is not directly falsifiable. That does not mean that the hypothesis is necessarily false but only that it does cannot be afforded the same status as the established theories of modern physics, such as the general theory of relativity, that are falsifiable. However falsification of a theory does not necessarily mean that it has to be abandoned, often it can be saved by changing the limits of its applicability. Newtonian mechanics has been falsified but it is still alive and well because within the solar system it works very well, except at very small distances, and there are only a very few examples where it does not predict the behaviour of matter to an accuracy similar to that which its effects can be observed.[181] Physicists and cosmologists want their model of an aspect of nature, which they describe by a theory, to be reality. However a model of reality is just that a model and it will only be accurate within certain limits. Some cosmologists are extreme in their criticism of hypotheses of the multiverse. Ellis and his co-authors have stated that:[174]

> 'The multiverse idea is not testable or provable in the usual scientific sense – existence of the hypothesized ensemble remains a matter of faith rather than proof.'

Even some of those who support the multiverse concept, like Weinberg, do so with some reluctance:[182]

> 'Theories based on anthropic calculation certainly represent a retreat from what we had hoped for: the calculation of all fundamental parameters from first principles.'

The anthropic principle is at present the only observable principle that can be verified but that in itself does not constitute a proof since it is only one result. The problem is that there is such a variety of multiverse hypotheses that it would be surprising if one of them did not satisfy the anthropic principle. If Popper's falsification criterion for physics is to be relaxed, is there some other criterion that is capable of discerning physics from metaphysics? The British philosopher, Colin Howson, with his background in probability studies, believes that Bayesian methodology is an alternative to the more rigid Popperism.[183] Bayesianism dates back to Thomas Bayes (ca. 1701–1761) who formulated a special case of what is now known as Bayes' theorem. Basically Bayes' theorem allows the inversion of the probability, $p(A \mid B)$, of some event A given that an event B is true, to obtain the probability $p(B \mid A)$:

$$p(B \mid A) = \frac{p(B)p(A \mid B)}{p(A)}, \qquad (6.7)$$

where $p(A)$ and $p(B)$ are the prior probabilities of A and B. However, as Kragh writes, it is doubtful that Bayesian arguments support the multiverse any more than Popperian ones do.[184] Scientists were little divided on the question of what constitutes science in the 1986 US Supreme Court case as to whether Creationism was science; Kragh quotes the definition provided by the anti-creationism coalition of scientists, which was supported by many Nobel Laureates including Weinberg, and American science organizations:[173]

> 'To be a legitimate scientific ''hypothesis'' an explanatory principle must be consistent with prior and present observations and must remain subject to continued testing against future observations. An explanatory principle that by its nature cannot be tested is outside the realm of science'

He wisely refrained from commenting on the relationship of this definition to the case of the multiverse.

6.5. Concluding Remarks

Although Stephen Hawking has avowed that philosophy is dead, philosophy is more important than ever. Modern physics asks philosophical questions. True it needs a philosopher with knowledge of modern physics to make any meaningful answers since modern physics is beyond everyday experience. The boundaries in knowledge are decreasing. Whether physicists like it or not, physics needs philosophy as much as it needs mathematics since its own philosophy is uncertain. Physics, classical or modern, philosophy, and mathematics are inseparable in the worldview.

Theists, at least of the Christian and Jewish religions, are mainly in retreat. Like classical physics, modern physics can be used by both atheist and theist in support of their position and it is more relevant to consider religious questions along side of the philosophical questions. Einstein's religious thoughts were inseparable from his philosophy and I think that is the more usual approach of most people.

Modern physics is a child of the twentieth century which has seen the rise in political systems that have shaped the world. The politics of the twentieth century have dominated people's lives and it is not surprising to see that domination spilling over to physics. Marxists in the twentieth century forced the philosophy of physics into the mould of dialectical materialism. However, it has been seen that the dialectic method has been used to argue both the orthodox Copenhagen version of quantum mechanics and Bohm's deterministic version.

Politics does not seem to have been affected by modern physics, though it has been obviously been affected by its technology. Politicians only accept science when it suits their purpose.

6.6. Notes

[1] Schrödinger (1958).
[2] Hawking and Mlodinow (2010).
[3] Stenholm (2011).
[4] See §2.6.
[5] Eddington (1928); Zimmer (1936).
[6] Filk and Müller (2010).
[7] Feynman (1965).
[8] Weiberg (1993).
[9] Stirrat and Cornwell (2013).
[10] This section relies heavily on the superb doctoral thesis of Imogen Clarke (2012).
[11] Mach (1900).
[12] The identification of Crookes and Lodge as the chemist and physicist in the quotation from Mach (1900) are due to Noakes (2008).
[13] Thomson still called these particles *corpuscles* until about 1912.
[14] Thomson (1909).
[15] Rowlands (2004).
[16] Russell (1961).

[17] Proposition XLXI, Problem IX, Book II: *The density and elastic force of a medium being given, to find the, velocity of the pulses,* Newton (1726). Newton's calculation of the speed of sound in a gaseous medium was too low by about 15% because his calculation tacitly assumed that sound propagation takes place isotropically whereas it takes place adiabatically without loss of heat. Laplace gave the correction for adiabatic propagation in 1802. This change would not have significantly affected Newton's attempt at calculating the density of aether if light propagated by same mechanism as sound, though the adiabatic correction depends slightly on the structure of the gas.
[18] Newton (1718).
[19] de Andrade *et al.* (2013).
[20] Lodge even arranged for a posthumous psychic experiment leaving a series of sealed envelopes to with a secret message which he would transmit after death. A panel of psychic experts failed to identify the messages in 1954 before they were opened. Lodge can be forgiven for a long time spent seeking proof of an existence after death for, like so many in the First World War, his son was killed and he sought the help of mediums to contact him. Naturally there are always the unscrupulous ready to take advantage of a person's grief. Rowlands (2004).
[21] Lodge (1933).
[22] Lodge (1913).
[23] Eddington (1928).
[24] Trimmer (1980).
[25] Heisenberg (1958).
[26] Galilei (1627).
[27] Dirac (1931).
[28] Dirac (1939).
[29] Wigner (1960).
[30] Dummett (1978).
[31] Tegmark (2014).
[32] Hilbert (1964).
[33] Sepkoski (2005).
[34] Barrow generously resigned his Lucasian Chair in 1669 in favour of Newton, who he saw as a rising star.
[35] Cohen (1980).
[36] Wittgenstein (1922).
[37] Wilson (2009). It is good that a number of others have seen the error in this example though many others have believed it illustrates that Newtonian mechanics can be indeterminate.
[38] Or at least they would have when I taught at university. Unfortunately there has been an unhealthy change in university education, with the growth in advanced computer programmes the emphasis on fundamentals has decreased considerably and many go straight to a computer and a commercial program rather than think about the problem first and obtain a first estimate of the solution. Computer programmes are only as good as their programmer many of whom are not completely familiar with the problem; also many computer programmes do not cope with multiple solutions or singularities and sometimes can give nonsense answers.
[39] Eddington (1933).
[40] Ruelle (1990).
[41] Broad *et al.* (1931).
[42] Jammer (1999).
[43] See §1.2.1.1.
[44] Stump and Kretzmann (1981).
[45] Published in the New York Times 28th January 1931.

[46] Quoted in Jammer (1999).
[47] Einstein (1930a). The quote comes from Schopenhauer's prize winning essay of 1839: *On Freedom of the Will.*
[48] Einstein (1930b).
[49] Einstein (1948).
[50] Einstein (1927).
[51] Einstein *et al.* (1935).
[52] Bohr (1935).
[53] Petersen (1963).
[54] Neumann (1932).
[55] Bohm (1952).
[56] Letter to Born on the 12th May 1952. Born and Einstein 1971.
[57] Myrvold (2003).
[58] The two famous papers of Bell: *On the problem of hidden variables in quantum mechanics* (1966) and *On the Einstein-Podolsky-Rosen paradox* (1964) are both to be found in Bell (1987).
[59] Published in 1966.
[60] Letter to Born on the 3rd March 1947. Born and Einstein (1971).
[61] Published in 1964.
[62] Aspect (1999).
[63] Yin *et al.* (2013).
[64] Feynman (2011).
[65] First published in English as *The Logic of Scientific Discovery*. (Popper 1959a).
[66] Popper (1982a).
[67] Popper (1982b).
[68] Popper (1959b).
[69] Jammer (1991).
[70] Lenin was the pseudonym that Vladimir Ilyich Ulyanov adopted in the early twentieth century for his political publications.
[71] Fostner (2008).
[72] Lenin (1938).
[73] Marek and Rissen (1977).
[74] Bohm had his US passport confiscated when he went to Brazil.
[75] Bohm (1951).
[76] Interview of Bohm by Wilkins on 6th June 1986, Niels Bohr Library & Archives, American Institute of Physics, College Park, MD, USA.
[77] Bohm (1957).
[78] Bohr (1949).
[79] Heisenberg (1927). Translation taken from Wick (1996).
[80] Jacobsen (2007).
[81] Rosenfeld (1953).
[82] Engels' *Anti-Dühring*, first published in 1878. First English translation Engels (1907).
[83] Freire and Lehner (2010).
[84] *in nuce*: in a condensed form OED.
[85] Olival Pessoa, Olival Freire, and Alexis De Greiff have written an excellent account of the Tausk affair. Pessoa *et al.* (2008). This paragraph gives a very brief account based on their paper.
[86] Hiley (2013).
[87] Bohm and Hiley (1993).
[88] Hiley (1997).
[89] Bohm (1980).

[90] Hanegraaff (1996).
[91] Bohr (1958).
[92] Smith (2013).
[93] Zukov (1979), Capra (1975).
[94] Scerri (1989).
[95] This paragraph is mainly based on Kragh (2007a).
[96] Dingle (1937).
[97] Bondi (1952).
[98] Whitrow and Bondi (1953).
[99] O'Neil (2002).
[100] Milton (2005).
[101] Eddington (1928).
[102] Kragh and Lambert (2007).
[103] Hubble (1926).
[104] A parsec (ps) is an astronomical unit of length and is the distance from the solar system that a star exhibits a parallax of 2 arc seconds when viewed from Earth at two points across its orbit around the Sun. A parsec is about 3.26 light-years.
[105] Lemaître (1927).
[106] Hubble estimated the constant of proportionality to be 500 km s^{-1}/Mps. Hubble (1929). The distances of the galaxies used by both Lemaître and Hubble in their calculations were grossly underestimated so that the present day value of Hubble's constant is approximately ten times smaller at 67 km s^{-1}/Mps (Ade *et al.*, 2014).
[107] Eddingtion (1931).
[108] Lemaître (1931).
[109] The term *primeval atom* was not used by Lemaître in his letter to *Nature* and is a translation of the French term *l'atome primitif* used by him.
[110] Kragh (2008).
[111] Godart and Heller (1985).
[112] Sorondo (2003).
[113] Hawking and Penrose (1970).
[114] Dickie *et al.* (1965).
[115] Ade *et al.* (2014).
[116] Guth (1997).
[117] Guth (2007).
[118] Gasperini and Veneziano (2003).
[119] Gasperini (2008).
[120] Guth (1997). Weinberg included this book in his list of the thirteen best popular science books of all time.
[121] Guth (1981).
[122] Linde (1982).
[123] Guth (1997, 2007).
[124] Kragh (2012).
[125] baryon comes from the Greek word for heavy.
[126] Bergh (1999).
[127] Bertotti *et al.* (2003).
[128] Freire and Wex (2010). In 1964 Irwin Shapiro showed that electromagnetic waves passing near a massive body are subjected to a relativistic time delay that can be calculated from the general theory of relativity.

[129] Dark matter cannot be generally distributed in the solar system or the orbits of the planets would be affected. However, since Newton and everyone else calculates the mass of the Sun and the planets from the universal law of gravitation (very slightly modified in modern times by the general theory of relativity), there is no philosophical prohibition on dark matter making up a very small portion of their mass but nothing like most of their mass which is the hypothesis for intergalactic dark matter.
[130] The existence, but not position, of Neptune was first predicted by Finnish born Anders Johan Lexell in 1781. Again in 1821 the French astronomer, Alexis Bouvard, noted the irregularities in the orbit of Uranus and the British mathematician and astronomer, John Couch Adams, made similar calculations for an unknown planet in about 1845. However, it was the calculations of the French mathematician, Urbain Jean Joseph Le Verrier published in 1846 that were used by Johann Gottfried Galle at the Berlin Observatory to discover Neptune. Afterwards it was realized by James Challis, the Director of the Cambridge Observatory that he had observed Neptune a month earlier but because of a lack of an accurate star chart, he did not recognize it as a planet at the time. There is still some confusion over the details of Neptune's discovery.
[131] Trimble (1987).
[132] Vanderburgh (2003).
[133] Newton (1726).
[134] Kroupa (2012).
[135] Popper (1957).
[136] Nadathur and Sarkar (2011).
[137] Singal (2014).
[138] Haas (1936).
[139] Translated quotation given by Overduin and Fahr (2003).
[140] Tryon (1973).
[141] Banerjee and Sen (1997).
[142] Because of their wave-like nature, quantum particles can penetrate barriers even if their energy is smaller than the barrier.
[143] Vilenkin (1982), Atkatz and Pagels (1982), Hartle and Hawkings (1983).
[144] Vilenkin (1984).
[145] Harlte and Hawking (1983).
[146] Gott and Li (1998).
[147] Allen (1997).
[148] See §4.2.
[149] I only consider arguments that directly make use of cosmology or, like the PSR, are used in the interpretation of the implications of cosmology.
[150] Craig and Smith (1995).
[151] Hawking (1966). See also Penrose (1965) and Hawking (1965).
[152] Hawking (1988).
[153] Weinberg and Polkinghorne (2001).
[154] Brandon Carter coined the term *anthropic principle* in 1974.
[155] Since the terms weak and strong anthropic principles have no recognized definition, Smolin (2004) prefers to call this statement an anthropic observation.
[156] Stoeger (2010).
[157] Roush (2003).
[158] Barrow and Tipler (1988).
[159] Gardner (1986).
[160] Stenger (2011).
[161] Rees (1999).

[162] Rees (2003).
[163] Tegmark *et al.* (2006).
[164] Ehrenfest (1918).
[165] Callender (2005).
[166] Tangherlini (1963).
[167] Penney (1965) and Einstein (1922a).
[168] Kant (1768).
[169] Cleve (1987).
[170] Adams (2008).
[171] Straus (1956).
[172] Rugh and Zinkernagel (2001).
[173] Kragh (2009).
[174] Ellis *et al.* (2004).
[175] Tegmark (2003).
[176] Carr (2008).
[177] Everettt did not go on to pursue a career in physics but worked on weapons systems and military modelling for the rest of his life.
[178] Polkinghorne (2001).
[179] Stoeger *et al.* (2006).
[180] Page (2008).
[181] The anomalous precession of the perihelion of Mercury due to relativity is only 43 arc seconds per century out of a total of 574 arc seconds per century. There is a precession of the perihelion of all the planets that is due to relativity but it is greatest for Mercury because it is closest to the Sun and therefore experiences the greatest warping of spacetime.
[182] Weinberg (2008).
[183] Howson and Urbach (2006).
[184] Kragh (2009). See also Palonen (2008).

Chapter 7

Popularization of Modern and Postmodern Physics

Newton and the era of classical physical was the culmination of a science revolution that had started with Copernicus. People were breaking away from the authority of the church. There was a thirst for knowledge of the new natural philosophy of Newton among the educated at the end of the seventeenth century, which led to its popularization. The conditions at the beginning of modern physics were very different. Literacy was becoming universal. In Britain more than 85% of the population were literate by 1900 and the 1902 Education Act created Local Education Authorities responsible for providing secondary as well as primary education for all. However that does not mean that there was such a general thirst for information about modern physics like there was for Newtonian natural philosophy at the end of the seventeenth century though there were some exceptions. At the beginning of the era of modern physics H.G. Wells wrote an article *Popularizing Science* for *Nature*, in which he among other things wrote that:[1]

> '[I]nstead of the current popular and yet philosophical books increasing, its volume appears if anything to dwindle, and many works ostensively addressed to the public by distinguished investigators, succeed in no notable degree, or fail to meet with appreciation altogether.'

Wells insisted that popular science should not be entertainment but written for an intelligent reader especially since 'the endowment of research is rapidly passing out of the hands of private or quasi-private organizations into those of the State'.

The written word, or lectures that have a published version, are the best way of popularizing physics. As William Henry Bragg wrote in the preface of his published version of his 1923 Christmas Lectures at The Royal Institution about the additions he had made to the lectures:[2]

> 'they were not very suitable for consideration at lectures, but are perhaps permissible in the book because the reader can omit them if he desires or read them more than once.'

It is the ability to go over the material again that makes the written word the best means of conveying science whether popular or otherwise. The BBC has broadcast programmes on science since 1924 only two years after it was formed, and soon understood the need to follow up a broadcast talk on radio, whether on science or any other serious subject, with a published version. From the earliest days at the BBC the magazine *The Listener* was 'a medium for the intelligent reception of broadcast programmes by way of amplification and explanation of those features which cannot ... be dealt with in the editorial columns of the Radio Times.'[3] It is regrettable that *The Listener*, which started in 1929, ceased to be published in 1991, but *The New Zealand Listener*[4] is still alive and well. Television provides entertainment and it is not a good medium for the popularization of physics precisely because sections of a programme cannot be repeated at one's own pace. Good popularization of physics should provide a philosophical understanding and not be simply an entertainment wedged between sitcom programmes. The high production costs of television programmes, Carl Sagan's (1934–1996) thirteen part *Cosmos* series for PBS in 1980 is said to have cost over $6 million with a further $2 million for its promotion, ensures that they must attract the maximum audience by making them mainly entertainment. Here I concentrate on popularizations that, though not relying on prior knowledge and presented in an engaging manner, do not dodge the important issues even if only a comparatively few are reached. The 'gee-whiz tour of the Strange But True' as Polkinghorne has called the entertainment only popular books on modern physics have been avoided.[5] Physics should be entertaining but it is not entertainment, hence popularization of physics on television is not discussed in particular. Such decisions on what constitutes popularization that is aimed at those who want to understand physics was not needed in the chapter on the popularization of Newtonian physics where even books for children were not seen primarily as entertainment. A sensible short discussion of the problems in the popularization of modern physics is the introduction to a paper by Helge Kragh and Kristian Nielsen.[6]

Apart from books and magazines, which are the major methods of popularizing physics, there are, as there were before the twentieth century, organizations whose purpose is to spread the knowledge of science. Most of those which existed in the nineteenth century have continued to function since such as The Royal Institution and the British Association for the Advancement of Science.[7] The Mechanics' Institutes, which had been so vigorous, lost their way by the end of the nineteenth century. The new education movement of the twentieth century was the Workers' Educational Association (WEA) which grew from, not the mechanics' institutes, but the University Extension Movement.[8]

The WEA was formed in 1903 by Albert Mansbridge (1876–1952) who was an office worker who had educated himself by attending university extension courses at Kings College London.[9] At first the association was only for men and had the unwieldy title of The Association to Promote the Higher Education of Working Men, two years later the association was extended to women and its title changed to the WEA. By 1914 the WEA had 179 branches in Britain. Mansbridge also toured Australia in 1913 and helped to establish the WEA in all six states, though not all survived. However courses in the sciences in general and physics in particular were not that well attended. In 1921 the physicist Norman Campbell wrote the small book *What is Science?* to encourage the study of science in the classes of the WEA and he wrote in the preface that:[10]

> 'there is a danger that the W.E.A. which stands for new ideas, will become the last stronghold of the reactionary doctrine that science and culture are antagonistic.'

Unfortunately Campbell was not successful in his attempt to stem the flow from science; as with the earlier mechanics' institutes the courses in science declined over the years. In 2015 the WEA offered 497 courses within 25 miles of London but there were only six science courses, two of which were in the life sciences leaving only four courses (less than 1% of the total) that would include topics in physics and cosmology discussed in this book but with plenty of other courses on such enlightening topics as Belly Dancing for Beginners or Paper Mache. Physics and science in general is failed by the WEA.

Another source of popularization of physics is the science museum.[11] However scientific and technological objects are easier to display than developments in theory and understanding. Thus when the Mercury space capsule *Friendship 7*, in which John Glenn orbited the Earth in February 1962, went on display in London's Science Museum for four days in May 1962, 25,000 people visited the exhibition and the queue to get into it stretched a quarter of a kilometre.[12] Museums are more adapted to applied than pure physics. London's Science Museum, which had its roots in the in the Great Exhibition of 1851, was originally housed in the South Kensington Museum, which opened in 1857 on part of the site of the present Victoria and Albert Museum. The Science Museum was initially modelled on the Musée Des Arts et Métiers in Paris and focused on contemporary science. It was not until 1914 that building commenced for the Science Museum in its present location but building was interrupted by the First World War. A geologist Henry Lyons (1864–1944) was appointed the first Director long before the first wing of the Science Museum in Exhibition Rd., South Kensington was officially opened in 1928.[13] Lyons was much influenced by the pure science exhibition of the Royal Society at the British Empire

Exhibition at Wembley in 1924, which showed many of the developments in modern physics, and he wanted to have a similar gallery in the Science Museum which he tentatively named *Properties of Matter and Physical Phenomena.*[14] One of the few exhibits of pure science in the Science Museum were photographs taken during the 1919 expedition to verify Einstein's general theory of relativity by observing stars close to the Sun during an eclipse.[15] However the Science Museum came under pressure to be aligned with industry, and in the later 1920s more emphasis was placed on industrially orientated applied physics than pure science. Lyons retired from the Science Museum in 1933 and was replaced by Ernest Mackintosh (1880–1957) a retired colonel in the British Army. Mackintosh embraced industrial exhibitions. However one notable exhibition *Atomic Tracks* in 1937 was of pure science and displayed the cloud chamber of Charles Wilson (1869–1959) with which the tracks of X-rays and alpha and beta particles were first visualized in 1911 and some eighty cloud track photographs. The first 1911 published photograph of the cloud formed by alpha particles is shown in Fig. 7.1. Thus the Science Museum before the World War II did not contribute that much to the popularization of modern physics. After the war one of the first exhibitions in the newly reopened Science Museum was *Atomic Energy and Uranium* in 1946. There was a historical introduction to the exhibit which contained things such as atom models of Bohr and some of the Wilson photographs from the 1937 exhibition but the main point of the exhibition was to present practical details of the use of uranium[12] From the 1960s the historical approach became more in favour as historians of science began to be appointed as curators.[16]

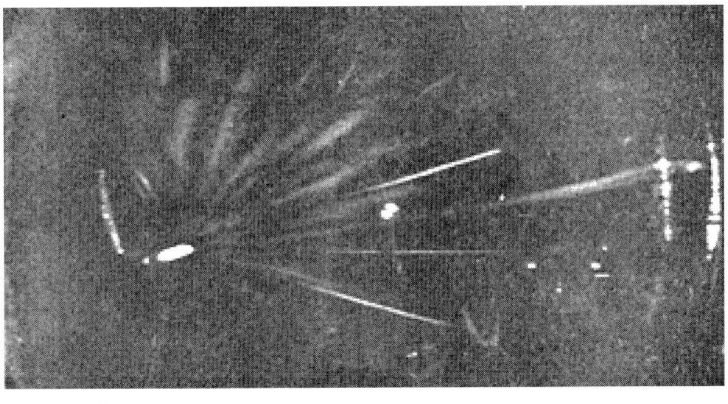

Fig. 7.1. Cloud formed on ions due to an α-particles emitted from a small piece of radium on the left (Wilson, 1911).

There is a general problem in popularizing physics after the nineteenth century in that modern physics does not supplant classical physics which still explains nearly all of what we observe. However classical physics is now not sexy despite it being what one must understand before going on to modern physics. Even in A-level physics in British schools most of what is taught is classical physics with a nod at a few topics in quantum mechanics and modern cosmology. For those who miss an exposure to physics at school there have been few interesting popular books on classical physics written in the twentieth or twentieth first century from which adults can start exploring physics, though there have been a few notable exceptions. In the 1930s the physicist Edward Andrade and the biologist Julian Huxley (1887–1975) wrote *Introduction to Science* which was published in four parts for children. It was found 'that a number of persons no longer young were of interest' and a new edition of the first three parts with the title *Simple Science* contained seven chapters, six of which were on classical physics, was published in 1934.[17] Another very successful book on science containing classical physics, *Science for the Citizen*, was written by the socialist biologist Lancelot Hogben (1895–1975) as a self-educator and published in 1938; the book was in five parts, the first three of which were on classical physics.[18] After the Second World War there were some books which started with Newtonian physics or earlier and then continued on to quantum mechanics and sometimes relativity and cosmology. Some of these books worked but many failed because of the sheer difficulty in covering so much history in a single volume. The German physicist, Max von Laue (1879–1960), succeeded for a special reason with his *History of Physics* written during the dark days of the Second World War. Von Laue was an opponent of the Nazis and his short *History of Physics* reads as a celebration of the enduring strength of science. As he wrote in the foreword for the English translation published in 1950:[19]

> '[T]he first draft of this book was written in 1943 and is accordingly a war product... The thought of the culture that is the common property of all nations and which was being so despicably mistreated was ever present throughout the period of composition.'

Physics and his Christian faith kept von Laue sane during the war. Another German born physicist, Rudolf Peierls (1907–1995), an assimilated Jew, was in Cambridge on a Rockefeller fellowship in 1933 when the Nazis came to power and could not return to Germany and he became a British national in 1940. Peierls reason for writing *The Laws of Nature*[20] published in 1956 was that after the war during frequent lectures on the atomic bomb he was often asked fundamental questions about physics which he could not answer satisfactorily in the time allotted.[21] The most famous of such books, by the extrovert Feynman, was not written as a popular book but as lecture notes for freshmen and sophomore classes at the California Institute of Technology, Pasadena taking a minor in physics during 1961–1963. The result was *The Feynman Lectures on*

Physics which became perhaps the most popular physics book ever written with more than 1.5 million copies sold in English and it has been translated into many languages.[22] A retrospective review in *Nature*, for the fiftieth anniversary of its publication in 1964, stated:[23]

> 'Feynman's physics is about simplicity, beauty, unity and analogy, presented with enthusiasm and insight that bursts from the page.'

In his Epilogue to the three volumes of *Feynman Lectures*[24] Feynman wrote:

> '[T]he main purpose of my teaching has not been to prepare you for some examination ....I wanted most to give you some appreciation of the wonderful world and the physicist's way of looking at it, which, I believe, is a major part of the true culture of modern times.'

Feynman also gave the Messenger Lectures for 1964 at Cornell University on *The Character of Physical Law*. These were recorded by BBC Television and subsequently broadcast; a transcription of the lectures was published in 1965.[25] This series of lectures gave Feynman's view of what lies at the heart of physical laws. He showed that 'one of the amazing characteristics of nature is the variety of interpretational schemes which is possible,' though he did 'not understand the reason why it is that the correct laws of physics seem to be expressible in such a tremendous variety of ways.'

It is impossible to mention all popular books on modern physics in this chapter, those presented are more-or-less a personal choice. Some of the books mentioned in Chapters 1 and 6 also fall under the general heading of popular modern physics these will not necessarily be mentioned again in this chapter. One such book must be singled out for particular mention Eddington's *Nature of the Physical World* based on his 1927 Gifford Lectures, in which he discussed 'some of the results of modern study of the physical world which give most food for philosophic thought.'[26] This book was extremely successful in both Britain and America and showed that it is possible to write a successful popular book that contains 'arguments of considerable difficulty.'

This chapter has been structured on the important topics in modern physics, but, since many popular accounts include more than one topic, it has not been possible to be completely consistent.

7.1. Invisible Rays

The first evidence that physics was changing at the end of the nineteenth century came with the discovery of the invisible rays, though this discovery was foreshadowed by Faraday whose experiments in the 1830s, where he passed a current through a partially evacuated glass tube that created a glow around the cathode and an arc discharge nearer the anode, announced the discovery of cathode rays in 1879. However, it was finally J.J. Thomson in 1897 who discovered that cathode rays are a stream of electrons, weighing about one two thousandth part of a hydrogen atom, which he first called corpuscles. X-rays discovered in 1895, were at first called *Röntgen Rays* after their discoverer Wilhelm Röntgen (1845–1923),[27] though he named them X-rays. In 1896 Henri Becquerel (1852–1908) was experimenting with potassium uranyl sulphate, which is phosphorescent and which he thought might emit X-rays. Wrapping a photographic plate in black paper he placed some crystals of potassium uranyl sulphate on the wrapped plate and exposed them to sunlight. On developing the plate, the silhouette of the crystals was visible. Becquerel tried to repeat the experiment later but the Sun did not come out and he placed the covered plate in a drawer. A couple of days later he developed the plate and to his surprise found an image of the crystals. After further experiments with non-phosphorescent uranium salts which also produced an image of the salt crystals, Becquerel reasoned that the penetrating rays must have come from the uranium salt itself. Becquerel received a joint Nobel Prize for Physics alongside Marie Skłodowska-Curie (1867–1934) and Pierre Curie (1859–1906) in 1903 for his work on radioactivity. Rutherford identified two radioactive rays, alpha and beta, and a third kind, gamma rays, were discovered by Paul Vilard (1860–1934). We now know that because of the particle-wave duality[28] that these rays are also particles. The alpha particle is particularly massive; it consists of two neutrons and two protons, and is identical to the nucleus of a helium atom.

7.1.1. *X-rays*

X-rays became a sensation with the public and almost immediately became a craze on both sides of the Atlantic. In 1896 the year after Röntgen had named them, the magazine *Photography* published the slightly naughty ditty X-actly So!

> The Roentgen Rays, the Roentgen Rays,
> What is this craze?
> The town's ablaze
> With the new phase
> Of X-rays' ways.

I'm full of daze.
Shock and amaze
For now-a-days
I hear they'll gaze
Thro' cloak and gown – and even stays,
These naughty, naughty Roentgen Rays.

X-rays were very quickly used in medicine and John Macintyre (1857–1928) set up the world's first radiology unit at Glasgow Royal Infirmary in 1896. Popular books were also quick off the mark with chapters on X-rays. Peter Atkinson, an electrical supplier, who also wrote and lectured on electrical topics, included a final chapter in the second edition of his book *Electricity for Everybody* writing in his introduction to the second edition published in 1897: [29]

> 'The development, within the last year of the wonderful X-rays, which were previously almost unknown, has made the addition of a chapter to the present edition necessary.'

Children were not left out in this rush to understand the latest developments in physics. The sixth and last part of the Christmas Lectures of 1896 for children at the Royal Institution, *Visible and Invisible Light*, given by Silvanus Thompson (1851–1916) in January 1897 contained a large section on X-rays.[30]

Most popular interest in X-rays was in their medical applications which fall outside of the scope of this book; in physics they were an important tool in understanding the structure of matter. The most important physicists who developed the use on X-rays for this purpose was the father and son team: William Henry Bragg and William Lawrence Bragg who in 1915 were jointly awarded the Nobel Prize in Physics 'for their services in the analysis of crystal structure by means of X-rays.' They are the only father and son to be awarded a Nobel Prize and Lawrence Bragg is the youngest recipient at 25 years of the prize for physics. Since crystals occur in nature their regular form has been discussed from the earliest times. In the seventeenth century ideas that the shape of crystals was due to the regular arrangement of globular corpuscles were current and expressed by Boyle and Hooke. In his *Micrographia* published in 1665 Hooke showed crystals of alum, see Fig.7.2, writing:[31]

> '[*Alum*] in this texture, composes a regular *Tetrahedron*, which is a very usual Figure of the *Crystals* of *Alum*, And (to hasten) there is no one Figure into which Alum is observ'd to be crystallized, but may by this texture of *Globules* be imitated, and by no other.'

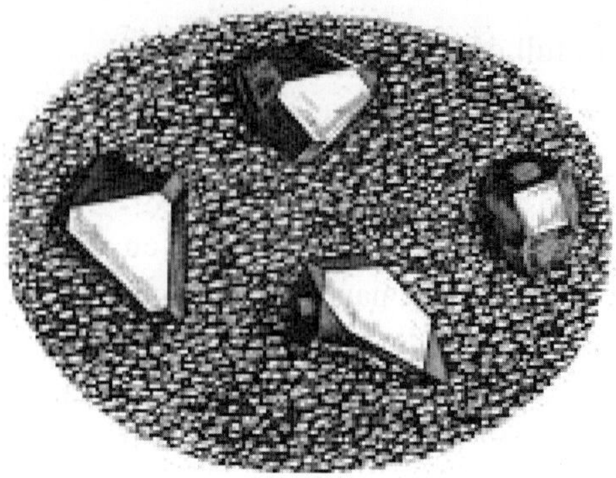

Fig. 7.2. Alum crystal, Plate 7, *Micrographia* (Hooke, 1665).

The geometry of crystals was studied in the nineteenth century by Auguste Bravais (1811–1863) who determined that 14 unique regular lattices, including the tetragonal, were possible in crystalline systems. Finally Evgraf Fedorov (1853–1919) in Russia and Arthur Schoenflies (1853–1928) in Germany independently hypothesized that the underlying form of crystals was due to the arrangement of atoms or molecules. X-rays enabled the arrangement of atoms to be deduced. Diffraction gratings, consisting of closely spaced lines, diffract light according to its wave length and enable its spectral lines to be examined. The grooves on a CD are spaced at 1.6 μm and are in the range used for normal diffraction gratings, which is the reason that a spectrum of light is reflected off them. It is the shortness of the wavelength of X-rays that gives them their penetrating power and the wave length of hard X-rays is less that 0.1 nm whereas the wave length of visible light is greater than 390 nm. Regular rows of atoms in crystals are close enough to act as diffraction gratings for X-rays and von Laue was awarded the Nobel Prize for Physics in 1914 for discovering this effect. The interference patterns observed in zinc blende (zinc sulphide) are shown in Fig. 7.3. Laue assumed that the crystal structure of zinc blende was simple cubic and that only a few definite wavelengths were present in the X-rays to deduce his interference patterns. Lawrence Bragg graduated from Cambridge University in 1911 and in the next year he gave a paper before the Cambridge Philosophical Society where he gave the correct law for X-ray interference using Laue's experiments on zinc blende.[32] Bragg made two assumptions:

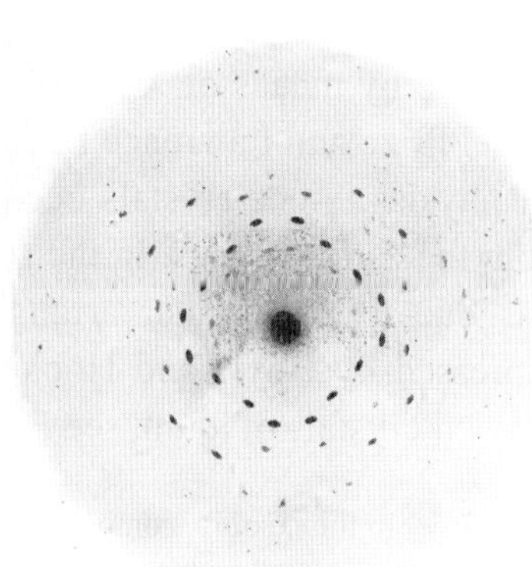

Fig. 7.3. Interference pattern for zinc blende crystal obtained by von Laue (Braggs, 1913).

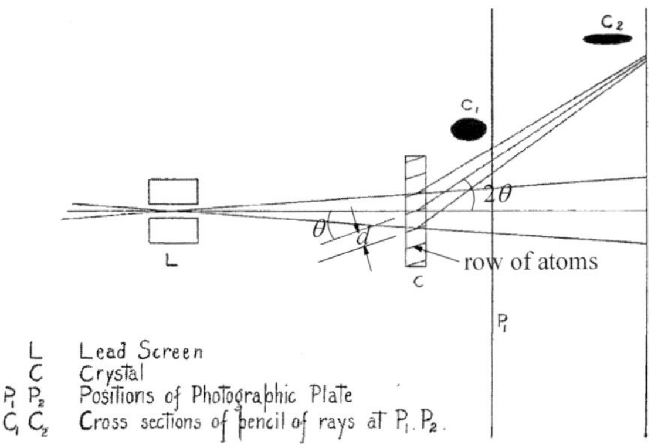

Fig. 7.4. Bragg's original diagram of x-ray diffraction with minor additions (Bragg, 1914).

(1) The X-rays had a continuous spectrum of wavelengths.
(2) The crystal structure of zinc blende was face centred cubic.

Lawrence Bragg's original diagram of the formation of the interference pattern is given in Fig. 7.4. Bragg's law derived by Lawrence in 1912 is expressed by

$$n\lambda = 2d\sin\theta, \tag{7.1}$$

where λ is the wavelength of the X-rays, n is an integer, d is the spacing between rows of atoms, and θ is the angle between the incident X-ray and the plane of the atoms. The diffracted X-rays undergo constructive interference when Bragg's law is satisfied. Since the incident X-rays form a fine pencil of rays, those that undergo constructive interference appear as black dots on an exposed plate whose position is given by Eq. (7.1) and if the wave length of the X-ray is known the crystal structure of the crystal can be obtained. It was an exciting discovery for the twenty two year old undergraduate but not one that lent itself to popularization. The father and son wrote a professional book, *X-rays and Crystal Structure*,[33] in 1915 but it was left at that time to others to inform the public. Frederick Soddy (1877–1956), who had been a collaborator of Rutherford, wrote an extended review of the Braggs' book for *Science Progress* with the subtitle *How an Important Question of Physical Science has been Solved*, which was reprinted in the *Scientific American Supplement* in 1916.[34] Soddy lauded the work writing:

> '[This work has] opened a new era of activity, not for one but for many sciences, and the first chapter of this new era in the science of crystallography.'

Although neither of the Braggs immediately delivered popular versions of their work, both were very much involved in publicizing physics and both spent many years at the Royal Institution where both were Directors at different times. One of the first things that W.H. Bragg did at the Royal Institution as its new Director was to deliver the 1923–1924 Christmas Lectures for Children taking for his title: *Concerning the Nature of Things*, which included showing how X-rays were used to understand crystal structure and helped in understanding atomic structure.[2] In 1924 John Reith (1889–1971), general manager of the newly formed BBC,[35] wanted to demonstrate how broadcasting could be used to promote education among the masses, especially in science. Bragg's Christmas Lectures were so well received that Reith considered him the ideal speaker for such a task and asked him to repeat them for the BBC.[36] The success of these radio broadcasts convinced Bragg that radio was a powerful medium for dissemination of scientific knowledge and he offered a new series on sound, which was accepted by Reith.

7.1.2. *Alpha, beta, and gamma rays*

The discovery of radioactive rays followed closely on the discovery of X-rays. In 1903 Pierre Curie lectured at the Royal Institution on radium and the radioactive rays it produced. On the other side of the Atlantic Rutherford, fresh from Cambridge, was appointed the Macdonald Research Professor of Physics at McGill University, Montreal in 1898 and continued his work on radioactivity there. Soddy also joined McGill University in 1900 as a demonstrator in the Chemistry Department and was soon collaborating with Rutherford. During his McGill years (1898–1907) Rutherford made major contributions to the understanding of radioactivity in which Soddy contributed in the early years. As well as writing professional papers, Rutherford sought to disseminate his research more widely by writing articles for *Harper's Monthly Magazine* and the *Scientific American.* In 1904 he wrote articles on radioactive radiation for both magazines in which he explained what was known about the decay of uranium, thorium, and radium.[37] Although alpha particles had only slight penetrating power they were the most important because of their large mass and energy. He sketched the decay of these three radioactive elements as it was known to him. For example, at the time Rutherford thought that thorium decayed to its final stable product in the four stages shown in Table 7.1. Thorium has an extremely long half life so that Rutherford was only seeing minute decay products and could not identify them precisely. The actual decay process is more complicated than Rutherford thought. He also gave estimates of the lives of the three radioactive elements for the *Scientific American* he expressed them as half-lives but for the general magazine *Harper's* Rutherford did not, presumably believing that those reading the *Scientific American* had more knowledge of science. In the same year an American electrical engineer William Hammer (1858–1934) capitalized on the radium craze and gave popular lectures, which he published as *Radium and Other Radio-Active Substances*.[38] After writing an article on the radioactivity of uranium for the *Scientific American* in 1903,[39] Soddy returned to England in 1904 where he accepted a position as lecturer at the University of Glasgow. In that year Soddy wrote an introductory text to radioactivity published for The Electrician Printing and Publishing Company for whom he had written a number of popular articles in their magazine *The Electrician* during the previous two years.[40]

Table 7.1. Rutherford's scheme for the decay of thorium (Rutherford, 1940a).

Rutherford's name	Modern name	Nuclide number
Thorium	Thorium	232
Thorium X	Radium	224
Thorium emanate	Radon	220
Matter causing excited activity	Polonium	216
Final product	Isotope of Lead	208

In 1905 Rutherford wrote another popular article in *Harper's* magazine which pointed out that radioactivity would cause heating of the Earth[41] so that William Thomson's estimates of the age of the Earth from the rate of cooling[42] would be too small. With no quantitative evidence Rutherford could not give an estimate of the age of the Earth in 1905 and it was left to Arthur Holmes (1890–1965) to estimate the age of the Earth in 1913 from radioactive decay as 1.6 billion years compared with Thomson's estimate of 20 to 40 million years.[43] The accepted age of the Earth today is 4.5 billion years. Even after Rutherford returned to England in 1907 and the Langworthy Chair of Physics at the University of Manchester, popular interest in his work on radioactivity continued in America. In 1908 his two papers in the Proceedings of the Royal Society with Hans Gieger (1882–1945) confirmed that an alpha particle was the nucleus of a helium atom.[44] An important topic of the first paper was also a method of counting alpha particles which later became the *Gieger Counter.* The *New York Times* and the *Chicago Tribune* both reported on these two papers in November 1908 but not *The Times* of London. Although Rutherford never claimed more than he could prove explicitly, this paper was an important step in understanding the atom and he was eager to disseminate this exciting new knowledge and in 1909 he wrote an article *Visualizing the Atom* for the *Scientific American*.[45]

Soddy was as interested in popularizing radioactivity as Rutherford. In 1909 Soddy published the *Interpretation of Radium*, based on six free popular lectures he had given at the University of Glasgow.[46] This book was extremely successful on both sides of the Atlantic and on the Continent and it went through many editions. Even in the first edition Soddy could see the potential development of radioactivity for energy:

> '[From] one grain of radium no less than 125,000,000 calories would be evolved.... By burning of a grain of coal … only about 500 calories are obtained…
>
> By these achievements of experimental science Man's inheritance had been increased, his aspirations had been ennobled to an extent beyond our present power to foretell. The real wealth of the world was its energy, and by these discoveries it, for the first time, transpired that the hard struggle for existence on the bare leavings of natural energy in which the race evolved was no longer the only possible or enduring lot of Man.'

The eleventh chapter of Soddy's the *Interpretation of Radium* was acknowledged by H.G. Wells as the inspiration for his novel: *The World Set Free* published in 1914 where he described 'atomic energy dominating every other source of power' and atomic bombs, which he conceived as devices 'that would continue to explode indefinitely.'[47]

It was not long after his 1908 papers that Rutherford obtained the evidence for the structure of the atom for which he was looking. Experiments at McGill University had shown that that alpha and beta particles were scattered slightly when they impinged upon thin metal foils. In 1909 Rutherford got his undergraduate student Ernest Marsden (1889–1970) to look for large angle scattering for his final year thesis. Surprisingly for all concerned, Marsden found there was occasional scattering at angles greater than 90^0. This scattering gave the proof that there was a tiny nucleus to an atom which when hit reflected the particle. In 1911 Rutherford gave the evidence in support of his theory that an atom consisted of a central charge concentrated at a point, and that the large deflections of alpha and beta particles were mainly due to their passage through a field of electrons surrounding a strong central charge.[48] J.J. Thomson's plum pudding model of an atom could not be true.[49] Bohr took Rutherford's atom and developed a quantum theory.

7.2. Quantum Mechanics

There were two phases in the popularization of quantum mechanics: from 1913 and Bohr's semi-classical atom to the new quantum mechanics engendered in

1925 by the three papers of Heisenberg, Born and Jordan[50] and after those papers which showed the indeterminacy of quantum mechanics.

7.2.1. *The structure of Bohr's 1913 atom*[51]

Bohr visited England in 1911–1912, after seeing Rutherford in Manchester, he conceived the idea of combining Rutherford's new hypothesis of the atom with Planck's quantum of energy. Back in Copenhagen Bohr wrote his famous three papers in 1913 that described the first Bohr atom.[52] By about 1915 Bohr's theory was largely accepted by most physicists though there were still some who resisted and proposed their own theories.[53] In Bohr's atom of 1913 there was a tiny positively charged nucleus surrounded by negative electrons in stationary orbits. However in classical electrodynamics a revolving electron would emit energy and the electron would collapse into the nucleus. Bohr was forced to postulate that an electron did not obey the classical laws of physics but he used classical laws to calculate the energy levels of the stationary orbits.

Bohr was not a popularizer and even fellow physicists found that his writing was at times difficult to understand. One of the first general accounts of the modern conception of the atom was given to a meeting of the Royal Society of Canada held at Montreal in May 1914 by Arthur Eve (1862–1948), an English physicist who was working at McGill University and for a time had been a colleague of Rutherford there.[54] Eve echoed Jeans' remarks at the 1913 British Association Meeting about Bohr:[55]

> 'His work is remarkable as leading to excellent numerical verification [of the spectral lines of hydrogen.]'

Rutherford also briefly discussed Bohr's contribution to the structure of the atom in a lecture at the National Academy of the Sciences delivered in Washington in 1914. This lecture was published in the *Popular Sciences Monthly* in 1915.[56] Rutherford mentioned the difficulty in Newtonian physics that a simple model of a hydrogen atom with an electron orbiting a nucleus faced writing that:

> 'Bohr ... concludes that the complexity of the spectrum [of hydrogen] is not due to the complexity of the atomic structure but to the variety of modes in which an electron can emit radiation... [T]here can be no doubt of the great interest and importance of this first attempt to deduce the structure of the simple atoms and to explain the origin of their spectra.'

Two other popular articles on Bohr's atom appeared in 1914: *The Content and Structure of the Atom*,[57] by the American physicist George Walter Stewart

(1876–1956) and the *Structure of the Atom*,[58] by Norman Campbell. The closing remarks in Campbell's article are worth quoting:

> 'There are only two alternatives open to the modern theoretical physicist: he may either suppose that the principles of the older mechanics are true, and that all the brilliant results which have followed from the application of the conceptions of Planck and Einstein to the most diverse phenomena are illusory and devoid of evidential value; or he may suppose that they are not true. Bohr's theory offers him the choice in its most striking form.'

Bohr's new atomic theory would probably have received more attention after 1913 had it not been for World War I. Denmark was neutral but in a particularly difficult position since it has a common frontier with Germany and also because, after the German–Danish wars of the nineteenth century, the sovereignty of Jutland had not been resolved. Not until the end of the war did the first non-specialist books on the atom appear. One of the first, if not the first, was *Die Atomtheori* by a German physicist Leo Graetz (1856–1941) published in 1918.[6] One of the earliest books in English to contain a description of Bohr's model was *The Atom* by an American physicist Albert Cushing Crehore (1865–1959) published in 1919 but its purpose was to promote his own model.[53] Then 1923 saw a host of popular books on Bohr's atom. Among these were *Atoms and Electrons*[59] written by John William Navin Sullivan (1886–1937) a popular science writer who, though he had attended University College London where he studied science, had left without taking a degree and *The ABC of Atoms*[60] lucidly written by that man of genius Bertrand Russell who, having recently married for the second time and without an academic position, needed the royalties from a popular book. *Atoms and Electrons* was first published for the People's Library series of Hodder and Stoughton and was a straightforward easy to read book; it had no illustrations but that is perhaps not a fault when writing about entities that do not lend themselves to explicit representation; classical physics is about the objects that we perceive and can readily be illustrated but the particles that make up an atom cannot be explicitly depicted. Russell's *The ABC of Atoms* was more demanding than Sullivan's and discussed the more complicated picture of atoms with the electrons moving in non planar elliptical orbits, which was developed by Bohr in the early 1920s. Edward Andrade (1887–1971) published *The Structure of the Atom*[61] in 1923, a book more suited to scientists who wanted an introduction to atomic physics rather than the general reader, though he did publish a small popular book, *The Atom*,[62] in 1927 with no mathematics which was more suitable. Norman Campbell also published a supplement to his early 1907 book *Modern Electrical Theory*[63] in 1923 with the same title as Andrade's book with which it is comparable. However the most significant publication in

English about Bohr's atom in 1923 was a book published in Copenhagen in 1922, which was translated as *The Atom and the Bohr Theory of its Structure* written by the Dutch physicist Hendrik Kramers (1894–1952) and the Danish science writer Helge Hoist (1871–1944).[64] As Rutherford explained in the foreword:[64]

> 'This volume, written by Dr. Kramers and Mr. Hoist, gives a simple and interesting account of our knowledge of atomic structure, with special reference to the work of Professor Bohr. Dr. Kramers is in an especially fortunate position to give a first-hand account of this subject, for he has been a valued assistant to Professor Bohr in developing his theories, and has himself made important original contributions to our knowledge in this branch of inquiry.'

This book was very successful and translated into four languages.[65] Although the book was written without mathematics in a style that is easy for the general reader to understand, it did not dodge difficult issues, for example, in writing about the jump of an electron from one stationary state to another we read:[64]

> 'During the transition from one stationary state to another we have no knowledge at all of the existence of the electron, indeed we do not even know whether it exists at that time or whether it perhaps is dissolved in the ether to be re-formed in a new stationary state.'

Niels Bohr was awarded the he Nobel Prize in Physics 1922 'for his services in the investigation of the structure of atoms and of the radiation emanating from them'. Bohr's visit to the USA during the fall and winter of 1923 was widely reported by the *New York Times*. In November he gave a series of six lectures on *The Atom and the Natural System of the Elements*, at Yale University, which were reported daily in the *New York Times*. In February 1924 the *New York Times* wrote:

> 'Since he came to this country last fall to lecture on his theory of the structure of the atom at Yale University and elsewhere, there has been a remarkable display of interest in his discoveries of marvelous convolutions within this infinitesimal particle of matter.'

In the second half of the 1920s, physicists who were not intimately concerned with the structure of the atom began to write popular accounts. Karl Darrow (1891–1982), a physicist first with Western Electric and then Bell Laboratories and Secretary of the American Physical Society for twenty-six years, was an outstanding writer of popular reviews of physics. Bernard Cohen considered that his popular books on modern physics ranked with those of de Boglie and constituted 'the high point of the exposition and explanation of modern physical science.'[66] In 1926 Darrow published *Introduction to Contemporary Physics*, a well received popular book, which outlined the experimental and theoretical developments in the structure of the atom; an updated second edition in 1939 was equally well received.

7.2.2. *The new quantum mechanics after 1925*

To understand the Copenhagen interpretation of quantum mechanics one must totally abandon the deterministic reality of the world that we perceive with our senses. In 1927 Heisenberg published his uncertainty paper where he showed that the position and momentum of an electron can only be known to an accuracy that depended upon Planck's constant.[67] In September 1927, Heisenberg was in Leeds explaining his work to the British Association. The science editor of the New York Times, Waldemar Kaempffert (1877–1956) reported on Heisenberg's address writing:[68]

> '[Heisenberg gave] a brief exposition of a concept which will make it necessary to modify belief in what we are pleased to call "common sense' and "reality."... The scientific world is faced with an upheaval as great as that brought about by Einstein.'

One of the first popular expositions of the new quantum mechanics was contained in Eddington's very popular *The Nature of the Physical World*, published in 1928, which was translated into eight languages and still is in print.[26] *The Nature of the Physical World* was based on his Gifford Lectures of 1927 with some additions to take account of Heisenberg's 1927 paper but, even before his 1927 lectures, Eddington was expressing 'that our precise knowledge of certain aspects of the behaviour of atoms leaves their intrinsic nature just as transcendental and inscrutable as the nature of the mind, so the difficulty of interaction of matter and mind is lessened.'[69] Eddington was a Quaker first and foremost and welcomed the uncertainty in the new quantum mechanics which reinforced his strong belief in free-will. To him the spiritual world was real:[69]

> 'I venture to say that the division of the external world into a material world and a spiritual world is superficial, and that the deep line of cleavage is between the metrical and non metrical aspects of the world.'

In his *The Nature of the Physical World*, Eddington gives a full discussion of the *Uncertainty Principle* which, since this name had not yet been given in 1927 when Eddington reworked his Gifford Lectures for publication, he called the *Principle of Indeterminacy.* Eddington summed up the general implications of the principle thus:[26]

'The principle of indeterminacy, like the principle of relativity, represents the abandonment of a mistaken assumption which we never had sufficient reason for making. Just as we were misled into untenable ideas of the aether through trusting to an analogy with the material ocean, so we have been misled into untenable ideas of the attributes of the microscopic elements of world-structure through trusting to analogy with gross particles.'

He makes it plain that that the new quantum mechanics has important applications apart from those in physics and wrote:

'[W]e may note that science thereby withdraws its moral opposition to freewill. Those who maintain a deterministic theory of mental activity must do so as the outcome of their study of the mind itself and not with the idea that they are thereby making it more conformable with our experimental knowledge of the laws of inorganic nature.'

Quakers are mystics. The last chapter of Eddington's book is *Science and Mysticism.* He saw that the new quantum mechanics had removed the scientific materialism challenge of physics to religion only to have the challenge replaced by 'the deadlier moral materialism which despises it.'

Paul Heyl (1872–1961), a physicist at the National Bureau of Standards in Washington, was also a popularizer of physics. In 1930 he wrote *New Frontiers of Physics*, in which the longest chapter was *Matter and its Structure* on quantum mechanics.[70] However, unlike Eddingon he did not point out the big difference between the early quantum theory and the new indeterminist quantum mechanics or discuss the wider significance of modern physics. Heyl also published a number of popular articles in the *Scientific American* in the early 1930s and in one of these he did discuss indeterminism but, unlike Eddington, he did not see that it had philosophical significance writing:[71]

'The importance of the principle of indeterminacy is undeniable, but we must be careful not to read too much into it. In some quarters it has been regarded as overthrowing the philosophy of determinism: This, I think, is going farther than is warranted.'

The physicist grandson of Charles Darwin, Charles Galton Darwin (1887–1962) published *The New Conceptions of Matter*, in 1931 based on lectures he had given to the Lowell Institute of Boston, Massachusetts.[72] However there are problems with his book.[73] Darwin emphasized wave mechanics over corpuscular interpretations of quantum mechanics despite the dual nature being well proved. As William Henry Bragg wrote about the mechanics of an electron:[74]

'On Mondays, Wednesdays and Fridays we use the wave theory; on Tuesdays, Thursdays and Saturdays we think in streams of flying energy quanta or corpuscles.'

Whereas Darwin presented quantum mechanics as almost completely described by wave mechanics; he thought that when 'the scaffolding of mathematical formalism' was removed one could 'gain something of an intuitive view of what the world is really like.'

In 1934 Ernst Zimmer (1887–1965) published in German *The Revolution in Physics* which was published in English in 1936.[75] Zimmer was a school teacher who had attended the universities of Freiberg and Berlin, where he was a student of Planck who wrote a very brief introduction to his book. This popular book is the only one from the early 1930s that at all rivals that of Eddington in both style and content, though Zimmer was not as quite as relaxed in his writing as Eddington. In common with *The Nature of the Physical World*, Zimmer's book ends with a more general chapter relating the wider implications of modern physics to philosophy. Unlike Eddington's book, Zimmer's deals entirely with quantum theory, except for a very brief mention of relativity in the introductory chapter on classical physics in many respects this division is appropriate since relativity is nearer to classical physics than to quantum theory.

The popularizations of Russian born physicist George Gamow (1904–1968) are very different from most of the rest. In 1938 Gamow wrote a story about the adventures of a bank clerk Mr C.G.H. Tompkins[76] with which he explained the features of relativity and quantum mechanics. *Harpers Monthly* and other similar magazines rejected the story. Fortunately when Gamow was in England the next year he told Charles Galton Darwin about his story Darwin suggested that he send it to C.P. Snow, who was then editor of the popular science magazine, *Discovery*, which Gamow did, and the first dream of Mr Tompkins *The Toy Universe* was published in 1939 followed by six other dreams about relativity and quantum mechanics. The series was very successful and the next year Cambridge University Press published a book version, *Mr Tompkins in Wonderland*. Four other Mr Tompkins' stories were published, the last in 1967.[77] The only similar work to that of Gamow at the time that I know is an article *What is a Quantum?* by Heyl published in the *Scientific American* in 1930, which is a dialogue between a scientist and a visitor. The article works and is good but it has not the magic of the Mr Tompkins' stories.

In 1937 de Broglie published *La Physique Nouvelle et les Quanta*. Bernard Cohen wrote of de Broglie that he[78]

> 'is a representative of that great French tradition of first-rank scientists who can write of their own field for ordinary men and women. His prose is brilliant and his genius at presenting difficult abstractions in an intelligible way is probably unmatched.'

Unfortunately English speakers, because of the war, had to wait until 1953 for a translation[79] and that, as Bernard Cohen stated, was somewhat leaden.[80] What I think distinguishes de Broglie's book from others is that, in his review of classical physics, he very carefully outlined its unsubstantiated assumptions based on commonsense understanding of the world around us which produced classical laws that very accurately describe the world of everyday experience but nonetheless are shadows of reality not reality itself. De Broglie thought that quantum physics merited the attention of all not just specialists because it 'incontestably broadened our horizons and introduced a number of new ways of thinking, deep traces of which will without doubt remain in the future expansion of human thought.' By introducing quantum theory in this way removes some of its mysterious nature and makes its development and that of relativity, which he mentions in a brief chapter, more logical. Cohen's 1953 judgement that 'the reader may be safely assured that he will not find another book to match this one for clarity and patient explanation' is still true.

Abraham d'Abro, a successful popular physics writer, was one of the few non-professional physicists who wrote popular books on quantum mechanics before the Second World War.[81] In 1939 he published *The Decline of Mechanism*, on quantum mechanics with an emphasis on the change from the mechanical interpretation of reality of Newtonian physics epitomized by Laplace.[82] This approach is refreshing. D'Abro emphasized that the possibility that quantum mechanics meant the end of complete determinism in physics began to be realized long before Heisenberg's uncertainty principle quoting Poincaré's doubts when he returned from the 1911 Solvay Conference:

> 'Newton realized (or thought he realized – today we are beginning to wonder) that the state of a mobile system, or more generally of the universe, depending only on its immediately preceding state; and that all change in Nature occurred continuously ... well, it is this fundamental belief that is being questioned today.'

However, d'Abro left the door open to a determinist interpretation and referred to Einstein's view that quantum mechanics was incomplete. I believe that d'Abro's book ranks with that of Zimmer. Both Bernard Cohen and the Irish mathematician and physicist, John Synge (1897–1995) gave generally favourable reviews in 1940 while pointing out some minor inaccuracies in the history of physics.[83] Although it was long for a popular book, almost 1,000 pages, it obviously sold well because a second edition with the more easily understood title, *The Rise of the New Physics* in two volumes, was published in 1951 by Dover.

There were few books published on modern physics in Britain during the Second World War because of paper restrictions and after the war the publication of popular quantum mechanics books did not quickly pick up. One of the first post-war popular books on quantum mechanics was *The Strange Story of the Quantum* published in 1947 by Banesh Hoffman (1906–1986) a mathematician and physicist who co-authored a few papers with Einstein between the wars.[84] The title of his book and chapter headings such as *Tweedledum and Tweedledee* on the duality of corpuscular and wave mechanics interpretation of quantum mechanics were obviously chosen to entice the popular reader but the content is fairly standard apart from a little light hearted banter. The book must have been a success because a second edition was published in 1959. However, after the Second World War and its two atomic bombs, the general public was more interested in nuclear physics and atomic energy and the number of popular books on quantum mechanics decreased. The New Age Culture's interest in quantum mechanics in the late 1970s[85] led to more popular books on quantum mechanics being published but there were less of them than were published on relativity. From about 1980 there was an exponential growth in popular books on quantum theory and, from the beginning of superstring theory in 1984, there have been many popular books on string theory. There are in fact so many that not all of them can be mentioned here.

The physicist John Polkinghorne wrote two popular books: *Particle Play*[86] and *The Quantum World*[5] published in 1979 and 1984 before he entered the priesthood and concentrated on books on the relationship between science and religion. These two short books give straight forward introductions to particle physics and quantum mechanics, the first does have a brief chapter on scattering experiments and how particles are detected but otherwise they are very similar to many other popularizations. One irritation, common with some other popularizations, is the use of twee chapter titles that give little indication of what the chapter contains.

Feynman's success as a physicist and as a popularizer of physics was in no little way due to the fact that he danced to his own beat. He wrote so that others might discover the fun he had trying to understand the apparently absurd workings of Nature according quantum physics. His major achievement was in understanding the relativistic quantum theory of electrodynamics (QED) for which he received the Nobel Prize in 1965 along with Sin-Itiro Tomonaga and Julian Schwinger. In 1985 Feynman published a semi-popular book *QED The Strange Theory of Light and Matter*, which served three different sets of readers: the intelligent general reader who wanted to know what QED is about, students of quantum mechanics, and established physicists who appreciated the Feynman touch.[87] In this popular book Feynman showed how the probability of interaction of photons of light and electrons could be deduced using formalized diagrams, now known as *Feynman diagrams*, and were first introduced by him at the 1948 Pocono Conference on quantum physics[88] and the first published Feynman diagram appeared in 1949.[89] In 2005 the US Postal Service issued a stamp honouring Richard Feynman containing eight Feynman diagrams.

Two quite prolific popular writers on quantum mechanics from the 1980s onwards Paul Davies and John Gribbin have similar backgrounds having both worked for a time with Fred Hoyle. Gribbin was Hoyle's research student from 1968 to 1971 and Davies did postdoctoral research with him between 1970 and 1972. While Gribbin has done little research since the 1970s, Davies has had a fine research record in physics though, since about the turn of the millennium, he has concentrated more on astrobiology rather than physics. Davies has also been active in popularizing modern physics on radio and television. Two of his programmes for BBC Radio 3: *The Ghost in the Atom* and *Desperately Seeking Superstrings* have been successfully produced in book form.[90] What was refreshing, about these programmes and the subsequent books, was that Davies managed to persuade a number of important figures in these fields to give interviews, nothing is better than listening to or reading an account of a development from the people actually involved in it. *New Physics* edited by Davies also consists of contributions from well known physicists.[91]

7.3. The Theory of Relativity

> 'Who would imagine that this simple law [the constancy of the velocity of light in a vacuum] has plunged the conscientiously thoughtful physicist into the greatest intellectual difficulties?' Einstein (1916b)

7.3.1. *Special theory of relativity*

Einstein's special theory of relativity[92] published in 1905 had little popular impact at the time. Even the first book-length account of the special theory, in English published by Ebenezer Cunningham, did not appear until 1914.[93] The special theory of relativity was not mentioned in general scientific magazines like *Nature* and the *Scientific American* until around 1910[94] but there was still no mention of Einstein's special theory of relativity in the British or American newspapers. In the Netherlands there was special interest in Einstein's relativity since it was closely related to the work of Lorentz who was professor of theoretical physics at Leiden University. The first two public discussions of the theory of relativity there were given in 1909 and 1912 at the inaugural lectures of Johannes van der Waals Jr. (1873–1971) when he succeeded his father as professor of physics at Amsterdam University and Paul Ehrenfest, who had succeeded Lorentz at Leiden University in 1912.[95]

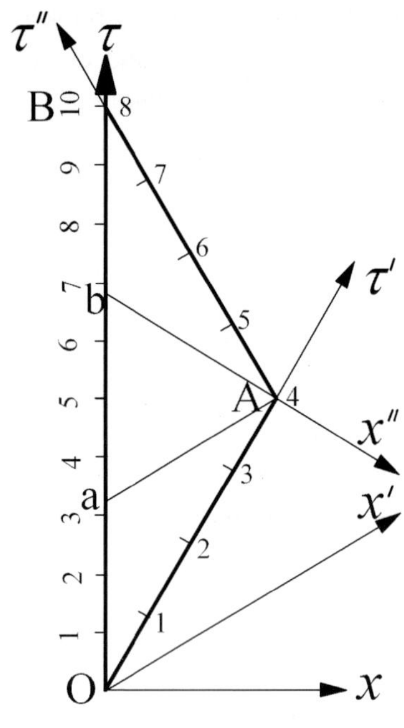

Fig. 7.5. Clock and twins' paradox.

One aspect of the special theory of relativity that did capture the general public's imagination was the relativity of time. In his original paper Einstein discussed a thought experiment with two clocks: C_1 is stationary in a reference frame Σ and clock C_2 is initially coincident with C_1 but then moves away at constant velocity v for time $\Delta t/2$, as measured by clock C_1, and then returns at the same velocity.[92] The world line of clock C_2 (OAB) in a Minkowski diagram is shown in Fig. 7.5 for a velocity v = 0.6; the inertial frames for the outward and inward journeys x', τ' and x'', τ'' are also shown. Einstein showed that the time registered by clock C_2 would be $\Delta t(1-\sqrt{1-v^2/c^2})$ slower than clock C_1[96] or, for the example illustrated in Fig. 7.5, 20% slower. There are problems with this world line at O, A, and B where there are singularities but let us leave these, and the general problem of thought experiments, for a moment. Einstein also remarked that a clock at the Equator must go slower than one at the North or South Pole. Paul Langevin (1872–1946) made the problem more interesting by substituting twins for the two clocks in 1911:[97] one twin remains at home while the other goes on a journey and when he returns finds the twin who has remained at home is older than he is. This version has become known as the Twins' Paradox. First let us put Langevin's twin paradox into practical perspective. Suppose one twin remains on the Earth and the other sets off on a spacecraft, ignoring the fact that the twin on the Earth is not stationary relative to an inertial frame at a velocity of 39,896 km/hr, which was the speed of the fastest manned spacecraft to have escaped from Earth and 0.00037 times the velocity of light. The longest space journey a human could possibly take is about eighty years, forty on the outward journey and forty on the return; on such a journey the space travelling twin would be 1.7 seconds younger than the stay-at-home twin. In practical terms the difference is imperceptible; for the difference in age at the twin's return to be a year, the twin would have to travel at nearly 240,000,000 km/hr.

Surprisingly, given the very slight practical possible difference in age, the Twins' Paradox simmered after Langevin first formulated it in 1911 until it boiled over in 1956 when Herbert Dingle, an emeritus professor of history and philosophy of science at University College London, expressed it as 'a new factor' in space travel in an article in *Nature*; it was the Sputnik Age and space travel had just become a possibility. The reality of the Twins' Paradox was a major controversy for the next eighteen months.[98] The controversy cannot be dealt with in detail here and those interested will find Hasok Chang's paper very informative.[99] Dingle, like Eddington was a Quaker, but came to physics comparatively late at the age of twenty-five. At first he was mostly orthodox in his understanding of relativity. In 1922 he wrote a short popular book *Relativity for All* with the stated aim 'to make the idea definite and intelligible to the ordinary mind.'[100] By 1939 Dingle was starting to have doubts about aspects of the theory of special relativity and the reality of time dilation.[101] Dingle continued his crusade against common interpretations of the special theory until he died and by 1961 was rejecting it almost entirely.[99] The heat went out of the controversy once editors started refusing to publish further contributions by Dingle. What was Dingle's motivation in continuing his battle? Chang's explanation was that Dingle wanted physical explanations of relativity rather than a mathematical demonstration; he wanted to know what slowed moving clocks down.[99]

In 1971 Joseph Hafele (1933–2014) and Richard Keating (ca. 1931–2005) decided to put the time difference due to relative motion to an experimental test rather than a thought experiment to try to end 'one of the most enduring scientific debates of [the]century … the relativistic clock paradox.'.[102] By 1971 extremely accurate portable caesium beam atomic clocks were available, and they took four such clocks around the world by commercial flights, both in an easterly and a westerly direction.[103] Since the clock at Washington is not in an inertial frame because of the Earth's rotation, the time on a hypothetical clock of an underlying non-rotating inertial fame calculated from the Washington clock was used as the reference. They had to account for the different speeds, directions, latitudes, and altitudes of the flight but their method can be understood assuming a flight at a constant speed, v, and altitude, h, at the latitude of Washington whose radius is R. Relative to the hypothetical reference clock, the clock at Washington, using the same approximation as Einstein,[92] must be slow by

$$t - t_0 = \frac{t_0}{2}\left(\frac{R\Omega}{c}\right)^2, \tag{7.2}$$

where t_0 is the time registered by the hypothetical reference clock and Ω is the rotational speed of the Earth. If v is the velocity of the flights and positive for the easterly one then the clock on that aircraft must be slow relative to the Washington clock by

$$t - t_0 = \frac{t_0}{2}\left(\frac{2R\Omega v + v^2}{c^2}\right). \quad (7.3)$$

Since the aircraft flew through the weak gravitational field of the Earth there are other time changes which are predicted by the general theory of relativity. Before publishing his general theory of relativity, Einstein in 1907 gave the time differences caused by weak gravitational fields by differences in gravitational potential Φ[104]

$$t - t_0 = \left(\frac{\Phi}{c^2}\right)t_0. \quad (7.4)$$

The gravitational potential of an aircraft whose altitude is h compared with one at Washington is gh, where g is the acceleration due to gravity, hence the gravitation field causes a clock at altitude h to show a time difference with the one at Washington of

$$t - t_0 = \left(\frac{gh}{c^2}\right)t_0. \quad (7.5)$$

The predictions and the measured time differences are shown in Table 7.2.[105]

Table 7.2. Predicted and experimental time differences in nanoseconds between the clocks on the aircraft and that at Washington.

Effect	Direction of flight	
	Easterly	Westerly
Special theory relativity	−184 ± 18	96 ± 10
Gravitational field	144 ± 14	179 ± 18
Theoretical total	−40 ± 23	275 ± 21
Experimental	−59 ± 10	273 ± 7

It can be seen that the experimental and theoretical time differences agree within the limits of accuracy giving confidence in the theory of relativity but the contribution from the general theory of relativity is the same order of magnitude as that from the special theory. The average flight time for all round trips was 44.9 hr so that the time difference in the times of the flights of 332 ns represents only 2×10^{-10} %, an extremely small percentage of the flight time.

Now we need to briefly address the problems with Einstein's thought experiment of the two clocks which is to do with thought experiments in general.

The word experiment in the term *thought experiment* seems to imply that we are dealing with reality but epistemologically there is no difference between such *experiments* than any other theoretical argument.[106] Thought experiments are performed in a model world and can only be as real as that model world. In the case of Einstein's thought experiment the model world of the special theory of relativity is one devoid of matter. His result, apart from any possible particular problems with his experiment, is true but for the model world not reality. I do not for a moment believe that Einstein was not aware of this fact, the problem is with the interpretation others have given it. The real experiment cannot be performed in empty space, in the experiment of Hafele and Keating we see that the Earth makes it necessary to consider a contribution caused by the general theory of relativity to get an accurate value. The particular problem with the thought experiment is with the infinite accelerations that are required at points O, A, and B in the Minkowski diagram shown in Fig. 7.5 of which that at A where the travelling twin reverses his direction relative the stay-at-home twin is the most serious. As the travelling twin reaches A the age of the stay-at-home twin appears to the travelling twin to be given by the point *a* but, immediately after he has reversed direction, his age appears to be given by point *b*, the stay-at-home twin appears to age by *ab* instantaneously. The specially theory can accommodate constant acceleration and it might be thought that substituting a constant high retardation on his journey before the turning point A is reached that the problem with acceleration could be removed but the world line of the stay-at-home twin, viewed by the travelling one, involves speeds greater than light.[99] In conclusion to this discussion, Einstein's demonstration that the travelling clock is slower than the travelling one on return is true but I am sure that he never intended the actual time difference to be an accurate measure of a real experiment.

Books and magazines were not the only media used in the 1920s to popularize relativity, in 1922 Hanns Kornblum (1878–1970) made a German full length silent feature film, more than two hours long, with the English title *The Basics of Einstein's Theory of Relativity*, describing the special theory of relativity using trick cinematography to illustrate the relativity of extension and time; there is also a final scene showing the deflection of starlight during the eclipse observed by the Eddington expedition.[107] The response was generally enthusiastic apart from a few anti-relativists. In Britain, Punch greeted the film on its Charivaria page with:[108]

> 'In Berlin they are preparing a film explaining the Theory of Relativity. We felt it would not be long before a serious attempt would be made to compete with the Charlie Chaplin Films.'

Unfortunately the film has been lost but in 1922 the *Scientific American* published a few film clips, the one in Fig. 7.6 shows a train, with clocks at both ends, travelling along a viaduct which also has clocks.[109] The white line at the bottom of the train represents light propagating from one end of the train towards the other. The clocks on the train are correctly shown as going slow compared with the ones on the viaduct. Physicists like von Laue did criticize some points in the film at the time but it is no wonder that it was a sensation.

In the USA Fleischer Studios released the silent film *The Einstein Theory of Relativity* in 1923 which relied on footage from the German film and had two versions one of which ran for about 20 minutes and another 50 minute version.

Fig. 7.6. Film sequence from Die Grundlagen der Einsteinschen Relativitätstheorie, courtesy Scientific American.

7.3.2. *General theory of relativity*

Einstein's first papers on the theory of general relativity were not generally immediately known in Britain or the United States even by physicists because of the First World War. However, when the deflection of light from stars near the Sun during an eclipse measured during the Eddington-Dyson expedition[110] were shown to be predicted by the theory of general relativity, Einstein became an immediate world celebrity. *The Times* announced the British Solar Eclipse Expedition to measure the deflection of starlight during the eclipse on the 29th May 1919, in April. On the 7th of November *The Times* reported the decision on the previous day by the joint Royal and Astronomical Societies that Einstein's general theory of relativity had accurately predicted the deflection of starlight with the headlines:

REVOLUTION IN SCIENCE

NEW THEORY OF THE UNIVERSE

NEWTONIAN IDEAS OVERTHROWN

There were a further 35 articles in *The Times* that mentioned Einstein after that day until the end of the year, which is nearly one article for every day *The Times* was published. The articles were very varied, one unattributed article was in the form of a short story about a priest and a monk, a time-traveller 'who has studied these poets [Newton and Einstein], as you rightly call the eminent geometers, chatting about space and time on a motor-omnibus.' The story reminds me somewhat of the Father Brown stories of G.K. Chesterton (1874–1930) who at the time wrote *Our Note Book*, a weekly comment in the *Illustrated London Weekly*, where he twice mentioned Einstein in 1920 and on occasions in later years. The *New York Times* also published many articles after the Royal and Astronomical Societies confirmed that Einstein's theory successfully predicted the deflection of starlight by the Sun.

Einstein himself wrote the first popular version of both the special and the general theories of relativity for 'those readers who, from a general scientific and philosophical point of view, are interested in the theory, but who are not conversant with the mathematical apparatus of theoretical physics,' in 1916.[111] Because of the war, the first English translation of Einstein's *Relativity* did not appear until 1920 and then went through fifteen editions and is still in print. In April 1921 Einstein went to the USA for the first time and delivered lectures at Columbia and Princeton Universities. His Stafford Little Lectures at Princeton were published as a popular book on relativity *The Meaning of Relativity* which went through five editions and is still in print.[112] Even in 1929 when Einstein wrote a popular article for both *The Times* and the *New York Times* explaining both the special and general theory of relativity as well as the new field theory for the structure of space-time, the excitement was such that the department store Selfridges in London pasted the six pages of his article in its widow and large crowds gathered to read it.[113]

Eddington, who was one of the first to learn about Einstein's general theory of relativity in Britain,[114] was also one of its early popularizers and in 1920 published *Space, Time and Gravitation.*[115] Eddington's popularization has also remained in print until the present day and has had French, German and Spanish editions. Through the *Scientific American* in 1920, the philanthropist, Eugene Higgins offered a prize of US$5,000, a princely sum for the time,[116] for the best 3,000 word essay on the theory of relativity. Not surprisingly there were three hundred entries for the prize including some from well known physicists such as de Sitter and Andrade but it was won by Lyndon Bolton (1860–1942), who like Einstein was on the staff of a patent office.[117] Bolton's essay was mainly on the special theory of relativity, which is not surprising seeing the limitation on its length but his essay was really neither very clear nor very informative. Although Bolton had no formal training in physics he had read mathematics at Cambridge. Methuen, who had published many books on relativity, approached him to write a book on relativity but his book[118] attracted few reviews.

The early 1920s saw an avalanche of popular books on relativity. Bolton listed thirteen books on relativity in English.[118] In 1922 Cunningham, who reviewed six such books and pamphlets in English, French, and German in *Nature*, wrote:[119]

> 'Another collection of books and pamphlets reminds us of the hold the theory of relativity has on the public imagination.'

The Austrian physicist Hans Thirring (1888–1976) published *The Ideas of Einstein's Theory* in 1921 which was a translation of *Die Idee der Relativitätstheorie* published in Berlin in the same year.[120] In a review essay for the *Scientific American* in 1950, James Newman (1907–1966) an American mathematician who was a regular reviewer, listed what he considered the twenty of the best popular science books, *The Ideas of Einstein's Theory* is one of the two on relativity that he listed and the other one was Bertrand Russell's *The ABC of Relativity*[121] first published in 1925, I agree with both his choices.[122] As well as being an excellent and simple explanation of relativity, in Thirring's last chapter he is one of the few popularizers of relativity, apart from Einstein himself, who pointed out that:[120]

> 'since our "world-space" is curved, we must take into account the possibility that it is a closed space, *i.e. our universe is perhaps finite, although it is certainly unbounded.*'

Russell's *ABC of Relativity* has been in print since 1921 and is generally recognized as one of the best, though that was not necessarily everybody's opinion when it was first published. At the beginning of a review in the *New York Times*, Benjamin Harrow, a professor of biochemistry at Columbia University, wrote:[123]

> 'If this is the ABC of relativity the sooner we stop thinking about Einstein the better.'

but Harrow did not like any book on relativity:

> 'I know of no popular book on relativity which is even as good as Russell's, which I am afraid is simply another way of saying that they are all bad.'

However, Harrow did like the last chapter of Russell's on the philosophical consequences of relativity.

The only other popular book on relativity that I shall mention from the early days of relativity is Sullivan's *Three Men Discuss Relativity*[124] published in 1925 and that is simply because he was one of the few in the twentieth century similar to the eighteenth and nineteenth century non-scientist popularizers of science. Sullivan's book is in the form of a dialogue between a mathematical physicist, a philosopher, and an ordinary intelligent person. There is a long history of books on science in the form of dialogues between people. Galileo's the *Two New Sciences* published in 1638 is one that worked well but by the twentieth century the genre was showing its age, especially as the three chosen by Sullivan were very boring characters. Nevertheless many must not have been of my opinion because the book sold very well.[125]

Although there were popular books on the theory of relativity published after 1925 they did not create so much attention as those in the early 1920s.

7.3.3. *Anti-Einsteinians*[126]

Einstein's theory of relativity is even more revolutionary than Newton's natural philosophy was and in addition it is outside of common sense experience, therefore it is not surprising that the theory was opposed by some. What might have caused some of the opposition is way the theory of relativity has been presented to the general public. The third line in *The Times'* headline *Newtonian ideas overthrown* is true but somewhat misleading. Newton's concept of absolute space and time was overthrown but Newtonian mechanics is still very accurate for predicting motion within the solar system and its effect cannot be noticed by ordinary people as J.J. Thomson stated in 1919.[127] Numbers showing the size of the effect of relativity are seldom presented in popular accounts because it somewhat lessens their appeal, one notable exception was Eddington's the *Nature of the Physical World* where the contraction of length for a rod travelling at the orbital speed of the Earth of about 19 miles per second (30 km/s) is given as 0.0000005%.[26]

The anti-Einsteinians were different to the anti-Newtonians. While the latter opposition was at first religious and only became secular in the nineteenth century, most opposition to Einstein's theories was secular from the beginning, though there were a few such as Arvid Reuterdahl (1876–1933), the Dean of the Engineering and Architecture Faculty at the Catholic College of St. Thomas in St. Paul, Minnesota in the early 1920s, who did have theist motives.[128] Reuterdahl was instrumental in forming the Academy of Nations in 1921 which, though it had no mention of the theory of relativity, was a coordinating body to refute it. Another difference between the two oppositions was, while there were no notable natural philosophers who were anti-Newtonian, the anti-Einsteinians included professional physicists. The anti-Einsteinians even included two German Nobel Laureates: Philipp Lenard (1862–1947), who received the Nobel Prize in 1905 for his research on cathode rays, and Johannes Stark (1874–1957), who received the Nobel Prize in 1919 for his research on the Doppler effect in anode rays. Both of these Nobel Laureates were anti-Semitists and supporters of the Deutsche Physik national movement in the 1930s. However, though anti-Semitism was a factor among many anti-Einsteinians, some were Jews like Oskar Kraus (1872–1942) who was a Czech philosopher. Kraus had intended to present a lecture at the anti-Einsteinian meeting in the Berlin Philharmonic Hall in 1920 organized by the anti-Semite Paul Weyland (1888–1972) but withdrew when he learned the anti-Semitic nature of the event.

One of the leading anti-Einsteinians was Ernst Gehrcke (1878–1960), an experimental physicist who was a Privatdozent[129] at the Friedrich-Wilhelms-Universität, Berlin from 1904 to 1921 when he was appointed to a professorship; Gehrcke was a critic of the theory of relativity from 1911 until his death. Gehrcke is known to have made a few anti-Semitic statements but his objections to relativity were not based on his anti-Semitism.[128] One objection that Gehrcke, and many other physicists, had to relativity was that it entailed abandoning the luminous aether. It is hard to imagine now how central was the aether to the physics of the late nineteenth and early twentieth century, the great William Thomson said at the Baltimore Lectures in 1884:[130]

> 'One thing we are sure of, and that is the reality and substantiality of the luminous ether.'

Oliver Lodge was one who, while not rejecting the theory of relativity, maintained a belief in the aether which he did not think was destroyed by it. One of the problems with aether is that it meant different things to different people. Einstein had a philosophical concept of an aether, but it did not have a physical reality, and in a lecture at Leyden in 1920 said:[131]

> '[A]ccording to the general theory of relativity space is endowed with physical qualities; in this sense, therefore, there exists an ether. According to the general theory of relativity space without ether is unthinkable; for in such space there not only would be no propagation of light, but also no possibility of existence for standards of space and time (measuring-rods and clocks), nor therefore any space-time intervals in the physical sense. But this ether may not be thought of as endowed with the quality characteristic of ponderable media, as consisting of parts which may be tracked through time. The idea of motion may not be applied to it.'

As Eddington said at the twenty-third Swarthmore Lecture given to the London Yearly Meeting of Quakers in 1929:[132]

> 'Among leading scientists today I think about half assert that the æther exists and the other half deny its existence, but as a matter of fact both parties mean exactly the same thing, and are divided only by words.'

There were certainly philosophical problems with relativity, without the Newtonian absolutes; space and time concepts were difficult to grasp and the increasing reliance on mathematical formulation also seemed to distance it from reality. Gehrcke's publications were not simply nonsense as some were and they provoked a response from Max Born, Hans Thirring, Hermann Weyl, and even Einstein himself.[128] Eventually by 1920 Gehrcke abandoned physical and philosophical arguments, especially in explaining why the theory of relativity had such a hold over the general public, and turned to psychological reasons and the power of mass suggestion. Soon Gehrcke stopped criticizing Einstein publicly but he still privately believed mass suggestion explained the acceptance of the theory of relativity.

As with anti-Newtonianism, non-academics propounded their own nonsensical world views. One typical exponent was Arthur Patschke (1865–1934) a German designer of steam turbines who seems to have much in common with Thomas Mackintosh who wrote *Electrical Theory of the Universe* in 1838.[133] The English translation of the title of Patschke's magnum opus was *Electromechanics: Uniform Explanation and Mechanics of Natural Forces*, the uniform explanation was that the aether caused all the forces of nature. In 1921 Patschke demolished the theory of relativity in his book with the English title *Overthrow of the Einsteinian Theory of Relativity; Introduction to the Uniform Explanation and Mechanics of Natural Forces*.

The international network that had existed to promote opposition to the theory of relativity gradually petered out in the late 1920s but opposition still lingers in such dark places as the American website *Conservapedia* run by extreme right-wing Christians who object to anything outside their narrow world.

7.4. Cosmology

Although the word cosmology has been in use since the sixteenth century and its earlier form cosmography since the fifteenth, nobody called themselves a cosmologist until about 1830[134] and cosmology as a university discipline did not exist until after the Second World War.[135] However speculations have been made about the beginning and development of the universe since earliest times.

Modern cosmology, founded on the general theory of relativity began in the 1920s. An excellent survey of cosmology since the earliest times until the present day has been written by Helge Kragh and which, though not written specifically as a popular book, is one of the best on general cosmology.[135]

Modern cosmology grew quickly during the 1920s and many of its popular books became best sellers. Although not strictly cosmology, one of the first was Dingle's *Modern Astrophysics* published in 1924 which was a little too early to contain many of the new exciting discoveries.[136] However, he could include Eddington's proposal of the mechanism for the vast energy of the stars in Eddington's address as President of the Mathematics and Physics Section of the British Association in Cardiff in 1920:[137]

> 'If 5 per cent, of a star's mass consists initially of hydrogen atoms, which are gradually being combined to form more complex elements, the total heat liberated will more than suffice for our demands, and we need look no further for the source of a star's energy.'

However, he did not add Eddington's typical following remark:

> 'It is difficult to assert, but perhaps more difficult to deny, that this is going on. Sir Ernest Rutherford has recently been breaking down the atoms of oxygen and nitrogen, driving out an isotope of helium from them; and what is possible in the Cavendish laboratory may not be too difficult in the Sun.'

Dingle was also able to write about the Doppler shifts in galaxies and stars but was too early to mention the Hubble expansion of the universe. Eddington published a short popular book *Stars and Atoms* in 1927 based on an Evening Discourse given at the meeting of the British Association in Oxford in August 1926. The general theme is the advance in knowledge of atoms and radiation which led to advances in astronomy and the reciprocal part that the study of matter in the extreme conditions in stars and nebulae has played in the progress of atomic physics. Eddington stated that he made his selection in his book from the advances that could be easily explained. The book, though much more readable than Dingle's, covered much the same ground.

James Jeans was as popular a writer as Eddington and in the late 1920s and early 30s published many popular books on cosmology. His first was a short book *Eos or, The Wider Aspects of Cosmology* published in 1928, which sprung from his Truman Wood Lecture to the Royal Society of Arts earlier in the same year and was very popular being reprinted only two months after it was first published.[138] *Eos* was an expansive look at the universe and its immensity and the possibility of planets of stars other than the Sun that could support life. Jeans had been appointed a research associate of the Mount Wilson Observatory in 1923 and in his book he published some splendid photographs of nebulae obtained with the 100 inch telescope. This book was followed with a larger *The Universe around Us* in 1929, which had many more photographs again mainly from Mount Wilson and was even more popular.[139] Jeans is unique among the writers on popular cosmology of the time in describing in *The Universe around Us* the creation of the universe in terms somewhat similar to Lemaître's primeval atom of 1931[140] though Jeans did not limit himself to one single event and wrote:

> '[It] leads us to contemplate a definite event, or series of events, or continuous process, of creation of matter at some time not infinitely remote. In some way matter which had not previously existed, came, or was brought, into being.... If we want a concrete picture, we may think of the finger of God agitating the ether.'

Other writers of popular cosmology between the wars were reticent about the early universe probably because there was no accepted version. The most popular of Jeans books on cosmology was *The Mysterious Universe*, published in 1930 which had an initial print run of 10,000 copies and was reprinted twice in the month in which it was first published.[141] *The Mysterious Universe* contained two chapters on modern physics with the chapter on relativity linked to aether and explaining that you cannot have both.

Hubble had published his famous 1929 paper giving the Hubble constant for the expansion of the nebulae before the publication of *The Mysterious Universe* but, while Jeans did mention that 'the speeds at which individual nebulae are receding from us are roughly proportional to their distances from us,' he did not discuss the implications at any length. Hubble's general conclusion was already known by Lemaître and others by 1929[142] and one would have thought that Hubble's conclusion would have been headline news in the newspapers but it was not until 1931 that *The Times* had a short article in February noting that Einstein accepted the work of Hubble and it reported his Rhodes Lectures at Oxford University in May where at the second lecture he spoke about the expanding universe.[143] Edmund Bowen (1898–1980), a Fellow of University College, had the foresight to save the blackboard (see Fig. 7.7) which Einstein had used in explaining his estimate of the diameter in light years (Ly) and age of the universe in years (y) both of which are reasonably close to today's estimate. A humorous ditty *Hubble Bubble* was written for Punch on the occasion of Einstein's lecture by Charles Graves (1856–1944), a regular contributor, but though it mentioned Hubble's 'Constant for red displacement,' the constant was not linked to the expansion of the universe.[144] Hubble gave the 1936 Rhodes Lectures with the general title *The Observational Approach to Cosmology*.[145] The first popular book to discuss the expansion of the universe was Eddington's *The*

Expanding Universe published in 1933. When Eddington gave the Messenger Lectures at Cornell University in 1934, which were published as *New Pathways in Science*, the expanding universe was one of its topics. This lecture was not missed by *Time*, which put Eddington on the cover on 16th April 1934 with the title: *His universe expanded into popularity.*[146]

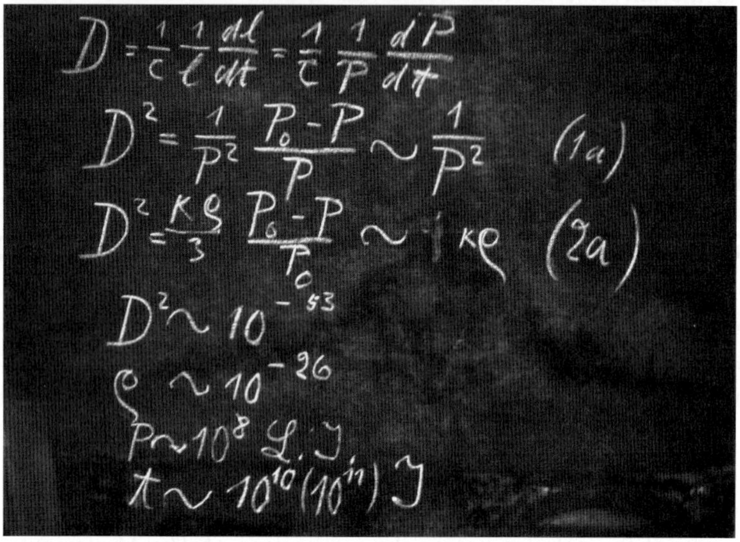

Fig. 7.7. The blackboard Einstein used at his second Rhodes Lecture in Oxford in 1931, courtesy Museum of the History of Science, Oxford.

Jeans gave the Christmas Lecture at the Royal Institution in the winter 1933/34 and published the lectures as *Through Space and Time.*[147] The theme of his lectures was the sky around us, I know that the Christmas Lectures were meant for primarily for children but as Jeans wrote in his preface:

> '[T]he lecturer will be confronted with an eager and critical audience, ranging in respect of age from under eight to over eighty, and in respect of scientific knowledge from the aforesaid child under eight to staid professors of science and venerable Fellows of the Royal Society, each of whom will expect the lecturer to say something that will interest him.'

and I would have thought he would have made more mention of some of the exciting developments in cosmology which had occurred since his *The Universe around Us* of 1929.

The most successful popular book on cosmology is Stephen Hawking's *A Brief History of Time*, first published in 1988, which explains the modern conception of the birth and expansion of the universe and has sold over ten million copies. Since Hawking's book is so well known, it will not be reviewed here. Cosmology after the 1950s and the Big Bang theory has been covered in Chapter 6 mostly using popular books as accessible references and is not repeated here but the Steady State theory, which was a valid theory until the 1960s is presented.

7.4.1. *The Steady State theory*

What was to become known as the Big Bang Theory was not the only cosmological theory. In the late 1940s and 1950s the Steady State theory was the topical theory.[148] The main supporters of the Steady State theory were Fred Hoyle, Hermann Bondi, and Thomas Gold (1920–2004). They believed in an expanding universe but considered that the application of the general theory of relativity with a cosmological constant unsatisfactory since it was manipulation of the constant that enabled the Hubble expansion to be modelled. Bondi and Gold also believed in the *perfect cosmological principle*, which assumes that the universe observed not only looks the same from any point in space but also looks the same at any point in time. The argument given by Bondi and Gold for the *perfect cosmological principle* is at best weak.[149] They make the point that a physical law should be the same anywhere and at anytime with which all would agree but then add that 'as conversely the structure of the universe depends upon the physical law, it follows that there may be a stable position.' Thus the *perfect cosmological principle* may be true but at best it is a possible hypothesis not a principle. Because they all believed in an expanding universe it could only remain the same if there was continuous creation of matter and Hoyle modified the general theory of relativity to allow the continuous creation by adding a creation field to Einstein's equations for which there was no scientific evidence.[150] The matter in continuous creation must be *creatio ex nihilo*. Surprisingly creation out of nothing may be possible. Edward Tryon was the first to suggest that *creatio ex nihilo* could be possible in 1973 because when matter of mass m is created it has a positive mass energy mc^2 but negative gravitational energy $-GmM/R$, where M is the mass of the universe contained within the Hubble radius $R = H$.[151] In 1994 Nathan Rosen showed that, by using Einstein's pseudotensor and assuming a Friedmann-Lemaître-Robertson-Walker universe, the two terms did indeed cancel one another.[152] However in 1997 Nilanjan Banerjee and S. Sen showed that the result depended on the model used and the two energies did not necessarily exactly cancel one another.[153]

In the 1940s there were observational problems with the Big Bang Theory because Hubble's constant was some ten times larger than its present value and gave the age of the universe as less than the independently calculated age of the solar system. There was no such problem with the Steady State model and both models had their supporters. Fred Hoyle presented his Steady State theory to the public in a 20 minute broadcast by the BBC Third Programme in 1949 entitled *Continuous Creation.*[154] As already noted in this chapter often the magnitude of what is being discussed by scientists is not given, in Hoyle's case he did give the order of magnitude in a graphic way and explained that continuous creation 'means that in a volume equal to one-pint milk bottle about one atom is created in a thousand million years,' However, there is room for a lot of milk bottles in the observable universe and as Hoyle stated about 10^{20} tonnes of matter is created every second but it is still too small to be detected.[155] Hoyle's talk was so successful that he was offered a series in the following year despite the fact that he was not yet a professor as was usually preferred by the BBC for their science programmes. The resulting lecture series was quickly published in 1950 as *The Nature of the Universe*, which closely followed the broadcast series.[155] Hoyle's last chapter was his personal view, which in the broadcast talks occupied only five minutes at the end of the talk, on the expanding universe and his Steady State theory. This last five minutes caused a lot of trouble. As already noted in Chapter 6, Hoyle was a staunch atheist and in that last five minutes, expanded into a chapter in his book, he ranted against those 'threatening to choke the life out of science,' such as secrecy, nationalism, which were indeed threats in 1950, and Marxists and Christians. Hoyle had worked on radar during the war and his scripts for the BBC still had to be vetted by the official censor in 1950 before they could be broadcast.

Hoyle published a larger popular book on astronomy and cosmology in 1955 *Frontiers of Astronomy*, which had a similar format to his *The Nature of the Universe.*[156] In the chapter on the expanding universe Hoyle addressed a possible problem for the Steady State theory with Olbers' Paradox: why is the night sky dark?[157] Bondi had previously in his book *Cosmology* addressed Olbers' Paradox at length.[158] As Hoyle explained, the intensity of the light of a star falling on the Earth is inversely proportional to the square of the distance of the star, r, assuming that on average stars give out the same light; if the spatial density of stars is the same the number of stars increases with the square of the distance from the Earth and therefore the total intensity of starlight is simply the sum of the number of stars provided they are older than r/c, where c is the speed of light. Thus the night sky should glow as brightly as the average star. Olbers' solution to the paradox was that it is dark because the interstellar medium or matter is not perfectly transparent and filters out the light from very distant stars.[159] Olbers' Paradox was more important to the Steady State theory than to the Big Bang theory because the Steady State theory implied an infinite universe, whereas the Big Bang universe is finite, and it had to be resolved. Of course Olbers did not know about the expansion of the universe. Because the universe is expanding the distance between galaxies increases with distance so there is a diminution in the light reaching Earth from the more distant ones and Hoyle stated that: '*The sky is dark at night because the Universe expands*'.[160] More modern analysis states that the extra-galactic background light is reduced by about a factor of two because of the expansion of the universe but that the major factor is the finite lifetime of the galaxies.[161] Absorption of extra-galactic background light is not thought to be a big factor. Hence Olbers' Paradox was one of the nails in the Steady State coffin.

In the 1950s the Steady State theory was as valid a hypothesis as the Big Bang theory and consequently created as much general interest. In 1959 The BBC Third Programme invited Hermann Bondi, William Bonnor (1920–2015), and Raymond Lyttleton (1911–1995) to talk about the rival views of cosmology. This series of three talks was so successful that the BBC invited them to repeat their talks again later in the same year and to give a fourth talk where the three discussed their different views under the chairmanship of Gerald Whitlow; these talks were published in book form in 1960.[162] In some ways Bonnor and Lyttleton were an odd choice to include with Bondi who, of course, put the Steady State view of the universe, because neither of them believed in the Big Bang. Bonnor believed in the relativistic theories of the universe without a cosmological constant but did not accept the Big Bang theory, since these theories implied a singularity, which at the time was thought to be at about eight billion years ago, and implied 'an act of creation in the finite past'. Bonnor was a staunch atheist and said of the possibility of invoking a 'creation of the universe' that:

> 'It seems to me highly improper to introduce God to solve our scientific problems.'

Lyttleton thought that the general theory of relativity was suspect because it could be made to give almost any result by 'introducing certain extra terms.' His alternative was 'an electric universe' where the proton instead of having equal but opposite charge to an electron has a very slightly larger charge (only 2×10^{-18} larger) which would give a repulsive force that would cause the expansion of the universe. Lyttleton's universe like Bondi's was Steady State. What is most curious is that in March 1959 Lyttleton and Bondi submitted a joint paper to the Royal Society which contains the same model presented by Lyttleton in the BBC Third programme.[163] Bondi did not mention this different version of the Steady State theory in his talk but, in the discussion between all four participants in the final broadcast of the series, Lyttleton did mention their joint work. All hinged on the possible difference in charge between a proton and an electron, the problem was that even in 1959 estimate of the maximum difference was already less than the value that Lyttleton and Bondi were using in their model. In 1960 Lyttleton and Bondi quietly dropped their electric universe theory.[164]

The detection of the cosmic microwave background radiation from 1964 which was predicted by the Big Bang theory combined with a sharp decrease in Hubble's constant which increased the estimates of the age of the universe both spelt the end of the Steady State theory. Hoyle continued to believe in the Steady State theory and produced modifications to it, but the theory effectively died with him. In the chapter *Cosmic Opera* in Gamow's Mr Tompkins book published in 1965, he made light hearted fun of many of those involved in cosmology – George Lemaître, James Jeans, Herman Bondi, Thomas Gold, the radio astronomer Martin Ryle (1918–1984) and most of all Fred Hoyle, who he made the librettist of his cosmic opera (it was Gamow's wife Barbara who wrote the libretto) the most apt verse of the opera for this discussion, set to the Thomas Arne's music that was used for *Rule Britannia*, is:[165]

> The universe, by Heaven's decree,
> Was never formed in time gone by,
> But is, has been, shall ever be—
> For so say Bondi, Gold and I.
> Stay, O Cosmos, O Cosmos, stay the same!
> We the Steady State proclaim!

Gamow published a popular account of his version of the Big Bang in 1952 where the elements formed out of a 'primordial mixture of nuclear particles,' for which he used the archaic word *Ylem*,[166] by '*nuclear cooking*, which lasted not more than an hour of time, [when] conditions throughout the universe closely approximated those existing in the centre of an exploding atomic bomb.'[167] Gamow ended his book with the observation:

> 'Indeed, it took less than an hour to make the atoms, a few hundred million years to make the stars and planets, but five billion years to make man!'

7.5. String Theory and Postmodern Physics

Modern physics has been very successful in giving predictions that have been verified by experiments or observations both at the very small and large scale to an extreme accuracy but it has not explained everything. Lee Smolin, a physicist much concerned with the directions of physics in the twenty-first century, has listed five great problems that remain to be answered:[168]

(1) Quantum gravity, the combination of quantum theory with gravity.
(2) Explanation of the foundations of quantum theory.
(3) Unification of the elementary particles and forces.
(4) Explanation how nature chose the constants of the standard quantum model.
(5) Explanation of the nature of dark matter and energy.

The main candidate for answering these questions has been string theory which has gone through several versions since it was first proposed. There are many popular books explaining the development of string theory with Steven Weinberg nominating Brian Greene's *The Elegant Universe*[169] as one of his thirteen best popular books on science for all time.[170] However, after more than forty years since its first appearance in the early 1970s, it is still by no means certain that the answer is going to come from string theory. Two of the main gainsayers to string theory Lee Smolin and Peter Woit have written the popular books *The Trouble with Physics*[168] and *Not Even Wrong*[171] on their objections to string theory. Woit has compared the arrogance of some string theorists and 'the way superstring theory research is pursued in physics departments and the way postmodernism has been pursued in humanities departments.' Roger Penrose has also effectively defined postmodernism in physics in his public lectures, *Fashion, Faith, and Fantasy in the New Physics*, which he gave at Princeton in 2003.[172] To Penrose string theory is *Fashion*, quantum mechanics applied at all levels is *Faith*, and inflationary cosmology is *Fantasy*. Some of the reasons why string theory has come to a position of ascendancy are not peculiar to string theory but are the result of postmodernism in universities

I and my older colleagues, who have experienced research and teaching in universities before the postmodern era, have also recognized a form of postmodernism in our own discipline and it is unfortunately quite general in all disciplines. In common with most, we did not have to follow the fashion in our research in our early careers and we had academic freedom to pursue whatever we chose. In our research in we mostly developed theories backed by experiment realizing that without experimental evidence one could waltz with the fairies. We were comparatively free to introduce courses that we considered beneficial to our discipline and students and made sure that they had an adequate experimental component. During our early careers deans and vice-chancellors were elected from faculty members or if appointed it was primarily for their research experience. Finance for research and teaching was often limited especially in the

1950s and 60s but what was available was shared roughly according to need without the necessity of spending much of our time in submitting proposals to committees of this and that. Regrettably in the postmodern era this position has completely changed. Everything has become judged by its cost and not its effectiveness for research and teaching. Our deans and vice-chancellors are now appointed for their managerial skills and their research standing counts for little. New teaching courses can only be introduced after a long vetting process by committees. Laboratory classes are expensive and have been reduced or cut out entirely and experimental research is often discouraged for the same reason. Because of the increase in numbers and size of universities, the finance available for each researcher has became very limited and has to be bid for with research proposals that take up to a third of our time to prepare and have a very low success rate. To maximize one's chances of funding it became necessary to work in fashionable areas, which were not necessarily the best, or in those areas which were deemed by bureaucrats to be in the national interest. Many fashionable research areas have been pursued long after it was apparent that they were unproductive. In the leading article of a recent issue of the *Foundations of Physics* the effect of the fashionability of string theory has been noted:[173]

> 'String theory monopolizes the best jobs in the academic career market, so that young talented researchers have no choice but following the trends if they want to have a chance of an academic career.'

String theory is unfortunately in part a product of postmodernist universities.

Before the first superstring revolution in 1984 there were few popular mentions of string theory. It is fitting that one of the first such mentions in the *Scientific American* in 1975 was by John Schwarz, one of the founders of string theory, who described how hadrons could be modelled as strings.[174]

7.5.1. *The superstring revolutions*

The original string theory did not incorporate all the symmetries that are possible. Supersymmetry allows a common description of fermions, which have integer valued spin, and bosons, which have spin of half integer units and superstring theories which have ten spacetime dimensions give this symmetry. The first superstring revolution began in 1984 when Green and Schwarz realized such a theory might allow the incorporation of gravity with the other three fundamental forces in a single theory which hopefully might lead to a *Theory of Everything*.[175] In the next two years popular articles appeared in the *Scientific American* explaining supersymmetry and superstring theory and included one by

Michael Green.[176] It was found that there were five superstring theories with eleven space-time dimensions each of which predicts both quantum mechanics and gravitation. A conceptual problem with a string theory in more than three space dimensions is that the extra dimensions must be hidden in some way. The most accepted reason why the extra dimensions remain hidden is that they are curled up into spaces of the order of Planck's length. In his article Green looked at the problem from a different perspective, instead of considering how the extra space dimensions were curled up so that we could not see them, he considered the situation just after the Big Bang when all the space dimensions must have been curled up and asked why did three space dimensions unfold. In 2012 three physicists in Japan published a Monte Carlo simulation for a type IIB superstring model of ten spacetime dimensions where the Big Bang is modelled as a symmetry breaking event which is found to cause three dimensions to unfold and expand.[177]

One of the first popular books on superstring theory *Superstrings* was the book form of a broadcast on BBC Radio 3 in 1987.[178] In this book Paul Davies gave a short introduction to superstring theory which was followed by reports of interviews with some leading physicists including Edward Witten, who was to lead the second superstring revolution, as well as Richard Feynman, who thought that 'superstring stuff is crazy,' and Sheldon Glashow, who likened superstring theorists to medieval theologists.

The superstring theory was able to unify the strong, weak, and electromagnetic forces at high energy that would exist at the Big Bang, and created a euphoric belief that a theory of everything was about to be grasped. Stephen Hawking devoted the penultimate chapter in *A Brief History of Time* in 1988 to explaining how string theory could lead to a unification of physics. In 1992 Steven Weinberg published his lyrical *Dreams of a Final Theory*[179] which was

> 'about a great intellectual adventure, the search for the final laws of nature... [Al]though we do not known what the final laws might be ... we think we are beginning to catch glimpses of the outline of a final theory.'

Weinberg is a reductionist and he hopes 'to find a few simple general laws that would explain why nature is the way it is,' but as he explained he is 'a compromising reductionist.'[180] He also thinks that 'a final theory will rest on the principles of symmetry;' and that superstring theory, which has the necessary symmetries, is the strongest candidate for a final theory. Such a belief is common in string theorists but is not universal. Roger Penrose is one who has reservations both about the existence of a final theory or that superstring theory is a highly likely candidate if there is one.[181] Penrose was not convinced that 'symmetry is indeed fundamental'. Both Weinberg's book and Penrose's review are now more than twenty years old and in hindsight some of the issues raised have now been answered. A large part of Weinberg's agenda in writing *Dreams* was to support the funding of the Superconducting Super Collider (SSC) which had been planned for Ellis County in Texas. One of the objectives of the SSC was to discover the Higgs particle which was thought to hold the secret of the symmetry between photons and bosons at the extremely high temperature that prevailed just after the Big Bang and gets broken at lower temperatures. Another Nobel Laureate, Leon Lederman, also wrote a popular book in support of the SSC entitled *The God Particle* but, though he admitted 'a connection, of sorts to another book, a *much* older one,' he gave the first reason why he named the Higgs particle, the *God Particle*, 'the publisher wouldn't let us call it the Goddamn Particle'.[182] Penrose made the remark that:

> 'Perhaps the symmetry of the current electroweak theory is not fundamental, and there is no proper Higgs particle. I would myself hope that this is the case.'

because of the problems it entailed with spontaneous symmetry breaking by requiring inflation after the Big Bang. Of course we now know that the SSC was not built but that Higgs particle was found nevertheless with the Large Hadron Collider at CERN in 2012.[183]

The second superstring revolution began when Witten announced at the annual Strings Conference in 1995 that the five consistent superstring theories were not independent theories but limiting cases of a deeper more comprehensive eleven-dimensional theory which became known as the M-theory. At first M-theory was not directly connected with the vibrations of membranes and Witten stated that there was 'no evidence for membrane excitations' such as might be expected 'if a consistent membrane theory with eleven-dimensional supergravity as its low energy limit really does exist.'[184] However, though Witten at first linked M with magic, mystery, or matrix, he was soon converted to the idea of membranes and higher dimensional branes. It is very difficult to write popular accounts of the developments in string theory because the mathematics behind it is not just a detail but an integral part. Unfortunately Witten himself has not written any popular books on M-theory. The temptation in writing popular works on physics is sensationalism; Witten has not fallen to this temptation so in the popular articles that he has written you will find no mention of M-theory being a theory of everything neither does he not think that the end of physics is imminent and he wrote in regard to M-theory in 1998 that:[185]

'Physicists and mathematicians are likely to spend much of the next century to come to grips with this theory.'

As already mentioned Brian Greene's *The Elegant Universe* is universally acclaimed as the best popular book on string theory, though it is now more than seventeen years old it still gives the best insight and the only other popular book that will be mentioned in this section is Lisa Randall's *Warped Passages*[186] which is one of Weinberg's thirteen best popular science books of all times, though it has not received such acclamation as *The Elegant Universe*. Lisa Randall is one of the few notable women physicists and even fewer women popular postmodern physics writers. So many of the popular books on string theory go over the same topics with minor variations, Randall's book is attractive because it is mainly concerned with a topic that is glossed over in others and that is a closer examination of the possible forms of the extra dimensions to our familiar three dimensional world. Her research into extra dimensions is not limited to string theory but obviously has an application to it and she has a refreshingly frank view of the position of string theory writing that:

'String theory … is based solely on theoretical questions and ideas and isn't even completely mathematically formulated yet, so we can't yet be certain of its predictions. As for me, I'm an agnostic on this subject – I don't know what string theory will ultimately be or whether it will solve the questions of quantum mechanics and gravity it sets out to address. But string theory has been a rich resource for new ideas, some of which I have exploited in my own research on extra dimensions of space. These ideas exist independently of string theory, but string theory gives us a good reason to think some of their underlying assumptions could be right.'

The only slightly negative aspect of Randall's book is the inclusion of an introductory section to each chapter narrated by two characters Athena and Ike, which seem to me to show contempt for the intelligence of the reader, but these are easily skipped. Only Gamow's Mr Tompkins has been able to handle a surrealist approach to modern physics successfully.

7.5.2. *String theory atheists*

Many have an almost religious belief in string theory. Lisa Randall admits that she is an agnostic. The Nobel Laureate Sheldon Glashow is a confirmed string theory atheist who is concerned that[187]

'For the first time ever, it is possible to see how our noble search could come to an end, and how Faith could replace Science once more.'

However, the two most vocal atheists are Lee Smolin and Peter Woit who strongly object to superstring theory that, since it has made no predictions, cannot be falsified. Of course philosophy too has its fashions and Popper is not fashionable at the moment.

String theorists believe that their theory has the potential to be the theory of everything and realize Weinberg's dream. The Standard Model of quantum theory does not pretend to be a candidate for a theory of everything not simply because it only applies to the very small and does not contain gravity but also because it contains so many constants which cannot be deduced. For string theory to be a candidate for a theory of everything it cannot rely on arbitrarily assigned constants it has to be unique. However the hope of a unique string theory has receded and there is a vast landscape of theories that require a belief in a multiverse. Observations of supernova in the twentieth century have showed that the universe is expanding entailing a positive cosmological constant. String theories that give a positive cosmological constant to agree with the observation have been found in fact there are an embarrassingly large number, some 10^{500}, of them. Instead of a theory of everything we have a theory of anything. Smolin started his book with the five great problems and assessed that of these five only it has only succeeded in solving the third problem unification of particles and forces.

It is more than thirty years since the first superstring revolution and still there is not a consensus among physicists whether string theory can explain the physical world. Is physics drifting into metaphysics? No doubt time will tell.

7.6. Concluding Remarks

Except for singular events, such as the 1919 Eddington/Dyson verification of the general theory of relativity or the discovery of Higgs boson with the CERN LHC in 2012, the great theoretical achievements of modern physics do not seem to have generated as much general excitement, as did those of Newtonian physics. There seems to have been more inclusiveness in the involvement of society in physics in the eighteenth and nineteenth centuries. In the early nineteenth century working class men sought to obtain education in the mechanics' institutes, which initially had many courses on natural philosophy and astronomy, but by the end of the century clerks and shop assistants began to dominate the audiences and the sciences were largely replaced by the liberal arts. In the twentieth century the WEA was even quicker to abandon the sciences and by the twenty-first century there were very few courses in the sciences. However there is still a strong market for popular books on physics and cosmology; commentators might say that Stephen Hawking's *A Brief History of Time*, which has sold more than 10 million copies, is the most purchased and least read popular science book but if only one copy in ten was read that still is a tremendous number readers and the other 90% must have had enough interest to purchase it. It would appear that those interested in the liberal arts and ephemeral subjects are more likely to attend courses whereas those interested in the sciences still read books.

7.7. Notes

[1] Wells (1894).
[2] Bragg (1925).
[3]This quotation appeared on the front page of the first issue of the *Listener*, not as part of an editorial but as an unattributed quotation in an advertising insert giving the good wishes of the Amplion radio manufacturer for its loud speakers.
[4] The *New Zealand Listener*, which now covers television as well as radio, was launched in 1939.
[5] Polkinghorne (1984).
[6] Kragh and Nielsen (2013).
[7] The British Association changed its name in 2009 to the British Science Association (BSA) and its annual meeting is known as the British Science Festival.
[8] Rowbottom (1981).
[9] These notes on Mansbridge and the early WEA are taken from Jennings (2004).
[10] Campbell (1921).
[11] Chapter 6 of Imogen Clarke's doctoral thesis has an excellent account of the early days at the Science Museum (Clarke, 2011).
[12] Gouyon (2014).
[13] The parts of the Eastern Wing of the Science Museum were used before the official opening.
[14] Clarke (2011).
[15] See §1.4.2.1.
[16] Introduction to Morris (2010).
[17] Andrade and Huxley (1934).
[18] Hogben (1938).
[19] Laue (1950).
[20] Peierls (1955).
[21] Peierls was the first in Britain to estimate what was the critical mass of uranium U235 a stable isotope with a natural half life of over 700 million years. His answer one kilogram surprised him by being so small. Although Peierls' estimate was too small by a factor of fifty, it did show that the critical mass was small enough to be obtained in a bomb. Peierls joined the Manhattan Project in the USA in 1943.
[22] Feynman (2011). The Feynman lectures are now freely available on the Web.
[23] Phillips (2013).
[24] There are more manageable sized extracts are available: Feynman (2005a,b).
[25] Feynman (1965).
[26] Eddington (1928).
[27] Röntgen was the first to systematically study X-rays but he was not the first to have observed their effects.
[28] See §1.5.2.

[29] Atkinson (1897).
[30] *The Times*, 11 January 1897.
[31] Hooke (1665).
[32] Bragg (1914).
[33] Bragg and Bragg (1915).
[34] Soddy (1916).
[35] Reith was the foundation General Manger of the British Broadcasting Company from 1922-1927 and he became the first Director General of the British Broadcasting Corporation when it was formed in 1927.
[36] Friday (1974).
[37] Rutherford (1904a,b).
[38] Hammer (1903).
[39] Soddy (1903).
[40] Soddy (1904).
[41] Rutherford (1905).
[42] See §2.3.
[43] Holmes (1913).
[44] Rutherford and Geiger (1908a,b).
[45] Rutherford (1909).
[46] Soddy (1909).
[47] Wells (1914).
[48] Rutherford (1911).
[49] See 4.1.7.
[50] Heisenberg (1925); Born and Jordan (1925); Born *et al.* (1925).
[51] This section is partly based on the excellent paper by Kragh and Nielsen (2013).
[52] See §1.2.2.4.
[53] Kragh (2011).
[54] A summary of this address was published in *Science*. (Eve, 1914).
[55] See §1.2.2.4.
[56] Rutherford (1915).
[57] Stewart (1914).
[58] Campbell (1914).
[59] Sullivan (1923a).
[60] Russell (1923).
[61] Andrade (1923).
[62] Andrade (1927).
[63] Campbell (1923).
[64] Kramers and Holst (1923).
[65] Kragh and Nielsen (2013) give a detailed description of this important book.
[66] Cohen (1940).
[67] Heisenberg (1927).
[68] Kaempffert (1927).
[69] Eddington (1925).
[70] Heyl (1930).
[71] Heyl (1931).
[72] Darwin (1931).
[73] Navarro (2009).

[74] Bragg (1921).
[75] Zimmer (1936).
[76] Gamow chose the first three constants in Table 1.1, as the initials of Tompkins.
[77] The easiest of the Mr Tompkins stories to obtain is *Mr Tompkins in Paperback*, which is a combination of *Mr Tompkins in Wonderland* and *Mr Tompkins Explores the Atom.* (Gamow, 1965).
[78] Cohen (1954).
[79] de Broglie (1954).
[80] Cohen (1953).
[81] I have not been able to find anymore biographical details about this obviously very talented writer of popular books on physics.
[82] d'Abro (1939).
[83] Cohen (1940); Synge (1940).
[84] Hoffman (1947).
[85] See §6.2.5.
[86] Polkinghorne (1979).
[87] Feynman (1985a).
[88] Kaiser (2005).
[89] Feynman (1949).
[90] Davies and Brown (1986, 1988).
[91] Davies (1989).
[92] Einstein (1905a).
[93] Cunningham (1914).
[94] The most noteworthy early article on Einstein's special theory of relativity was published by Max Planck in the *Scientific American* in 1910.
[95] Summaries of both of these lectures were printed in the *Algemeen Handelsblad* though this was simply usual practice and does not indicate interest by the general public (van Besouw, 2013).
[96] Einstein did not give this result exactly in his 1905 paper but its approximation $(v/c)^2\,\Delta t/2$, which holds if v is small compared with the velocity of light.
[97] Langevin (1911).
[98] Dingle (1956).
[99] Chang (1993).
[100] Dingle (1922).
[101] Dingle (1939).
[102] Hafele and Keating (1972b).
[103] They used mainly flights on Boeing 707s but used the Concorde across the Atlantic.
[104] Einstein (1907).
[105] Hafele and Keating (1972a).
[106] Norton (2004).
[107] Wazeck (2010).
[108] Charivaria Punch Historical Archives, Vol. 162, 12th April 1922, p. 281.
[109] Anon.(1922).
[110] See §1.4.2.1.
[111] Einstein (1916b).
[112] Einstein (1922a).
[113] Pais (1994).
[114] See §1.2.1.3.
[115] Eddington (1920a).

[116] Equivalent today to about US$150,000.
[117] Bolton (1921a).
[118] Bolton (1921b).
[119] Cunningham (1922).
[120] Thirring (1921).
[121] Russell (2009).
[122] Newman (1950).
[123] Harrow (1926).
[124] Sullivan (1925).
[125] Bowler (2009).
[126] Milena Wazeck's book and a little known paper by Varadaraja Raman have been used in this section and are recommended for further information. Wazeck (2014), Raman (1972).
[127] Thomson (1919).
[128] Wazeck (2013).
[129] The title acknowledges permission to teach and supervise PhD students independently.
[130] Thomson (1904).
[131] Einstein (1922b). Einstein gave his lecture in German, it is the published translation given here.
[132] Eddington (2013).
[133] See §3.3.
[134] The term cosmologist is first recorded to have been used in the OED in1792 but that was for the author of Genesis in the Bible.
[135] Kragh (2007).
[136] Dingle (1924).
[137] Eddington (1920b).
[138] Jeans (1928).
[139] Jeans (1929).
[140] See §6.3.
[141] Jeans (1930).
[142] See §6.3.
[143] *The Times* 6th February and 18th May 1931.
[144] Punch 27th May 1931.
[145] Hubble (1937).
[146] Einstein featured on the cover of *Time* many times. Publication of *Time*, and its practice of having a well known person on its cover, did not start until 1923. The first time Einstein was on the cover was in 1929.
[147] Jeans (1934).
[148] Helge Kragh's *Cosmology and Controversy* Kragh (1996) gives an excellent account of the controversy as to whether the Steady State or the Big Bang theory was correct.
[149] Bondi and Gold (1948).
[150] Hoyle (1948).
[151] Tryon (1973).
[152] Rosen (1994).
[153] Banerjee and Sen (1997).
[154] Milton (2005).
[155] Hoyle (1950).
[156] Hoyle (1955).
[157] See §6.3.

[158] Bondi (1952).
[159] Jaki (1967).
[160] Hoyle's italics.
[161] Wesson (1986).
[162] Bondi *et al.* (1960).
[163] Lyttleton and Bondi (1959).
[164] Kragh (1997).
[165] Gamow (1965).
[166] The term was used by Roger Bacon in the thirteenth century. (Kragh, 2007).
[167] Gamow (1952).
[168] Smolin (2006).
[169] Greene (1999).
[170] Weinberg 3rd April 2015, *The Guardian*, London.
[171] Woit (2006). The title of Woit's book comes from a quote from Wolfgang Pauli 'It is not even wrong.' Pauli was a blunt speaker and Peierls (1960) in his Pauli's memoir of recounted that Pauli had uttered the phrase when "a friend showed him the paper of a young physicist which he suspected was not of great value.'
[172] Penrose (2016).
[173] De Haro *et al.* (2013).
[174] Schwartz (1975).
[175] Green and Schwarz (1984).
[176] Green (1986).
[177] Kim *et al.* (2012).
[178] Davies and Brown (1988).
[179] Weinberg (1992).
[180] Weinberg does not believe that there necessarily has to be a final theory and people like Popper (1983) who do not believe in the 'idea of an ultimate explanation' may be right.
[181] Penrose (1993).
[182] Lederman (1993).
[183] See §1.5.6.
[184] Witten (1995).
[185] Witten (1998).
[186] Randall (2005).
[187] Glashow (1986).

Chapter 8

Modernism and Postmodernism

Modernism in the arts has been mirrored by similar developments in physics and they have both represented nature as it had not been perceived before. Postmodernism, which developed in the second half of the twentieth century, has many different interpretations but I believe again has similarities to the developments in physics. It is difficult to define postmodernism concisely but a good introduction to modernism and its development into postmodernism has been given by Pericles Lewis in *The Cambridge Introduction to Modernism.*[1] The flavour of postmodernism can be tasted from an example given by Lewis: the film of Akira Kurosawa (1910–1998), *Rashamon*. In this period film, set in the eighth century, a samurai is killed in the forest and a bandit arrested for his murder. We see and hear four different versions of the event from the wife, bandit, woodcutter, and, through a medium, the samurai himself. This film, which is one of the greatest of all time, certainly enthralled me when I first saw it in the 1950s and it still enthrals me. It is the general consensus that indeterminacy is a mark of postmodernism. According to Dick Higgins, artists around 1958 started asking the ontological postmodernistic questions: [2]

> 'Which world is this?
> What is to be done in it?
> Which of my selves is to do it?'

The two cultures debate came to head in the twentieth century, but it had its beginnings in the nineteenth century and the intense publicity given to relativity caused it to be revived in the 1920s well before the better known Snow-Leavis affair. In this case it was a minor spat between the science popularizer J.W.N. Sullivan and the literary writer John Middleton Murry (1889–1957). The exchange was initiated by a letter Sullivan wrote Murry in 1922 in which he asserted that Einstein's relativity was bringing about a new Renaissance.[3] Murry responded with an article in *The Times* on *Literature and Science* which began with:[4]

'A friend wrote to me the other day that "the sceptre has passed from literature to science"'

Sullivan replied with an article *Science and Literature*.[5] There were many who realized that there was a symbiotic relationship between science and the arts even if both sides did not fully understand each other. However during the twentieth century, physics and science in general gained a position of cultural ascendancy unfortunately at the expense of the arts, though this has since been eroded by the anti-culture of technology and business. There was a slight lessening of the suspicion between the arts and the sciences towards the end of the twentieth century. The *Society for Literature and Science* was founded in 1985 to foster the perceived need 'to develop a grand unified theory of literature and science' and was later broadened to become the *Society for Literature, Science, and the Arts*.[6] Although articles in its journal *Configurations* appear to consist of mainly interdisciplinary studies rather than examples of influence across the two aspects of culture, its heart is in the right place. Another sign of rapprochement between science and the arts has been the election of a few science writers, such as Paul Davies and John Gribbin, to the Royal Society of Literature. On the science side, physics courses have been provided for non-physicists students at universities, often under the alliterative generic title of Physics for Poets with at least two books published with that title. *Physics for Poets* by Robert March, a Professor of Physics and Liberal Studies at the University Wisconsin, is a very successful example of such a book and it has had five editions since it was originally published in 1970.[7] Leon Lederman and Christopher Hill have also used the generic title for their *Quantum Physics for Poets*, published in 2011.[8] Unfortunately despite books such as these, enrolments in science courses by liberal arts students have decreased in recent times. Shirley Tilghman, the President of Princeton University, spoke of the problem of declining participation in science courses by liberal arts students in an address to the Council of Independent Colleges in 2010.[9] This decline is not due to the quality of science teaching in liberal arts colleges because sixteen Nobel Laureates out of seventy during the last twenty years in America had attended liberal arts colleges, as Tilghman noted. Princeton has long done its bit to help bridge the cultural divide and in the early 1940s its graduate college organized speakers on a wide range of topics on Wednesdays under the auspices of the Dean of the Graduate School, Luther Eisenhart (1876–1965) during the time that Richard Feynman was a student. One such talk on poetry illustrates the difficulties faced when determining whether there has been an influence across the divide. At the talk Eisenhart asked Feynman to compare poetry with theoretical physics Feynman got up and said:[10]

> 'Yes, it's very closely related. In theoretical physics, the analog of the word is the mathematical formula, the analog of the structure of the poem is the interrelationship of the theoretical bling-bling with the so-and-so.'

He continued making 'perfect analogies' between physics and poetry for everything but then said:

> 'It seems to me that no matter what you say about poetry, I could find a way of making up an analog with any subject, just as I did for theoretical physics. I don't consider such analogs meaningful.'

There is a definite danger of seeing influences in everything, especially as some people in the humanities have only a vague idea of modern physics and often know little more than the names of the protagonists though many physicists are just as bad in their appreciation of the arts. Einstein's *Theory of Relativity* is particularly prone to misinterpretation in fact he wished later that he had called it the *Theory of Invariance*. Relativity was and still is often misunderstood as relativism so much so that Bertrand Russell felt it necessary in 1925 to make certain that there was no misunderstanding in his *ABC of Relativity* writing:[11]

> 'The 'subjectivity' concerned in the theory of relativity is a physical subjectivity, which would exist equally if there were no such things as minds or senses in the world. Moreover, it is a strictly limited subjectivity. The theory does not say that everything is relative; on the contrary, it gives a technique for distinguishing what is relative from what belongs to a physical occurrence in its own right.'

People grossly overestimated the understanding of relativity in the 1920s and still do. In *Time and Western Man* published in 1920 Wyndham Lewis (1882–1959) the modernist artist and writer was probably being satirical when he wrote:[12]

> 'Within a few years of the arrival of Einstein upon the European scene, the layman, I suppose, knows more about Relativity physics than any layman has ever known about the newtonian cosmology, either during Newton's lifetime or since.'

as he made clear from his remarks further on in his book:

> 'The physical or philosophical theory in the ascendant at any moment is humbly and reverently picked up, in an abridged, and usually meaningless form, by the majority of people.... Then they will say in conversation, "It appears that there is no such thing as time"; or "Everything is relative, Einstein says. I always thought it was."'

In literature Einstein and the theory of relativity are also often seen as synonymous with modern physics probably because its foundations were laid by one man and they burst into popular view in 1919 promising a new beginning after the war to end all wars. Quantum mechanics was developed relatively slowly by many and did not have the same publicity in the 1920s. Names matter for publicity and when the Higgs boson was called the *God Particle* by Lederman it caught the public's attention, leading confusion similar to that caused by relativity.[13] In what Sean Carroll, a cosmologist at Caltech, has called 'the worst paragraph I've ever read about the Higgs boson' CBS news described it thus:[14]

> 'The Higgs boson is often called "the God particle" because it's said to be what caused the "Big Bang" that created our universe many years ago. The nickname caught on so quickly (even though scientists and clergy alike do not care for it) partly because it's a great explanation of what it's supposed to do — the Higgs boson is what joins everything and gives it matter.'

People have always been interested in the stars and their mention certainty does not necessarily show an influence of astronomy and only if the influence can be shown to stem from the new cosmology has it anything to do with modern physics. Cigoli's depiction of the craters in the moon, which he saw through Galileo's telescope, on which the Madonna rests her foot in the fresco on the dome of Santa Maria Maggiore in Rome (shown in Fig. 5.4), certainly is a prime example of the influence of astronomy in the seventeenth century but the mention of a telescope in the twentieth century does not necessarily imply any influence of astronomy. Modernity matters. In the case where someone other than the creator of the work pontificates on apparent analogies it is important, and often difficult, to determine whether the analogies are true or imagined. However if it is the creator of the work who mentions some aspect of physics or says that he was influenced, it is less important whether it is absolutely true since here we have someone who believes that different aspects of culture influence one another even if he is wrong in the detail. There is such influence and many modern physicists have been ready to see it. Feynman, for example, 'recognized that the creative arts, music, literature, and science all participate in the common endeavour of telling stories about the universe,'[6] but it is necessarily to find the chapter and verse of the influence.

The imaginative arts do not have a monopoly on imagination as the poet and essayist Ralph Waldo Emerson (1803–1882) recognized in his essay *Poetry and Imagination* written in 1872 where he wrote:[15]

> 'Science does not know its debt to imagination. Goethe did not believe that a great naturalist could exist without this faculty.'

Einstein fully recognized the importance of imagination, when George Viereck[16] (1884–1962) asked him, during an interview in 1929, whether he made his discoveries through intuition or inspiration and Einstein replied:[17]

> 'Both. I sometimes feel I am right, but do not know it. When two expeditions of scientists went to test my theory I was convinced they would confirm my theory. I wasn't surprised when the results confirmed my intuition, but I would have been surprised had I been wrong. I'm enough of an artist to draw freely on my imagination, which I think is more important than knowledge. Knowledge is limited. Imagination encircles the world.'

In modern physics imagination has been even more important than during the classical period because its reality is not that of common experience and often can only be portrayed mathematically. Imagination in the arts and physics may initially be similarly unlimited but in physics it must give insights into the physical world that are testable by more than aesthetics.

Physics was at the heart of modernity but was only one of the competing influences on the modernist movement. Two world wars and the totalitarian politics of communism and fascism took their toll either directly as in the Soviet Union or Nazi Germany where art had to serve the regime, or indirectly where the younger writers who reached adulthood after the First World War, such as George Orwell (1903–1950) and W.H. Auden (1907–1973) became politicized. The dissemination of modern physics in Britain and America was interrupted by the First World War and the only widely popularized parts before the war were X-rays and radioactive rays. Bohr's 1913 model of the atom and Einstein's theory of relativity were not widely known until well after the end of the war and any claims that these parts of modern physics influenced the imaginative arts before then need to be scrutinized closely.

8.1. Literature

The term modernism was first applied to early twentieth century literature in 1908 by Rolfe Arnold Scott-James (1878–1959) a literary critic who was a close friend of Wyndham Lewis.[18] In Britain the literary figures mainly associated with modernism in the early twentieth century were two American expatriates Ezra Pound (1885–1972), whose 1928 injunction to 'make it new' is a key element in modernism, and T. S. Eliot (1888–1965), an Irishman James Joyce (1882–1941), the English Virginia Woolf (1882–1941) and David Herbert Lawrence (1885–1930), more recognizable as D. H. Lawrence, all of whom had connections with others of this generation. All these writers reached their twenties well before the First World War, the futility of which had a large influence on the modernist movement. Of the same generation William Carlos Williams (1883–1963), a paediatrician by day and a poet by night, was one of the founders of modernistic literature in the US.

Einstein's first visit to the US in the spring of 1921 was enthusiastically received and acted as a catalyst for modernism. William Carlos Williams was moved to write a modernistic paean *St. Francis Einstein of the Daffodils* which in its final stanza refers to Einstein's four dimensions of spacetime:

> It is Einstein
> out of complicated mathematics
> among the daffodils —
> spring winds blowing
> four ways, hot and cold,
> shaking the flowers!

The American poet and anthologist Louis Untermeyer (1885–1977) parodied Williams paean to Einstein in a mock anthology of poems supposedly written by well known poets in his *Rhyme and Relativity* written shortly afterwards, parts of which were first published in *Vanity Fair* and the avant-garde magazine, *Broom*.[19] There are seventeen poems reflecting, as stated in the introduction, 'Relative Motion, Substitutes for Gravitation, The Michelson-Morley Experiments, Time as a Fourth Dimension, Deflected Light-Rays, non-Euclidean Warps in Space and The Shifting of Spectral Lines toward the Red.' Untermeyer definitely had a good knowledge of the physics of the time despite having not completed high school, much better than that of most modern critics as Elizabeth Cornell has recently pointed out.[20] She picks out, *Einstein among the Coffee-Cups*, the parody of T.S. Eliot's style, as one with some of the most interesting allusions to the physics of the time. The title of the parody comes from the well known line 'I have measured out my life with coffee spoons' in Eliot's *The Love Song of J. Alfred Prufrock*. However, though Eliot up to the early 1920s occasionally used a scientific vocabulary and gave considerable thought to the relationship of science to poetry, he did not make use of scientific images in his poetry.[21] Neither did Ezra Pound, a prime driver of modernism, use scientific images, though he favoured the language of mathematics, physics and chemistry. To Pound science represented modernity. In the introduction to his 1910 book on literary criticism *Spirit of Romance*, Pound explained that, though 'there are a number of sciences connected with the study of literature, ... the Art ... is not, and never will be, a science.'[22] Pound believed that poetry should not be romantic but should reflect ordinary life:[23] 'Science is unpoetic only to minds jaundiced with sentiment and romanticism.'

It is difficult to write succinctly about the influence of modern physics on all modernist writers and here only those whose involvement is easy to show have their own individual section. Some writers such as Samuel Beckett (1906–1989) have so many layers in their work that a simple study of one of the layers, the influence of modern physics, becomes misleading and only a comprehensive study of their work, which cannot be undertaken here, really makes real sense. In many cases the modernist writer's love of puns, allusions and ambiguity make the detection of real influence from modern physics extremely difficult. For example Beckett's *Murphy*, first published in 1930, has undoubtedly references to modern physics but some apparent references are simply the use of words associated with modern physics to make puns such as: 'The horse leech's daughter is a closed system. Her quantum of wantum cannot vary' is an allusion to thermodynamics and the fact that in a closed system the lust of the horse leech's daughter, like energy cannot vary. The quantum of wantum is simply a good pun bringing in an allusion to modern physics but really having no significance. The reader can get some idea of Beckett's use of modern physics from Katherine Ebury's *Modernism and Cosmology*.[24]

The distinction between modern and postmodern literature that occurred around the middle of the twentieth century, is not clear cut and not very important since many writers span the two. To me Beckett's play *Waiting for Godot* originally written in French in 1949 and given its first performance in English in 1956 represented a change I could appreciate. Beckett recognized the play as postmodern himself and gave a succinct description of the difference between modernist and postmodernist literature in an interview with a reporter for *The New York Times*, when it was on Broadway in 1956.[25] He said that:

> 'Joyce was a superb manipulator of material.... He was making words do the absolute maximum of work. The kind of work I do is one in which I'm not master of my material. The more Joyce knew the more he could. He's tending toward omniscience and omnipotence as an artist. I'm working with impotence, ignorance. I don't think impotence has been exploited in the past.'

One would have thought that postmodernist writers, if they made used of physics, would at least use some postmodernist physics but that is usually not the case. Many modernists on the other hand quickly incorporated elements of modern physics into their works even if they all did not completely understand it and had been making it new long before Ezra Pound gave his exhortation. The postmodernist writer Thomas Pynchon has incorporated many diverse themes into his novels including physics but not postmodernist physics and has mainly been concerned with thermodynamics of the nineteenth century. Kinsley Amis' *Time's Arrow* is similarly concerned with entropy. It is not that postmodernists must include postmodernist physics if they include physics in their works but one would have thought that postmodernist physics would sit more comfortably with their ethos and have had as much or more influence than modern physics did on the modernists.

8.1.1. *James Joyce* (1882–1941)

James Joyce was one of the most influential writers of the early twentieth century and the writer Harold Nicholson (1886–1968) gave an apt description of him as the 'Einstein of English fiction.'[26] Unlike Eliot, Joyce did not have a significant knowledge of physics, classical or modern, but he still appreciated its cultural importance. After school Joyce entered University College, Dublin in 1899, reading modern languages but was disappointed by the narrowness of the syllabus and, not applying himself, he only obtained a pass degree in 1902. In that same year Joyce enrolled in the Cecilia Street Medical School, Dublin and attended a few lectures in biology, chemistry, and physics, but finding these subjects disagreeable, he quickly left.[27] Surprisingly Joyce then applied for admission to the faculty of medicine at the University of Paris but he was not accepted by the university. That was an end to any formal studies in the sciences by him. In a letter to his patroness Harriet Weaver (1876–1961), Joyce wrote about his problem with understanding chemistry, 'I never could learn it or understand in the least what it is about.'[28] His own brother Stanislaus (1884–1955) commented 'whatever method there is in Jim's life is highly unscientific.'[29] While Joyce was in Copenhagen in 1936 the Danish poet and writer Tom Kristensen (1893–1974) asked him if he believed in Giambattista Vico's (1668–1744) *Scienza Nuova*, which was not about science in the modern sense, as both Kristensen and Joyce knew, but about knowledge of the history of the ideas, the customs, and the deeds of mankind.[27] However, Joyce deliberately interpreted the question as "did he believe in science?" and replied 'I don't believe in any science' and added that 'my imagination grows when I read Vico.'[30] One must conclude that Joyce had little aptitude for physics or science. I agree with Andrzej Duszenko that 'there is no evidence that Joyce read systematically on the subject of new developments in physics;'[31] Jeffrey Drouin has provided a comprehensive table of 'items that Joyce read or might have read' that contains not one article or book on science.[32]

Joyce did not fully develop his postmodernist style until *Finnegans Wake* published in 1939, which fully embraced indeterminacy in its language which was to become one of the elements of postmodernism.[1] His first book *Dubliners* (1914) was a collection of short stories that heralded a new modernistic style. Joyce introduced his alter ego, Stephen Dedalus, and a stream of consciousness in *Portrait of an Artist as a Young Man* (1916), the complex mosaic of his writing increased in *Ulysses* (1922) and in *Finnegans Wake*, he thought that he could not 'use words in their ordinary connections' remarking to the Swiss sculptor August Suter (1887–1965) that 'Je suis au bout de I'anglais' (I am at the end of English).[27] As Samuel Beckett wrote:[33]

> '[T]his stuff is not written in English. It is not written at all. It is not to be read – or rather it is not only to be read. It is to be looked at and listened to. His writing is not *about* something; it is *that something itself.*'

Finnegans Wake borders on the impenetrable. Einstein is difficult, for most other than physicists and mathematicians, but he is difficult through necessity because only the language of mathematics can illuminate modern physical nature; Joyce is difficult because he needs his own special language to express human nature. Stuart Gilbert (1883–1969) began an appraisal of *Work in Progress*, Joyce's name for *Finnegans Wake* during the seventeen years he took to complete it, with a quotation from G.K. Chesterton:[34]

> 'Great poets are obscure for two opposite reasons; now, because they are talking about something too large for anyone to understand and now, again, because they are talking about something too small for anyone to see.'

Human nature moulds culture that in turn defines humanity. Joyce wanted to portray all aspects of culture and in *Finnegans Wake* he made use of modern physics. However some Joycean critics imply an understanding of modern physics and its relationship to classical physics that is equal to that of Eddington or Jeans, I think that Joyce himself would have been very amused. Joyce made use of modern physics in *Finnegans Wake*, just as he did with his language incorporating neologisms, portmanteau words, puns and many different languages into his work. Although *Finnegans Wake* abounds with references to relativity and quantum mechanics, physics was used by Joyce in his own way for its effect not its meaning, which it is doubtful that he really understood, as shown by Duszenko and others.[35] In *Finnegans Wake* Joyce represented Einstein as overthrowing Newtonian physics using the symbol of the apple (pomme) to link the fall of Newtonian physics to the fall of Adam and Eve (fammy of levity) from grace writing:[36]

> 'For then was the age when hoops ran high. Of a noarch and a chopwife; of a pomme full grave and a fammy of levity; or of golden youths that wanted gelding; or of what the mischievmiss made a man do.'

and a few lines on 'Het wis if ee newt [Newton]' is Joyce's clue to the meaning.[31] However, this analogy is not entirely true because Einstein did not cause a fall from grace of Newtonian physics which is still alive and well. Newtonian physics is based on Galilean relativity[37] which is the limit of Einstein's relativity when velocities are small compared with the speed of light and is still true. Joyce reinforced his criticism of Newton and the *Principia* further on in *Finnegans Wake* writing: 'heaving alljawbreakical expressions out of old Sare Isaac's universal of specious aristmystic unsaid.'[31]

Although *Ulysses* was completed after the euphoric announcement of the vindication of Einstein's general theory of relativity in 1919, I do not believe that there is any evidence that Joyce made use of modern physics in it apart from a few mentions of X-rays and, though Joyce was not necessarily limited by historical accuracy, the action in *Ulysses* was set by him during Bloomsday on the 16th June 1904 and to include either relativity or quantum mechanics would have been anachronistic. The Ithaca episode was greatly influenced by science as Joyce acknowledged but the physics is classical. In a letter to his artist friend Frank Budgen (1882–1971) in February 1921, Joyce wrote that he was writing the Ithaca episode:[27]

> 'in the form of a mathematical catechism. All events are resolved into their cosmic physical, psychical &c equivalents, e.g. Bloom jumping down the area, drawing water from the tap, the micturition in the garden, the cone of incense, lighted candle and statue so that not only will the reader know everything and know it in the baldest coldest way, but Bloom and Stephen thereby become heavenly bodies, wanderers like the stars at which they gaze.'

It is significant that of the four sciences named: mathematics, astronomy (cosmic), physics and psychology it is mathematics that is the most prominent in Joyce's *Notesheets* for *Ulysses* in what has been called 'mathematical doodling'.[38] To confound our interpretation of what Joyce really thought, he delighted in confusing his readers and in the *Ithaca* episode one of the ways he did it was by intentional errors.[39] As Joyce wrote in Episode 9:[40]

> 'Bosh! Stephen said rudely. A man of genius makes no mistakes. His errors are volitional and are the portals of discovery.'

And in the *Eumaeus* episode Bloom is portrayed as correcting an error he makes:[40]

'[It] is one thing for instance to invent those rays Rontgen did or the telescope like Edison, though I believe it was before his time Galileo was the man, I mean,'

However, when in the *Ithaca* episode he makes two errors in examples concerning heat we have no means of knowing whether these are both deliberate errors. One of the errors is quite minor: the calculation of the thermal units 'needed to raise 1 pound of water from 50 degrees to 212 degrees Fahrenheit,' the second error is more fundamental. The statement that '[t]he cold of interstellar space, [is] thousands of degrees below freezing point or the absolute zero of Fahrenheit, Centigrade or Reaumur,' if not a deliberate error, reveals a limited knowledge of physics. However, *Ithaca* does contain many quite detailed references to stars and their constellations, which are correct. Drouin has argued that the *Wandering Rocks* and *Ithaca* episodes show evidence of Joyce's knowledge of the theory of relativity that he probably gained from reading Dora Marsden's (1882–1960) series of articles on *The Science of Signs* in *The Egoist*. Joyce almost certainly read at least some of these articles but they are on philosophy and make no mention of the ideas of relativity theory which were not well known at the time.[41] Drouin argues that Marsden learnt of the theory of relativity from her former professor, Samuel Alexander (1859–1938), at Owens College, Manchester. Alexander was one who knew of Einstein's work before most and wrote *Space, Time, and Deity* based on his Gifford Lectures at Glasgow during 1916–1918 where he mentioned it, but his was a work of philosophy and he admitted that as far as the theory of relativity was concerned he was a 'layman' and wrote:[42]

'Now I understand the essential spirit of the doctrine to be that Space and Time are not independent of each other but united in the one four-dimensional world, and with this the result of our empirical or metaphysical inquiry is in agreement.'

As far as Marsden's rambling *The Science of Signs*, I agree with her biographer Les Garner:[43]

'If the "British Public" struggled with Joyce's prose they had an almost impossible task in unraveling Dora's own philosophical ramblings in *The Egoist*. In particular her [writing] tended to be based on unsubstantiated assertion. It lacked realism and was expressed in inexorable prose.'

In the *Notesheets* for *Ithaca* there is mention of two nineteenth century mathematicians Nicholai Lobaschevsky[44] (1792–1856) and Bernhard Riemann (1826-1866) who developed non-Euclidean geometry. Einstein referred to Lobaschevsky and used Riemann geometry in his general theory of relativity but in the nineteenth century their geometry was purely mathematical and not specifically connected to physics. Although Joyce did not make use of these notes in *Ithaca* and it looks like he was perhaps contemplating contrasting Milly's straight Euclidean spine with perhaps Reuben J. Dobbs' curved Lobaschewsky/Riemann one but the only mention of curvature of the spine in *Ulysses* is:[40]

> 'Stand up straight, girl [Dilly Dedalus], he said. You'll get curvature of the spine.'

8.1.2. *Virginia Woolf* (1882–1941)

The Bloomsbury Group was important for the exchange of ideas among the modernists. The Queen of this group was Virginia Woolf one of the foremost of the modernists. The Group crystallized around the time of Virginia's marriage to Leonard Woolf[45] (1890–1969) in 1912 and had close links to Cambridge University and the Cambridge Conversazione Society better known as The Apostles because there were twelve founding members in 1820. Leonard Woolf was an apostle as were the philosophers Bertrand Russell and George Edward Moore (1873–1958) who, though not in the inner circle, had close connections with the Bloomsbury Group.

Virginia Woolf was educated at home by her farther Leslie Stephen (1832–1904), who was ambivalent about women's education, and she did not go to university[46] like her brothers Thoby (1880–1906) and Adrian (1993–1948) who both went to Trinity College, Cambridge.[47] Virginia was self-educated apart from a few classes in Greek at King's College, London and had no training in physics. However, Virginia had a lifetime fascination with telescopes and astronomy, she read the popular works of Jeans and Eddington and George Thomson's[48] (1892–1975) popular book *The Atom*, was in the Woolfes' library, and its is also most likely that Virginia read Bertrand Russell's *ABC of Atoms* and *ABC of Relativity*.[49]

Gillian Beer is a sound commentator on the indubitable influence of modern physics on Virginia Woolf but, does not see the influence of physics in everything, as some people have done since, such as Elizabeth Nielsen whose extravagant claims Woolf herself rejected in 1938.[50] Woolf saw modernist literature as corresponding to the modern developments in philosophy and physics one of which was the realization that reality was insubstantial, as Eddington discussed in his introduction to *The Nature of Physical World*.[51] Woolf even coined her own word in 1923 for the process of reducing the substance of reality *insubstantise*, which Arnold Bennett (1867–1931) had accused her of doing in her novel *Jacob's Room*, There is, of course, a difference in the meanings of the reality that Woolf portrayed in *Jacob's Room* and the physical reality that Eddington was writing about and there is certainly no direct connection between the two since *The Nature of Physical World* was published later in 1928. However there may be an echo of Eddington's introduction to *The Nature of Physical World* in *The Waves* published in 1931.[52] In *The Waves*, Woolf's most experimental novel, Bernard says:

> 'I begin to doubt the fixity of tables, the reality of here and now, to tap my knuckles smartly upon the edges of apparently solid objects and say, "Are you hard?"'

As Beer pointed out, Eddington poetically summed up reality as like 'a child which cannot survive without its nurse illusion.'[53] In her novels Woolf was trying to strip illusion from reality. There does seem to have been a shift in Woolf's approach to reality between *To the Lighthouse*, published in 1927 before the publication of *The Nature of Physical World*, where she concentrated on the philosophical aspects, and *The Waves* where she was interested in physical reality.[54] In *To the Lighthouse*, Lily Briscoe asks Andrew Ramsay 'what his father's books were about' and receives the reply 'Subject and object and the nature of reality.' Lily does not understand so Andrew gives a Berkeleyan idealist's response: 'Think of a kitchen table when you're not there.' George Moore and Bertrand Russell rejected the modern idealism that was current in Cambridge at the turn of the century and in 1903 Moore had written an essay refuting idealism.[55] In this essay Moore discussed the chain argument: if A is B, B is C, and C is D, it follows that A is D, pointing out that this argument is only as strong as its weakest link and if any link in the chain is false, the argument fails. Moore visited Virginia and Leonard Woolf in 1920 and discussed Berkeley's philosophy with them but Virginia confessed to her diary that she did not follow his argument.[56] As Lorraine Sim pointed out in *To the Lighthouse*, the musings of Mr Ramsay that 'his splendid mind' was 'like the alphabet … ranged in twenty-six letters all in order,' has a rhythm that is not unlike that of the chain argument of Moore and, since it follows quite closely on Andrew's Berkeleyan response, it may represent Woolf's musing on idealistic reality.[57]

Whereas Joyce used modern physics for effect, Woolf made use of its meaning. Even when the mention of modern physics occurs in a throwaway line, such as in *Mrs Dalloway* where Mr Bentley thought that the aeroplane shooting away in the distance was a symbol of man's determination 'to get outside his body, beyond his house, by means of thought, Einstein, speculation, mathematics, the Mendelian theory,' it has meaning as well as effect.[58]

8.1.3. *D. H. Lawrence* (1885–1930)

Lawrence differed from the other modernists discussed above in that he came from a working class family; his father went down a coalmine in the Nottingham district when he was only ten years old, but his mother came from a distressed lower middle class family.[59] After attending Nottingham High School and a few months working as a clerk in a factory, Lawrence became a pupil teacher in Eastwood and from 1906 to 1908 he attended the Day Training College at University College Nottingham to qualify as a teacher. He then taught at the Davidson Road elementary school in Croydon, South London until he quit teaching in 1911 for a career in writing. Lawrence was not totally ignorant of physics and science generally since a testimonial from the Davidson Road School referred to him as having 'to a great extent influenced the science teaching of the whole school.'[60] Lawrence's interest in modern physics seems to have begun about the time in 1921 when he wrote to his friend Samuel Koteliansky (1880–1955), a Russian-born translator and publishing manager who Lawrence had first met in 1914 and affectionately called Kot, asking him to 'lend me or send [to] me a simple book on Einstein's Relativity.'[61] A few days later Lawrence received his book on relativity, its title is not revealed but it has been suggested that it was the 1920 English translation of Einstein's own popular book: *Relativity: The Special and General Theory* [60] and Lawrence's letter to Koteliansky, on the day after he received the book, certainly implies that its author was Einstein. What Lawrence disliked about Newtonian physics was its mechanical action and complete separation of subject and object. Lawrence's 1915 novel *The Rainbow*, holds out some promise of a new beginning for humanity at the rainbow's end which stretches four generations from old Tom Brangwen, and the evils of the industrial revolution, to his granddaughter Ursula: [62]

> [who] 'saw in the rainbow the earth's new architecture, the old, brittle corruption of houses and factories swept away, the world built up in a living fabric of Truth, fitting to the over-arching heaven.'

As Michael Whitworth has noted, Lawrence showed his distaste for a mechanical view of nature that marked the industrial revolution.[21] In the novel, the woman physicist Dr Frankstone expresses the view that there is no 'special order of life,' which she sees as no 'different in kind and distinct from everything else in the universe,' that is mechanical. Even more explicit are the thoughts of Winifred Inger who:

'worshipped the impure abstraction, the mechanisms of matter. There, there, in the machine, in service of the machine, was she free from the clog and degradation of human feeling. There, in the monstrous mechanism ... held all matter, living or dead, in its service.'

Although in 1915 I doubt that Lawrence was aware of the new physics that was soon to burst in the newspapers, he would have seen his concluding passage to *The Rainbow* describing Ursula's vision of 'the earth's new architecture' a fitting announcement of it.

After reading the book on relativity that Koteliansky sent him and finding that it fitted with his philosophy, Lawrence wrote the long essay *Fantasia of the Unconscious* making use of it.[63] Lawrence was interested in what he called *subjective science*, which he saw as being 'established on data of living experience and of sure intuition' and really had nothing to do with science. Lawrence was selective in his approach to science and only accepted what appeared to agree with his own sure intuition. Thus he was not really influenced by relativity; his quick assessment in the letter he wrote to Koteliansky the day after receiving the book:[61]

'Einstein isn't so metaphysically marvellous, but I like him for taking out the pin which fixed down our fluttering little physical universe.'

indicates that Lawrence liked relativity because he thought that it removed mechanical action. In *Fantasia* Lawrence repeats the idea of removing the pin holding down the universe twice. In some statements about relativity it appears that Lawrence at least partially understood the significance of relativity:

'Mr. Einstein's Theory of Relativity does not supersede the Newtonian Law of Gravitation or of Inertia. It only says, "Beware! The Law of Inertia is not the simple ideal proposition you would like to make of it. It is a vast complexity. Gravitation is not one elemental uncouth force. It is a strange, infinitely complex, subtle aggregate of forces." And yet, however much it may waggle, a stone does fall to earth if you drop it.'

However Lawrence also confused relativity with relativism writing:

'I feel inclined to Relativity myself. I think there is no one absolute principle in the universe. I think everything is relative.'

I end this section on Lawrence with a short poem *Relativity* taken from his *Pansies* published in 1929.

I like relativity and quantum theories
because I don't understand them
and they make me feel as if space shifted about

like a swan that can't settle,
refusing to sit still and be measured;
and as if the atom were an impulsive thing
always changing its mind.

In this poem Lawrence demonstrated that, even if as he said he did not fully understand relativity and quantum theory, he could convey their essence poetically.

8.1.4. *Aldous Huxley* (1894–1963)

Aldous Huxley had both science and the arts in his DNA. His grandfather on his father's side was Thomas Huxley and his granduncle on his mother's side was Matthew Arnold and his great grandfather was Thomas Arnold, the famous headmaster of Rugby School. Aldous Huxley went to Balliol College, Oxford and read English language and literature, his father Leonard Huxley (1860–1933), a writer and biographer, had also gone to Balliol College and read classical moderations and *literae humaniores*.[64] However Aldous' elder brother Julian (1887–1975) and his younger half-brother, the Nobel Laureate Andrew Huxley (1917–2012), both studied and worked in the life sciences.[65] Thus in the Huxleys at least there was a spectrum of culture and it is not surprising that Aldous Huxley felt compelled to enter into the two cultures debate with *Literature and Science*[66] However Aldous Huxley's novels show how science can enhance literature in practice better than his *Literature and Science*. It is difficult to find examples of literature enhancing science. It is doubtful that literature can directly influence science but it can influence the way science is presented particularly in popular presentations. Two such presentations of Eddington have already been discussed in Chapter 7. In a review of *The Nature of the Physical World* and *Science and the Unseen World*, Evander Bradley McGilvary (1864–1953) wrote:[67]

> 'Rare is the man who is an acknowledged master in a great scientific field and who also has mastery of the language in which he writes. Professor Eddington is such a man, and these books are books that only such a man can write.'

With his background it is not surprising that it was the life sciences that influenced Aldous Huxley the most but he was influenced by physics as well. Huxley received most of his knowledge of modern physics through the popularizer J.W.N. Sullivan who became a close friend.[3] Sullivan published many articles on relativity between 1919 and 1921 during which time Huxley was a subeditor of the Athenaeum under John Murry and would have read some of them. However Huxley first mentions of modern physics were satirical such as those in his first novel *Crome Yellow* published in 1921 in which there is an explicit satirical mention of Einstein when Priscilla Wimbush says 'This Einstein theory. It seems to upset the whole starry universe. It makes me so worried about my horoscopes.' Einstein used trains to illustrate the special theory of relativity and the first chapter with its train and Denis Stone, who emphasizes his name to the porter 'STONE' (stein) and Denis' worry at loosing two hours in the journey all point to a satirical take on relativity and there are more examples in a similar vein.[3] David Bradshaw believes that Huxley's flippant remarks in *Crome Yellow* about relativity rankled Sullivan.[3] Huxley believed the modernist creed that 'it should theoretically be possible to make poetry out of anything whatsoever of which the spirit of man can take cognizance'[68] and came to believe that physics could be used in literature as had Wordsworth at the end of the eighteenth century.[69] Sullivan was the cause of Huxley's change of heart[3] and he re-worked his 1920 article on *The Subject-Matter of Poetry* in 1923 writing:[68]

> 'The men who do feel passionately about abstractions, the men to whom ideas are as persons — moving and disquietingly alive — are very seldom poets. They are men of science and philosophers, preoccupied with the search for truth and not, like the poet, with the expression and creation of beauty. It is very rarely that we find a poet who combines the power and the desire to express himself with that passionate apprehension of ideas and that passionate curiosity about strange remote facts which characterize the man of science and the philosopher. If he possessed the requisite sense of language and the impelling desire to express himself in terms of beauty, Einstein could write the most intoxicating lyrics about relativity and the pleasures of pure mathematics. And if, say, Mr. Yeats understood the Einstein theory – which, in company with most other living poets, he presumably does not, any more than the rest of us – if he apprehended it exultingly as something bold and profound, something vitally important and marvellously true, he too could give us, out of the Celtic twilight, his lyrics of relativity.'

After Sullivan stayed a few months with Huxley in Florence in early 1924, Huxley became even more receptive to modern physics. Sullivan, as well as Eddington, thought that modern physics countered the materialism of the nineteenth century and gave some support to mysticism. It is no coincidence that Huxley's third novel, *Those Barren Leaves*, recommenced while Sullivan was staying with him and published in 1925, had mysticism for its theme.[3] Huxley also modelled Calamy on Sullivan, even to the extent of almost paraphrasing his enthusiasm for the new physics expressed in Sullivan's article *The New Scientific Horizon* published in 1921.[70] Cardan, the foil for Calamy, and a generation younger says:[71]

> 'I made my entry in the late fifties – almost a twin to The Origin of Species. . . . I was brought up in the simple faith of nineteenth-century materialism; a faith untroubled by doubts and as yet unsophisticated by that disquieting scientific modernism which is now turning the staunchest mathematical physicists into mystics.'

However after Huxley renewed his acquaintance with D.H. Lawrence in 1925, the influence of Sullivan waned. In *Point Counter Point*, Rampion was modelled on Lawrence and he returned to treating physics flippantly and wrote in *Along the Road* that 'Science has "explained" nothing; the more we know the more fantastic the world becomes and the profounder the surrounding darkness.'[72] Both Lawrence and Sullivan died in the 1930s and in the end it was Sullivan's influence that prevailed.[3] In 1944 Huxley published *Time Must Have a Stop* with Sullivan reappearing as De Vries and saying:[73]

> 'This was the revolution that had changed the whole course of scientific thinking, brought back idealism, integrated mind into the fabric of Nature, put an end forever to the Victorians' nightmare universe of infinitesimal billiard balls.'

8.1.5. *Lawrence Durrell* (1912–1990)

The style of Lawrence Durrell's *Alexandria Quartet*, published between 1957 and 1960 was influenced by the theory of relativity, as he tells us in a note to the second novel *Balthazar*:[74]

> 'Three sides of space and one of time constitute the soup-mix recipe of a continuum. The four novels follow this pattern.
>
> The three first parts [*Justine*, *Balthazar*, and *Mountolive*] however, are to be deployed spatially … and are not linked in a serial form. They interlap, interweave, in a purely spatial relation. Time is stayed. The fourth part [*Clea*] alone will represent time and be a true sequel.
>
> The subject-object relation[75] is so important to relativity that I have tried to turn the novel through both subjective and objective modes.'

In the novels of the *Quartet* there are no direct references to modern physics apart from Pursewarden's study of science in *Balthazar*:

> 'This for some reason annoyed Justine who took him to task for wasting his time in these studies. He defended himself by saying that the Relativity proposition was directly responsible for abstract painting, atonal music, and formless (or at any rate cyclic forms) in literature… He added: "In Space and Time marriage we have the greatest Boy meets Girl story of the age.'

I do think that abstract painting and literature have been influenced by the theory of relativity but, as you will see in §8.3, not atonal music and I assume with this regard that Durrell is confusing the theory of relativity with relativism. The introduction of relativity into *Balthazar* is in keeping with the character of Pursewarden who was modelled on Wyndham Lewis.

In an interview shortly after he had finished the *Quartet*, Durrell said:[76]

> '[I]n the concept of the space-time continuum you've got an absolutely new concept of what reality might be, do you see? Well, this novel is a four-dimensional dance, a relativity poem. Of course, ideally, all four volumes should be read simultaneously, as I say in my note at the end; but as we lack four-dimensional spectacles the reader will have to do it imaginatively, adding the part of time to the other three, and holding the whole lot in solution in his skull. I call it a continuum.'

In a later interview in 1968 he explained:[77]

> 'I'm not a scientist but you don't need to be to grasp the by product of the relativity idea philosophically … to grasp the four or five principles behind it. An infant of two can. But I took it seriously in the sense that it is our cosmology. Unconsciously we are living by it. The state of our cinema, the form of the novel the cyclic forms are dominated by the philosophical ideas which are, God knows, fifty years old now.'

Although his equating the understanding of the philosophy of relativity to that of a two-year old infant shows how little many outside of the sciences really understand the nature of modern physics, Durrell clearly saw the link between modern physics and modernism in general.

Anne Zahlan suggests that the *Alexandria Quartet* marks Durrell's passage from modernism to postmodernism and her last sentence of her article is worth quoting:[78]

> 'In Clea, last of the Alexandrian narratives, Lawrence Durrell leaves modernism behind and prepares to cross the border into a postmodernist space that opens into "the totally unpredictable".'

8.1.6. *Vladimir Nabokov* (1899–1977)

Vladimir Nabokov's *Lolita*, first published in Paris in 1955 because it contained paedophilia, was an early postmodernist novel though Nabokov hated to be pigeonholed. Two other later novels also featured sexual perversions: *Ada or Ardor* (1969) and *Look at the Harlequins!* (1974) but these also had modern/postmodern physics themes that were central to them. Lolita gave the impression that Nabokov was a sexual adventurer but he showed by having incest in *Ada* and in *LATH*[79] that the sexual mores themselves were not important. As he said in an interview in *Time* on the occasion of the publication of *Ada*:[80]

'If I had used incest for the purpose of representing a possible road to happiness or misfortune, I would have been a best-selling didactician dealing in general ideas. Actually I don't give a damn for incest one way or another. I merely like the "bl" sound in siblings, bloom, blue, bliss, sable.'

Barton Johnson argues convincingly that "bl" became Nabokov's signal for incest.[81] The probable real father of Vadim Vadimovich N., the fictional writer of the autobiography *LATH*, a doppelgänger of Vladimir Vladimirovich Nabokov, was Count Starov who was also the probable father of Vadim's first two wives: Iris Black and Anna Blagovo.[81]

Katherine Hayles has given a fine discussion of Nabokov's use of the modern physics' conception of symmetry in *Ada* and *LATH* that has been used here.[82] A major source of Nabokov's modern physics themes was Martin Gardner (1914–2010), an American popular mathematics and science writer, who the narrator, Van Veen in *Ada*, mentions:[83]

'"Space is a swarming in the eyes, and Time a singing in the ears," says John Shade, a modern poet, as quoted by an invented philosopher ("Martin Gardiner") in The Ambidextrous Universe, page 165. Space flutters to the ground, but Time remains between thinker and thumb, when Monsieur Bergson uses his scissors. Space introduces its eggs into the nests of Time: a 'before' here, an 'after' there – and a speckled clutch of Minkowski's 'world-points.' A stretch of Space is organically easier to measure mentally than a 'stretch' of Time. The notion of Space must have been formed before that of Time (Guyau in Whitrow).'

Gardner's *The Ambidextrous Universe*, first published in 1964, examined the effect of mirror symmetry on physics and life.[84] There are three types of possible symmetry in the Standard Model of particle physics:

- C, symmetry in electric charge, or physical laws independent of the sign of the charge.
- P, parity or mirror-like symmetry in physical laws.
- T, symmetry in time, or physical laws, which are independent of the direction of time.

Prior to 1956 all these symmetries were believed to hold in particle physics. However in 1956, Chien-Shung Wu (1912–1997) showed that the β decay of radioactive cobalt violated parity. The loss of parity did not cause the concept of symmetry to be abandoned and physicists assumed that even if parity was lost that symmetry could be maintained in the combined CP symmetry and T. This symmetry did not last long. In 1964 James Cronin and Val Fitch (1923–2015) found that the decay of neutral K-mesons violated CP symmetry.[85] Shortly after this discovery Eugene Wigner wrote a popular article on symmetry and its violation in which he wrote:[86]

> 'Physicists are left with the belief that the very last mirror, the CPT mirror, is a true mirror. This belief is not based on nature's innate preference for symmetry; it is based on the stubborn fact that we cannot formulate equations of motion in quantum field theory that lack this symmetry and still satisfy the postulates of Einstein's special theory of relativity.'

In part 4 of *Ada*, set in 1922, Van Veen lectures on *The Texture of Time* which considers the symmetry of time and speculates that:[83]

> 'The direction of Time, the ardis of Time, one-way Time, here is something that looks useful to me one moment, but dwindles the next to the level of an illusion obscurely related to the mysteries of growth and gravitation. The irreversibility of Time (which is not heading anywhere in the first place) is a very parochial affair: had our organs and orgitrons[87] not been asymmetrical, our view of Time might have been amphitheatric and altogether grand, like ragged night and jagged mountains around a small, twinkling, satisfied hamlet.'

The first quotation from *Ada* contains other interesting facts. There was an exchange between Gardner and Nabokov. In his first edition of *The Ambidextrous Universe*, Gardner quoted the lines: 'Space is a swarming in the eyes, and Time a singing in the ears,' crediting them to John Slade without explaining that the quoted poem occurred in Nabokov's *Pale Fire* and that John Slade was a fictitious poet; Nabokov continued the joke by writing that they were 'quoted by an invented philosopher ("Martin Gardiner")' deliberately misspelling his name as well as that of his own fictitious poet.[88] Van Veen's conception of time is nearer to that of Bergson than Einstein; as Van Veen tells us: 'We reject without qualms the artificial concept of space-tainted, space-parasited time, the space-time of relativist literature.' However, he was also influenced by Jean-Maire Guyau (1854–1888) who, in his posthumously published major work, argued that the conception of space must have preceded that of time.[89]

Nabokov also extended the concept of symmetry in physics to cosmology with two slightly asymmetrical worlds Terra and Antiterra.[90] Demonia or Antiterra was the world of Van and Ada. Van, who started a dissertation *Terra: Eremitic Reality or Collective Dream?*, 'wondered what really kept him alive on terrible Antiterra, with Terra a myth … [and] even the deepest thinkers, the purest philosophers … were emotionally divided in their attitude toward the possibility that there existed "a distortive glass of our distorted glebe"...'[91]

Symmetry is even more central to Nabokov's last novel *LATH* than *Ada*. Vadim Vadimovich had feared *dementia paralytica* since he was a young man. His 'motley madness' is manifest in his inability to mentally transpose left to right and conceive of the symmetry of parity though, as he tells his first wife Iris:

> 'In actual, physical life I can turn as simply and swiftly as anyone. But mentally, with my eyes closed and my body immobile, I am unable to switch from one direction to the other.'

Vadim also mentions what Gardner called the Ozma Problem.[92] In the continuing conversation between Vadim and Iris, she says:

> 'what we had been trying to do was to solve a stupid philosophical riddle – on the lines of what does "right" and "left" mean in our absence, when nobody is looking, in pure space,'

Gardner defined the Ozma Problem slightly differently posing the question: can we communicate the meaning of left and right to people on a distant planet? Such a communication must be made through referring to a physical law where the symmetry of parity is violated. If classical physics were exactly true all physical laws would be perfectly symmetrical. Communication only became possible because Wu's experiments in the β decay of radioactive cobalt of 1956 showed that parity could be violated. The Ozma Problem has had a long history. Kant addressed the problem in his *Prolegomena* of 1783,[93] in 1890 William James in describing the sides of a cube remarked that 'there remains no form of words by which we can describe to another person which of the remaining sides is right and which is left,'[94] and, as Gardner acknowledged, Richard Feynman discussed the problem in terms of modern physics in his famous *Feynman Lectures on Physics*[95] written in the early 1960s. In *LATH* Vadim's mental inability to transpose left and right finally becomes a physical reality:

> 'What used to happen so often in thought, now had happened for keeps: I could not turn. To make that movement would mean rolling the world around on its axis and that was as impossible as traveling back physically from the present moment to the previous one. Maybe I should not have panicked, should have waited quietly for the stone of my limbs to regain some tingle of flesh. Instead, I performed, or imagined performing, a wild wrenching movement – and the globe did not bulge. I must have hung in a spread-eagle position for a little while longer before ending supine on the intangible soil.'

Vadim had had his 'great seizure' and was paralyzed.

8.2. Visual Art

The date of the beginning of what is considered the modern era depends upon the discipline and even for a particular discipline is not precise. Modern art is usually considered to date from about the last third of the nineteenth century with impressionism the first modern movement. Although the first impressionist paintings shocked the public, impressionism did not completely reject traditional Western art with its single view-point perspective and though the style was not completely realistic it gave a true representation of the psychological impact of the subject matter. In the beginning of the twentieth century fauvism and expressionism saw a departure from realism and with the cubists there was an experimental departure from a single view-point perspective and a decreation as well as creation that tick all the boxes for modernism. Hence in this section I discuss the relation between modern physics and twentieth century art from the cubists onwards.

The postmodern artists are less likely to describe themselves as painter or sculptor and their work is often in a combination of media; the mode of expression can to be more important than what is being expressed and there is a tendency to produce very large art works. In 1969 the married couple Christo and Jeanne-Claude Javacheff,[96] both born on the same day in 1935 wrapped 2.5 km of cliffs in Little Bay, Sydney, Australia in white fabric to create a work of purely aesthetic impact that was surely postmodern. The prefix post, though it implies it, does not necessarily mean a chronological order modern – postmodern. In part postmodernism is a reaction to modern art, which is perceived as elitist, and often is populistic, an example is the pop-art of Andy Warhol (1928–1987) with its multi-images of Marilyn Monroe and Campbell Soup tins. Conceptual art is a new development. Christo's cliff wrapping and installations are art where the actual work is planned and may be executed by a number of assistants, of course many artists through the ages have used assistants but they have been orchestrated by the artists, conceptual art in theory could be constructed in the absence of the artist.

There is no precise definition of postmodernist art each art critic having his own version. One improvement in my opinion is the demise of the rules of art and the formation of isms. There are a few surveys of the term but I have found the one by Dan Karlholm to be the most useful.[97]

8.2.1. *Cubism and the fourth dimension*

Cubism flourished between 1907 and 1920 though, of course. it continued to influence new artists. During the early stages of Cubism the only elements of modern physics that were widely known were X-rays and the radioactive rays: alpha, beta, and gamma. Most importantly outside of scientific circles little was known about the theory of relativity or quantum mechanics in the early days of Cubism. The Cubists rejected realism and a single view-point perspective and were drawn to the nineteenth century developments in geometry and in particular to the fourth dimension. From the 1880s until 1919, the term *fourth dimension* was used to denote a supposed fourth dimension to space that existed but was not visible. The fourth dimension was linked to the astral plane of spiritualists and theosophists and related to X-rays and aether by them. Not until after Einstein's theory of relativity became widely known was the fourth dimension linked to time by the general public. The original popularization of the fourth dimension in the 1880s had nothing directly to do with physics but it was associated with it and is part of the story. Hyperspace is not necessarily curved. In non-Euclidean geometry space is curved and the curvature can be constant or variable like Riemann's *n*-dimensional space. The conclusion of Linda Dalrymple Henderson, who, in her heroic book first published in 1983, convincingly reconstructed the influence of the fourth dimension on modern art, is that from 1900 to 1930 the fourth dimension was an influence on Cubism and almost all of the other major art movements.[98]

The fourth dimension was developed as a part of pure mathematics but both the mathematicians Carl Friedrich Gauss (1777–1855) and Nikolai Lobachevsky, who described non-Euclidean geometry in the early nineteenth century, speculated that may be there was space on a large scale that was non-Euclidean but they tested their hypothesis without success.[99] The popularizers of the fourth dimension all believed in its reality. Howard Hinton (1853–1907), an English mathematician, who was a science teacher at Uppingham School in Rutland from 1877 to 1887,[100] published in 1880 *What Is the Fourth Dimension?* in the Dublin University Magazine and it was reprinted in other places in the early 1880s.[101] Hinton's major works on the fourth dimension were *A New Era of Thought* (1888) and *The Fourth Dimension* (1904). Bertrand Russell gave a favourable review of *The Fourth Dimension*, and summed up with: [102]

> 'The merit of speculations on the fourth dimension – a merit which the present work possesses in full measure – is chiefly that they stimulate the imagination, and free the intellect from the shackles of the actual.'

Although he did add the caveat:

> 'If Cheshire cats were common, if there were many well-attested cases of the appearance of ghosts in closed rooms, if Messrs. Maskelyne & Cooke could cause the disappearance of a left-hand glove, and its reappearance immediately afterwards as a right-hand glove,[103] the fourth dimension would begin to present scientific attractions which as- yet are lacking to it.'

A related publication in the 1880s was Edwin Abbott's (1838–1926) *Flatland.*[104] Whereas Hinton wrote about a four dimensional world that could not be perceived, Abbott under the pseudonym, A Square, wrote a romance of many dimensions, the first part of which was about a flatland of two dimensions and the second part was about other hyperdimensional worlds. In the second part a

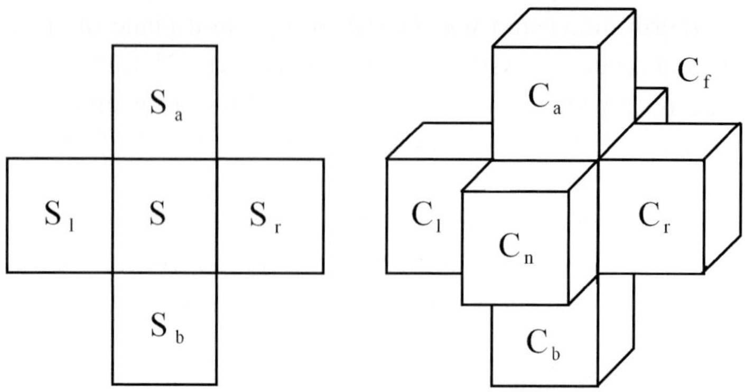

Fig. 8.1. Schematic representation of (a) a cube in two dimensions and (b) a tesseract in three dimensions (after Hinton, 1904).

Sphere from a three dimensional world visits the Square who views him as a Circle. The Sphere takes the Square to his three dimensional world. The Square on becoming accustomed to the three dimensional world inquires about the 'land of Four Dimensions.' After the Square goes through what the Sphere has taught him about one to three dimensions, he extends the argument to four and thereby defining the *tesseract* the term coined by Hinton in *A New Era of Thought* to describe a four dimensional hypercube:[105]

> 'In One Dimension, did not a moving Point produce a Line with two terminal points?
> In Two Dimensions, did not a moving Line produce a Square with four terminal points?
> In Three Dimensions, did not a moving Square produce did not this eye of mine behold it that blessed Being, a Cube, with eight terminal points?
> And in Four Dimensions shall not a moving Cube alas, for Analogy, and alas for the Progress of Truth, if it be not so shall not, I say, the motion of a divine Cube result in a still more divine Organization with sixteen terminal points?'

A tesseract obviously cannot be illustrated but Hinton schematically represented it in three dimensional perspective and this is shown in Fig. 8.1b a cube is also shown schematically represented in two dimensions in Fig. 8.1a. The square and the cube are moved, as described in the quotation from *Flatland*, in one two and three dimensions; a tesseract is moved in a fourth dimension, resulting in sixteen terminal points. However by 1930 the fourth dimension in space was so forgotten that when the *Scientific American* published an article on the fourth dimension its editors had so many inquiries, from confused readers who were used to time as the fourth dimension, that they had to respond.[106]

The English art critic Frank Rutter (1876–1937) reported that according to the influential French art dealer Léonce Rosenberg (1879–1947) Cubism obtained its name in 1908 when a painting by Georges Braque (1882–1963) was being

Fig. 8.2. Pablo Picasso, *Head of a Woman in a Mantilla*, ca. 38 × 30 cm, 1909 (location unknown).

carried into the Salon des Indépendants and a member of the Hanging Committee exclaimed 'Encore des Cubes! assez de cubisme'. Braque and his fellow painters readily accepted the term.[107] It was Georges Braque and Pablo Picasso (1881–1973) who together created Cubism around 1908. The inspiration for Cubism did not come at first from the geometry of the fourth dimension, but from the facetted forms and limited palette of the late paintings of Paul Cézanne (1839–1906) and African primitive art. Picasso's *Les Demoiselles d'Avignon* painted in 1907 was one of the major steps towards Cubism.[108] Picasso 'abolished curves, sharpened the angles of planes, and laid the foundation of Cubism in his painting of the Lady in a Mantilla', which is better known as *Head of a Woman in a Mantilla*, painted in 1909 and shown in Fig. 8.2.

The Cubists learnt about the fourth dimension largely from Maurice Princet (1875–1968) a French insurance actuary with an interest in mathematics. Princet became known to Picasso and his Bateau-Lavoir group[109] by 1905 or 1907.[110] Picasso himself denied ever discussing the fourth dimension with Princet but he was renowned for denying all kinds of influences.[111] It was the analytically inclined artist Jean Metzinger (1883–1956), who was most influenced by Princet in the Bateau-Lavoir group.[98] According to Metzinger one source of the fourth dimension for Princet was the German mathematician Victor Schlegel (1843–1905) who wrote a number of articles on *n*-dimensional geometry in the 1880s and even constructed models of the projections of hyperpolyhedra in three dimensions.[98] Another source of the fourth dimension was Esprit Jouffret (1837–1904) a French artillery officer who, after he retired, wrote two popular books on the fourth dimension. Henderson states that Jouffret's 1903 book, *Traité Élémentaire de Geométrie à Quatre Dimensions* was known to Marcel Duchamp (1887–1968) and other Cubists.[112] Jouffret gave two-dimensional projections of four dimensional polyhedra that appear superficially similar to some Cubists' paintings. Henderson particularly notes the similarity of one Jouffret's projections with Picasso's Cubist portrait of Ambroise Vollard painted in 1910; she shows the portrait in black and white where the similarity might seem close but it vanishes when colour is added. Jouffrey used hatching in his two-dimensional projections which also has been compared to the hatching on the breast and nose of one the demoiselles in Picasso's *Les Demoiselles d'Avignon*. Such inessential detail cannot be used as evidence of influence without other strong supporting evidence; the mind sees what it wants to see. However I do believe that the geometry of the fourth dimension did have a general influence on modern art and on Cubism in particular.

There is a difficulty in discussing art in terms of movements because really each great artist has his own individual movement. The Czech artist František Kupka (1971–1957) resisted being identified with any particular group. In 1906 he settled in Puteaux a suburb of Paris which came to be the home of the Puteaux Cubists with whom he interacted. Kupka was a Theosophist and was interested in the hyperspace philosophy through Hinton. In the period between 1909 and 1912 he painted the two *Planes by Colour* paintings shown in Plate 17. Hinton used coloured sections to illustrate the intersections of four dimensional hyperpolyhedra with three dimensional planes and Kupka's *Planes by Colour* paintings might have been partially inspired by them. However Henderson suggested that both of the *Planes by Colour* paintings were inspired by Kupka's interest in X-rays and particularly drew attention to the nose of the woman in Plate 17b and compared it to a similar nose in another painting of Kupka, *Portrait of the Musician Follot* of the same period that, though not including *Planes by Colour* in its title, is clearly of the same genre, she calls these 'X-ray noses'.[113] The noses in both cases are the same colour as other planes of colour in the paintings and I think my explanation is more plausible but it may only be a case of *my mind* seeing what it wants to see.

8.2.2. *Dada and De Stijl*

Dada[114] was more than an art movement and included all the imaginative arts. According to Dadaist legend it was founded in 1916 at the small bar in Zurich, Switzerland called the Cabaret Voltaire by two Germans Hugo Ball (1886–1927) and Richard Huelsenbeck (1892–1974), two Rumanians, Tristan Tzara (1896–1963) and Marcel Janco (1895–1984), and the Alsatian, Hans Arp (1886–1966) all of whom had left their native countries because of their opposition to the war. Dada was a revolt of young men against complacent society; it was aimed at liberating the individual. While neutral Switzerland drew many avant-garde artists, others went to the United States including Marcel Duchamp and the French born Francis Picabia (1879–1953) who arrived in New York in 1915 and together with the American Man Ray (1890–1976) formed the Arensberg Circle[115] which was effectively an independent Dada group and because of the war it took a couple of years before the two groups were aware of each other.

Dada and other avant-garde movements adopted the Fourth Dimension with open arms. Theo van Doesburg (1883–1931) was a co-founder of the Dutch artistic movement *De Stijl* (The Style), which grew out of Dada in 1917, and it was also the name of van Doesburg's magazine. In 1923 van Doesburg wrote in a pamphlet *What is Dada?????????*, which he sold on a Dada campaign, that claimed 'Dada ... [was] one of the strongest manifestations of the fourth dimension, transposed onto the subject.'[116] In the same pamphlet he called Bergson, Einstein, and Charlie Chaplin, Dadaists and on an advertising poster he wrote: 'Dada has always existed. The holy virgin was already a Dadaist.' Van Doesburg also applied fourth dimensional concepts to *De Stijl* architecture. After van Doesburg met the Russian El Lissitzky (1890–1941) in 1921, Henderson asserted that he began to appreciate Einstein's theory of relativity. In 1922 van Doesburg published an article in *De Stijl*, *The Will to Style*, which discussed the relevance of relativity for art.[98] However, there is really no hard evidence that van Doesburg understood the theory of relativity even in a rudimentary form.[117] Van Doesburg certainly was very interested in the 'combining of the impetus of space and time' as he wrote in his *De Stijl* article but he then gives the example of Viking Eggeling (1880–1925) and Hans Richter (1888-1976) 'using film technique in the painting of pure form gives … the artistic solution of the dichotomy of static and dynamic, of spatial and temporal elements', which does not seem to me to be referring to the spacetime of the theory of relativity.[118] Van Doesburg is equally vague in his *De Stijl* article of 1924 *Towards a Plastic Architecture*, confusing classic constructional strength calculations with non-Euclidean geometry writing:

> 'The two-dimensional spatial composition *fixed* in a ground-plan will be replaced by an exact *constructional calculation* – a calculation by means of which the supporting capacity is restricted to the simplest but strongest supporting capacity. For the purpose Euclidean mathematics will be of no use – but with the aid of calculation that is non-Euclidean and takes into account the four dimensions everything will be very easy.'

I have quoted from Doesburg's article at some length to demonstrate the complete nonsense that he wrote. It is reported that van Doesburg was an important source for the influential book, *Time and Architecture: The Growth of a New Tradition*, first published in 1941 by the architectural critic Sigfried Giedion (1888–1968),[98] but he too, though he discusses the interconnectedness of space and time, does not explicitly discuss the application of the theory of relativity to architecture.

8.2.3. *The Russian avant-garde*

Kazimir Malevich (1878–1935) began the Suprematism avant-garde movement in pre-revolutionary Russia based on geometrical forms. Lissitzky, who had studied architectural engineering in the Technische Hochschule, Darmstadt, Germany, was attracted to Suprematism because of its use of geometric forms, and developed a variant of it he called *Proun*.[119] Lissitzky knew at least a little about non-Euclidean geometry and the special theory of relativity. In 1925 he wrote an essay in German about space and art, whose title in English was *Art and Pangeometry*, The mathematician Lobachevsky coined the word *Pangeometry*, which he used for the title of his major work on non-Euclidean geometry. Lissitzky's essay is mainly about the 'variability of our conception of space and the corresponding forms of A[rt]'[120] He did briefly mention relativity but Lissitzky suggested that artists 'have based their ideas, in a flexible superficial way, on the most modern scientific theories, without acquiring any knowledge of them (multidimensional spaces, the theory of relativity, the universe of Minkovsky, and so on).' I think this is a true assessment of most of the use of modern physics in all forms of art. Lissitzky was most likely to have been familiar with Minkowski's geometrical representation of spacetime[121] and was particularly taken by the before and after light cones of a world-point and the corresponding hyperboloid sheets geometrical representation of spacetime, depicted by Minkowksi and reproduced in Fig. 8.3, because it appears in many of his art works around 1923 including the lithograph of the *New Man* (*Neuer*) one of a series or ten lithographs that Lissitzky produced for a 1923 revival of the Futurist opera of 1913 *Victory Over the Sun* by Mikhail Matyushin and Aleksei Kruchenykh, which is shown in Plate 18.[122]

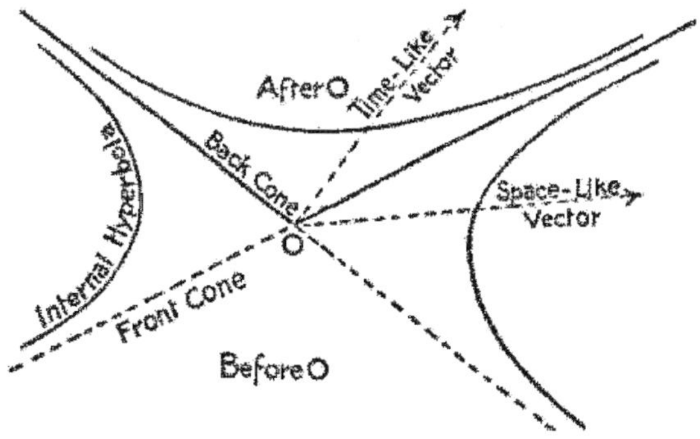

Fig. 8.3. Before and after light cones and hyperbolae shown in Fig. 2 (Minkowski, 1908).

An avant-garde artist, who probably really understood the basics of the theory of relativity, was the Russian born sculptor Naum Gabo (1890–1977); his family name was Pevsner but he changed his name during the First World War to avoid confusion with his older brother Antoine Pevsner who was also an artist. Initially Gabo was under family pressure to study medicine and in 1910 he enrolled at Munich University, where he attended courses in anatomy, physics, and chemistry for two years. Sommerfeld was the Professor of Theoretical Physics at Munich during those two years, and in 1911 he gave a colloquium on Einstein's special theory of relativity, which Gabo very probably attended making him one of the few non-physicists to be aware of spacetime and relativity in those early years.[123] Gabo also attended courses in civil engineering at the Technische Hochschule in Munich, which stood him well in his later Constructivist sculpture. As with very many young men, the First World War disrupted his studies at Munich, and Gabo and his two brothers Antoine and Alexei spent the war in Oslo (called Christinia until 1924) where Gabo experimented in pre-Constructivist sculpture made by joining together flat planar elements of cardboard at first and later of wood and metal. After the 1917 Revolution the brothers returned to Russia and in 1920 Gabo and Antoine Pevsner issued their Constructivist *Realistic Manifesto* in which they stated:[124]

> 'Space and time are re-born to us today.... The realization of our perceptions of the world in the forms of space and time is the only aim of our pictorial and plastic art.'

In the first sentence of this quotation Gabo may have been referring to the Einsteinian conception of spacetime, which I think undoubtedly inspired him, but in the second he was probably thinking about kinetic art. In the same year as Gabo issued his manifesto he produced his *Kinetic Construction* (*Standing Wave*) which was a vertical metal rod excited to vibrate by an electromagnet. As if to emphasize the second sentence quoted from his manifesto, Gabo renamed this work *Kinetic Construction* (*Time as a New Element of Plastic Art*) when he re-exhibited in Berlin in 1922.

Fig. 8.4. Naum Gabo, *Model for Construction in Space: Two Cones*, 1928. May be an earlier state of the one now in the Tate Gallery (Martin *et al.*, 1937).

Although Gabo was obviously influenced by the spirit of the theory of relativity, there is little direct evidence in his sculptures. Gabo did carve sculptures but most were constructed as were his first ones and many were models for large monuments, such as his 26 metre tall sculpture erected outside the Bijenkorf Department Store in Rotterdam, but few of his projects were actually constructed and most exist only as models. These models were often constructed in plastic, as was his 1928 *Model for a Construction in Space: Two Cones* shown in Fig. 8.4 with celluloid cones. The two cones in this sculpture are reminiscent of Minkowski's diagram reproduced in Fig. 8.3. *Construction in Space: Soaring* made in 1929/30 also had two celluloid cones and was probably similarly inspired.

8.2.4. *Surrealism*[125]

Like many of the early twentieth century avant-garde movements Surrealism developed out of Dada and was also a movement of both the visual and literary arts. In 1924 two rival manifestoes were issued, one by the Alsatian poet Yvan

Goll (1891–1950), the other by the French writer André Breton (1896–1966) was the more influential. In Breton's 1924 manifesto, Surrealism was defined as:[126]

> 'Psychic automatism in its pure state, by which one proposes to express – verbally, by means of the written word, or in any other manner – the actual functioning of thought. Dictated by the thought, in the absence of any control exercised by reason, exempt from any aesthetic or moral concern.'

With such a definition, Surrealism owes more to Freud than Einstein, though even the theory of relativity helped to demolish the concept of a mechanical world with its attendant materialism, and Salvador Dali (1904–1989) was probably influenced by the loss of absolute time in his paintings depicting melting watches such *Persistence of Memories* painted in 1931 shown in Plate 19 However, Dali was suitably ambiguously surrealistic when he wrote:[127]

> 'Salvador Dali in 1935 no longer contents himself with serving you, out of this agonizing and colossal question of Einsteinian space-time, with anthropomorphism … he serves you cheese out of it, for you may rest assured that Salvador Dali's famous soft watches are nothing but the soft, extravagant, and solitary paranoiac-critical Camembert of time and space.'

His meticulous academic style of painting was anything but automatic. According to Henderson, Breton and the Surrealists accepted a time/space duality in their approach to the fourth dimension because of its mystical and irrational associations.[98] In his *La Conquéte de l'Irrationnel*, Dali expressed a concern for the fourth dimension and in his 1954 painting *Crucifixion (Corpus Hypercubicus)* shown in Plate 20 the cross is a Hinton tesseract as illustrated in Fig. 8.1b.

Surrealist art showing influences of modern physics did not start appearing until the late 1920s for a good reason. Not until Heisenberg developed his uncertainty principle did modern physics have a theory that could be called surrealistic. Quantum mechanics was a bigger influence on Surrealism than the theory of relativity simply because it was more surrealistic. Austrian born Wolfgang Paalen (1905–1959) and other Surrealists were drawn to modern physics. After Paalen had moved to Mexico to avoid the Second World War, he left the Breton Surrealists to form his own group announcing this move in the first issue his art review *Dyn* in 1942 with what was in effect a manifesto.[128] The surrealism of Paalen was based on the 'achievements and methods of science' In the second issue of *Dyn*, Paalen published an article *Art and Science* where he made it plain that the science he was writing about was the 'new Quantum Physics [that] is compelled to abandon the rigorous determinism that until now was held to be the very foundation of physics.'[129] Paalen gave his mantra for his new image in the last issue of *Dyn*:[130]

'No more painting with a subject, but no picture without a theme.'

The frontispiece for his essay *The New Image*[131] in the first issue of *Dyn* was his *Figure Pandynamique*, which Paalen had painted in 1940 and is shown in Plate 21, which clearly has no recognizable subject but does have theme, which Gavin Parkinson has suggested is the complementarity of the particle-wave duality in quantum mechanics, and I think he may be correct.[132] There are other Paalen paintings from the *Dyn* era that use this same theme.

8.2.5. *Postmodernism*

The major difference between modern and postmodern art, where the artist is influenced by physics in its execution, is that whereas with modern art one can usually see evidence of the physics in the work of art, yet often the artists themselves did not write about the relation between physics and their work in detail, with postmodern art there is often little direct evidence of physics visible in the work, yet the artists often write at length on how physics has been incorporated into the work. Also the modernists were most interested in the philosophical implications of modern physics whereas the postmodernists often seem more interested in the technical aspects of physics.

I start this discussion of postmodernist artists with an example that, while it does not strictly involve physics but technology, is interesting as far as the Two Cultures debate. is concerned. In 1962 the National Aeronautics and Space Administration (NASA) began its Art Program to present and record NASA's work in an accessible form and Andy Warhol was one of the artists who have been commissioned by it. Warhol was commissioned in 1987 and chose to depict the first manned mission to the Moon of 1969 showing Buzz Aldrin standing by the flag planted by Neil Armstrong (1930–2012). Wahol completed two series of silk screen prints: one shown in Plate 22 with a magenta space suit and a second with a yellow one but he died before he could produce a version with multiple images similar to those of his Marilyn Munroe series. The image shown in Plate 22 was a composite produced from a number of photographs of the Moonwalk, Armstrong and the Lunar Module are reflected in the face plate of the space helmet. The US Flag flying high is an anti-Newtonian feature that gave some credibility to conspiracy theories about the Manned Moon Mission. Since there is no atmosphere on the Moon there can be no wind to make a flag fly and a normal flag would hang limp about its mast. Obviously this was not the image that the Americans wanted to present, so to make the flag apparently fly high a rod, fixed horizontally, was inserted in a sleeve along the top of the flag and is the reason why the top of the flag seems unnaturally straight.

While it is questionable whether I should have included any of Warhol's works, many of the works of Matthew Ritchie, a prolific postmodernist artist born in London and now working in New York, are inspired by modern or postmodern physics. Ritchie's installations tend to be very large and not easy to represent in a book. I have chosen to show in Plate 23 his very large oil painting *M Theory* from 2000. From the title we know that this is about M-theory but to me there are no elements in the painting that make me say yes I can see that. Contrast this panting with that of Paalen shown in Plate 21, its title *Figure Pandynamique* gives no clue to what it represents but Gavin Parkinson recognized that it could be the complementarity of the particle-wave duality in quantum mechanics. I am not sure that I would have identified it as such without Parkinson's observations, but I can see that he could well be right. In Ritchie's case he gave a very detailed story that he attached to *M Theory* in an interview conducted by Robin Clark of the San Francisco Museum of Modern Art:[133]

> 'The formal structure of *M Theory* is similar to most of my paintings from that period. A hard ground, in this case a pale amethyst field, is overlaid with complex pseudo-solids, each containing multiple color zones. Here, the crystalline red structure refers to "the landscape," a concept in string theory that allows for the separation of very distinct physics regimes. The rapidly painted marks on top of this tessellated surface indicate the smaller, rapidly inflated "bubble universes" proposed by theoretical physicist Andrei Linde. The red finger prints refer to the thermal signature of the big bang, sometimes called "the fingerprint of the universe." The scribbled texts on the surface are string theory equations describing the relational terms through which our universe and probably many other variants can exist. any of these terms such as surface, action, harmony, curvature, line sheet, and even the idea of a vibrating, energetic string itself seem to have a direct resonance with the history of painting.'

Now even with Ritchie's explanation I cannot see what he sees in his painting. That does not mean I think it is a poor painting or that Ritchie is not being truthful in what he wrote. Paalen's *Figure Pandynamique* I like and also I think that I understand it. Ritchie's *M Theory* I also like but I only understand it aesthetically. However *M*-theory is a work influenced by postmodern physics but the influence is Ritchie's personal artistic interpretation as really all influences across different aspects of our culture must be. Ritchie's *Main Sequence* body of work 1997–2003, which includes *M-theory*, contains many works with images inspired by physics. For example in 2008 he painted a series of four large paintings representing the four fundamental forces of physics.

Some postmodernist artists have studied physics at university first before they became full time artists. One of the most successful of these is Julian Voss-Andreae, a German-born sculptor also now working in the USA, who began painting but changed directions to study physics. He obtained a degree from the Free University of Berlin and started a doctorate in quantum physics before going back to his love of art and studying sculpture in the USA.[134] His first sculptures were more or less models of protein molecules, though sometimes imaginatively coloured. However since 2005 it has been the philosophical implications of quantum physics that have fascinated him and become the main influence on his art. A typical sculpture of Voss-Andreae is *Quantum Man*; three different views of which are shown in Plate 24. The construction of *Quantum Man* was appropriately inspired by quantum physics and consists of a large number of vertical stainless steel plates, which represent wave fronts of the walking quantum man's wave duality. As Voss-Andreae noted,[135] the quantum wavelength[136] of a walking man is about equal to the Planck length (1.6×10^{-35} m) which is unimaginably small, the interatomic distance of an iron atom is many magnitudes larger at about 4×10^{-10} m, thus a more realistic spacing of the stainless steel plates is impossible. *Quantum Man* does have a high aesthetic impact and his appearance varies from solid (Plate 24a) to transparent (Plate 25c) and almost invisible (Plate 24b) depending on where you stand.

8.3. Music

Musical theory is part of classical physics because it needs a planet like Earth that has an atmosphere, it has a human scale and does not exist at either the elementary particle or the cosmic scale. Arnold Schöenberg (1874–1971) is often called the Einstein of music because of his tonal revolution at the beginning of the twentieth century. Although this revolution occurred at the same time as the revolution that gave us modern physics it did not have anything directly to do with it. Up until about 1907 Schöenberg explored chromatic harmony but by 1909 Schöenberg had finally arrived at atonality in the first of his three pieces for piano (Opus 11). Between 1921 and 1923 Schöenberg had organized his atonality into the twelve tone system where no tone has importance over another. In *Opinion or Insight* written in 1926, Schöenberg wrote on the emancipation of dissonance:[137]

> 'Few of today's composers still find a few tonal triads sufficient … most of them have noted what has been done by the works of Wagner, Strauss, Mahler, Reger, Debussy, Puccini, etc., and they have drawn conclusions about harmony, whose outcome is recognizable as the emancipation of the dissonance… Tonality's origin is found … in the laws of sound. But there are other laws that music obeys, apart from these and the laws that result from the combination of time and sound: namely those governing the working of our minds… *Tonality is not an end in itself, but a means to an end…* [C]onsonance and dissonance differ

> not as opposites do, but only in point of *degree*...There is no reason in physics or aesthetics that could force a musician to use tonality in order to represent his idea.'

Schöenberg's innovation that heralded modern music was an aesthetic innovation not directly related to physics.[138] Modern physics is not relevant to musical theory. Although modern and postmodern physics have not influenced musical theory they have provided a rich source of inspiration for composers.

8.3.1. *Modern physics, the theory of relativity, and black holes*

The keystone of relativity is the principle of the constancy of the velocity of light relative to any inertial frame and it was the Michelson-Morley experiments of 1887 that first established this principle and prepared the way for Einstein's theory of relativity.[139] Michelson was professor of physics at the Case School of Applied Science and Morley was professor of chemistry at the Western Reserve University both of which were located in Cleveland Ohio. The two institutions have since combined to become the Case Western Reserve University who commissioned Philip Glass to compose music in celebration of the one hundredth anniversary of the Michelson-Morley experiments of 1887. Glass described his one-movement work *The Light* as a portrait of the two men and the moment that heralded beginning of modern physics. Although the two men did not at the time realize the full import of their results, it was an important moment in physics.

Earlier in 1975 Glass had composed an opera *Einstein on the Beach* in collaboration with the avant-garde director Robert Wilson. This five hour avant-garde opera used Einstein as primarily a representative of someone who has changed our thinking rather than concentrate completely on the substance of his work. This opera subsequently became part of Glass' trilogy of similar operas: *Satyagraha* (1979) and *Akhnaten* (1983). The particular title of the Einstein opera has little significance and was chosen by Wilson simply because he had seen a photograph of Einstein standing on a beach.[140] Glass' music is minimalist and repetitive and the lyrics, if they can be called that since they include counting and the tonic sol-fa, also tend to be repetitive and are meant to create ambiguity and allow the audience to supply their own interpretation.[141] It is doubtful that Einstein himself would have liked Glass' music. Einstein was generally unimpressed by modern music; his favourite composers were Mozart and Bach, though he did like the Swiss-born American composer Ernst Bloch (1880–1959) and a few others modernists, writing in 1950:[142]

'My knowledge of modern music is very restricted. But in one respect I feel certain: True art is characterised by an irresistible urge in the creative artist. I can feel this urge in Ernst Bloch's work as in a few later musicians.'

Einstein on the Beach has four acts connected by intermezzi or knee plays as Glass calls them. Some of the motifs of the acts can be identified with relativity. There are scenes with early twentieth century trains which figured so much in explanations of the special theory of relativity and some feature clocks. Other scenes reference Einstein's development of the atom bomb. The court scenes might be interpreted as judging Einstein's role in the development of the atomic bomb since an Einstein-like figure plays the violin in one scene but for the fact that in the court scene the defendant turns into Patty Hearst toting a machine-gun. So the opera is truly postmodern and everyone has to supply their own interpretation.

Another opera *Einstein* first performed in 1974, which is strictly outside of the scope of the present book, concentrated entirely on Einstein's involvement with the United States Manhattan Project to build an atomic bomb. The music was by Paul Dessau (1894–1979), a German composer who wrote the music to accompany many Brecht plays and the librettist was Karl Mickel (1935–2000). Unfortunately the libretto was largely fictitious because Einstein's involvement with the Manhattan Project was minimal though important.[143]

The works described above were inspired by modern physics but more by its singers not their songs. Hèctor Parra is a Catalan composer who is both inspired by modern physics and deeply interested in attempting to interpret it in music. In *Stress Tensor* composed in 2009, Parra goes to the heart of Einstein's general theory of relativity. In his 1916 paper giving the Foundation of the General Theory of Relativity, Einstein in his introduction stated, emphasizing with italics, that:[144]

'*The general laws of nature are to be expressed by equations which hold good for all systems of co-ordinates, that is, are co-variant with respect to any substitutions whatever.*'

Then in his section on the mathematical formulation of covalent equations, Einstein introduced tensors and stated:[144]

'If ... a law of nature is expressed by equating all the components of a tensor to zero, it is generally covariant. By examining the laws of formulation of tensors, we acquire the means of formulating generally covariant laws.'

Tensors were not generally used by physicists in the early twentieth century, though the concept can be traced back to the Gauss, they did not even become well known to many mathematicians until 1900,[145] and Einstein had to include a section on tensor analysis in his paper so that it could be understood by physicists. In general relativity the energy–momentum tensor (sometimes called the stress–energy–momentum tensor or stress–energy tensor) is the source of the gravitational field in Einstein's theory of general relativity. Parra's title *Stress Tensor*[146] is a direct reference to Einstein's tensor analysis and as Parra wrote in the introduction of his score:[147]

> 'There is a magical aspect to the musicality inherent in humans that I would, in aesthetic terms, put in parallel with the beauty that physicists – Einstein in particular – found in nature: the stress tensor – of energy and momentum[148] (all musical materials) – that is the source of the gravitational field, the curvature of the universe.'

Just as the artist Ritchie cannot recreate M-theory in his painting of the same name nor can Parra recreate in *Stress Tensor* the tensor analysis of general relativity in music but what he has done, as he stated in the programme notes to his CD, is to make 'possible the appreciation of [its] architectural beauty, a shared characteristic of Art and Science.'

Parra is not only interested in physics but also in the life sciences and in Caressant l'Horizon, a tone poem premiered by the Ensemble Intercontemporain at the Philharmonie de Paris in 2011, he combined these interests. *Caressant l'Horizon* or caressing the horizon contrasts the cosmic force of a black hole where the gravitational pull at its event horizon becomes so great that nothing not even light can escape and the evolution of life in our oasis in the solar system where spacetime is virtually flat and enables us to exist. Parra imagines 'what we would physically experience if the unimaginably intense gravitational waves which arise when two black holes collide actually penetrated us.'[149] During composing the music for *Caressant l'Horizon* Parra made a number of sketches to aid him conceive the gravitational waves generated by the coalescence of two black holes, one of which is shown in Plate 25. Here we have a case of the artistic conception occurring before the actual event. On the 14th September 2015 two black holes of mass 29 and 36 times that of the Sun were observed to merge as discussed in §1.2.1.3. The predictions of the spatial strain were accurately predicted from the general theory of relativity and are shown in Plate 26. Parra's sketch in Plate 25 and his music contain the essence of that black hole merger.

8.3.2. *Postmodernism and hyperspace*

Three postmoderns have combined in an exciting project to create a truly postmodernist opera *Hypermusic Prologue* first performed in the Pompidou

Centre in Paris in 2009.[150] The three postmoderns were: Hèctor Parra who composed the music, Lisa Randall the physicist who wrote the libretto, and Matthew Ritchie the artist who designed the sets. There are just two singers in this opera: a soprano physicist-composer and her lover a baritone. As in many operas the love between them is threatened because while the soprano loves the baritone she also has a love of physics and wants to explore its world. After a classic lovers quarrel, the soprano leaves our familiar four dimensional spacetime (three of space and one of time) to explore the world of the five dimensional spacetime (four of space and one of time). I do not think there is another opera where the librettist has cast herself as the heroine because it is Lisa Randall who is exploring her own fifth dimension. The baritone is left confined to our four dimensional spacetime. However, this is not a tragic opera and it ends, with the hope that finally they will find not just each other but reality at last, as the two sing a duet:

> Is it true?
> We can reach deeper?
> Find more?

The concept of a fifth dimension – as a solution to the hierarchy problem caused by the extreme, smallness of the fundamental energy scale of gravity compared with the three other fundamental forces – was first mooted by Kaluza in the early twentieth century and Klein suggested the fourth space dimension was curled up so that it was invisible but the idea was abandoned when its predictions did not match experiments. What Lisa Randall and her fellow American physicist Raman Sundrum have done is to take the idea of five dimensional spacetime and propose that the fifth dimension is not curled up but extremely warped and invisible to us. The fifth dimension is bounded by two four dimensional spacetime branes: one the weak brane, which contains our universe and all interactions between particles except gravitational ones, the other brane is the gravity brane where most gravitational interactions take place. In their first model RS1 the fifth dimension is finite but in their later model RS2 the there is only one brane and the fifth dimension is infinite.[151] The Randall-Sundrum model RS1 solves the hierarchy problem and does not have the draw backs of the Kaluza-Klein five dimensional spacetime.

To simulate the postmodern physics in the music of this opera, Parra created an extensive fabric of sound to represent spatial extension, time, mass, and energy. For example sounds in our own weak brane where the baritone is trapped are weaker and less spectrally rich than those of the soprano as she approaches the gravity brane along the fifth dimension because mass and energy increase as she moves towards it. The result is exciting music.

8.4. Concluding Remarks

Physics has gone through two great revolutions. The first was heralded by Newton in the seventeenth century; the second at the beginning of the twentieth century strictly cannot be linked to one person but it would be churlish not to personify it with the name of Einstein since his theory of relativity marked the beginning of the end to a strictly mechanical universe. Both revolutions have had a marked effect on all the imaginative arts. Imagination has to be new, Ezra Pound's catch cry 'make it new' still rings true. Modern physics did not initiate modernism but I do not think it is chance that both occurred at about the same time. We humans are social animals and though individuals are not governed completely by society it does have a significant effect on us. In this general sense the arts affect science as well as science affecting the arts.

Postmodern physics is significantly different to modern physics but it is not a revolution, which I think partly explains the reason why postmodern physics has apparently not had such a large affect on the arts. There are other reasons. One I think is that society has become seduced by technology. When now people talk about science they are more usually actually talking about technology. Culture should not be affected by technology which should be society's servant not savant. Another reason may be that it is just too soon to make judgements on the influence of postmodern physics.

8.5. Notes

[1] Lewis (2007).
[2] Higgins (1978).
[3] Bradshaw (1996).
[4] *The Times* 26th May 1922.
[5] *Nation and Athenaeum*, 24 June 1922.
[6] Gossin (2003).
[7] March (1970). March's book won the Science writing Award of the American Institute of Physics in 1971.
[8] Lederman and Hill (2011).
[9] Tilghman (2010).
[10] Feynman (1985b).
[11] Russell (2009).
[12] Lewis (1993).

[13] See §7.5.1.
[14] Sean Carroll www.preposterousuniverse.com March 18th 2013.
[15] Emerson (2010).
[16] Vierech was a poet and a journalist who interviewed many famous and infamous people between the wars. He was born in Germany of a German father and an American mother; the family immigrated to the US when he was a child. Vierech was a Nazi sympathizer and was imprisoned in the USA between 1942 and 1947.
[17] Philadelphia Saturday Evening Post , October 26th, 1929.
[18] Scott-James (1908).
[19] Untermeyer (1922).
[20] Cornell (2015).
[21] Whitworth (2001).
[22] Pound (1910).
[23] Pound (1973), original published in twelve parts in the *New Age* between 1911 and 1912.
[24] Edbury (2014).
[25] Interview of Samuel Beckett by Israel Shenker, New York Times, 6th May 1956.
[26] Nicholson (1931).
[27] Ellmann (1982).
[28] Joyce (1966).
[29] Joyce (1971).
[30] Beckett (1928) was the first to point out the influence that Vico and Scienza Nuova had on Joyce in the writing of *Work in Progress* (*Finnegans Wake*).
[31] Duszenko (1994a).
[32] Drouin (2010).
[33] Beckett (1929).
[34] Gilbert (1929).
[35] Duszenko (1994a,b).
[36] Joyce (1964).
[37] See §1.2.1.
[38] Herring (1972).
[39] McCarthy (1984).
[40] Joyce (1969).
[41] Drouin (2014).
[42] Alexander (1920).
[43] Garner (1990).
[44] Misspelled by Joyce, Lobatschewsky.
[45] Introduced to Virginia by her brother Thoby who was a fellow Apostle.
[46] Women were admitted to Girton College Cambridge from 1869 but could not obtain a full degree until 1936. At the University of London women could obtain a degree from 1880.
[47] Gordon (2004b).
[48] George Thomson was the son of J.J. Thomson.. He shared the Nobel Prize for physics in 1937.
[49] Henry (2003).
[50] Beer (1996). This section relies heavily on Gillian Beer's essay.
[51] See the introduction to Chapter 1.
[52] Woolf (1931).
[53] Eddington (1928).
[54] Woolf (1927).

[55] Moore (1903).
[56] Woolf (1978).
[57] Sim (2010).
[58] Woolf (1925).
[59] These early biographical details taken from Worthen (2004).
[60] Salter (1984).
[61] Lawrence (2002).
[62] Lawrence (1995).
[63] Lawrence (1922).
[64] The bent towards the humanities in Aldous and his father may have come from Leonard's Australian mother Henrietta Ann Heathorn whose father was Henry Heathorn an advocate of Bathurst, Australia.
[65] Andrew had initially read natural sciences at Trinity College, Cambridge before he switched to physiology.
[66] Huxley (1963).
[67] McGilvary (1930).
[68] Huxley (1923).
[69] See §5.3.1.1. Huxley repeated part of the quotation of Wordsworth in *Lyrical Ballads* in his *Subject-Matter of Poetry*.
[70] Published in the *Nation and Athenaeum* and also reprinted in Sullivan's *Aspects of Science*, Sullivan (1923b).
[71] Huxley (1994).
[72] Huxley (1928).
[73] Huxley (1998).
[74] Durrell (1976).
[75] To a physicist this use of subject-object seems strange and is in fact due to the Galilean transformation not the Lorentzian transformation and comes from the fact that nothing is absolutely at rest, see §1.2.1.1.
[76] Durrell (1959–1960).
[77] Durrell (1998).
[78] Zahlan (1999).
[79] Vadim's own abbreviation for *Look at the Harlequins!*
[80] A Letter from the Publisher, *Time*, 23rd May 1969.
[81] Johnson (1984).
[82] Hayles (1984).
[83] Nabokov (2000).
[84] Gardner (2005).
[85] For which they received the Nobel Prize for Physics in 1980.
[86] Wigner (1965).
[87] Orgitrons are not mentioned elsewhere in *Ada* and one assumes that they are symmetrical anti-organs similar to the symmetrical antiparticles of physics.
[88] Gardner explained the joke in his third edition of *The Ambidextrous Universe*. Gardner (2005).
[89] Guyau (1890). Nabokov seems to have read about Guyau's ideas in Gerald Whitrow's (1912–2000) *The Natural Philosophy of Time*. Whitrow (1961).
[90] Menahem Paz (2008) believes that Nabokov employed Everett's alternative universes interpretation of quantum mechanics in his conception of Terra and Antiterra (see §6.4.5.4) but I

see no evidence of Everett's ideas in *Ada* and think it is more likely that Nabokov was extending the concept of symmetry.

[91] Distractive glebe is a deliberate misquotation of 'In this distractive globe' from Hamlet Act I, Scene V, where Hamlet replies to his father's ghost. Nabokov was found of misquotations and mistranslations.

[92] Gardner's name *The Ozma Problem* derived from the SETI experiment begun in 1960 by Frank Drake to search for extraterrestrial life through interstellar radio waves. Drake was a fan of Frank Baum (1856–1919) and his series of books about Oz, the most well known being the first *The Wizard of Oz* published in 1900. The program was named after Princess Ozma who appears in all of the books except the first and was inspired by the supposed radio communication with Oz by Baum to discover the events in Oz. The Ozma Problem is described in Chapter 18, Gardner (2005).

[93] Kant (1783). See §4.1.4 and Bennett (1970).

[94] James (1950).

[95] Feynman *et al.* (2011).

[96] Jeanne-Claude died in 2009.

[97] Karlholm (2009).

[98] Henderson (1983).

[99] Jammer (1993).

[100] Hinton lost his position at Uppingham School because, under an assumed name, he went through a bigamist marriage. He left England in 1887 first for Japan and from 1893 to 1900 he was on the faculty of Princeton University. From then on he worked for the US Patent Office.

[101] Hinton (1980).

[102] Russell (1904).

[103] This feat is impossible in three dimensions but can be accomplished in four dimensions.

[104] Abbott (1884).

[105] This link to Hinton and the fourth dimension was very possibly because Abbott knew of Hinton through a life-long friend Howard Candler who was the mathematics master at Uppingham School at the same time that Hinton was there and Abbott and his family frequently visited Candler. Banchoff (1990).

[106] Henderson (2009).

[107] Rutter (1926).

[108] Henderson (2013).

[109] The creaky house of Picasso at the time he was painting *Les Demoiselles d'Avignon* was affectionately known as Le Bateau-Lavoir (the wash boat) because in stormy weather it creaked and swayed like the wash boats on the Seine.

[110] According to Miller (2001) Picasso's model and one time girl friend Alice Géry (1884–1975), who was Princet's mistress, introduced the two men in 1905. Henderson (1913) suggests that Princet became known to Picasso's group at least by the middle of 1907 when Miller states that Princet frequented cafés with Picasso's group in spring 1907.

[111] Miller (2001).

[112] Henderson (1983). Poincaré's book *La Science et l'Hypothèsee* (1905) is also often mentioned as another source but only one chapter in this book is on space and geometry and the details about the fourth dimension in it are slight and very general.

[113] Henderson (1988).

[114] The word Dada may be related to the French word meaning hobbyhorse but was chosen to be meaningless.

[115] Walter and Louise Arensberg were wealthy art collectors.

[116] Quoted in Henderson (1983).
[117] Hatch (2010) is more reserved in his comments on Doesburg's knowledge of the theory of relativity.
[118] Eggleling and Richter had visions of a new universal language for film that could be understood by all cultures and wrote a pamphlet *Universelle Sprache* in 1920, which they sent to influential people including Einstein but it had nothing to do with the theory of relativity. Elder (2007).
[119] In Cyrillic script Проун is an acronym for Проекты Утвержденного Нового, which means New Projects Approved.
[120] Lissitzky (1968).
[121] Minkowski (1908).
[122] Hatch (2010).
[123] Hammer and Lodder (2000).
[124] Gabo and Pevsner (1920).
[125] Gavin Parkinson's book published in 2008 is the first to deal in detail with the influence of modern physics on Surrealism.
[126] Breton (1969) he also issued a second manifesto in 1929.
[127] Dali (1935).
[128] Paalen (1942a).
[129] Paalen (1942b).
[130] Paalen (1944).
[131] Paalen (1942c).
[132] Parkinson (2004).
[133] Clark and Ritchie (2015).
[134] Alaina Levine (2014) reviews three other artists who have studied physics before turning to art: Paul Friedlander, Robert Lang, and Thomas Babinec.
[135] Voss-Andreae (2011).
[136] See §1.2.2.4.
[137] Schöenberg (1975).
[138] There have been attempts to link atonality with the indeterminism of modern physics, see for example Delaere and Daly (1990), but they are not convincing.
[139] See §1.2.
[140] Broadhurst (2012).
[141] Owens (1977).
[142] Einstein (1950).
[143] Einstein signed a letter written by Leó Szilárd that was sent to President Roosevelt in 1939 Roosevelt warning him that the Nazis were working an atomic bomb and urging him to initiate similar work in the United States. Einstein was reluctant to support the building of the Bomb and in an 1947 interview with *Newsweek* said 'had I known that the Germans would not succeed in developing an atomic bomb, I would have done nothing.' He himself was not involved with the Manhattan Project.
[144] Einstein (1916a).
[145] The landmark paper was Méthodes de calcul différentiel absolu et leurs applications, published in *Mathematische Annalen* in 1900 by two Italian mathematicians Gregorio Ricci-Curbastro and Tullio Levi-Civita, (absolute differential calculus is now called tensor analysis).
[146] It is a little unfortunate that Parra has chosen to abbreviate one of the more usual terms because the term stress tensor is used for the tensor description of the state of stress in a solid in classical physics.
[147] Parra (2011).

[148] The quotation was in French and Parra used the word *impulsion*, which I have translated as momentum but there is a close connection in classical physics between impulse and momentum; an impulse causes a corresponding change in momentum. Also in Spanish *impulso* can be translated as either impulse or momentum.
[149] Parra's 2011 programme notes to *Caressant l'Horizon*.
[150] Parra (2013).
[151] Randall and Sundrum (1999a, 1999b).

Epilogue

The science discussed in this book has been limited to physics so while much of what has been written applies to science in general, not necessarily all of it does. Physics is not synonymous with science and, since I have been arguing for the unity of culture which embraces all intellectual and artistic endeavours, by only writing about physical science I appear to be introducing my own division of science and culture. I have limited what I have written to the affect of physics because the book would otherwise be unmanageable and also physics is what is most familiar to me. However despite this limitation, what I have written about physics here is mostly typical of science in general.

The knowledge revolution that began in the sixteenth century was, especially in England, as much about the replacing of the scholastic tradition by Erasmian humanism as it was about science. There was a change in the categories of knowledge in the seventeenth century from the Aristotelian division of knowledge between certain, such as mathematics or philosophy, and knowledge that is not certain but probable or a matter of opinion such as rhetoric. During the sixteenth and seventeenth century this simple division in knowledge was replaced by a family of categories whose truth ranged from absolute certainty through various degrees of probability to speculation. Scholasticism emphasized dialectic logic whereas humanism was more in tune with the pragmatism in English science, in the rest of Europe science was less based on pragmatism and less affected by the new humanism. In the seventeenth century the new scientific ideas were shared by the educated in England and were an integral part of their culture. The new science was taken to the English universities where at first it was not a formal part of the education but more of a pastime. Oxford took the lead with the establishment of chairs in astronomy and geometry with the endowments in 1620 by the classical scholar Henry Savile (1549–1622). The sciences were Savile's first interest as a youth and his classical scholarship only took precedence from the 1590s. Earlier Savile had lectured on both geometry and astronomy and was the first to teach the Copernican system at an English university. These Savilian chairs were quickly followed by natural philosophy in 1621 and anatomy in 1624. After the establishment of the new chairs, geometry, astronomy, and natural philosophy became part of the curriculum. The library at Oxford, which had been stripped during the sixteenth century in the Reformation, was refounded by Thomas Bodley in 1602, one floor of the new library was devoted to mathematics and the frieze in the reading room showed men of the scientific revolution such as Copernicus as well as past academic and religious figures. Although in Oxford University the prime place in the curriculum was given to the humanities, mathematics and physics became integral to an undergraduate education. Cambridge University lagged behind Oxford in establishing chairs in mathematics and the sciences but they too required geometry, astronomy, and natural philosophy for a degree. So in England during the seventeenth century the sciences became part of the curriculum and were accepted as a part of intellectual life. Education was wide and the idea of specialization was foreign. None would have understood the idea of two cultures.

In England the latitudinarians saw that natural philosophy could supplement religious revelation. While latitudinarians did not form a majority in the English church, they contained some of the most able minds and influenced a wide section of the educated population. Thus natural philosophy became an accepted part of the culture of society and religion was the first facet of society to make use of it. During its first decades the Royal Society itself promoted the use of natural philosophy in the design argument for the existence of God. Most religious figures did not object to natural philosophers trying to determine the laws of nature but some like Bishop Berkeley objected to trying to understand their metaphysical implications.

Natural philosophy was used by deists and atheists to argue both for and against Christianity. Advocates of materialism also made use of natural philosophy and in the nineteenth century John Tyndall's Belfast Address in 1874 shocked the Christian world. Thermodynamics revitalized the use of physics in religious debate during the nineteenth century. In the twentieth century the use of physics in religious debate continued but modern physics has not seemed to have the same impact that classical physics did. Although the modern cosmology and the Big Bang theory gave the application of physics to religion another burst of life and even belatedly brought the Roman Catholic Church to realize that physics could have a positive a part to play in their religion. However, the dialogue between believers and non-believers using modern cosmology has largely been played out between professionals and has not involved the general public to any great extent probably because most people no longer think religion important.

Today only a few scientists become celebrity figures, not that many desire to be one, but more fame is given to the ability to get a ball into a hole than to explaining the nature of the universe. In the seventeenth century Newton was a celebrity, though very few understood his *Principia*, there was a clamour among the educated for knowledge of the new natural philosophy. Initially this knowledge was supplied by churchmen but coffee-houses soon proved to be more convivial and lectures in them on Newton's new natural philosophy became popular. Most lecturers had benefited from the newly developed university courses on natural philosophy but ordinary folk like Benjamin Martin, who started life as a ploughboy, also educated themselves and became lecturers. The popular lectures on natural philosophy were expensive in the early eighteenth century but by the middle of the century lecturers like Martin reduced their cost and they became more affordable to artisans. During the eighteenth century popular lectures were organized by the lecturers themselves but by the nineteenth century lectures, institutes, and associations became important in disseminating natural philosophy. The mechanics' institutes catered for artisans and initially the Royal Institution also attracted a wide audience from the aristocrats to the working class. The British Association was primarily aimed at scientists but it did also cater for all who had an interest in science. The renamed British Science Association and the Royal Institution have continued to function up until the present, though the Royal Institution has faced recent financial problems that caused it even to consider selling its historic home in Albemarle Street, London. The original intention of the mechanics' institutes was to educate artisans in the physical sciences but by the second half of the nineteenth century their intentions became more industrially orientated and eventually in the twentieth century they no longer fulfilled their original intention. Initially in the twentieth century the Workers Education Association replaced the function of the mechanics' institutes but they too have now almost abandoned teaching science completely.

Books were expensive in the eighteenth century and beyond the reach of most artisans. One of the important early functions of the mechanics' institutes in the nineteenth century was to provide libraries. However those, who could afford them, bought popular books on physical science. Women as well as men have been interested in the physical sciences and popular books were published specifically for women in the eighteenth century. Children also had their own books and Faraday introduced the Christmas Lectures at the Royal Institution for children in the early nineteenth century.

Einstein became the celebrity of the second revolution in physics almost overnight in November 1919 when his general theory of relativity accurately predicted the deflection of starlight passing close to the Sun. Newspapers on both sides of the Atlantic were full of Einstein and his story until the end of that year. Quantum mechanics did not quite have the same sudden impact on the general public as relativity. Cosmology, because it comments on the origins of the universe has had an impact on the mind of most, and many very successful popular books were published before the Second World War. When in the swinging sixties the Big Bang theory emerged as the most probable scenario for creation, cosmology gained in popularity. The emergence of mystical black holes and dark matter further cemented the popularity of cosmology and Stephen Hawking, with his book *A Brief History of Time* that put them altogether became another celebrity in physics for the general populace.

Before the nineteenth century philosophy and physics were not completely separate and the term natural philosophy was not entirely synonymous with what today is called physics. Natural philosophy was part of philosophy. The real difference between Descartes' *Principles of Philosophy* and Newton's *Mathematical Principles of Natural Philosophy* was not that one was mainly philosophy and the other just physics but that Newton used mathematical principles. During the nineteenth century the modern day professional specialization began and scientists started to separate themselves from philosophers but almost immediately some scientists felt the need to understand the philosophy of their discipline. William Whewell who had coined the word *scientist* in 1833 also in effect renamed in 1837 what had been the philosophy part of natural philosophy the *philosophy of science* in his *History of the Inductive Sciences*. What is clear is that however we categorize philosophy it must be influenced by physics and physics itself cannot be independent of the other sciences. Even Kant in his *Critique of Pure Reason* admitted that all knowledge begins with experience even if it is possible to have *a priori* pure reason. The philosophical implications of modern physics, especially quantum mechanics and string theory, have captured the attention of both philosophers and physicists

As I have already noted, Einstein himself saw that scientists and artists both made use of imagination and what would be surprising is if sometimes images in the imaginative arts did not come from physics. Here I have only highlighted a few of the more notable examples of this use of common images. Postmodernist physics seems to be a symbiotic relationship with the postmodernist arts. Like philosophy, music and the visual arts also have a tangible relationship with physics distinct from any imaginative relationship. What I have not much shown is the influence of our wider culture on physics not because I believe that there is none but because it is much more difficult to demonstrate such influence.

In the background of my mind, while writing this book, has been the question why a rancorous dichotomy developed between science and the humanities in the twentieth century. That the protagonists of the most well known example of Snow and Leavis should both be Cambridge men is completely inexplicable to me. Collegiate life in Cambridge or Oxford fosters an appreciation and understanding of all the facets of culture, I would have thought that Cambridge would have been the last place in the world where the childish argument of 'what

I do is better and more important than what you do' would take place. For such an argument to start, one has first to have people who work in well defined areas, which are quite separate. In the seventeenth century there was a unity to education. Specialization gradually became necessary after the seventeenth century because of the burgeoning sciences and the humanities, and today a unified education is practically impossible. Because science and its offshoot technology have an economic advantage over the humanities, they have prospered at the expense of the humanities, especially since the Second World War. This dichotomy has led to some in the humanities being resentful of the success of the sciences and to some in the sciences being disdainful of the humanities because of their own success and financial supremacy. I believe that a wide knowledge in both science and the humanities marks an educated person. A liberal arts course tacked on to a science degree or the reverse does not work well. The huge numbers of university students now make a collegiate life practically impossible for all but, with the example of Snow and Leavis, it does not seem to work for all in any case. It is difficult in the postmodern world to be an educated person fully aware of our culture. Although I show that many do overcome this difficulty and do develop our rich culture.

I hope, I have to some extent, demonstrated that the physical sciences have had an influence on the humanities since they became separated and that there is only one human culture even if some still believe in two separate cultures.

Bibliography

Abbott, B.P. *et al.* (2016) Observation of gravitational waves from a binary black hole merger, *Phys. Rev. Lett.*, 166, 061 102:1-16.

Abbott, E.A. (1884) *Flatland: A Romance of Many Dimensions*, Seely & Co., London.

d'Abro, A. (1939) *The Decline of Mechanism* (*in Modern Physics*), Van Nostrand, New York.

Adams, F.C. (2008) Stars in other universes: stellar structure with different fundamental constants, *J. Cosmol. Astropart. Phys.*, 08010:1-28.

Adams, R.M. (1954) The text of "Paradise Lost:" emphatic and unemphatic spellings, *Mod. Philology*, 52:84-91.

D'Agostino, S. (2004) The Bild conception of physical theory Helmholtz, Hertz, and Schrödinger, *Phys. Persp.*, 6:372-389.

Airy, G.B. (1859) On the apparent projection of stars upon the Moon's disk in occultations, *Mon. Not. Roy. Astron. Soc.*, 19:208-211.

d'Alembert (1751) *Preliminary Discourse to the Encyclopedia*, Trans. Schwab, R.N., and Rex, W.E. (1995), University of Chicago Press, Chicago.

Alexander, S. (1920) *Space, Time and Deity*, 2 Vols., The Humanities Press, New York.

Algarotti, F. (1739) *Sir Isaac Newton's Philosophy Explained for the Use of the Ladies: In Six Dialogues on Light and Colour*, Trans. Carter, E., Printed for E. Cave, London.

Allen, C. (1997) The principle of sufficient reason and the uncaused beginning of the universe, *Dialogue – Can. Philos.*, 36:555-562.

Amaldi, U., de Boer, W., and Fürstenau, H. (1991) Comparison of grand unified theories with electroweak and strong coupling constants measured at LEP, *Phys. Lett.*, 260:447-455.

de Andrade, E.M.P., Faber, J., and Rosa, L.P. (2013) A spontaneous physics philosophy on the concept, ether throughout the history of science: Birth, death and revival, *Found. Sci.*, 18:559-577.

Andrade, E.N. da C. (1923) *The Structure of the Atom*, Bell and Sons, London.

Andrade, E.N. da C. (1927) *The Atom*, Thomas Nelson and Sons, London.

Andrade, E.N. da C., and Huxley, J. (1934) *Simple Science*, Basil Blackwell, Oxford.

Anon. (1894) Strange light on mars, *Nature*, 50:319.

Anon. (1922) Relativity in the films, *Sci. Am.*, 127:92.

Anstey, P.R. (2005) Experimental versus speculative natural philosophy, In: *The Science of Nature in the Seventeenth Century*, Eds. Anstey, P.R., and Schuster, J.A., Springer, Dordrecht, The Netherlands.

Appleby, J.H. (1990) Erasmus King: Eighteenth-century experimental philosopher, *Ann. Sci.*, 47:375-392.

Aspect, A. (1999) Bell's inequality test: more ideal than ever, *Nature*, 398:189-190.

Aston, N. (2004) From personality to party: the creation and transmission of Hutchinsonianism, *Stud. Hist. Philos. Sci.*, 35:625–644.

Astore, W.J. (2004) Dick, Thomas (1774–1857), *Oxford Dictionary of National Biography*, Oxford University Press, Oxford.

Atkatz, D., and Pagels, H. (1982) Origin of the universe as a quantum tunneling event, *Phys. Rev. D*, 25:2065-2073.

Atkinson, P. (1897) *Electricity for Everybody*, 2nd edition, The Century Co., New York.

Auerbach, J. (2001) The Great Exhibition and historical memory, *J. Vic. Cult.*, 6:89-112.

Austin, W.H. (1970) Isaac Newton on science and religion, *J. Hist. Ideas*, 31:521-542.

Axtell, J.L. (1965) Locke's Review of the "Principia", *Notes Rec. Roy. Soc. Lond.*, 20:152-161.

Bacon, F. (1620) *Novum Organum*, Trans. Spedding, J. (1858) *The Collected Works of Francis Bacon*, Vol. 4, Longmans, London.

Baker, T. (1700) *Reflections upon Learning wherein is Shewn the Insufficiency Thereof, in its Several Particulars, in Order to Evince the Usefulness and Necessity of Revelation*, 2nd edition, Printed for A. Bosvile, London.

Banchoff, T.F. (1990) *From Flatland to Hypergraphics: Interacting with Higher Dimensions*, *Interdiscipc. Sci. Rev.*, 15:364-372.

Banerjee, N. and Sen, S. (1997) Einstein pseudotensor and total energy of the universe, *Pramana*, 49:609-615.

Banfield, M. (2007) Darwinism, doxology, and energy physics: The new sciences, the poetry and the poetics of Gerard Manley Hopkins, *Vict. Poet.*, 45:175-194.

Barnett, J.R., Miller, S., and Pearce, E. (2006) Colour and art: A brief history of pigments, *Opt. Laser Technol.*, 38:445–453.

Barrow, J.D., and Tipler, F.J. (1986) *Anthropic Cosmological Principle*, Oxford University Press, Oxford.

Barton, R. (1987) John Tyndall, pantheist: a rereading of the Belfast Address, *Osiris*, 3:111-134.

Bausch, A., and Meyers, S. (2002) New experimental limit on the validity of local position invariance, *Phys. Rev. D*, 65:081101.

Beardley, J. (1872) *The Earth Proved not be a Sphere, nor the Surface of Water upon it Convex*, F. Pittman, London.

Beckett, R.B. (1970) *John Constable's Discourses*, Suffolk Rec. Soc., Vol XIV, Ipswich.

Beckett, S. (1929) Dante… Bruno. Vico…Joyce, In *Our Exagmination Round His Factification for Incamination of Work in Progress*, Beckett, S., Brion, M., Budgen, F., Gilbert, S., Jolas, E.J., Llona, V., McAlmon, R., McGreevy, T., Paul, E., Rodker, J., Sage, R., and Williams, W.C., with letter of protest, Slingsby, G.V.L., and Dixon, V., Faber and Faber, London.

Beech, M. (1989) The great meteor of 18th August 1783, *J. Brit. Astron. Assoc.*, 99:130-134.

Beer, G. (1991) Helmholtz, Tyndall, Gerard Manley Hopkins: Leaps of the prepared imagination, In *Comparative Criticism*, Vol. 13, *Literature and Science*, Ed. Shaffer, E.S., Cambridge University Press, Cambridge.

Beer, G. (1996) Physics, sound, and substance: Later Woolf, In *Virginia Woolf: The Common Ground*, Edinburgh University Press, Edinburgh.

Bell, J.S. (1987) *Speakable and Unspeakable in Quantum Mechanics*, Cambridge University Press, Cambridge.

Bender, R., and Saglia, R.P. (2007) Supermassive black holes in local galaxies, *C. R. Phys.*, 8:16-25.

Bennett, J. (1970) The difference between right and left, *Am. Philos. Quart.*, 7:175-191.

Bentley, R. (1699) *The Folly and Unreasonableness of Atheism Demonstrated from the Advantage and Pleasure of a Religious Life, the Faculties of Humane Souls, the Structure of Animate Bodies, & the Origin and Frame of the World*, 4th edition, Printed for H. Mortlock, London.

Bergh, van den, S. The early history of dark matter, *Publ. Astron. Soc. Pac.*, 111:657-666.

Berkeley, G. (1710) Principles of Human Knowledge, In *The Works of George Berkeley, Bishop of Cloyne,* Vol. 2, Eds. Luce, A.A. and Jessop, T.E., (1949) Thomas. Nelson, London.

Berkeley, G. (1713) Three Dialogues Between Hylas and Philonous, in Opposition to Sceptics and Atheists, In *The Works of George Berkeley, Bishop of Cloyne*, Vol. 2, Eds. Luce, A.A. and Jessop, T.E., (1949) Thomas. Nelson, London.

Berkeley, G. (1721) De Motu, In *The Works of George Berkeley, Bishop of Cloyne*, Vol. 4, Eds. Luce, A.A. and Jessop, T.E., (1951) Thomas. Nelson, London.

Berington, S. (1750) *Dissertations on the Mosaical Creation, Deluge, Building of Babel, and Confusion of Tongues*, Printed for the author, London.

Berman, M. (1972) The early years of the Royal Institution 1799–1810: A re-evaluation, *Soc. Sci. Stud.*, 2:205-240.

Bernstein, J. (1997) Heaven's net: The meeting of John Donne and Johannes Kepler, *Am. Scholar*, 66:175-195.

Bertotti, B., Iess, L., and Tortora, P. (2003) A test of general relativity using radio links with the Cassini spacecraft, *Nature*, 425:374-376.

van Besouw, J. (2013) *Einstein and the Popular Image of Science in the Netherlands*, Master's Dissertation, Utrecht University.

Bindman, D. (1977) *Blake as an Artist*, Oxford University Press, Oxford.

Blackmore, J.T. (1972) *Ernst Mach-His Life, Work & Influence*, University of California Press, Berkeley.

Blake, W. (1904) *Jerusalem*, Eds. Maclagan, E.R.D., and Russell, A.G.B., A.H. Bullen, London.

Blake, W. (1907) *Milton*, Eds. Maclagan, E.R.D., and Russell, A.G.B., A.H. Bullen, London.

Blaikie, W.G. (1880) *The Personal Life of David Livingstone*, J. Murray, London.

Bohm, D. (1951) *Quantum theory*, Prentice Hall, New York.

Bohm, D. (1952) A suggested interpretation of the quantum theory in terms of "hidden" variables, *Phys. Rev.*, 52:166-193.

Bohm, D. (1957) *Causality and Chance in Modern Physics*, Routledge & Kegan Paul, London.

Bohm, D. (1980) *Wholeness and the Implicate Order*, Routledge & Kegan Paul, London.

Bohm, D., and Hiley, B.J. (1993) *The Undivided Universe*, Routledge, London.

Bohr, N. (1913) On the constitution of atoms and molecules. Part I. *Phil. Mag.* 26:1-24; Part II. Systems containing only a single nucleus.26:476-502; Part III. Systems containing several nuclei. *Phil. Mag.*, 26:857-875.

Bohr, N. (1914) On the effect of electric and magnetic fields on spectral lines. *Phil. Mag.*, 27:506-524.

Bohr, N. (1918) *On the Quantum Theory of Line Spectra. Part I. On the General Theory; Part II. On the Hydrogen Spectrum*. Høst & Son, Copenhagen.

Bohr, N. (1921) Atomic structure, *Nature*, 107:104-107.

Bohr, N. (1935) Can quantum-mechanical description of physical reality be considered complete? *Phys. Rev.*, 48:696-702.

Bohr, N. (1949) Discussion with Einstein on epistemological problems in atomic physics, In *The Library of Living Philosophers, Vol 7, Albert Einstein: Philosopher-Scientist*, pp.199-241, Ed. Schilpp, P.A., Open Court, Chicago.

Bohr, N. (1958) Biology and atomic physics, In *Atomic Physics and Human Knowledge*, Ed. Bohr, N., Wiley, New York.

Bolton, L. (1921a) Relativity. The winning essay for the Eugene Higgins five thousand dollar prize, *Sci. Amer.*, 124:105-115.

Bolton, L. (1921b) *An Introduction to the Theory of Relativity*, Methuen, London.

Bondi, H. (1952) *Cosmology*, Cambridge University, Press, Cambridge.

Bondi, H., Bonnor, W.B., Lyttleton, R.A., and Whitrow, G.J. (1960) *Rival Theories of Cosmology*, Oxford University Press, Oxford.

Bondi, H., and Gold, T. (1948) The steady state theory for the expanding universe, *Mon. Not. Roy. Astron. Soc.*, 108:252-270.

Booth, S.A., and van Helden, A. (2001) The Virgin and the telescope: The moons of Cigoli and Galileo, In *Galileo in Context*, Ed. Renn, J., Cambridge University Press, Cambridge.

Born, M. (1926) Quantenmechanik der Stoβvorgänge, *Z. Phys.*, 38:803-827.

Born, M., and Einstein, A. (1971) *The Born-Einstein Letters, with Commentaries by Max Born*, Trans. Born, I., Macmillan Press, London.

Born, M., Heisenberg, W., and Jordan, P. (1925) Zur Quantenmechanik II, *Z. Phys.*, 35:557-615; English trans., On quantum mechanics II, In *Sources of Quantum Mechanics*, pp. 321-386, Ed. Van der Waerden, B.L. (1967) North-Holland Publishing Co., Amsterdam.

Born, M., and Jordan, P. (1925) Zur Quantenmechanik I, *Z. Phys.*, 34:858-888; English trans. On quantum mechanics I, In *Sources of Quantum Mechanics*, pp. 277-306, Ed., Van der Waerden, B.L., (1967) North-Holland Publishing Co., Amsterdam.

Bowen, I.S. (1927) The origin of the nebulium spectrum, *Nature*, 120:473.

Bowler, P.J. (2009) *Science for All: The Popularization of Science in Early Twentieth-Century Britain*, University of Chicago, Chicago.

Boyle, R. (1660) *New Experiments Physico-Mechanical, touching the Spring of the Air, and its Effects*, Printed for Thomas Robinson, Oxford.

Boyle, R. (1691) *The Sceptical Chymist or Chymico-Physical Doubts & Paradoxes*, Printed for J. Crooke, London.

Bradshaw, D. (1996) The Best of Companions: J. W. N. Sullivan, Aldous Huxley, and the New Physics, *Rev. Eng. Stud.*, 47:188-206, 352-368.

Bragg, W.H. (1921) *Electrons and Ether Waves, Being the Twenty-Third Robert Boyle Lecture*. Humprey Milford, Oxford University Press, London.

Bragg, W.H. (1925) *Concerning the Nature of Things*, G. Bell & Sons, London.

Bragg, W.H., and Bragg, W.L. (1915) *X-rays and Crystal Structure*, G. Bell & Sons, London.

Bragg, W.L. (1914) The diffraction of short electromagnetic waves by a crystal, *Proc. Cambridge. Philos. Soc.*, 17:43-57.

Brauner, B. (1924). Einstein and Mach, *Nature*, 113:927.

Brentano, F. (1987) *On the Existence of God*, Ed. and Trans. Kranz, . S.F., Nijhoff, Dordrecht.

Breton, A. (1969) *Manifestoes of Surrealism*, Trans. Seaver, R., and Lane. H.R., University of Michigan Press, Ann Arbor, MI.

Brewster, D. (1834) On a new analysis of solar light, indicating three primary colours, forming coincident spectra of equal length, *Trans. Roy. Soc. Edin.*, 12:123-136.

Brewster, D. (1853) *A Treatise on Optics*, New edition, Longman, Brown, Green & Longmans, London.

Broad, C.D., Eddington, A.S., and Braithwaite, R.B. (1931) Symposium: Indeterminacy and Indeterminism, *P. Aristotelian Soc. Sup. Vols.*, 10:135-196.

Broadhurst, S. (2012) Einstein on the Beach: A study in temporality, *Perform. Res.*, 17:34-40.

de Broglie, L. (1953) *The Revolution in Physics: A Non-Mathematical Survey of Quanta*, Trans. Niemeyer, R.W., Noonday Press, New York.

Brougham, H.P (1825) *Practical Observations on the Education of the People Addressed to the Working Class and their Employers*, Longmans, London.

Brown, C.G. (2001) *The Death of Christian Britain*, Routledge, Abington.

Brown, D. (2012) William Rowan Hamilton and William Wordsworth: The poetry of Science, *Stud. Romanticism*, 51:475-501.

Buchdahl, G. (1951) Sources of scepticism in atomic theory, *Brit. J. Philos. Sci.*, 10:120-134.

Buffon, (G.-L. Leclerc) (1785) *Natural History, General and Particular*, Trans. Smellie, W., 2nd edition, Vol. 1, Printed for W. Strahan and T. Cadell, London.

Burnet, T. (1684) *The Theory of the Earth Containing an Account of the Original of the Earth, and of All the General Changes Which It Hath Already Undergone, or is to Undergo, Till the Consummation of All Things*, Printed for W. Kettilby, London.

Burtt, E.A. (1951) *The Metaphysical Foundations of Modern Science*, Humanities Press, New York.

Bush, R. (2004) Eliot, Thomas Stearns (1888–1965), *Oxford Dictionary of National Biography*, Oxford University Press.

Caballero-Nieves, S.M. *et al.* (2009) The ultraviolet spectrum and physical properties of the mass donor star in HD 226868 = Cygnus X-1, *Astrophys. J.*, 701:1895–1905.

Cahan, D. (2012) Helmholtz and the British scientific elite: From force conservation to energy conservation, *Notes Rec. Roy. Soc.*, 66:55–68.

Callender, C. (2005) Answers in search of a question: 'proofs' of the tri-dimensionality of space, *Stud. Hist. Philos. M. P.*, 36:113–136.

Campbell, G. (2004) Milton, John (1608–1674), *Oxford Dictionary of National Biography*, Oxford University Press, Oxford.

Campbell, N.R. (1914) The structure of the atom, *Nature*, 92:586-587.

Campbell, N.R. (1921) *What is Science?* Methuen & Co. Ltd., London.

Campbell, N.R. (1923) *Modern Electrical Theory, Supplementary Chapter XVII, The Structure of the Atom*, Cambridge University Press, Cambridge.

Campion, N. (2009) Astronomy and political theory, The role of astronomy in society and culture, *Proc. IAU Sym.* No. 260, pp. 595-602.

Candelas, P., Horowitz, G.T., Strominger, A., and Witten, E. (1985) Vacuum configurations for superstrings, *Nucl. Phys. B*, 258:46-74.

Caneva, K.L. (1997) Colding, Orsted, and the meanings of force, *Hist. Stud. Phys. Biol. Sci.*, 28:1-138.

Cantor, G.N. (1970) Thomas Young's lectures at the Royal Institution, *Notes Rec. Roy. Soc.*, 25:87-112.

Capon, L.J., Ed. (1959) *The Adams-Jefferson Letters: The Complete Correspondence Between Thomas Jefferson and Abigail and John Adams*, Vol. 2, University of North Carolina Press, Chapel Hill, NC.

Capra, F. (1975) *The Tao of Physics*, Fontana, London.

Carr, B. (2008) Introduction and Overview, In *Universe or Multiverse?* Ed. Carr, B., Cambridge University Press, Cambridge.

Cat, C. (2012) Into the 'regions of physical and metaphysical chaos': Maxwell's scientific metaphysics and natural philosophy of action (agency, determinacy and necessity from theology, moral philosophy and history to mathematics, theory and experiment), *Stud. Hist. Philos. Sci.*, 43:91–104.

Catcott, A. (1756) *Remarks on the Second Part of the Lord Bishop of Clogher's Vindication of the Histories of the Old and New Testaments*, London.

Cavendish, M. (1653a) *Poems and Fancies*, Printed for J. Martin, and J. Allestrye, London.

Cavendish, M. (1653b) *Philosophy and Fancies*, Printed for J. Martin, and J. Allestrye, London.

Chalmers, T. (1817) *A Series of Discourses on Christian Revelation Viewed in Connexion with Modern Astronomy*, Kirk and Mercein, New York.

Chamberlen, P. (1715) *A Philosophical Essay upon Actions on Distant Subjects*, 3rd edition, Printed by H. Parker.

Chandrasekhar, S. (1987) *Truth and Beauty: Aesthetics and Motivation in Science*, University of Chicago Press, Chicago.

Chang, H. (1993) A misunderstood rebellion; The twin-paradox controversy and Herbert Dingle's vision of science, *Stud. Hist. Philos. Sci.*, 24:741-790.

Chang, H., and Li S.W. (2005) The absolute and its measurement: William Thomson on temperature, *Ann. Sci.*, 62:281-308.

Christensen, T. (1993) *Rameau and Musical Thought in the Enlightenment*, Cambridge University Press, Cambridge.

Clapeyron, É. (1834) Mémoire sur la puissance motrice de la chaleur, *J. École Polytech.*, 23:153-190.

Clarke, I. (2012) *Negotiating Progress: Promoting 'Modern' Physics in Britain, 1900–1940*, Doctoral Thesis, University of Manchester.

Clarke, S. (1705) *A Demonstration of the Being and Attributes of God: More Particularly in Answer to Mr. Hobbs, Spinoza, and Their Followers*, James Knapton, London.

Clarke, S. (1706) *A Discourse Concerning the Unchangeable Obligations of Natural Religion*, James Knapton, London.

Clarke, S. (1717) *A Collection of Papers Which Passed Between the Late Learned Mr Leibnitz and Dr Clarke in the Years 1715–1716*, James Knapton, London.

Clausius, R. (1851) On the moving force of heat, and the laws regarding the nature of heat itself which are deducible therefrom, *Phil. Mag.*, (4) 2:1-21, 102-119.

Clausius, R. (1865) Ueber verschiedene für die Anwendung bequeme Formen der Hauptgleichung der mechanischen Wämetheorie, *Ann. Phys. Chem. Paris*, 125:353-401, Trans. Hirst, T.A. (1867) On several convenient forms of the fundamental equations of the mechanical theory of heat, In: *Mechanical Theory of Heat with Applications to the Steam Engine and to the Physical Properties of Bodies*, John van Voorst, London.

Clausius, R. (1868) On the second fundamental theorem of the mechanical theory of heat, *Phil. Mag.*, (4) 35:405-419.

Le Clerc, J. (1706) *The Life and Character of Mr John Locke, Author of the Essay Concerning Human Understanding*, Trans. from French, T.F.P, Printed for John Clark, London.

Clerke, A.M. (1902) *A Popular History of Astronomy during the Nineteenth Century*, 4th edition, Adam & Charles Black, London.

Clerke, A.M. (1905) *The System of the Stars*, 2nd edition, Adam & Charles Black, London.

Clerke, A.M. (rev. McConnell, A.) (2004) Perry, Stephen Joseph (1833–1889), *Oxford Dictionary of National Biography*, Oxford University Press, Oxford.

Cobb, A.D. (2012) Is John F. W. Herschel an inductivist about hypothetical inquiry? *Persp. Sci.*, 20:409-439.

Cohen, I.B. (1940) A. d'Abro – the decline of mechanism in modern physics, *Isis*, 32:380-382.
Cohen, I.B. (1953) A classic of modern science, *Yale Rev.*, 43:283-284.
Cohen, I.B. (1979) Notes on Newton in the art and architecture of the Enlightenment, *Vista. Ast.*, 22:523-537.
Cohen, I.B. (1980) *The Newtonian Revolution.* Cambridge University. Press, Cambridge.
Cohen. I.B. (1997) *Science and the Founding Fathers*, W.W. Norton & Co., New York.
Colclough, D. (2004) Donne, John (1572–1631), *Oxford Dictionary of National Biography*, Oxford University Press, Oxford.
Colding, A. (1864) On the history of the principle of the conservation of energy, *Phil. Mag.*, 27: 56-64.
Connell, P. (2009) Newtonian physico-theology and the varieties of Whiggism in James Thomson's The Seasons, *Huntington Lib. Q.*, 72:1-28.
Conrad, J. (1921) Henry James, an appreciation, In *Notes on Life and Letters*, Doubleday, Page & Co., Garden City, NY.
Conrad, J. (1994) *The Secret Agent*, Penguin Books, London.
Conrad, S. (2004) Wilson, James (1742–1798), *Oxford Dictionary of National Biography*, Oxford University Press, Oxford.
Coppola, A. (2008) Retraining the virtuoso's gaze: Behn's Emperor of the Moon, the Royal Society, and the spectacles of science and politics, *Eighteenth-Cent. Stud.*, 41:481-506.
Cornell, E. (2015) Louis Untermeyer's poetic engagement of the popularization of Einstein's relativity theory, *South Cent. Rev.*, 32:48-66.
Cotterell, B. (2010) *Fracture and Life*, Imperial College Press, London.
Cotterell, B., and Kamminga, J. (1992) *Mechanics of Pre-industrial Technology*, Cambridge University Press, Cambridge.
Coulomb, C.A. (1785) Premier mémoire sur l'électricité et le magnétisme, *Histoire de l'Académie Royale des Sciences*, pp. 569-577.
Cowan, C.L., Reines, F., Harrison, F.B., Kruse, H.W., and McGuire, A.D. (1956). Detection of the Free Neutrino: a Confirmation, *Science,* 124:103-104.
Craig, W.L., and Smith, Q. (1995) *Theism, Atheism and Big Bang Cosmology,* Oxford University Press, Oxford.
Creese, M.R.S. (2004) Somerville, Mary (1780–1872), *Oxford Dictionary of National Biography*, Oxford University Press, Oxford.
Crowe, M.J. (1986) *The Extraterrestrial Life Debate (1750–1900)*, Cambridge University Press, Cambridge.
Cunningham, E. (1914) *The Principle of Relativity*, Cambridge University Press, Cambridge.
Cunningham, E. (1922) *Prof. Eddington's Romanes Lecture*, Oxford University Press, Oxford.
Dahm, J.J. (1970) Science and apologetics in the early Boyle Lectures, *Church Hist.*, 39:172-186.
Dali, S. (1935) La conquéte de l'irrationnel, In *The Collected Writings of Salvador Dali,* Trans. Finkelstein, H. (1998) Cambridge University Press, Cambridge.
Dalton, J. (1808) *A New System of Chemical Philosophy*, R. Bickerstaff, London.
Damon, S.F. (2013) *A Blake Dictionary: The Ideas and Symbols of William Blake*, Updated edition, Eaves, M., Dartmouth College Press Hanover, N.H.
Darwin, C. (1909) *The Foundations of the Origin of Species: Two Essays Written in 1842 and 1844*, Ed. Darwin, F., Cambridge University Press, Cambridge.
Darwin, C.G. (1931) *The New Conceptions of Matter*, Bell and Sons, Ltd., London.

Davidson, G (1900) The apparent projection of stars upon the bright limb of the Moon at occultation, and similar phenomena at total solar eclipses, transits of Venus and Mercury, etc., *Proc. Calif. Acad. Sci.*, 1:63-102.

Davies, P.C.W. (Ed.) (1989) *The New Physics*, Cambridge University Press, Cambridge.

Davies, P.C.W., and Brown, J. (Eds.) (1986) *The Ghost in the Atom: A Discussion of the Mysteries of Quantum Physics*, Cambridge University Press, Cambridge.

Davies, P.C.W., and Brown, J. (Eds.) (1988) *Superstrings A Theory of Everything?* Cambridge University Press, Cambridge.

Daw, J.W. (2011) *The Victorian Poetic Imagination and Astronomy: Tennyson, De Quincey, Hopkins and Hardy*, Doctoral Thesis, University of Sussex.

Delaere, M., and Daly, P.H. (1990) Mutations in systems in the natural sciences and music in the first half of the twentieth century, *Int. Rev. Aesth. Sociol. Mus.*, 21:3-28.

Derham, W. (1713) *Physico-Theology: Or, a Demonstration of the Being and Attributes of God, from His Works of Creation*, Printed for W. Innys, London.

Derham, W. (1719) *Astro-Theology: Or, a Demonstration of the Being and Attributes of God, from a Survey of the Heavens*, 3rd edition, Printed by W. and J. Innys, London.

Desaguliers, J.T. (1717) *Physico-Mechanical Lectures, or an Account of What is Explain'd and Demonstrated in the Course*, Printed for the Author, London.

Desaguliers, J.T. (1727) An account of a machine for measuring any depth in the sea, with great expedition and certainty; shewn to the Royal Society, by J. T. Desaguliers, LLD and RSS contriv'd by the Rev. Mr. Stephen Hales, FRS and himself, *Philos. T. R. Soc. Lond.*, 35:559-562.

Desaguliers, J.T. (1728) *The Newtonian System of the World, The Best Model of Government an Allegorical Poem*, Printed for J. Roberts, London.

Desaguliers, J.T. (1734) *A Course of Experimental Philosophy,* Vol. 1, Printed for J. Senex, W, Innys, J. Osborn, and T, Longmans, London.

Desaguliers, J.T. (1744) *A Course of Experimental Philosophy,* Vol. 2, Printed for W, Innys, London.

Descartes, R. (1637, 1644) *Discours de la Méthode, and Meditationes*, Trans., Haldane, E.S., (2003) *Discourse on Method, and Meditations*, Dover, New York.

Descartes, R. (1644) Principia Philosophiae , Trans. Cottingham, J., Stoothoff, R., and Murdock, D. (1985) In *Philosophical Writings of Descartes,* Vol. 1, Cambridge University Press, Cambridge.

Desmond, A., Moore, J., and Browne, J. (2004) Darwin, Charles Robert (1809–1882), *Oxford Dictionary of National Biography*, Oxford University Press.

Dick, T. (1826) *The Christian Philosopher, or the Connection of Science and Philosophy with Religion*, 1st American edition, G. & C. Carvill, New York.

Dick, T. (1838) *Celestial Scenery, Illustrating the Perfections of Deity and a Plurality of Worlds*, Harper & Brothers, New York.

Dick, T. (1844) *The Sidereal Heaven,* Harper & Brothers, New York.

Dick, T. (1845) *The Practical Astronomer*, Seeley, Burnside, and Seeley, London.

Dick, T. (1846) *The Solar System*, The Religious Tract Society, London.

Dickie, R.H., Peebles, P.J.E., Roll, P.G., and Wilkinson, D.T. (1965) Cosmic black-body radiation, Astrophys. J., 142:414-419.

Dingle, H. (1922) *Relativity for All*, Little, Brown and Company, Boston.

Dingle, H. (1924) *Modern Astrophysics*, W. Collins Sons & Co. Ltd., London.

Dingle, H. (1937) Modern Aristotelianism, *Nature*, 139:784-786.

Dingle, H. (1939) The relativity of time, *Nature*, 144:888-890.

Dingle, H. (1956) Relativity and space travel, *Nature*, 177:782-784.

Dirac, P.A.M. (1928a) The quantum theory of the electron, Part I, *Proc. Roy. Soc. Lond.*, 117:610-624.

Dirac, P.A.M. (1928b) The quantum theory of the electron, Part II, *Proc. Roy. Soc. Lond.*, 118:351-361.

Dirac, P.A.M. (1931) Quantised singularities in the electromagnetic field, *Proc. Roy. Soc. Lond.*, 133:60-72.

Dirac, P.A.M. (1939) The relation of mathematics to physics, *Proc. Roy. Soc. Edin.*, 59:122-129.

Dirac, P.A.M. (1978) The monopole concept, *Int. J. Theor. Phys.*, 17:235-247.

Dirac, P.A.M. (1987) The inadequacies of quantum field theory, In *The Dirac memorial Volume*, Eds. Kursunoglu, B., and Wigner, E.P., Cambridge University Press, Cambridge.

Ditton, H. (1705) *The General Laws of Nature and Motion with their Applications to Mechanics*, Printed for J. Seller, C. Price, and J. Senca, London.

Donne, J. (1611a) *Ignatius his Conlave: or the Enthronization of Loyola in Hell*, Printed for R. More, London.

Donne, J. (1611b) *An Anatomy of the World Wherein, by Occasion of the Untimely Death of Mistris Elizabeth Drury the Frailty and the Decay of this Whole World is Represented*, Printed for S. Macham, London.

Doone, J. (1635) *Poems by J.D.*, Printed for J. Marriot, London.

Donne, J. (1651) *Essays in Divinity*, Ed. Jessop, A. (1855), John Tupling, London.

Downing, L. (2005) Berkeley's natural philosophy and philosophy of science, In *The Cambridge Companion to Berkeley*, Ed. Winkler, K., Cambridge University Press, Cambridge.

Drake, S. (1973) Galileo's Experimental Confirmation of Horizontal Inertia: Galileo's Experimental Confirmation of Horizontal Inertia, Unpublished Manuscripts (Galileo Gleanings XXII), *Isis*, 64:291-305.

Drouin, J.S. (2010) *"The Einstein of English Fiction": James Joyce, the New Physics, and Modernist Print Culture*, Doctoral dissertation, The City University of New York.

Drouin, J.S. (2014) *James Joyce, Science, and the Modernistic Print Culture: The Einstein of English Fiction*, Routledge, New York.

Drysdale, W. (1798) *Sacred Scripture Theory of the Earth from Its first Atom to Its Last*, Printed for the author, London.

Ducheyne, S. (2006) Reid's adaptation and radicalization of Newton's natural philosophy, *Hist. Eur. Ideas*, 32:173-189.

Ducheyne, S. (2011) Kant and Whewell on bridging principles between metaphysics and science, *Kant-Stud.*, 102:22-45.

Duck, M. (1987) The Bezold–Brücke phenomenon and Goethe's rejection of Newton's Opticks, *Am. J. Phys.*, 55:793-796.

Dummett, M. (1978) *Truth and Other Enigmas*, Harvard University Press, Cambridge, MA.

Durrell, L. (1976) *Balthazar*, Faber and Faber, London.

Durrell, L. (1959–1960) Lawrence Durrell, The Art of Fiction No. 23, Interviewed Andrewski, G., and Mitchell, J,, *Paris Rev.*, No. 22.

Durrell, L. (1998) Investigating a nightingale, Interviewed by S. Gray, In *Lawrence Durrell: Conversations*, Ed. Ingersol, E.G., Associated University Presses, Cranbury, NJ.

Duszenko, A. (1994a) The Relativity Theory in "Finnegans Wake", *J. Joyce Quart.*, 32:61-70.

Duszenko, A. (1994b) The Joyce of Science: Quantum Physics in "Finnegans Wake", *Irish Uni. Rev.*, 24:272-282.

Dyson, F.W., Eddington, A.S., and Davidson, C. (1920) A determination of the deflection of light by the sun's gravitational field, from observations made at the total eclipse of May 29, 1919, *Philos. T. R. Soc. Lond.*, 220:291-333.

Edbury, K. (2012) "In this valley of dying stars": Eliot's cosmology, *J. Mod. Lit.*, 35:139-157.

Edbury, K. (2014) *Modern Cosmology: Absurd Lights*, Palgrave Macmillan, Basingstoke.

Eddington, A.S (1916) Gravitation and the principle of relativity, *Nature*, 98:328-330.

Eddington, A.S. (1920a) *Space, Time and Gravitation*, Cambridge University Press, Cambridge.

Eddington, A.S. (1920b) The Internal Constitution of the Stars, In *Report of the British Association for the Advancement of Science*, John Murray, London.

Eddington, A.S. (1923) *The Mathematical Theory of Relativity*, Cambridge University Press, Cambridge.

Eddington, A.S. (1925) The domain of physical science, In *Science, Religion, and Reality*, Ed. Needham, J., Macmillan & Co., New York.

Eddington, A.S. (1928) *Nature of the Physical World*, Macmillan, New York.

Eddington, A.S. (1931) The end of the world: from the standpoint of mathematical physics, *Nature*, 127:447-453.

Eddington, A.S. (1933) Physics and philosophy, *Philos.*, 8:30-43.

Eddington, A.S. (2013) Science and the unseen world, In *A.S. Eddington and the Unity of Knowledge*, Ed. Heine, V., Cambridge University Press, Cambridge.

Ehrenfest, P. (1918) In what way does it become manifest in the fundamental laws of physics that space has three dimensions? *KNAW Proc.*, 20:200-209.

Eibenberger, S., Gerlich, M., Arndt, M., Mayor, M., and Tüxen, J. (2013) Matter-wave interference of particles selected from a molecular library with masses exceeding 10000 amu, *Phys. Chem. Chem. Phys.*, 15:14696-14700.

Einstein, A. (1905a) Zur Elektrodynamik bewegter Körper, Trans. Perrett, W., and Jeffery, G.B., On the electrodynamics of moving bodies, In Lorentz, H.A., Einstein, A., Minkowski, H. and Weyl, H. (1952) *The Principle of Relativity*, Dover, New York.

Einstein, A. (1905b) Über einen die Erzeugung und Verwandlung des Lichtes betreffenden heuristischen Gesichtspunkt, *Ann. Phys.*, 17:132-148.

Einstein, A. (1907) Über das Relativitätsprinzip und die aus demselben gezogenen Folgerungen, *Jahrbuch der Radioaktivität und Elektronik,* 4:411-462, Trans. Beck, A. (1987) On the relativity principle and the conclusions drawn from it, In *The Collected Papers of Albert Einstein*, Vol. 2, pp. 433-484, Princeton University Press, Princeton,

Einstein, A. (1915) Die Feldgleichungen der Gravitation, Königlich Preuβische Akademie der Wissenschaften (Berlin), *Sitzungsberichte*, pp. 844-847.

Einstein, A. (1916a) *Die Grundlage der allgemeinen Relativitätstheorie*, Trans. Perrett, W., and Jeffery, G.B. The foundation of the general theory of relativity, In Lorentz, H.A., Einstein, A., Minkowski, H. and Weyl, H. (1952) *The Principle of Relativity*, Dover, New York.

Einstein, A. (1916b) *Über die spezielle und die allgemeine Relativitätstheorie, gemeinverständlich, F. Vieweg,* Braunschweig. Trans. Lawson, R.W. (1920) *Relativity: The Special and General Theory*, Methuen & Co., London.

Einstein, A. (1917) Kosmollogische Betrachtungen zur allgemaeinen Relativitätstheorie, *Sitzungsberichte der Preussischen Akademie der Wissenschaften*, 8:142-152, Trans. Perrett, W., and Jeffery, G.B., Cosmological considerations on the general theory of relativity, In

Lorentz, H.A., Einstein, A., Minkowski, H. and Weyl, H. (1952) *The Principle of Relativity*, Dover, New York.

Einstein, A. (1922a) *The Meaning of Relativity*, Princeton University Press, Princeton.

Einstein, A. (1922b) *Ether and Relativity*, Methuen & Co. Ltd, London.

Einstein, A. (1930a) What I believe, *Forum*, 84:193-194.

Einstein, A. (1930b) Religion and Science, In *Ideas and Opinions – Albert Einstein*, Ed. Seelig, C. (1954) pp. 36-40, Crown Publishers, New York.

Einstein, A. (1931) Zum kosmologischen Problem der allgemeinen Relativitätstheorie, *Sitzber. K. Preuss. Aka.*, 235-237.

Einstein, A. (1948) Religion and Science: Irreconcilable? In *Ideas and Opinions – Albert Einstein*, Ed. Seelig, C. (1954) pp. 49-52, Crown Publishers, New York.

Einstein, A. (1950) In *Albert Einstein, the Human Side: New Glimpses from His Archives*, Eds. Dukas, H., and Hoffman, B. (1979) Princeton University Press, Princeton.

Einstein, A., Podolsky, B., and Rosen, N. (1935) Can quantum-mechanical description of physical reality be considered complete? *Phys. Rev.*, 47:777-780.

Elder, R.B. (2007) Hans Richter and Viking Eggeling: The dream of universal language and the birth of the absolute film, In *Avant-Garde Film*, Eds. Graf, A., and Scheunemann, D., Rodopi, New York.

Ellis, G.F.R., Kirchner, U., and Stoeger, W.R. (2004) Multiverses and physical cosmology, *Mon. Not. Roy. Astron. Soc.*, 347:921-936.

Ellmann, R. (1982) *James Joyce*, New and Revised Ed., Oxford University Press, Oxford.

Emerson, R.W. (2010) *Collected Works of Ralph Waldo Emerson*, Vol. *8 Letters and Social Aims*, Eds. Johnson, G.M., and Myerson, J., Harvard University Press, Cambridge, MA.

Engels, F. (1940) *Dialectics of Nature*, Trans. Dutt, C. with a preface by Haldane, J.B.S., International Publishers, New York.

Engels, F. (2001) *Anti-Dühring*, Electric Book Company.

Eve, A.S. (1914) Modern views on the constitution of the atom, *Science*, 40:115-121.

Eve, A.S. (1939) *Rutherford*, Cambridge University Press, Cambridge.

Fara, P. (2003) *Newton: The Making of Genius*, Columbia University Press, New York.

Fara, P. (2004a) Heavenly bodies: Newtonianism, natural theology and the plurality of worlds debate in the eighteenth century, *J. Hist. Astron.*, 35:143-160.

Fara, P. (2004b) John Theophilus Desaguliers (1683–1744), In *Oxford Dictionary of National Biography*, Oxford University Press, Oxford.

Faraday, M. (1822) Description of an electro-magnetical apparatus for the exhibition of rotary motion, *Q. J. Sci. Lit. Arts*, 12:283-285.

Faraday, M. (1846) Thoughts on ray-vibrations, *Phil. Mag.*, 28:345-350.

Faraday, M. (1852) On the physical character of the lines of magnetic force, *Phil. Mag.*, 3:401-428.

Faraday, M. (1861) *A Course of Six Lectures on the Chemical History of a Candle, to which is added A Lecture on Platinum*, Harper & Brothers, New York.

Ferguson, A, (1769) *The Institutes of Moral Philosophy for the Use of Students in the College of Edinburgh*, Printed for A. Kincaid and J. Bell, Edinburgh.

Ferguson, A. (1792) *The Principles of Moral and Political Science; Being Chiefly a Retrospect of Lectures Delivered in the College of Edinburgh*, Printed for A. Strahan and T. Cadell, London, and W. Greech, Edinburgh.

Ferguson, J. (1754) *An Idea of the Material Universe, Deduced from a Survey of the Solar System*, Printed for the author, London.

Ferguson, J. (1756) *Astronomy Explained upon Sir Isaac Newton's Principles and Made Easy for Those Who Have Not Studied Mathematics*, Printed for the author, London.
Ferguson, J. (1760) *Lectures on Select Subjects in Mechanics, Hydrostatics, Pneumatics, and Optics with the Use of the Globes, the Art of Dialling, and the Calculation of the mean times of the New and Full Moons and Eclipses*, Printed for A. Millar, London.
Ferguson, J. (1768) *The Young Gentleman and Lady's Astronomy, Familiarly Explained in Ten Dialogues between Neander and Edosia*, Printed by Boulter Grieson, Dublin.
Ferguson, J. (1773) *Select Mechanical Exercises: Shewing How to Construct Different Clocks, Orreries, and Sun-Dials, on Plain and Easy Principles with Several Miscellaneous Articles*, Printed for W. Strahan and T. Caddel, London.
Ferré, FR. (1980) Einstein on religion and science, *Amer. J. Theol. Phil.*, 1:21-28.
Feynman, R.P. (1949) Space-time approach to quantum electrodynamics, *Phys. Rev.*, 76:769-789.
Feynman, R.P. (1965) *The Character of Physical Law*, MIT Press, Cambridge, MA.
Feynman, R.P. (1985a) *QED The Strange Theory of Light and Matter*, Princeton University Press, Princeton.
Feynman, R.P., (1985b) with Leighton, R. "*Surely You're Joking, Mr. Feynman*!": *Adventures of a Curious Character*, W.W. Norton and Company, New York.
Feynman, R.P. (2005a) *Six Easy Pieces: Essentials of Physics Explained by its Most Brilliant Teacher*, Basic Books, New York.
Feynman, R.P. (2005b) *Six Not-So-Easy Pieces: Einstein's Relativity, Symmetry, and Space-Time*, Basic Books, New York.
Feynman, R.P., Leighton, R.B., and Sands, M. (2011) *The Feynman Lectures on Physics*, 3 Vols., Basic Books, New York.
Fiedler, I. (1989) A technical evaluation of the Grande Jatte, *Art. Inst. Chicago Mus.*, 14:173-179.
Finley, G.E. (1967) Turner: An early experiment with colour theory, *J. Warburg Courtauld*, 30:357-366.
Finley, I. (1991) Pigment into light: Turner, and Goethe's theory of colours, *Eur. Rom. Rev.*, 2:39-60.
Finn, S.R. (2006) *Thomas Hobbes and the Politics of Natural Philosophy*, Continuum, London.
Finocchiaro, M.A. (1989). *The Galileo Affair: A Documentary History*, University of California Press Berkeley, CA.
Fitzmaurice, J. (2004) Cavendish, Margaret, duchess of Newcastle upon Tyne (1623–1673), *Oxford Dictionary of National Biography*, Oxford University Press, Oxford.
Fixsen, D.J. (2009) The temperature of the cosmic microwave background, *Astrophys. J.*, 707:916-920.
O'Flaherty, N. (2010) The rhetorical strategy of William Paley's Natural theology (1802): Part 2, William Paley's Natural theology and the challenge of atheism, *Stud. Hist. Phil. Sci.*, 41:128-137.
Floto, F. (1999) Stephen Hales and the cohesion theory, *Trends Plant Sci.*, 4:209-209.
Fontenelle, B. de (1686) *Entretiens sur la Pluralité des Mondes*, Trans. Behn, A. (1688) *A Discovery of New Worlds*, Printed for W. Canning, London.
Force, J.E. (1984) Hume and the relation of science to religion among certain members of the Royal Society, *J. Hist. Ideas*, 45:517-536.
Force, J.E. (1987) Hume's interest in Newton and science, *Hume Stud.*, 13:166-216.
Force, J.E., and Popkin, R.H. (1990) *Essays on the Context, Nature, and Influence of Isaac Newton's Theology*, Kluwer, Dordrecht, The Netherlands.

Forstner, C. (2008) The early history of David Bohm's quantum mechanics through the perspective of Ludwik Fleck's thought-collectives, *Minerva*, 46:215-229.

Freire, O., and Lehner, C. (2010) "Dialectical materialism and modern physics", An unpublished text by Max Born, *Notes Rec. Roy. Soc.*

Friday, J. (1974) The Braggs and broadcasting, *P. R. Inst. G.B.*, 47:59-85.

Friedman, A. (1922) Über die Krümmung des Raumes, *Z. Phys.*, 10:377–386.

Friedman, M. (2012) Newton and Kant: Quantity of matter in the Metaphysical Foundations of Natural Science, *South. J. Phil.*, 50:482-503.

Frost, I. (1846) *Two Systems of Astronomy*, Simpkin, Marshall, & Co., London.

Fyfe, A. (2003) Science for Children, In *Science for Children*, Ed. Fyfe, A., Vol. 1, pp.xi-xxii, Thoemmes Press.

Fyfe, A. (2005a) Science for sixpence (in green covers), *Notes Rec. Roy. Soc.*, 59:209-210.

Fyfe, A. (2005b) Conscientious workmen or booksellers' hacks? The professional identities of science writers in the mid-nineteenth century, *Isis*, 96:192-223.

Gabo, N., and Prevsner, A. (1920) The Realistic Manifesto, In *The Tradition of Constructivism*, Ed. Bann, S. (1974) Viking Press, New York.

Galilei, G. (1610) *The Sidereal Messenger of Galileo Galilei and a Part of the Preface to Kepler's Dioptrics Containing the Original Account of Galileo's Astronomical Discoveries*, Trans. Carlos, E.S., Rivingtons, London.

Galilei, G. (1623) *The Assayer*, Selections Trans. Drake, S. (1957) In *Discoveries and Opinions of Galileo*, pp. 231-280, Doubleday, Garden City, NY.

Galilei, G. (1632) *Dialogue Concerning the Two Chief World Systems,* Trans. Drake, S., 2nd edition (1967) University of California Press, Berkeley.

Galilei, G (1638) *Galileo. [Discourses on the] Two New Sciences*, Trans. Drake, S. (1974) University of Wisconsin Press, Madison.

Gamow, G. (1952) *The Creation of the Universe*, Viking Press, New York.

Gamow, G. (1965) *Mr Tompkins in Paperback*, Cambridge University Press, Cambridge.

Garber, D. (1985) Leibniz and the Foundations of Physics: The Middle Years, In *Natural Philosophy of Leibniz*, Eds. Okruhlik, K., and Brown, J.R., D, Reidel Pub. Co., Dordrecht.

Garber, D. (2009) *Leibniz: Body, Substance, Monad*, Oxford University Press, Oxford.

Gardner, M. (1986) Wap, Sap, Pap, and Fap, *New York Review of Books* 23:22–25.

Gardner, M. (2005) *The New Ambidextrous Universe*, 3rd edition, Dover, New York.

Garner, L. (1990) *A Brave and Beautiful Spirit: Dora Marsden, 1882–1960*, Brookfield, Aldershot.

Gascoigne, J. (2004) Clarke, Samuel (1626–1701), *Oxford Dictionary of National Biography*, Oxford University Press, Oxford.

Gasperini, M. (2008) *The Universe Before the Big Bang*, Springer-Verlag, Berlin.

Gasperini, M., and Veneziano, G. (2003) The pre-big bang scenario in string cosmology, *Phys. Rep.*, 373:1-212.

Gay-Lussac, J.L. (1802) Recherches sur la dilatation des gaz et des vapeurs, *Ann. Chim.*, 43:137–175.

Georgi, H., and Glashow, S.L. (1974) Unity of all elementary-particle forces, *Phys. Rev. Lett.*, 32:438-441.

Georgi, H., Quinn, H.R., and Weinberg, S. (1974) Hierarchy of interactions in unified gauge theories, *Phys. Rev. Lett.*, 33:451-454.

Gellmann, M. (1964) A schematic model of baryons and mesons, *Phys. Lett.*, 8:214-215.

Gibbons, G.W., and Hawking, S.W. (1971) Evidence for black holes in binary star systems, *Nature*, 232:466-467.

Gilbert, S. (1929) Prolengomena to Work in Progress, In *Our Exagmination Round His Factification for Incamination of Work in Progress*, Beckett, S., Brion, M., Budgen, F., Gilbert, S., Jolas, E.J., Llona, V., McAlmon, R., McGreevy, T., Paul, E., Rodker, J., Sage, R., and Williams, W.C., with letter of protest, Slingsby, G.V.L., and Dixon, V., Faber and Faber, London.

Gilbert, W. (1600) *On the Magnet*, Trans. Thompson, S.P., Ed. Price, D.J. (1958) The Collector's Series in Science. Basic Books, New York.

Glashow, S.L. (1961) Partial-symmetries of weak interactions, *Nucl. Phy.*, 22:579-588.

Glashow, S.L. (1986) Does elementary particle physics have a future? In *The Lesson of Quantum Theory*, Eds. de Boer J., Dal, E., and Ulfbeck O., Elsevier, Amsterdam.

Godart, O., and Heller, M. (1985) *Cosmology of Lemaître*, History of Astronomy Series, Vol. 3, Pachart Publishing House, Tucson.

Gödel, K. (1931) Über formal unentscheidbare Sätze der Principia Mathematica und verwandter Systeme, I. *Monatsh. Math. Phys.*, 38:173-198.

Goethe, J.W. (1810) *Zur Farbenlehre*, Cotta, Türbingen, Trans. Eastlake, C.L. (1810) *Goethe's Theory of Colours*, John Murray, London.

Gold, B.J. (2002) The consolation of physics: Tennyson's thermodynamic solution, *PMLA*, 117:449-464.

Golinski, J. (2004) Nicholson, William (1753–1815), *Oxford Dictionary of National Biography*, Oxford University Press, Oxford.

Good, G. (1987) John Herschel's optical researches and the development of his ideas on method and causality, *Stud. Hist. Phil. Sci.*, 18:1-41.

Gordon, A. (2004a) Gregory, Olinthus Gilbert (1774–1841), *Oxford Dictionary of National Biography*, Oxford University Press, Oxford.

Gordon, L. (2004b) Woolf, Virginia (1882–1941), *Oxford Dictionary of National Biography*, Oxford University Press, Oxford.

Goss, E.M. (2011) "Almost unmade": Hopkins and the body apocalyptic, *Victorian Poetry*, 49:83-103.

Gossin, P. (2003) Literature and the modern physical sciences, In *The Modern Physical and Mathematical Sciences*, Ed. Nye, M.J., The Cambridge History of Science, Vol. 5, Cambridge University Press, Cambridge.

Gott, J.R., and Li, L.-X. (1998) Can the Universe create itself? *Phys. Rev. D*, 58, 023501.

Gouk, P. (1986) Newton and music: From the microcosm to the macrocosm, *Int. Stud. Philos. Sci.*, 1:36-59.

Gouyon, J.-B. (2014) Making science at home: visual displays of space science and nuclear physics at the Science Museum and on television in postwar Britain, *Hist. Technol.*, 30:37-60.

Graves, R.P. (1859) *Life of Sir William Rowan Hamilton*, Vol. 1, Hodges, Figgis & Co., Dublin.

Green, B. and Butler, D. (2002) From acoustics to Tonpsychologie, In *The Cambridge History of Western Music Theory*, Ed. Christensen, T., Cambridge University Press, Cambridge.

Green, M. (2012) London Cafes: The Surprising History of London's Lost Coffeehouses, *Telegraph*, 20th March 2012.

Green, M.B. (1986) Superstrings, *Sci. Amer.*, 255:48-60.

Green, M.B., and Schwartz, J.H. (1984) Superstring field-theory, *Nucl. Phys. B*, 243:475-536.

Greene, B. (1999) *The Elegant Universe*, Jonathon Cape, London.

Gregory, O (1799) *Astronomical and Philosophical, for the Amusement and Instruction of British Youth*, 2nd edition, Printed for T. Conder, London.

Gregory, O. (1802) *A Treatise on Astronomy*, Printed for G. Kearsley, London.

Gregory, O. (1806) *A Treatise on Mechanics*, Printed for G. Kearsley, London.

Guerrini, A. (1986) The Tory Newtonians: Gregory, Pitcairne, and their circle, *J. Brit. Stud.*, 25:288-311.

Guth, A.H. (1981) Inflationary universe: A possible solution to the horizon and flatness problems, *Phys. Rev. D*, 23:247-356.

Guth, A.H. (1997) *The Inflationary Universe*, Jonathan Cape, London.

Guth, A.H. (2007) Eternal inflation and its implications, *J. Phys. A-Math. Theor.*, 40:6811-6826.

Guyau, M. (1890) *La Genèse de l'Idée du Temps*, Ancienne Librairie Germer Bailiére et Co., Paris.

Haas, A. (1936) An attempt to a purely theoretical derivation of the mass of the universe, *Phys. Rev.*, 49:411.

Hafele, J.C. and Keating, R.E. (1972a) Around-the-world atomic clocks: Predicted relativistic time gains, *Science*, 177:166-168.

Hafele, J.C. and Keating, R.E. (1972b) Around-the-world atomic clocks: Observed relativistic time gains, *Science*, 177:168-170.

Hakfoort, C. (1992) Science deified: Wilhelm Ostwald's energeticist world-view and the history of scientism, *Ann. Sci.*, 49:525-544.

Hales, S. (1727) *Vegetable Staticks: Or, an Account of some Statical Experiments on the Sap in Vegetables: Being an Essay towards A Natural History of Vegetation*, Printed for W. and J. Innys, and T. Woodward, London.

Hall, B. (2009) Effecting a transition: How to fill the gap in Kant's system of critical philosophy, *Kant-Stud.*, 100:187-211.

Halley, E. (1705) *Miscellanea Curiosa*, Printed for J. Wale, London.

Hammer, M., and Lodder, C. (2000) *Constructing Modernity: The Art & Career of Naum Gabo*, Yale University Press, New Haven, CT.

Hammer, W.J. (1903) *Radium, and Other Radio-Active Substances: Polonium, Actium, and Thorium*, Van Nostrand, New York.

Hanegraaff, W.J. (1996) *New Age Religion and Western Culture: Esotericism in the Mirror of Secular Thought*, E.J. Brill, Leiden.

de Haro, S., Dieks, D., 't Hooft, G., and Verlinde, E. (2013) Forty years of string theory reflecting on the foundations, *Found. Phys.*, 43:1-7.

Harris, J. (1698) *The Atheistical Objections Against the Being of a God and His Attributes Fairly Considered and Fully Refuted in Eight Sermons, Preach'd in the Cathedral-Church of St. Paul, London*, Printed for R. Wilkin, London.

Harris, J. (1729) *Astronomical Dialogues between a Gentleman and a Lady wherein the Doctrine of the Spheres, Use of the Globes, and the Elements of Astronomy and Geography are Explain'd; in a Pleasant, Easy, and Familiar Way*, 2nd edition, Printed for A. Bettesworth and J. Batley, London.

Harrison, E. (1987) *Darkness at Night: A Riddle of the Universe*, Harvard University Press, Cambridge, MA.

Harrison, P. (1995) Newtonian science, miracles, and the laws of nature, *J. Hist. Ideas*, 56:531-553.

Harrow, B. (1926) Bertrand Russell on relativity, *New York Times*, 10th January, p. 15.

Hartle, J.B., and Hawking, S.W. (1983) Wave function of the Universe, *Phys. Rev. D*, 28:2960-2975.

Harvey, A., and Schucking, E. (2000) Einstein's mistake and the cosmological constant, *Am. J. Phys.*, 68:723-727.

Hatch, J.G. (2010) Some Adaptations of Relativity in the 1920s and the Birth of Abstract Architecture, *Nexus Network J.*, 12:131–147.

Hauksbee, F. (1709) *Physico-Mechanical Experiments on Various Subjects*, Printed for the Author, London.

Hawking, S.W. (1965) Occurrence of singularities in open universes, *Phys. Rev. Lett.*, 15:689-690.

Hawking, S.W. (1966) Singularities in the universe, *Phys. Rev. Lett.*, 15:444-445.

Hawking S.W. (1988) *A Brief History of Time*, Bantam Books, London.

Hawking, S.W. (2002) *Gödel and the end of physics*, Public lecture at the Texas A&M University.

Hawking, S.W., and Mlodinow, L. (2010) *The Grand Design*, Bantam Books, London.

Hawking, S.W., and Penrose, R. (1970) The singularities of gravitational collapse and cosmology, *Proc. Roy. Soc. Lond.* A, 314:529-548.

Hayles, N.K. (1984) *The Cosmic Web: Scientific Field Models and Literary Strategies in the Twentieth Century*, Cornell University Press, Ithaca, NY.

Hays, J.N. (1981) The rise and fall of Dionysius Lardner, *Ann. Sci.*, 38:527-542.

Hays, J.N. (2004) Lardner, Dionysius (1793–1859), *Oxford Dictionary of National Biography*, Oxford University Press, Oxford.

Hegel, G.W.F. (1805–1830) *Lectures on the History of Philosophy*, Vol. 3, Trans. Haldane, E.S. (1892) K. Paul, Trench, Trübner & Co., London.

Hegel, G.W.F. (1817) *Philosophy of Nature*, Vol. 1, Trans. Petry, M.J. (1970), Allen & Unwin, London.

Heisenberg, W. (1925) Über quantentheoretische Umdeutung kinematischer und mechanischer Beziehungen, *Z. Phys.*, 33:879-893, Ed. Van der Waerden, B.L. (1967) Quantum-theoretical re-interpretation of kinematic and mechanical relationships, In *Sources of Quantum Mechanics*, pp. 261-276, North-Holland Publishing Co., Amsterdam.

Heisenberg, W. (1927) Über den anschaulichen Inhalt der quantentheoretischen Kinematik und Mechanik. *Z. Phys.*, 43:172-198.

Heisenberg, W. (1958) *Physics and Philosophy*, Harper, New York.

Heisenberg, W., and Pauli, W.E.P. (1929) Zur Quantendynamik der Wellenfelder, *Z. Phys.*, 56:1-61.

Heisenberg, W., and Pauli, W.E.P. (1930) Zur Quantentheorie der Wellenfelder, II, *Z. Phys.*, 59:168-190.

Helmers, S. (2001) Painting as rhetorical performance: Joseph Wright's an experiment on a bird in the air pump, *JAC*, 21:71-95.

von Helmholtz, H. (1867) On integrals of the hydrodynamical equations, which express vortex-motion, *Phil. Mag.*, 33:485-512.

von Helmholtz, H. (1878) The Facts in Perception, Eds. Cohen, R.S., and Elkana, Y. (1977) In *Hermann von Helmholtz Epistemological Writings*, pp. 115–185, D. Reidel, Dordrecht.

von Helmholtz, H. (1893) On the relation of optics to painting, In *Popular Lectures on Scientific Subjects by Hermann von Helmholtz*, pp. 73-138, Trans. Atkinson, E., Longmans, Green, and Co., London.

von Helmholtz, H. (1895) *On the Sensations of Tone as a Physiological Basis for the Theory of Music*, Trans. Ellis, A.J., 3rd edition, Longmans Green & Co., London.

Henderson, L.D. (1983) *The Fourth Dimension and Non-Euclidean Geometry in Modern Art*,, Princeton University Press, Princeton.

Henderson, L.D. (1988) X rays and the quest for invisible reality in the art of Kupka, Duchamp, and the Cubists, *Art J.*, 47: 323-340.

Henderson, L.D. (2009) Einstein and the 20th century: A romance in many dimensions, In *Einstein for the 21st Century: His Legacy in Science, Art, and Modern Culture*, Eds. Gallson, P.L., Holton, G., and Scweber, S.S., Princeton University Press, Princeton.

Henderson, L.D. (2009) The image and imagination of the fourth dimension in twentieth-century art and culture, *Configurations*, 17:131-160.

Henry, H. (2003) *Virginia Woolf and the Discourse of Science: The Aesthetics of Astronomy*, Cambridge University Press, Cambridge.

Herbert, G.B. (1987) Hobbes' phenomenology of space, *J. Hist. Ideas*, 48:709-717.

Herring, P.F. (1972) Joyce's *Ulysses Notesheets in the British Museum*, University Press of Virginia, Charlottesville.

Herschel, J.F.W. (1830) *Preliminary Discourse on the Study of Natural Philosophy*, Printed for Longman, Orme, Brown, Green, and Longmans & Taylor.

Herschel, J.F.W. (1841) Art. VI – 1. History of the inductive sciences. By the Rev. William Whewell. 2. The philosophy of the inductive sciences. By the Rev. William Whewell., *Q. Rev*. 68:177-238.

Hertz, H. (1894) *The Principles of Mechanics*, Trans. Jones, D.E., and Walley, J.T., Macmillan & Co., London, (1899).

Hetherington, N.S. (1983) Mid-nineteenth-century American astronomy: Science in a developing nation, *Ann. Sci.*, 40:61-80.

Heyl, P.R. (1930) *New Frontiers of Physics*, Appleton & Co., New York.

Heyl, P.R. (1931) The perspective of modern physics, *Sci. Am.*, 45:168-170.

Hiebert, E.N. (1966) The uses and abuses of thermodynamics in religion, *Daedalus*, 95:1046-1080).

Higgins, D. (1978) *A Dialectic of Centuries: Notes towards a Theory of the New Arts*, VT: Printed Editions, New York.

Hilbert, D. (1964) On the infinite, Trans. Putnam, E., and Massey, G.J., In *Philosophy of Mathematics*, Eds. Benacerraf, P., and Putnam, H., Cambridge University Press, Cambridge.

Hiley, B.J. (1997) David Joseph Bohm, *Biog. Mem. Fellows Roy. Soc.*, 43:106-131.

Hinton, C.H. (1980) *What is the Fourth Dimension?* In *Scientific Romances, Vol. 1, Speculations on the Fourth Dimension*, Selected Writings of Charles H. Hinton, Dover, New York.

Hinton, C.H. (1888) *A New Era of Thought*, Swan Sonnenschein, London.

Hinton, C.H. (1904) *The Fourth Dimension*, Swan Sonnenschein, London.

Hirsch, R.F. (1979) The riddle of the gaseous nebulae, *Isis*, 70:196-212.

Hoare, M.R. (2005) *The Quest for the True Figure of the Earth: Ideas and Expeditions in Four Centuries of Geodesy*, Ashgate Publishing Ltd., Aldershot.

Hobbes, T. (1651) *Leviathan or the Matter, Forme and Power of a Commonwealth Ecclesiasticall and Civil*, Printed for A. Crooke, London.

Hoffman, B. (1947) *The Strange Story of the Quantum*, Harper, New York.

Hogben, L.T. (1938) *Science for the Citizen: A Self-Educator Based on the Social Background of Scientific Discovery*, Allen and Unwin, London.

Hollis, H.P. (rev. by Brück, M.T.) (2004) Clerke, Agnes Mary (1842–1907), *Oxford Dictionary of National Biography*, Oxford University Press, Oxford.

Holmes, A. (1913) *Age of the Earth*, Harper Brothers, London.

Homer, W.I. (1964) *Seurat and the Science of Painting*, MIT Press, Cambridge, MA.

t' Hooft, G. (2013) On the foundations of superstring theory, *Found. Phys.*, 43:46-53.

Hooke, R. (1665) *Micrographia or Some Physiological Descriptions of Minute Bodies made by Magnifying Glasses with Observations and Inquiries thereupon*, Printed for J. Martin and J. Allestry, London.

Horne, G. (1756) *An Apology for Certain Gentlemen in the University of Oxford.* Printed for C. Rivington and E. Withers, London.

Howard, D. (2014) Einstein and the development of twentieth-century philosophy of science, In *Cambridge Companion to Einstein*, Eds. Janssen M., and Lehner, C., Cambridge University Press, Cambridge.

Howard, J. (2004) 'Physics and fashion': John Tyndall and his audiences in mid-Victorian Britain, *Stud. Hist. Phil. Sci.*, 35:729–758.

Howason, C., and Urbach, P. (2006) *Scientific Reasoning: The Bayesian Approach*, Open Court, Chicago.

Hoyle, F. (1948) A new model for the expanding universe, *Mon. Not. Roy. Astron. Soc.*, 108:372-382.

Hoyle, F. (1950) *The Nature of the Universe*, Blackwell, Oxford.

Hoyle, F. (1955) *Frontiers of Astronomy*, Heinemann, London.

Hubble, E. (1926) Extra-Galactic Nebulae, *Astrophys. J.*, 64:321-369.

Hubble, E. (1929) A relation between distance and radial velocity among extra-galactic nebulae, *Proc. Natl. Acad. Sci.*, 15:168-173.

Hubble, E. (1937) *The Observational Approach to Cosmology*, Oxford University Press, Oxford.

Hudson, J.W. (1851) *The History of Adult Education*, Longmans, London.

Huelsenbeck, R. (1920) En avant Dada, In *The Dada Painters and Poets*, Ed. Motherwell, R., George Wittenborn, Inc., New York.

Hueta, S. (2007) Oval domes: history, geometry and mechanics, *Nexus Net*. J., 9:211-248.

Huggins, W. (1868) Further observations on the spectra of some of the stars and nebulae, with an attempt to determine there from whether these bodies are moving towards or from the earth, also observations on the spectra of the sun and of comet II, *Philos. T. R. Soc. Lond.*, 158:529-564.

Huggins, W. and Miller W.A. (1864a) On the Spectra of Some of the Fixed Stars, *Philos. T. R. Soc. Lond.*, 154:413-435.

Huggins, W. and Miller W.A. (1864b) On the Spectra of Some of the Nebulae, *Philos. T. R. Soc. Lond.*, 154:437-444.

Hume, D. (1739, 1740) *A Treatise of Human Nature: Being an Attempt to Introduce the Experimental Method of Reasoning into Moral Subjects*, 3 Vols., Printed for John Noon, London.

Hume, D. (1748) *Philosophical Essays concerning Human Understanding*, Printed for A. Millar, London.

Hume, D (1751) *An Enquiry Concerning the Principles of Morals*, Printed for A. Millar, London.

Hume, D. (1777) *The Life of David Hume, Esq. Written by Himself*, Printed for W. Strahan; and T. Cadell, London.

Hutcheson, F. (1725) *An Inquiry into the Original of Our Ideas of Beauty and Virtue; in Two Treatises. I. Concerning Beauty, Order, Harmony, Design. II. Concerning Moral Good and Evil*, Printed for William and John Smith, Dublin.

Hutchins, R. (2004) Proctor, Richard Anthony (1837–1888), *Oxford Dictionary of National Biography*, Oxford University Press, Oxford.

Hutchinson, J. (1724) *Moses's Principia of the Invisible Parts of Matter of Motion of Visible Forms and their Dissolution and Reformation*, Printed for J. Bettenham, London.

Hutchinson, J. (1748-9) *The Philosophical and Theological Works of the Late Truely Learned John Hutchinson*, 12 Vols., Printed for J. Hodges, London.

Hutchison, K. (1981) W. J. M. Rankine and the rise of thermodynamics, *Brit. J. Hist. Sci.*, 14: 1-26.

Huxley, A.L. (1923) Subject-matter of poetry, In *On the Margin notes and Essays*, Chatto & Windus, London.

Huxley, A.L. (1928) *Along the Road: Notes and Essays of a Tourist*, Chatto & Windus, London.

Huxley, A.L. (1963) *Literature and Science*, Hope & Row, New York.

Huxley, A.L. (1994) *Those Barren Leaves*, Introduction Bradshaw, D., Vintage Books, London.

Huxley, A.L. (1998) *Time Must Have a Stop*, Dalkey Archive Press, Chicago.

Huxley, T.H. (1964) *Science and Education*, Citadel Press, New York.

Huygens, C. (1698) *Cosmotheoros: Or, Conjectures Concerning the Inhabitants of the Planets*, English trans., Printed for R. Urie, Glasgow.

Iltis, C. (1971) Leibniz and the vis viva controversy, *Isis*, 62:21-35.

Inge, W.R. (1933) *God and the Astronomers*, Longmans, Green and Co., London.

Issitt, J. (2002) Jeremiah Joyce: science educationist, *Endeavour*, 26:97-101.

Issitt, J. (2006) *Jeremiah Joyce: Radical, Dissenter, and Writer*, Ashgate Publishing Ltd., Aldershot, England.

Jacob, M.C. (1976) *The Newtonians and the English Revolution 1689–1720*, Cornell University Press, Ithaca, N.Y.

Jacobsen, A.S. (2007) Léon Rosenfeld's Marxist defense of complementarity, *Hist. Stud. Phys. Biol. Sci.*, 37(Suppl.):3-34.

Jaki, S.L. (1967) Olbers', Halley's, or whose paradox? *Am. J. Phys.*, 35:200-210.

James F.A.J.L. (2002) 'Never talk about science, show it to them': the lecture theatre of the Royal Institution, *Interdiscipl. Sci. Rev.*, 27:225-228.

James F.A.J.L. (2004a) Faraday, Michael (1791–1867), *Oxford Dictionary of National Biography*, Oxford University Press, Oxford.

James, W. (1950) *Principles of Psychology*, Holt and Company, New York.

Jammer, M. (1991) Sir Karl Popper and his philosophy of physics, *Found. Phys.*, 21:1351-1368.

Jammer, M. (1993) *Concepts of Space: The History of Theories of Space in Physics*, Dover, Mineola, N.Y.

Jammer, M. (1999) *Einstein and Religion*, Princeton University Press, Princeton, NJ.

Jeans, J.H. (1914) Discussion on Radiation, In *Report of the British Association for the Advancement of Science,* John Murray, London, pp. 376-386.

Jeans, J.H. (1928) *Eos or, The Wider Aspects of Cosmology*, Kegan Paul, London.

Jeans, J.H. (1929) *The Universe Around Us*, 2nd Ed., Cambridge University Press, Cambridge.

Jeans, J.H. (1930) *The Mysterious Universe*, Cambridge University Press, Cambridge.

Jeans, J.H. (1934) *Through Space and Time*, Cambridge University Press, Cambridge.

Jennings, B. (2004) Mansbridge, Albert (1876–1952), *Oxford Dictionary of National Biography*, Oxford University Press, Oxford.

Johns, C.M.S. (1992) Art and science in eighteenth-century Bologna: Donato Creti's astronomical landscape paintings, *Z. Kunstgeschichte*, 55:578-589.

Johnson, D.B. (1984) Inverted reality in Nabokov's Look at the Harlequins, *St. 20 C. Lit.*, 8(2): Art. 10.

Johnson, W. (1995) Swift's Gulliver, no friend to science, *Int. J. Mech. Sci.*, 37:899-917.

Johnson, W. (2004) Pemberton, Henry (1694–1771), *Oxford Dictionary of National Biography*, Oxford University Press, Oxford.

Johnson, W., and Chandrasekar, S. (1990a) Voltaire's contribution to the spread of Newtonianism – I. Letters from England: Les Lettres Philosophiques, *Int. J. Mech. Sci.*, 32:423-453.

Johnson, W., and Chandrasekar, S. (1990b) Voltaire's contribution to the spread of Newtonianism – II. Eléments de la Philosophie de Neuton: The elements of the philosophy of Sir Isaac Newton, *Int. J. Mech. Sci.*, 32:521-546.

Jones, H.B. (1868–1869) Obituary Notices of Fellows Deceased: Michael Faraday (1791–1804), *Proc. Roy. Soc. Lond.*, 17:i-lvxiii.

Jones, S. (1793) *Rudiments of Reason; or the Young Experimental Philosopher*, 3 Vols., Printed for E. Newbery, London.

Joule, J.P. (1843) On the calorific effects of magneto-electricity, and on the mechanical value of heat, *Phil Mag.*, (2)23pp. 263-276, 347-355, 435-443.

Joule, J.P. (1845) On the existence of an equivalent relation between heat and the ordinary forms of mechanical power, *Phil Mag.*, (3)23:205-207.

Joyce, UJ. (1964) *Finnegans Wake*, Faber and Faber, London.

Joyce, J. (1966) *Letters of James Joyce*, Vol. I, Ed. S. Gilbert, Viking Press, New York.

Joyce, J. (1969) *Ulysses*, Penguin, Harmondsworth.

Joyce, S. (1971) *The Complete Dublin Diary of Stanislaus Joyce*, Cornell University Pres, Ithaca.

Kaempffert, W. (1927) *New York Times*, 2nd September 1927.

Kaiser, D. (2005) Physics and Feynman's diagrams, *Am. Sci.*, 93:156-165.

Kant, I. (1746-1749) Thoughts on the true estimation of living forces, In *Kant: Natural Science*, Ed. Watkins, E. (2015), Cambridge University Press, Cambridge, pp. 1-155.

Kant, I. (1755) Allgemeine Naturgeschichte und Theorie des Himmels, Trans. Johnston, I. (2008) *Universal History and Theory of the Heavens*, Richer Resources Publ., Arlington, Virginia.

Kant, I. (1768) Concerning the ultimate foundation of the differentiation of regions of space, In *Kant Selected Pre-critical Writings and Correspondence*, Trans. Kerferd, G.B., and Walford, D.E. (1968) Manchester University Press, Manchester.

Kant, I. (1783) *Prolegomena to any Future Metaphysics that Will Be Able to Present Itself as a Science*, Ed. Gray-Lucas, P. (1953), Manchester University Press, Manchester.

Kant, I. (1786) *Metaphysical Foundations of Natural Science*, Ed. M. Frierdman, Cambridge University Press, Cambridge.

Kant, I. (1787) *Critique of Pure Reason*, 2nd Ed. Trans. Meiklejohn, J.M.D. (2011) Pacific Publishing Studios, Seattle, Wash.

Kargon, R. (1966) Thomas Hariot, the Northumberland Circle and early atomism in England, *J. Hist. Ideas*, 27:128-136.

Karlholm, D. (2009) Surveying contemporary art: Post-War, Postmodern, and then what? *Art Hist.* 32:712-733.

Kavanaugh, B.T. (1886) *The Electric Theory of Astronomy*, Cranston and Stowe, Cincinnati.

Keats, P.B. (1820) *Keats: Poems Published in 1820*, Ed. Robertson, M. (1909) Clarendon Press, Oxford.

Keill. J. (1698) *An Examination of the Reflections on the Theory of the Earth together with a Defence of Remarks on Mr. Whiston's New Theory of the Earth*, Printed for H. Clemens, Oxford.

Keill, J. (1721) *An Introduction to the True Astronomy: Or, Astronomical Lectures Read in the Astronomical School of the University of Oxford*, Printed for B. Linton, London.

Kennard, E.H. (1927a) Zur Quantenmechanik einfacher Bewegungstypen, *Z. Phys.*, 44:326-352.

Kennard, E.H. (1927b) Note on Heisenberg's indetermination principle, *Phys. Rev.*, 31:344-348.

Kepler, J. (1619) Harmonices Mundi, Trans Aiton, . E.J., Duncan, A.M., and Field, J.V. (1997) *Mem. Am. Philos. Soc.*, Vol. 209.

Khachatryan, V., Sirunyan, A.M., Tumasyan, A., et al. (2015) Precise determination of the mass of the Higgs boson and tests of compatibility of its couplings with the standard model predictions using proton collisions at 7 and 8 TeV, *Euro. Phys. J. C*, 75:UNSP 212.

Kim, S.-W., Nisimura, J., and Tsuchiya, A. (2012) Expanding (3+1)-dimensional universe from a Lorentzian matrix model for superstring theory in (9+1) dimensions, *Phys. Rev. Lett.*, 108:011601.

King, E. (nd) *A Catalogue of the Experiments Made by Mr. King in his Course of Natural Philosophy*, London.

King-Hele, D.G. (1984) *Shelley, His Thought and Work,* 3rd edition, Macmillan, London.

King-Hele, D.G. (1992) Shelley and science, *Not. Rec. Roy. Soc. Lond.*, 46:253-266.

Kinoshita, T. (1990a) High precision tests of quantum electrodynamics – an overview, In *Quantum Electrodynamics*, Ed. Kinoshita, T., World Scientific Publishing, Singapore, pp. 1-14.

Kinoshita, T. (1990b) Theory of the anomalous magnetic moment of the electron, In *Quantum Electrodynamics*, Ed. Kinoshita, T., World Scientific Publishing, Singapore, pp. 218-237.

Kobe, D.H. (1998) Copernicus and Martin Luther: An encounter between science and religion, *Am. J. Phys.*, 66:190-196.

Kollerstrom, N. (1999) The path of Halley's comet, and Newton's late apprehension of the law of gravity, *Ann. Sci.*, 56:331-356.

Kragh, H.S. (1996) *Cosmology and Controversy: The Historical Development of Two Theories of the Universe*, Princeton University Press, Princeton.

Kragh, H.S. (1997) The electrical universe: grand cosmological theory versus mundane experiments, *Phys. Perspect.*, 5:199-231.

Kragh, H.S. (2004) *Matter and Spirit in the Universe: Scientific and Religious Preludes to Modern Cosmology*, Imperial College Press, London.

Kragh, H.S. (2007a) The controversial universe: A historical perspective on the scientific status of cosmology, *Phys. Philos.*, http://hdl.handle.net/2003/24422.

Kragh, H.S. (2007b) *Conceptions of Cosmos. From Myths to the Accelerating Universe: A History of Cosmology*, Oxford University Press, Oxford.

Kragh, H.S. (2008) Pierre Duhem, entropy, and Christian faith, *Phys. Perspect.* 10:379–395.

Kragh, H.S. (2009) Contemporary history of cosmology and the controversy over the multiverse, *Ann. Sci.*, 66:529-551.

Kragh, H.S. (2011) Resisting the Bohr atom: The early British opposition, *Phys. Perspect.* 13: 4-35.

Kragh, H.S. (2012) Karl Popper on physical cosmologies, *J. Hist. Astron.*, 43:347-356.

Kragh, H.S., and Lambert, D (2007) The context of discovery: Lemaître and the origin of the primeval-atom, *Ann. Sci.*, 64:445-470.

Kragh, H.S., and Nielsen, K.H. (2013) Spreading the gospel: A popular book on the Bohr atom in its historical context, *Ann. Sci.*, 70:257-283.

Kramers, H.A., and Hoist, H. (1923) *The Atom and the Bohr Theory of its Structure: An Elementary Presentation*, Trans. Lindsay, R.B., and Lindsay, R.T., Gyldendal, London.

Kroupa, P. (2012) The dark matter crisis: falsification of the current standard model of cosmology, *Publ. Astron. Soc. Aust.*, 29:395–433.

Lamb, W.E., and Retherford, R.C. (1947) Fine structure of the hydrogen atom by a microwave method, *Phys. Rev.*, 72:241–243.

Lambert, R.H. (1761) Cosmologische Briefe über die Einrichtung des Weltbaues, Trans. Jaki, S.L. (1976) *Cosmological Letters on the Arrangement of the World-Edifice*, Scottish Academic Press, Edinburgh.

Lange, F.A. (1879-1881) *The History of Materialism and Criticism of Its Present Importance*, Trans. E. C. Thomas, 3 Vols., 2nd edition, Trubner, London.

Langevin, P. (1911) L' evolution de l' espace et du temps, *Scientia*, 10:31-54.

Laplace, P.-S. (1796) *Exposition du Système du Monde*, Paris.

Von Laue, M. (1950) *History of Physics*, Academic Press, New York.

Lawrence, D.H. (1922) *Fantasia of the Unconscious*, Thomas Seltzer, New York.

Lawrence, D.H. (1995) *The Rainbow*, Wordsworth Editions, Ware, Hertfordshire.

Lawrence, D.H. (2002) *The Letters of D. H. Lawrence*, Vol. 4, June 1921–March 1924, Eds. Roberts, W., Boulton, J.T., and Mansfield, E., Cambridge University Press, Cambridge.

Leavis, F.R. (1962) *Two Cultures? The Significance of C.P. Snow*, with an Essay on Sir Charles Snow's Rede Lecture by Michael Yudkin, Chatto and Windus, London.

Lederman, L.M., and Hill, C.T. (2011) *Quantum Physics for Poets*, Prometheus Books, New York.

Lederman, L.M., with Teresi, D. (1993) *The God Particle: If the Universe is the Answer, What is the Question?* Delta, New York.

Lee, M. (2004) Birkbeck, George (1776–1841), *Oxford Dictionary of National Biography*, Oxford University Press, Oxford.

Leibniz, G.W. (1684) Nova methodus pro maximis et minimis, itemque tangentibus, que nee fractas, nee irrationales quantitates moratur, et singulare pro illis calculi genius, *Acta Eruditorum*, pp. 467-473.

Leibniz, G.W. (1686a) De geometria recondite et analysi indivisibilium atque infinitorum, *Acta Eruditorum*, pp. 292-300.

Leibniz, G.W. (1686b) Brevis demonstratio erroris memorabilis Cartesii et aliorum circa legem naturalem, secundum quam volunt a Deo eandem semper quantitatem motus conservari; qua et in re mechanica abutuntur, *Acta Eruditorum*, pp. 161-163, Trans. Loemker, L.E. (1956) *Gottfried Wilhelm Leibniz, Philosophical Papers and Letters*, 2 Vols., University of Chicago Press, Chicago, Vol. 1, pp. 455-463.

Leibniz, G.W. (1689) On Copernicus and the Relativity of Motion, Eds. and Trans. Ariew, R., and Garber, D. (1989) In *G.W. Leibniz Philosophical Essays*, Hackett, IN.

Leibniz, G.W. (1693) Supplementum geometriae dimensoriae, seu generalissima teragonismorum effectio per motum, *Acta Eruditorum*, pp.385-392.

Leibniz, G.W. (1696) Explanation of the New System concerning the Communication between Substances, to serve as a reply to the Memoir of M. Foucher inserted in the Journal des Savants of September 12, 1695, Trans. Duncan, G.M. (1908) In *The Philosophical Works of Leibniz*, 2nd edition, Tuttle, Morehouse & Taylor, New Haven.

Leibniz, G.W. (1697) Letter to Thomas Burnet, Ed. Gerhardt, G.I. (1978) In *Die philosophischen Schriften von G. W. Leibniz*, Vol. 3 p.217, Berlin 1875-1890, reprinted Olms, Hildesheim.

Leibniz, G.W. (1703) Explication de l'arithmétique binaire, *Mem. Acad. Roy.*, pp. 85–89.

Leibniz, G.W. (1710) Essais de Théodicée sur la Bonté de Dieu, la Liberté de l'Homme et l'Origine du Mal, Trans. Hubbard, E.M. (1985) *Theodicy: Essays on the Goodness of God, the Freedom of Man, and the Origin of Evil*, Open Court, La Salle, Ill.

Lemaître, G. (1927) Un univers homogène de masse constante et de rayon croissant, rendant compte de la vitesse radiale des nébuleuses extra-galactiques, *Ann. Soc. Sci. Bruxelles*, 47:49-59.

Lemaître, G.(1931) The beginning of the world from the point of view of quantum theory, *Nature*, 127:706.

Lenin, V.I. (1938) *Materialism and Empirio-Criticism*, Lawrence & Wishart Ltd., London.

Lester, J. (2002) Rameau and eighteenth-century harmonic theory, Ed. Christensen, T., In *The Cambridge History of Western Music Theory*, Cambridge University Press, Cambridge.

Levine, A.G. (2014) Right brain, left brain: physicists as artists, *APS News*, 23:(10).

Lewis, P (2007) *The Cambridge Introduction to Modernism*, Cambridge University Press, Cambridge.

Lewis, W. (1993) *Time and Western Man*, Ed. Edwards, P., Black Sparrow Press, Santa Rosa, USA.

Lightman, B.V. (2007) *Victorian Popularizers of Science: Designing Nature for New Audiences*, University of Chicago Press, Chicago.

Lindberg D.C. (1965) Galileo's experiments on falling bodies, *Isis*, 56:352-354.

Linde, A.D. (1982) A new inflationary universe scenario: A possible solution of the horizon, flatness, homogeneity, isotropy and primordial monopole problems, *Phys. Lett.*, 108B:389-393.

Lindsay, D.W. (1989) The order of Blake's large color prints, *Huntington Lib. Q.*, 52:19-41.

Lissitzky, El. (1968) A. and pangeometry, In *El Lissitzky, Life, Letters, Texts*, Ed, Lissitzky-Kuppers, S. Thames and Hudson, London.

Locke, J. (1690a) *Essay Concerning Human Understanding*, Printed for T. Basset, London.

Locke, J. (1690b) *Two Treatises of Government: In the Former, The False Principles and Foundation of Sir Robert Filmer, And His Followers, are Detected and Overthrown. The Latter is an Essay concerning The True Original, Extent, and End of Civil-Government*, Printed for A. Churchill, London.

Locke, J. (1699) Mr. Locke's reply to the Bishop of Worcester's answer to his second letter, In *The Works of John Locke in Nine Volumes*, 12th edition, (1824) Rivington, London, Vol. 3.

Lockyer, N. (1887) Researches on the spectra of meteorites: a report to the Solar Physics Committee, *Proc. Roy. Soc. Lond.*, 43:117-156.

Lockyer, N. (1919) Valedictory memories, *Nature*, 104:189-190.

Lodge, O. (1913) Presidential Address, *British Association for the Advancement of Science*, Birmingham.

Lodge, O. (1933) *My Philosophy*, E. Benn Ltd., London.

Lorentz, H.A. (1904) Electromagnetic phenomena in a system moving with any velocity less than that of light, Trans. Perrett, W., and Jeffery, G.B. (1952) In Lorentz, H.A., Einstein, A., Minkowski, H. and Weyl, H. (1952) *The Principle of Relativity*, Dover, New York.

Lyttleton, R.A., and Bondi, H. (1959) On the physical consequences of a general excess of charge, *Proc. Roy. Soc. Lond.*, 252:313-333.

McCarthy, P.A. (1984) Joyce's unreliable catechist: mathematics and the narration of "Ithaca", *ELH*, 51:605-618.

McCormmach, R. (1966) Ormsby MacKnight Mitchel's "Sidereal Messenger", 1846–1848, *Proc. Am. Philos. Soc.*, 110:35-47.

Mach, E. (1872) *The History and Root of the Principle of Conservation of Energy*, Trans. Jordain, P.E.B., Open Court, Chicago.

Mach, E. (1883) *The Science of Mechanics: A Critical and Historical Account of Its Development*, Trans. McCormack, T.J., Open Court, La Salle, IL.

Mach, E. (1886) *The Analysis of Sensations and the Relation of the Physical to the Psychical*, Trans. Williams, C.M. revised Waterlow, S., Open Court, Chicago.

Mach, E. (1895) The economical nature of physical inquiry, Trans. McCormack, T.J. In *Open Court. A Weekly Journal Devoted to the Religion of Science* 8:4263-4269, 4271-4275.

Mach, E. (1900) The propensity toward the marvellous, In *Open Court. A Weekly Journal Devoted to the Religion of Science*, 14:539-550.

McGilvary, E.B. (1930) Book reviews, *J. Philos*., 27:180-194.

McGuire, I.E. (1968) The origin of Newton's doctrine of essential qualities, *Centaurus*, 12:233-260.

Mackenzie, E. (1827) *Institutions for Education: The Trinity-House School, Historical Account of Newcastle-upon-Tyne: Including the Borough of Gateshead*, pp. 443-445.

McKie, D., and de Beer, G.R. (1951) Newton's apple, *Not. Rec. Roy. Soc. Lond.*, 9:46-54.

MacKillop, I. (2004) Leavis, Frank Raymond (1895–1978), *Oxford Dictionary of National Biography,* Oxford University Press, Oxford.

Mackintosh, T.S. (1838), *Electrical Theory of the Universe or the Elements of Physical and Moral Philosophy*, Abel Heywood, London.

McMullin, E. (2001) The impact of Newton's Principia on the philosophy of science, *Philos. Sci.*, 68:279-310.

Malcolm, N.R. (2004) Hobbes, Thomas (1588–1679), *Oxford Dictionary of National Biography*, Oxford University Press, Oxford.

Mandelbrote, S. (2004) Hutchinson, John (1674–1737), *Oxford Dictionary of National Biography*, Oxford University Press, Oxford.

Marcet, J.H. (1824) *Conversations on Natural Philosophy in which the Elements of that Subject are Familiarly Explained and adapted to Young People*, 4th edition, Longman, Jurst, Rees, Orme, Brown, and Green, London.

Marcet, J.H. (1809) *Conversations on Chemistry, in which the Elements of that Subject are Familiarly Explained*, Cooke & Co., New Haven.

March, R.H. (1970) *Physics for Poets*, McGraw-Hill, New York.

Marchitello, H. (2011) *The Machine in the Text: Science and Literature in the Age of Shakespeare and Galileo*, Oxford University Press, Oxford.

Marek, J., and Rissen, H. (1977) Lenin's relationship to the ideas of physicists, *Stud. Sov. Thought*, 17:63-80.

Martin, B. (1743) *A Course of Lectures in Natural and Experimental Philosophy Geography and Astronomy*. Printed by J. Newberry and C. Micklewright, Reading.

Martin, B. (1747) *Philosophia Britannia, or a New and Comprehensive System of the Newtonian Philosophy, Astronomy and Geography, A Course of Twelve Lectures*, Vols. 1 and 2, Printed by C. Micklewright and Co. for the author, Reading.

Martin, B. (1749) *A Panegyrick on the Newtonian Philosophy Shewing the Nature and Dignity of the Science*, Printed for W. Owen, London.

Martin, B. (1751) *A Plain and Familiar Introduction to the Newtonian Philosophy Design*, Printed for W. Owen, London.

Martin, J.L., Nicholson, B., Gabo, N. (1937) *Circle: International Survey of Constructive Art*, Faber and Faber, London.

Massimi, M. (2011) Kant's dynamical theory of matter in 1755, and its debt to speculative Newtonian experimentalism, *Stud. Hist. Philos. Sci.*, 42:525–543.

Maxted, I. (2004) Newbery, John (bap. 1713, d. 1767), *Oxford Dictionary of National Biography*, Oxford University Press, Oxford.

Maxwell, J.C. (1855, 1856) On Faraday's lines of force, Ed. Niven, Ed W.D. (1890), In *The Scientific Papers of James Clerk Maxwell*, Vol. 1, Cambridge University Press, Cambridge.

Maxwell, J.C. (1861a) On physical lines of force: Part I – The theory of molecular vortices applied to magnetic phenomena, *Phil. Mag.*, 21:161-175.

Maxwell, J.C. (1861b) On physical lines of force: Part II – The theory of molecular vortices applied to electric currents, *Phil. Mag.*, 21:338-348.

Maxwell, J.C. (1865). A dynamical theory of the electromagnetic field, *Phil. Trans. Roy. Soc. Lond.*, 155:459–512.

Maxwell, J.C. (1871) *Theory of Heat*, Longmans Green & Co., London.

Maxwell, J.C. (1878) Tait's "Thermodynamics", *Nature*, 17:257-259.

Mayer, A.M. (1875) Researches in Acoustics No. VI, *Phil. Mag.*, 49:352-365.

Mayer, J.R. (1842) Remarks on the forces of inorganic nature, Trans. Foster, G.C. (1862) *Phil Mag.*, (4)24:371-377.

Mazzotti, M. (2004) Newton for ladies: gentility, gender and radical culture, *Brit. J. Hist. Sci.*, 37:119–146.

Meadows, A.J. (1992) Astronomy and geology, terrible muses! Tennyson and 19th-Century science, *Not. Rec. Roy. Soc. Lond.*, 46:111-118.

Meckier, J. (2003) Conradian reminders in Aldous Huxley's Island: Will Farnaby's moksha-medicine experience and "the essential horror'. *Stud. Novel*, 35:44-67.

Meyer, W.F. (1904) *Encyklopädie der Mathematischen Wissenschaften mit Einschluss ihrer Anwendungen*, Vol. 2, Arithmetic und Algebra, Druck und Verlag von E. G. Teubner, Leipzig.

Milburn, J.R. (1983) The London evening courses of Benjamin Martin and James Ferguson, eighteenth-century lecturers on experimental philosophy, *Ann. Sci.*, 40:437-455.

Milburn J.R. (1985) James Ferguson's lecture tour of the English midlands in 1771, *Ann. Sci.*, 42:397-415.

Milburn, J.R. (2004) Martin, Benjamin (bap. 1705, d. 1782), *Oxford Dictionary of National Biography*, Oxford University Press, Oxford.

Mill. J.S. (1846) *A System of Logic, Ratiocinative and Inductive: being a Connected View of the Principles of Evidence and the Methods of Scientific Investigation*, Harper, New York.

Miller, A.I. (2001*) Einstein, Picasso: Space, Time, and the Beauty that Causes Havoc*, Basic Books, New York.

Miller, D.C. (1937) *Sound Waves Their Shape and Speed*, Macmillan, New York.

Milton, J. (1674) *Paradise Lost. A Poem in Twelve Books*, 2nd edition, Printed by S. Simmons, London.

Milton, J.R. (2004) Locke, John (1632–1704), *Oxford Dictionary of National Biography*, Oxford University Press, Oxford.

Milton, J.R. (2012) Laws of nature, In *The Cambridge History of Seventeenth-Century Philosophy*, Volume 1, Eds. Gaber, D., and Ayers, M. Cambridge University Press, Cambridge.

Michell, J. (1760) Conjectures concerning the cause, and observations upon the phaenomena of earthquakes; particularly of that great earthquake of the first of November, 1755, *Trans. Roy. Soc. Lond.*, 51:566-634.

Michell, J. (1784) On the means of discovering the distance, magnitude, &c. of the fixed stars, in consequence of the diminution of the velocity of their light, in case such a diminution should be found to take place in any of them, and such other data should be procured from observations, as would be farther necessary for that purpose, *Phil. Tran. Roy. Soc. Lond.*, 74:35-57.

Minkowski, H. (1908) Space and time, In A. Einstein, H.A. Lorentz, H. Minkowski and H. Weyl, (1952) *The Principle of Relativity: A Collection of Original Memoirs on the Special and General Theory of Relativity*, Dover, New York.

Mitchel, O.M. (1851) *The Orbs of Heaven, or The Planetary and Stellar Worlds: a Popular Exposition of the Great Discoveries and Theories of Modern Astronomy*, Office of the National Illustrated Library, London.

Mitchel, O.M. (1860) *Popular Astronomy: A Concise Elementary Treatise on the Sun, Planets and Comets*, Phinney, Blakeman and Mason, New York.

Mitchel, O.M. (1871) *The Planetary and Stellar Worlds: a Popular Exposition of the Great Discoveries and Theories of Modern Astronomy*, Oakley, Mason & Co., New York.

Mitton, S. (2005) *Conflict in the Cosmos: Fred Hoyle's Life in Science*, National Academic Press, Washington.

Montes, l. (2008) Newton's real influence on Adam Smith and its context, *Cambridge. J. Econ.*, 32:555–576.

Montmerle, T., Augereau, J.-C., Chaussidon, M., Gounelle, M., Marty, B., and Morbidelli, A. (2006) From suns to life: A chronological approach to the history of life on earth 3. Solar system formation and early evolution: the first 100 million years, *Earth Moon Planets*, 98:39-95.

Mooney, C., and Kirshenbaum, S. (2009*) Unscientific America: How Scientific Illiteracy Threatens our Future*, Basic Books, New York.

Moore, G.E. (1903) The refutation of idealism, *Mind*, 12:433-453.

De Morgan, A. (1872) *A Budget of Paradoxes*, Spottiswoode and Co., London.

Morrell, J., and Thackray, A. (1981) *Gentlemen of Science, Early Years of the British Association for the Advancement of Science*, Clarendon Press, Oxford.

Morris, P.J.T. (Ed.) (2010) *Science for the Nation: Perspectives on the History of the Science Museum*, Palgrave Macmillan, Basingstoke.

Morrison-Low, A,D, (2004) Brewster, Sir David (1781–1868), *Oxford Dictionary of National Biography*, Oxford University Press, Oxford.

Morton, A.Q. (1990) Lectures on natural philosophy in London, 1750–1765: S. C. T. Demainbray (1710-1782) and the 'Inattention' of his countrymen, *Brit. J. Hist Sci.*, 23:411-434.

Murry, A. (1799) *The Sequel to Mentoria or the Young Ladies Instructor in Familiar Conversations on a Variety of Interesting Subject, in Which are Introduced Lectures on Astronomy and Natural Philosophy*, Printed for C. Dilly, London.

Mussell, J. (2007) Nineteenth-century popular science magazines, *Journalism Stud.*, 8:656-666.

Myrvold, W.C. (2003) On some early objections to Bohm's theory, *Int. Stud. Philos. Sci.*, 17:8-24.

Nabokov, V.V. (2000) *Ada or Ardor*, Penguin Books, London.

Nadathur, S., and Sarkar, S. (2001) Reconciling the local void with the CMB. *Phys. Rev.*, D83:063506.

Navarro, J. (2009) "A dedicated missionary". Charles Galton Darwin and the new quantum mechanics in Britain, *Stud. Hist. Phil. Mod. Phys.*, 40:316-326.

Naylor, R.H. (1973) Galileo and the problem of free fall, *Brit. J. Hist. Sci.*, 7:105-134.

Naylor, R.H. (2003) Galileo, Copernicanism and the origins of the new science of motion, *Brit. J. Hist. Sci.*, 36:151–181.

Needham, J. and Wang, L. (1962) *Science and Civilisation in China*, Vol. 4. Physics and Physical Technology, Part I. Physics, Cambridge University Press, Cambridge.

O'Neil, J.C. (2002) How early is the doctrine of creatio ex nihilo? *J. Theol. Stud.*, 53:449-465.

Neumann, C. (1896) *Über das Newton'sche Princip der Fernwirkung*, B.G. Teubner, Leipzig.

Neumann, J. von (1932). *Mathematische Grundlagen der Quanten-Mechanik*, Trans. Beyer, R.T. (1955) *Mathematical Foundations of Quantum Mechanics*, Princeton University Press, Princeton, NJ.

Newbery, J. (1761) *The Newtonian System of Natural Philosophy, Being the Substance of Six Lectures Read to the Lilliputian Society, by Tom Telescope*, Printed for J. Newbery, London.

Newton, I. (1704) *Opticks or a Treatise of the Reflections, Refractions, Inflections and Colours of Light*, Printed for S. Smith and N.J.U. Wattford, London.

Newton, I. (1718) *Opticks or a Treatise of the Reflections, Refractions, Inflections and Colours of Light*, 2nd edition, Printed for W. and J. Innys, London.

Newton, I. (1726) *The Principia: Mathematical Principles of Natural Philosophy*, 3rd edition, Trans. Cohen, I.B., and Whitman, A. (1999) University of California, Berkeley, CA.

Newton, I. (1728) *The Chronology of the Ancient Kingdoms Amended*, Printed for J. Tonson, J. Osborn, and T. Longman, London.

Newton, I. (1756) *Four Letters from Sir Isaac Newton to Doctor Bentley Containing Some Arguments in Proof of a Deity*, Printed for R. and J. Dodsley, London.

Newton, I. (2008) *The Correspondence of Isaac Newton*, 7 Vols., Cambridge University Press, Cambridge.

Nicholson, H.G. (1931) *The New Spirit in Literature: An Essay Introductory to Twelve Talks*, BBC, London.

Nicolson, M.H. (1935) The "new astronomy" and English imagination, *Stud. Philol.*, 32:428-462.

Nicolson, M.H. (1950) *The Breaking of the Circle*, Northwestern University Press, Evanston, IL.

Nicolson, M.H. (1963) *Newton Demands the Muse: Newton's Opticks and the Eighteenth Century Poets*, Archon Books, Hamden, CT.

Nicolson, M.H., and Mohler, N.M. (1937) The scientific background of Swift's Voyage to Laputa, *Ann. Sci.*, 2:299-334.

Nicolson, M.H., and Rousseau, G.S. (1968) *"This Long Disease my Life," Alexander Pope and the Sciences*, Princeton University Press, NJ.

Noakes, R. (2008) The 'world of the infinitely little': connecting physical and psychical realities circa 1900, *Stud. Hist. Phil. Sci.*, 39:323–334.

Norton, J.D. (2004) Why thought experiments do not transcend empiricism, Ed. Hitchcock, C., In *Contemporary Debates in the Philosophy of Science*, Blackwell, Oxford, pp. 44–66.

Oerter, R. (2005) *The Theory of Almost Everything: The Standard Model, the Unsung Triumph of Modern Physics*, Pearson Education, London.

Ohm, G.S (1843) Ueber die Definition des Tones, nebst daran geknüpter Theorie de Sirene und ähnlicher tonbildender Vorrichtungen, *Ann. Phys.*, 59:513-565.

Oliver, P.M. (2002) *Selected Letters: John Donne*, Carcanet Press, Manchester.

Orange, A.D. (1971) The British Association for the Advancement of Science: The provincial background, *Sci. Stud.*, 1:315-329.

Overduin, J., and Fahr, H.-J. (2003) Vacuum energy and the economical universe, *Found. Phys Lett.*, 16:119-125.

Owens, C. (1977) "Einstein on the Beach": The primacy of metaphor, *October*, 4:21-32.

Ozawa, M. (2003) Universally valid reformulation of the Heisenberg uncertainty principle on noise and disturbance in measurement, *Phys. Rev. A*, 67, Art. 042105.

Oz-Salzberger, F. (2004) Ferguson, Adam (1723–1816), *Oxford Dictionary of National Biography*, Oxford University Press, Oxford.

Paalen, W. (1942a) Farewell au Surrealisme, *Dyn.*, 1:26.

Paalen, W. (1942b) Art and science, In *Form and Sense* (2013), Arcade Publishing, New York.

Paalen, W. (1942c) The new image, In *Form and Sense* (2013), Arcade Publishing, New York.

Paalen, W. (1944) On the meaning of Cubism, In *Form and Sense* (2013), Arcade Publishing, New York.

Page, D.N. (2008) Predictions and tests of multiverse theories, In *Universe or Multiverse?* Ed. Carr, B., Cambridge University Press, Cambridge.

Paine, T. (1794) *The Age of Reason being an Investigation of True and False Theology*, Barrois, Paris.

Pais, A. (1994) *Einstein Lived Here*, Clarendon Press, Oxford.

Paley, W. (1802) *Natural Theology or, Evidences of the Existence and Attributes of the Deity Collected from the Appearances of Nature*, Printed for R. Faulder, London.

Palonen, V. (2008) Bayesian considerations on the multiverse explanation of cosmic fine-tuning, preprint arXiv: 0802.4013.

Panofsky, E. (1956) Galileo as a critic of the arts: Aesthetic attitude and scientific thought, *Isis*, 47:3-15.

Parkinson, G. (2004) Surrealism and quantum mechanics: Dispersal and fragmentation in art, life, and physics, *Sci. Context*, 17:557-577.

Parkinson, G. (2008) *Surrealism, Art and Modern Science*, Yale University Press, New Haven, Conn.

Parra, H. (2011) *Stress Tensor*, New version commissioned by the Munich Biennale for Klangapuren Plus 2011, Tritó Edicions, Barcelona.

Parra, H. (2013) Hypermusic Prologue: A projective opera in seven planes, *Sonograma*, No. 018.

Patey, D.L. (1991) Swift's Satire on "Science" and the structure of Gulliver's travels, *ELH*, 58: 809-839.

Patrick, S. (1662) *A Brief Account of the New Sect of Latitude-Men Together with some Reflections upon the New Philosophy in Answer to a Letter from his Friend at Oxford*, London.

Pauli, W.E.P. (1925). Über den Zusammenhang des Abschlusses der Elektronengruppen im Atom mit der Komplexstruktur der Spektren, *Z. Phys.*, 31:765-783.

Paz, M. (2008) Vaniada, *Orbis Litterarum*, 63:285-303.

Peierls, R.E. (1955) *The Laws of Nature*, Allen & Unwin, London.

Peierls, R.E. (1960) Wolfgang Ernst Pauli. 1900–1958, *Biog. Mem. Fellows Roy. Soc.*, 5: 174-192.

Pemberton, H. (1728) *A View of Sir Isaac Newton's Philosophy*, Printed by S. Palmer, London.

Penney, R. (1965) On the dimensionality of the real world, *J. Math. Phys.*, 6:1607-1611.

Penrose, R. (1965) Gravitational collapse and space-time singularities, *Phys. Rev. Lett.*, 14:57-59.

Penrose, R. (1993) Nature's biggest secret, *New York Rev. Books*, 40:78-82.

Penrose, R. (2016) *Fashion, Faith, and Fantasy in the New Physics of the Universe*, Princeton University Press, Princeton, NJ.

Pessoa, O., Freire, O., and De Greiff, A. (2008) The Tausk controversy on the foundations of quantum mechanics: Physics, philosophy, and politics, *Phys. Perspect.*, 10:138-162.

Petersen, A. (1963) The philosophy of Niels Bohr, *Bull. Atomic Sci.*, 19(1):8-14.

Phillips, R. (1817a) The phœnomena called by the name of Gravitation proved to be proximate effects of the orbicular and rotary motions of the earth, *Phil. Mag.*, (1), 49: 430-442.

Phillips, R. (1817b) Further considerations on the doctrine that the phœnomena of terrestrial gravitation are occasioned by known terrestrial motions, *Phil. Mag.*, (1), 50:101-105.

Phillips, R. (1817c) On the new theory of the system of the universe, *Phil. Mag.*, (1), 50: 219-224.

Phillips, R. (1821) *The Proximate Causes of Material Phenomena, and the True Principles of Universal Causation Considered and Illustrated*, J. and C. Adlard, London.

Phillips, R. (2013) The Feynman lectures on physics, *Nature*, 504:30-31.

Pickard, W,F. (1981) The ascent of sap in plants, *Prog. Biophys. Mol. Biol.*, 37:181-229.

Pinkard, T. (2008) Hegel: A life, In *The Cambridge Companion to Hegel and Nineteenth-Century Philosophy*, Ed. Beiser, F.C., Cambridge University Press, Cambridge.

Planck, M. (1899) Über irreversible Strahlungsvorgänge, Sitzungsberichte der Königlich, Prussian Acad. Sci. Berlin, 5:440-480, Trans. Masius, M. (1914) In *The Theory of Heat Radiation*, P. Blakiston's Son & Co., Philadelphia.

Planck, M. (1901) Über das Gesetz der Energieverteilung im Normalspektrum, *Ann. Phys.*, 4:553-563.

Planck, M. (1915) *Eight Lectures on Theoretical Physics Delivered at Columbia University in 1909*, Trans. Wills, A.P., Reprinted (1998) Dover Publications, Mineola, NY.

Platts, J. (2006) Newton, Goethe and the process of perception: an approach to design, *Opt. Laser Technol.*, 38:205–209.

Poe, E.A. (1829) *Al Aaraaf, Tamerlane, and Minor Poems*, Hatch and Dunning, Baltimore.

Poe, E.A. (1848) *Eureka: A Prose Poem*, Geo. P. Putnam, New York.

Polkinghorne, J.C. (1979) *Particle Play: An Account of the Ultimate Constitution of Matter*, F.W. Freeman, Oxford.

Polkinghorne, J.C. (1984) *The Quantum World*, Longman, London.

Polkinghorne, J.C. (2001) Understanding the Universe, *Ann. New York Acad. Sci.*, 950:175-182.

Popper, K.R. (1940) What is dialectic? *Mind*, 49:403-426.

Popper, K.R. (1959a) *The Logic of Scientific Discovery*, Hutchinson & Co., London.

Popper, K.R. (1959b) The propensity interpretation of probability, *Brit. J. Philos. Sci.*, 10:25-42.

Popper, K.R. (1957) *The Poverty of Historicism*, Routledge, London.

Popper, K.R. (1982a) A critical note on the greatest days of quantum theory, *Found. Phys.*, 12:971-976.

Popper, K.R. (1982b) *Quantum Theory and the Schism in Physics*, Hutchinson, London.

Popper, K.R. (1983) *Realism and the Aim of Science*, Rowman and Littlefield, Totowa, NJ.

Posch, T. (2004) Hegel's criticism of Newton's physics: A reconsideration, *Oxford Conference on Hegel and British Thought.*

Pound, E. (1910) *The Spirit of Romance: An Attempt to Define Somewhat the Charm of the Pre-Renaissance Literature of Latin Europe*, J.M. Dent & Sons, London.

Pound, E. (1973) I gather the limbs of Osiris, In *Ezra Pound: Selected Prose 1909–1965*, Ed. Cookson, W., Faber and Faber, London.

Pound, R.V., and Rebka, G.A. (1959). Gravitational red-shift in nuclear resonance, *Phys. Rev. Lett.*, 3:439-441.

Preti, G. (2009) Schwarzschild radius before general relativity: Why does Michell-Laplace argument provide the correct answer? *Found. Phys.*, 39:1046–1054.

Priestley, J. (1767) *The History and Present State of Electricity with Original Experiments*, Printed for J. Dodsley, J. Johnson, B. Davenprt, and T. Cadell, London.

Priestley, J. (1777) *Disquisitions on Matter and Spirit*, Printed for J. Johnson, London.

Pritchard, C. (1870) Other worlds than ours, *Nature*, 2:161-162.

Proctor, R.A. (1868) *Half-Hours with a Telescope*, Hardwicke, London.

Proctor, R.A. (1870) *Other Worlds than Ours*, Longman, London.

Pumfrey, S. (2004) Hauksbee, Francis (bap. 1660–1713), *Oxford Dictionary of National Biography,* Oxford University Press, Oxford.

Pumfrey, S. (2011) 'Your astronomers and ours differ exceedingly': the controversy over the 'new star' of 1572 by Thomas Digges, *Brit. J. Hist, Sci.*, 44:29–60.

Quennell, P. (1956) *Hogarth's Progress*, Viking Press, New York.

Raman, V.V. (1972) Relativity in the early twenties: Many-sided reactions to a great theory, *Indian J. Hist. Sci.*, 7:119-145.

Rameau, J.-P. (1722) *Traité de l'Harmonie Reduite à Ses Principes Naturels*, Trans. Gossett, P (1971), Dover Publications, New York.

Randall, L. (2005) *Warped Passages: Unravelling the Mysteries of the Universe's Hidden Dimensions*, Ecco, New York.

Randall, L., and Sundrum, R. (1999a) Large mass hierarchy from a small extra dimension, *Phys. Rev. Lett.*, 83:3370-3373.

Randall, L., and Sundrum, R. (1999b) An alternative to compactification, *Phys. Rev. Lett.*, 83:4690-4693.

Rankine, W.J.M. (1851) On the centrifugal theory of elasticity, as applied to gases and vapours, In *Rankine, Miscellaneous Scientific Papers*, (1881) Ed. Millar, W.J., C. Griffin & Co., London.

Ray, M.W., Ruokokoski, E., Kandel, S., Möttönen, M., and Hall, D.S. (2014) Observation of Dirac monopoles in a synthetic magnetic field, *Nature*, 505:657–660.

Rayleigh, Lord (1889) On the transmission of light through an atmosphere containing small particles in suspension and on the origin of the blue of the sky, *Phil. Mag.*, 47:375-394.

Rees, M. (1999) *Just Six Numbers: The Deep Forces that Shape the Universe*, Weidenfeld and Nicholson, London.

Rees, M. (2003) Numerical coincidences and 'tuning' in cosmology, *Astrophys. Space Sci.*, 285:375-388.

Reid, F. (2005) Isaac Frost's Two Systems of Astronomy (1846): plebeian resistance and scriptural astronomy, *Brit. J. Hist. Sci.*, 38:161–177.

Reid, T. (1748) An essay on quantity; occasioned by reading a treatise, in which simple and compound ratio's are applied to virtue and merit, *Philos. T. R. Soc. Lond*, 45:505-520.

Reid, T. (1764) *An Inquiry into the Human Mind on the Principles of Common Sense*, Printed for A. Kincaid and J. Bell, Edinburgh.

Reid, T. (1788) *Essays on the Active Power of Man*, Printed for J. Bell, Edinburgh.

Reiss, A.G. et al. (1998) Observational evidence from supernovae for an accelerating universe and a cosmological constant, *Astron. J.*, 116:1009-1038.

De Regt, H.W. (1999) Ludwig Boltzmann's Bildtheorie and scientific understanding, *Synthese*, 119:113-134.

Richards, J.L. (1977) The evolution of empiricism: Hermann von Helmholtz and the foundations of geometry, *Brit. J. Philos. Sci.*, 28:235-253.

Rickles, D. (2014) *A Brief History of String Theory: From Dual Models to M-Theory*, Springer-Verlag, Berlin.

Robertson, J.L. (Ed.) (1908) *The Complete Poetical Works of James Thomson*, Oxford University Press, Oxford.

Robertson, J. (2004) Hume, David (1711–1776), *Oxford Dictionary of National Biography*, Oxford University Press, Oxford.

Roche, J.J. (2004) Lower, Sir William (c. 1570–1615), *Oxford Dictionary of National Biography*, Oxford University Press, Oxford.

Rogers, G.A.J. (1978) Locke's Essay and Newton's Principia, *J. Hist. Ideas*, 39:217-232.

Rood, O.N. (1879) *Modern Chromatics, with Applications to Art and Industry*, D. Appleton, & Co., New York.

Rosen, N. (1994) The energy of the universe, *Gen. Rel. Gravit.*, 26:319-321.

Rosenfeld, L. (1953) Strife about complementarity, *Sci. Prog.*, 41:393–410.

Rosenthal, J. (2010) The large novel and the law of large numbers; or, why George Eliot hates gambling, *ELH*, 77:777-811.

Roush, S. (2003) Copernicus, Kant, and the anthropic cosmological principles, *Stud. Hist. Philos. M. P.*, 34:5–35.

Rowbottom, S. (1981) Travellers in a strange country: Responses of working class students to the University Extension Movement — 1873–1910, *History Workshop*, No. 12, pp. 62-95.

Rowlands, P. (2004) Lodge, Oliver (1851–1940), *Oxford Dictionary of National Biography*, Oxford University Press, Oxford.

Ruelle, D. (1990) The Claude Bernard Lecture, 1989: Deterministic chaos: the science and the fiction, *Proc. R. S. Lond. A*, 427:241-248.

Ruff, W., and Rodgers, J. (1979) Kepler's Harmony of the World: A realization for the ear, *Am. Sci.*, 67:286-292.

Rugh, S., and Zinkernagel, H. (2001). The quantum vacuum and the cosmological constant problem, *Stud. Hist. Philos. M. P.*, 33:663–705.

Russell, B. (1900) *A Critical Exposition of the Philosophy of Leibniz*, Cambridge University Press, Cambridge.

Russell, B. (1904) The Fourth Dimension. By C. Howard Hinton, *Mind*, 13:573-574.

Russell, B. (1910) *Philosophical Essays*, Longmans and Green, London.

Russell, B. (1912) On the notion of cause, *P. Aristotelian Soc.*, 13:1-26.

Russell, B. (1914) On the nature of acquaintance, II Neutral monism, *Monist*, 24:161-187.

Russell, B. (1923) *The ABC of Atoms*, Kegan Paul, London.

Russell, B. (1961) *History of Western Philosophy*, George Allen & Unwin, London.

Russell, B. (1975) *Autobiography of Bertrand Russell*, Allen and Unwin, London.

Russell, B. (2009) *ABC of Relativity*, Routledge Classics, London.

Russell, J.L. (1964) Kepler's Laws of Planetary Motion: 1609–1666, *Brit. J. Hist. Sci.*, 2:1-24.

Russell, J.L. (1989) Catholic astronomers and the Copernican System after the condemnation of Galileo, *Ann. Sci.*, 46:365-386.

Rutherford, E. (1904a) Disintegration of radioactive elements, *Harper's Mon. Mag.*, 108: 279-284.

Rutherford, E. (1904b) The radiation and emanation of radium: Part I & II, *Sci. Am. Suppl.*, 58:24073-24075 and 24086-24088.

Rutherford, E. (1905) Radium – the cause of the earth's heat, *Harper's Mon. Mag.*, 110: 390-396.

Rutherford, E. (1909) Visualizing the atom: How the modern scientist counts and measures atoms, *Sci. Am. Suppl.*, 68:178-179 and 202-203.

Rutherford, E. (1911) The scattering of α and β particles by matter and the structure of the atom, *Phil. Mag.*, 21:669-688.

Rutherford, E. (1915) The constitution of matter and the evolution of the elements, *Pop. Sci. Mon.*, 87:1015-142.

Rutherford, E., and Geiger, H. (1908a) An electrical method of counting the number of alpha-particles from radio-active substances. *Proc. Roy. Soc. Lond. A*,141-161.

Rutherford, E., and Geiger, H. (1908b) The charge and nature of the alpha-particle, *Proc. Roy. Soc. Lond. A*, 81: 162-173.

Rutter, F. (1926) *Evolution in Modern Art: A Study of Modern Painting 1870–1925*, George Harrup, London.

Sahni, V. (2002) The cosmological constant problem and quintessence, *Class Quantum Gravit.*, 19:3435–3448.

Salam, A. (1968) Weak and electromagnetic interactions, In *Elementary Particle Physics: Relativistic Groups and Analyticity*, Proc. 8th Nobel Symp., 1968, Lerum, Sweden,

Salter, L.F. (1984) D.H. Lawrence and science, *Theoria*, 62:57-61.

Salusbury, T. (1665) Galileo's Mechanics, In *Mathematical Collections and Translations*, Vol. II.

Sambrook, J. (2004) Thomson, James (1700–1748), *Oxford Dictionary of National Biography*, Oxford University Press, Oxford.

Sarasohn, L.T. (2010) *The Natural Philosophy of Margaret Cavendish: Reason and Fancy during the Scientific Revolution*; John Hopkins University Press, Baltimore.

Saum, L.O. (1999) The proctor interlude in St. Joseph and in America: Astronomy, romance and tragedy, *Am. Stud. Int.*, 37:34-54.

Scerri, E.R. (1989) Eastern mysticism and the alleged parallels with physics, *Am. J. Phys.*, 57:687-692.

Schiebinger, L. (2003) *The Philosopher's Beard: Women and Gender in Science*, Cambridge History of Science, Vol. 4, Eighteenth Century Science, Ed. Porter, R., Cambridge University Press, Cambridge.

Schiebinger, L. (2006). Women of Natural Knowledge, *Cambridge History of Science*, Vol. 3, Early Modern Science, Ed. Park, K., and Daston, L., Cambridge University Press, Cambridge.

Schöenberg, A. (1975) Opinion or insight, In *Style and Idea: Selected Writings of Arnold Schöenberg*. Ed. Stein, L., Trans. Black, L., Faber and Faber, London.

Schofield, R.E. (2004) *The Enlightened Joseph Priestley*, Penn State Univversity Press, University Park, PA.

Schrödinger, E. (1926a) An undulatory theory of the mechanics of atoms and molecules, *Phys. Rev.*, 28:1048-1070.

Schrödinger, E. (1926b) Quantisierung als Eigenwertproblem, *Ann. Phys.*, 79:361-376; 79:489-527; 80:437-490, 81:109-139.

Schrödinger, E. (1926c) Über das Verhäitnis der Heisenberg-Born-Jordanschen Quantenmechanik zu der meinen, *Ann. Phys.*, 79:734-756.

Schrödinger, E. (1958) *Mind and Matter*, Cambridge University Press, Cambridge.

Schwarz, J.H. (1975) Dual-resonance models of elementary particles, *Sci. Am.*, 232:61-69.

Schweizer, P.D. (1982) John Constable, rainbow science, and English color theory, *Art Bull.*, 64:424-445.

Scott-James, R.A. (1908) *Modernism and Romance*, John Lane, London.

Secord, J.A. (1985) Newton in the nursery: Tom Telescope and the philosophy of tops and balls, 1761–1838, *Hist. Sci.*, 23:127-151.

Secord, J.A. (2000) *Victorian Sensation, Extraordinary Publication, Reception, and Secret Authorship of Vestiges of Natural History of Creation*, University of Chicago Press, Chicago.

Sepkoski, D. (2005) Nominalism and constructivism in seventeenth-century mathematical philosophy, *Hist. Math.*, 32:33–59.

Serres, M. (1982) *Hermes: Literature, Science, Philosophy*, Eds. Harari, J.V., and Bell, D.F., Johns Hopkins University Press, Baltimore.

Shapin, S. (1981) Of gods and kings: Natural philosophy and politics in the Leibniz-Clarke disputes, *Isis*, 72:187-215.

Shapin, S., and Barnes, B. (1977) Science, nature and control: Interpreting Mechanics' Institutes, *Soc. Stud. Sci.*, 7:31-74.

Shapiro, A.E. (1973) Kinematic Optics: A study of the wave theory of light in the seventeenth century, *Arch. Hist. Exact Sci.*, 11:131-266.

Shapiro, B.J. (1969) Law and science in seventeenth-century England, *Stan. Law Rev.*, 21:727-766.

Shapiro, B.J. (1991) Early modern intellectual life: humanism, religion and science in seventeenth century England, *Hist. Sci.*, 29:45-71.

Silliman, R.H. (1963) William Thomson: Smoke rings and nineteenth-century atomism, *Isis*, 54:461-474.

Sim, L. (2010) *Virginia Woolf: The Pattern of Ordinary Experience*, Ashgate, Farnham.

Singal, A.K. (2014) A discrepancy observed in the dipole anisotropy in the radio sky at 1.4 GHz and that in the CMBR – A threat to the cosmological principle? *National Conference on Contemporary Issues in High Energy Physics and Cosmology*, Conference Series 481: 012007.

Smith, B.T.L (2013) Indeterministic metaphors: The popular science books of Fritjof Capra and Gary Zukav, *Public Underst. Sci.*, 22:538–545.

Smith, C. (1976) Natural philosophy and thermodynamics: William Thomson and 'The Dynamical Theory of Heat', *Brit. J. Hist. Sci.*, 9:293-319.

Smith, C. (1977) William Thomson and the creation of thermodynamics: 1840–1855, *Arch. Hist. Exact Sci.*, 16:231-288.

Smith, C. (2004) North British network (act. c. 1845–c.1890), *Oxford Dictionary of National Biography*, Oxford University Press, Oxford.

Smith, C., and Norton, M.N. (1989) *Energy and Empire: A Biographical Study of Lord Kelvin*, Cambridge University Press, Cambridge.

Smith, R.W. (2002) Remaking Astronomy: Instruments and Practice in the Nineteenth Century, Chapter 8, The Modern Physics and Mathematical Sciences, Vol. 5, *Cambridge History of Science*, Cambridge University Press, Cambridge.

Smolin, L. (2004) Scientific alternatives to the anthropic principle, arXiv:hep-th/0407213.

Smolin, L. (2006) *The Trouble With Physics: The Rise of String Theory, the Fall of a Science, and What Comes Next*, Houghton Mifflin, New York.

Snobelen, S.D. (1998) On reading Isaac Newton's Principia in the 18th century, *Endeavour*, 22:159-163.

Snobelen, S.D. (2001) "God of Gods, and Lord of Lords": The theology of Isaac Newton's General Scholium to the Principia, *Osiris*, 16:169-208.

Snobelen, S.D. (2003) A time and times and the dividing of time: Isaac Newton, the Apocalypse and 2060 A.D. *Can. J. Hist.*, 38:537-551.

Snobelen, S.D. (2004) William Whiston, Isaac Newton and the crisis of publicity, *Stud. Hist. Philos. Sci.*, 35:573–603.

Snobelen, S.D. (2007) Isaac Newton: his science and religion, In *Science, Religion, and Society: An Encyclopedia of History, Culture, and Controversy*, 2 vols., Ed. Eisen, A. and Laderman, G., Armonk, New York, vol.1, pp. 355-369.

Snow, C.P. (1956) The two cultures, *New Statesman*, 6 October 1956.

Snow, C.P. (1959) *The Two Cultures and the Scientific Revolution*, Cambridge University Press, Cambridge.

Soddy, F. (1903) The radio-activity of uranium, *Sci. Am. Suppl.*, 55:22739-22740.

Soddy, F. (1904*) Radio-Activity: an Elementary Treatise, from the Standpoint of the Disintegration Theory*, The Electrician Printing and Publishing Company, London.

Soddy, F. (1909) *The Interpretation of Radium Being the Substance of Six Free Popular Experimental Lectures Delivered at the University of Glasgow, 1908*, John Murray, London.

Soddy, F. (1916) The problem of the X-rays: How an important question of physical science has been solved, *Sci. Am. Suppl.*, No 2138 pp. 404-405.

Somerville, M. (1831) *Mechanism of the Heavens*, John Murray, London.

Somerville, M. (2009) *Personal Recollections of Mary Somerville by Martha her Daughter*, Echo Library, Teddington.

Sorondo, M.S. (2003) Papal addresses to the Pontifical Academy of Sciences 1917-2002, *Ex Aedibvs Academicis in Civitate Vaticana.*

Sorrenson, R. (1996) Towards a history of the Royal Society in the eighteenth century, *Not. Rec. R. Soc. Lond.*, 50:29-46.

Spencer, H. (1867) *First Principles*, 2nd edition, Williams and Norgate, London.

Spergel, D.N. (2015) The dark side of cosmology: Dark matter and dark energy, *Science*, 347:1100-1102.

Spinoza, B. (1663) *Earlier Philosophical Writings: The Cartesian Principles and Thoughts on Metaphysics*, Trans. Haynes, F.A. (1963) Bobbs-Merrill, IN.

Spinoza, B. (1677) *Ethics*, Trans. Parkinson, G.H.R. (2000) Oxford University Press, Oxford.

Stanley, J.L. (1997) Marx's Critique of Hegel's Philosophy of Nature, *Sci. Soc*., 61:449-473.

Stanley, M. (2003) An Expedition to Heal the Wounds of War" The 1919 Eclipse and Eddington as Quaker Adventurer, *Isis*, 94:57-89.

Stenger, V.J. (2011) The *Fallacy of Fine-Tuning: Why the Universe is not Designed*, Prometheus Books, Amherst, N.Y.

Stenholm, S. (2001) *The Quest for Reality: Bohr and Wittgenstein Two Complementary Views*, Oxford University Press, Oxford.

Stephen, M.D., and Roderick, G.W. (1972) Science, the working classes and Mechanics' Institutes, *Ann. Sci.*,29:349-360.

Stephenson, B. (1994) *The Music of the Heavens: Kepler's Harmonic Astronomy*, Princeton University Press, Princeton, NJ.

Stern, R. (1994) British Hegelianism: A non-metaphysical view? *Eur. J. Philos*, 2:293-321.

Stern, R. (2009) *Hegelian Metaphysics*, Oxford University Press, Oxford.

Stewart, B., and Lockyer, N. (1868) The sun as a type of the material universe, Part II. The place of life in a universe of energy, *Macmillan's Mag.*, 18:319-327.

Stewart, G.W. (1914) The content and structure of the atom, *Science*, 40:661-663.

Stewart, L. (1986) The selling of Newton: Science and technology in early eighteenth-century England, *J. Brit. Stud.*, 25:178-192.

Stewart, L. (1999) Other centres of calculation, or, where the Royal Society didn't count: commerce, coffee-houses and natural philosophy in early modern London, *Brit. J. Hist. Sci.*, 32:133-153.

Stirrat, M., and Cornwell, R.E. (2013) Eminent scientists reject the supernatural: a survey of the Fellows of the Royal Society, *Evol. Edu. Outreach*, 6:33-37.

Stoeger, W.R. (2010) God, physics and the Big Bang, In *The Cambridge Companion to Science and Religion*, Ed. Harrison, P., Cambridge University Press, Cambridge.

Stoeger, W.R. Ellis, G.F.R., and Kirchner, U. (2006) Multiverses and cosmology: Philosophical issues, preprint, arXiv:astro-ph/0407329.

Straus, E.G. (1956) In *Helle Zeit – Dunkle Zeit*, Ed. Seelig, C., Europa-Verl., Zürich.

Strömberg, G. (1925) Analysis of radial velocities of globular clusters and non-galactic nebulae, *Astrophys. J.* 61:353-362.

Stump, E., and Kretzmann, N. (1981) Eternity, *J. Philos.*, 78: 429-458.

Sullivan, J.W.N. (1923a) *Atoms and Electrons*, Hodder and Stoughton, London.

Sullivan, J.W.N. (1923b) *Aspects of Science*, Jonathon Cape, London.

Sullivan, J.W.N. (1925) *Three Men Discuss Relativity*, W. Collins, London.

Susskind, L. (1969) Harmonic-oscillator analogy for the Veneziano model, *Phys. Rev. Lett.*, 23:545-547.

Susskind, L. (1970) Structure of hadrons implied by duality, *Phys. Rev. D*, 1:1182-1186.

Swededborg, E. (1734) Principia rerum naturlium sive novorum tentaminum phaenomena mundi elementaris philosophice explicandi, Trans. Rendell, J.R. (2010) *The Principia Or the First Principles of Natural Things V1: To Which Are Added the Minor Principia and Summary of the Principia*, Kessinger Publishing, Whitefish, MT.

Swift, J. (1710) *A Tale of a Tub, to which is Added the Battle of the Books and the Mechanical Operation of the Spirit*, Eds. Guthkelch, A.C., and Nichol Smith, D. (1920), Clarendon Press, Oxford.

Swift, J. (1726) Travels into Several Remote Nations of the World. In Four Parts. By Lemuel Gulliver, First a Surgeon, and then a Captain of several Ships. Vol. I., Printed for B. Motte, London.

Swift, D. (1755) *An Essay upon the Life, Writings, and Character of Dr Jonathan Swift*, Printed for C. Bathurst, London.

Synge, J.L. (1940) The Decline of Mechanism in Modern Physics. By A. d'Abro, *Bull. Math. Soc.*, 46:720-723.

Sysak, J. (1993) Coleridge's construction of Newton, *Ann. Sci.*, 50:59-81.

Tait, P.G. (1868) *A Sketch of Thermodynamics*, Edmonton & Douglas, Edinburgh.

Tait, P.G. (1876a) Force, *Nature*, 14 :459-463.

Tait, P.G. (1876b) *Lectures on Some Recent Advances in Physical Science*, Macmillan & Co., London.

Tait, P.G. (1898) On knots part 1, In *Scientific Papers*, Cambridge University Press, Cambridge.

Takabayasi, T. (1970) Internal structure of hadron underlying the Veneziano amplitude, *Prog. Theor. Phys.*, 43:117-1118.

Tangherlini, F.R. (1963) Schwarzschild field in n dimensions and the dimensionality of space problem, *Nuovo Cimento*, 27:636–651.

Tegmark, M. (2003) Parallel universes, In *Science and Ultimate Reality: From Quantum to Cosmos*, Eds. Barrow, J.D., Davies, P.C.W., and Harper, C.L., Cambridge University Press, Cambridge.

Tegmark, M., Aguirre, A., Rees, M.J., and Wilczek, F. (2006) Dimensionless constants, cosmology, and other dark matters, *Phys. Rev. D*, 73:023505.

Thomas, W.K., and Ober, W.U. (1989) *A Mind Ever Voyaging: Wordsworth at Work Portraying Newton and Science*, University of Alberta Press, Edmonton, Alberta.

Thomson. J.J. (1909) *Presidential Address*, British Association for the Advancement of Science, Winnipeg.

Thomson. J.J. (1919) Address of the President, Sir J. J. Thomson, O.M., at the Anniversary Meeting, December 1, 1919, *Proc. Roy. Soc. Lond.*, 96:311-322.

Thomson, W. (1848) On an absolute thermometric scale founded on Carnot's theory of the motive power of heat, and calculated from Regnault's observations, *Proc. Cambridge Philos. Soc.* 1: 66-71.

Thomson, W. (1852) On a Universal Tendency in Nature to the Dissipation of Mechanical Energy, *Phil. Mag.*, 4:304-306.

Thomson, W. (1862) On the age of the sun's heat, *Macmillan's Mag.*, 5:388-393.

Thomson, W. (1867) On vortex atoms, *Phil. Mag.*, 33:15-24. Meuthen & Co., London.

Thomson, W. (1904) *Baltimore Lectures on Molecular Dynamics and the Wave Theory of Light*, C. J. Clay and Sons, London.

Thurring, J.H. (1921) *The Ideas of Einstein's Theory: The Theory of Relativity in Simple Language*, Trans. Russell, R.A.B., Methuen, London.

Tilghman, SW.M. (2010) *The Future of Science Education in the Liberal Arts College*, Presented at Presidents Institute, Council of Independent Colleges.

Toland, J. (1704) *Letters to Serena*, Printed for Bernard Lintot, Fleet St., London.

Tondre, M. (2012) George Eliot's "Fine Excess": Middlemarch , energy, and the afterlife of feeling, *Nineteen. Cent. Lit.*, 67:204-233.

Topham, J.R. (1992) Science and popular education in the 1830s: The role of the "Bridgewater Treatises", *Brit. J. Hist. Sci.*, 25:397-430.

Topham, J.R. (2000) Scientific publishing and the reading of science in nineteenth-century Britain: a historiographical survey and guide to sources, *Stud. Hist. Philos. Sci.*, 31:559–612.

Trimble, V. (1987) Existence and nature of dark matter in the universe, *Ann. Rev. Astron. Astrophys.*, 25:425-472.

Trimmer, J.D. (1980) The present situation in quantum mechanics: a translation of Schrödinger's "cat paradox" paper, Proc. *Am. Philos. Soc.*, 124:323-338.

Turnbull, G. (1723) *De Scientiae Naturalis cum Philosophia Morali Conjunctione*, Printed for J Nicol, Aberdeen.

Turnbull, G. (1726) *De Pulcherrima Mundi cum Materialis tum Rationalis Constitutione*, Printed for J. Nicol, Aberdeen.

Turnbull, G. (1740) *The Principles of Moral Philosophy an Inquiry into the Wise and Good Government of the Moral World*, Printed for J. Noon, London.

Turnbull, H.W. Ed. (1960) *The Correspondence of Isaac Newton, Vol. II 1676–1687*, Cambridge University Press, Cambridge.

Turnbull, H.W. Ed. (1961) *The Correspondence of Isaac Newton, Vol. III 1688–1694*, Cambridge University Press, Cambridge.

Tylecote, M.P. (1957) *Mechanics' Institutes of Lancashire and Yorkshire before 1850*, Manchester University Press, Manchester.

Tyndall, J. (1854) On the vibrations and tones produced by the contact of bodies having different temperatures, *Phil. Trans. Roy. Soc. Lond.*, 144:1-10.

Tyndall, J. (1857) On the sounds produced by the combustion of gases in tubes, *Phil. Mag.*, (4)13:473-479.

Tyndall, J. (1863) *Heat Considered as a Mode of Motion*, Longman, Green, Longman, Roberts, and Green., London.

Tyndall, J. (1870) *Researches on Diamagnetism and Magne-Crystallic Action*, Longmans, Green, and Co., London.

Tyndall, J. (1896a) John Tyndall, The Belfast address, In *Fragments of Science: A Series of Detached Essays, Addresses and Reviews*, 2 Vols., Appleton, New York.

Tyndall, J. (1896b) On matter and force, In *Fragments of Science: A Series of Detached Essays, Addresses and Reviews*, 2 Vols., Appleton, New York.

Tryon, E.P. (1973) Is the universe a vacuum fluctuation? *Nature*, 248:396-397.

Ukers, W.H. (1922) *All about Coffee, The Tea and Coffee*, Trade Journal Company, New York.

Untermeyer, L. (1922) *Heavens*, Harcourt, Brace and Company, New York.

Usher, P. (2006) *Hamlet's Universe*, Aventine Press, Chula Vista.

Vanderburgh, W.L. (2003) The dark matter double bind: Astrophysical aspects of the evidential warrant for general relativity, *Philos. Sci.*, 70:812-832.

Veneziano, G. (1968) Construction of a crossing-symmetric, Regge-behaved amplitude for linearly rising trajectories, *Nuovo Cimento*, 57:190-197.

Vilenkin, A. (1982) Creation of universes from nothing, *Phys. Lett. B*, 117:25-28.

Vilenkin, A. (1984) Quantum creation of universes, *Phys. Rev. D*, 30:509-511.

Voltaire (1733) *Letters Concerning the English Nation*, Printed for C. Davis and A. Lyon, London.

Voltaire (1738) *The Elements of Sir Isaac Newton's Philosophy*, Rev., Corr., and Trans. Hanna, J., Printed for S. Austen, London.

Voltaire (1752) Micromégas, In *Candide and other Stories*, Ed. and Trans. Pearson, R. (2006) Oxford University Press, Oxford.

Voss-Andreae, J. (2011) Quantum sculpture: art Inspired by the deeper nature of reality, *Leonardo*, 44:14-20.

Vourlidas, A., Davis, A.C.J., Eyles, C.J. Crothers, S.R., Harrison, R.A., Howard, R.A., Moses, J.D., Socker, D.G. (2007) First direct observation of the interaction between a comet and a coronal mass ejection leading to a complete plasma tail disconnection, *Astrophys. J. Lett.*, 668:L79–L82.

Wade, I.O. (1950) *Voltaire's 'Micromegas': A Study in the Fusion of Science, Myth and Art*. Princeton University Press, Princeton.

Wakeham, N., Bangura,, A., Xu, X.-F., Mercure, J.-F., Greenblatt, M. and Hussey N.E., Gross violation of the Wiedemann–Franz law in a quasi-one-dimensional conductor, *Nat. Commun.* 2:396 doi: 10.1038/ncomms1406.

Walker, A. (1799) *A System of Familiar Philosophy in Twelve Lectures*, printed for the author, London.

Wallace, A.R. (1903) *Man's Place in the Universe a Study of the Results of Scientific Research in Relation to the Unity or Plurality of Worlds*, McClure, Phillips & Co., New York.

Wallis, J. (1677) A letter from Dr Wallis concerning a new musical discovery, *Philos. T. R. Soc. Lond.*, 12:839–842.

Wallis, J., and Wren, C. (1669) A Summary Account of the General Laws of Motion by Dr. John Wallis, and Dr. Christopher Wren, *Philos. T. R. Soc. Lond.*, 3:864-868.

Walmsley, J. (2004) Locke's Natural Philosophy in Draft A of the Essay, *J. Hist. Ideas*, 65:15-37.

Walpole, H. (1798) Walponia; Or Bon-Mots, Apophthegms, Observations on Life and Literature, with extracts from Original Letters, of the Late Horace Walpole, Earl of Orford, *Mon. Mag.*, Vol. 6, XC VII.

Walsh, F. (1743) *The Antediluvian World; or a New Theory of the Earth*, Printed by WS. Powell, Dublin.

Watson, E.C. (1947) Joule's only general exposition of the principle of conservation of energy, *Am. J. Phys.*, 15:385-390.

Watts, C.T. (Ed.) (1969) *Joseph Conrad's Letters to Cunninghame Graham*, Cambridge University Press, Cambridge.

Watts, C.T. (1977) *Conrad's Heart of Darkness: A Critical and Contextual Discussion*, Mursia International, Milano.

Wazeck, M. (2010) The 1922 Einstein film: cinematic innovation and public controversy, *Phys. Perspect.*, 12:163–179.

Wazeck, M. (2013) Marginalization processes in science: The controversy about the theory of relativity in the 1920s, *Soc. Stud. Sci.*, 43:163-190.

Wazeck, M. (2014) *Einstein's Opponents: The Public Controversy about Relativity in the 1920s*, Trans. Kobey, G.S., Cambridge University Press, Cambridge.

Webb, B. (1926) *My Apprenticeship*, Cambridge University Press Cambridge.

Weinberg, S. (1967) A model of leptons, *Phys. Rev. Lett*, 19:1264-1266.

Weinberg, S. (1992) *Dreams of a Final Theory*, Pantheon Books, New York.

Weinberg, S. (1993) *The First Three Minutes, a Modern View of the Origin of the Universe*, Basic Books, New York.

Weinberg, S (2001) A Universe with no designer, *Ann. New York Acad. Sci.*, 950:169-174.

Weinberg, S. (2008) *Living in the multiverse, In Universe or Multiverse?* Ed. Carr, B., Cambridge University Press, Cambridge.

Weinberg, S., and Polkinghorne J.C. (2001) An Exchange between Steven Weinberg and John Polkinghorne, *Ann. New York Acad. Sci.*, 950:183-190.

Weintraub, S. (2004) Snow, Charles Percy, Baron Snow (1905–1980), *Oxford Dictionary of National Biography*, Oxford University Press, Oxford.

Wells, H.G. (1894) Popularising Science, *Nature*, 50:300-301.

Wells, H.G. (1898) *War of the Worlds*, Heinemann, London.

Wells, H.G. (1914) *The World Set Free*, Macmillan & Co., London.

Wells, H.G. (1931) Preface to the time machine, In *The Time Machine*, Random House, New York.

Wesson, P.S. (1986) The extra-galactic background light: A modern version of Olbers' paradox, *Space Sci., Rev.*, 44:169-176.

West, W.K. (1978) Problems in the cultural history of the ellipse, *Technol. Cult.*, 19:709-712.

Westfall, R.S. (1980*) Never at Rest: A Biography of Isaac Newton*, Cambridge University Press, Cambridge.

Weston, T. (2012) Marx on the dialectics of elliptical motion, *Hist. Mater.*, 20:3–38.

Westphal, K.R. (2008) Philosophizing about nature: Hegel's philosophical project, philosophizing about nature, In *The Cambridge Companion to Hegel and Nineteenth-Century Philosophy*, Ed. Beiser, F.C., Cambridge University Press, Cambridge.

Wettersten, J. (1994) William Whewell: Problems of induction vs. problems of rationality, *Brit. J. Philos. Sci.*, 45:716-742.

Wheeler, J.A. (1957) On the nature of quantum geometrodynamics, *Ann. Phys.*, 2:604-614.

Whewell, W. (1833) *Astronomy and General Physics Considered with Reference to Natural Theology*, 2nd edition. William Pickering London.

Whewell, W. (1840) *Philosophy of the Inductive Sciences, Founded upon their History*, 2 Vols., J.W. Parker, London, J. & J.J. Deighton, Cambridge.

Whewell, W. (1847) Remarks on a review of the Philosophy of the Inductive Sciences, In *Philosophy of the Inductive Sciences, Founded upon their History*, Vol. 2, New Ed., J. W. Parker, London.

Whewell, W. (1849) *Of Induction, with Especial Reference to Mr J. Stuart Mill's System of Logic*, J.W. Parker, London.

Whewell, W. (1853) *Of the Plurality of Worlds: An Essay*, J. W. Parker and Son, London.

Whewell, W. (1860) *On the Philosophy of Scientific Discovery*, J. W. Parker and Son, London.

Whewell, W. (1831) Review of J. Herschel's preliminary discourse on the study of natural philosophy, *Quart. Rev.*, 45:374-407.

Whiston, W. (1698*) A Vindication of the New Theory of the Earth from the Exceptions of Mr. Keill and Others with an Historical Preface of the Occasions of the Discoveries therein contain'd, and some Corrections and Additions*, Printed for B. Tooke, London.

Whiston, W. (1702) *A Short View of the Chronology of the Old Testament*, Cambridge University Press, Cambridge.

Whiston, W. (1715) *Astronomical Lectures, Read in the Publick Schools at Cambridge*, J. Senex and W. Taylor, London.

Whiston, W. (1716) *Sir Isaac Newton's Mathematical Philosophy more easily Demonstrated with Dr Halley's Account of Comets Illustrated*, J. Senex and W. Taylor, London.

Whiston, W. (1725a) *A New Theory of The Earth*, Fourth Ed., Printed for S. Tooke and B. Motte, London.

Whiston, W. (1725b) *Astronomical Principles of Religion, Natural and Reveal'd*, Printed for J. Sennex, W. and J. Innys, J. Osborne, and T. Longman, London.

Whiston, W., and Hauksbee, F. (1714a) *A Course of Mechanical, Optical, Hydrostatical, and Pneumatic Experiments.*

Whiston, W., and Hauksbee, F. (1714b) *A Experimental Course on Astronomy: An Advertisement for the Course.*

White, A.N. (2004) Hopkins, Gerard Manley (1844–1889), *Oxford Dictionary of National Biography*, Oxford University Press, Oxford.

Whitehead, A.N. (1948) *Science and the Modern World*, The New American Library, New York.

Whitehurst, J. (1788). *An Inquiry into the Original State and Formation of the Earth; Deduced from Facts and Laws of Nature, to Which is Added an Appendix Containing some General Observations on the Strata in Derbyshire*, Printed for the author, London.

Whitrow, G.J. (1961) *The Natural Philosophy of Time*, Thomas Nelson, London.

Whitrow, G.J., and Bondi, H. (1953) Is physical cosmology a science? *Brit. J. Philos. Sci.*, 4:271-283.

Whitworth, M.H. (2001) *Einstein's Wake: Relativity, Metaphor, and Modernist Literature*, Oxford University Press, Oxford.

Whitworth, M.H. (2011) Natural science, In *T.S. Eliot in Context*, Ed. J. Harding, Cambridge University Press, Cambridge.

Whyte, L.L. (1958) R. J. Boscovich, S.J., F.R.S. (1711-1787), and the Mathematics of Atomism,. *Not. Rec. Roy. Soc. Lond.*, 13:38-48.

Wick, D. (1996) *The Infamous Boundary: Seven Decades of Heresy in Quantum Mechanics*, Copernicus, New York.

Wigelsworth, J.R. (2008) Bipartisan politics and practical knowledge: advertising of public science in two London newspapers, 1695–1720, *Brit. J. Hist. Sci.*, 41:517–540.

Wigner, E. (1960) The unreasonable effectiveness of mathematics in the natural sciences, *Commun. Pure Appl. Maths.*, 13:1-14.

Wigner, E. (1965) Violations of symmetry in physics, *Sci. Am.*, 213:28-37.

Will, C.M. (1988) Henry Cavendish, Johann von Soldner, and the deflection of light, *Am. J. Phys.*, 56:413-415.

Will, C.M. (2006) The confrontation between general relativity and experiment, *Living Rev. Relativ.*, 9:3-100.

Willmoth, F. (2004) Hodgson, James (bap. 1678–1755), *Oxford Dictionary of National Biography*, Oxford University Press, Oxford.

Wilson, C. (1985) The great inequality of Jupiter and Saturn from Kepler to Laplace, *Arch. Hist. Exact Sci.*, 33:15-290.

Wilson, C.T.R. (1911) On a method of making visible the paths of ionising particles through a gas, *Proc Roy. Soc. Lond. A*, 85:285-288.

Wilson, J. (2007) *Collected Works of James Wilson*, Vol. 1., Eds. Hall, K.L., and Hall, M.D., Liberty Fund, IN.

Winch, D. (2004) Smith, Adam (bap. 1723, d. 1790), *Oxford Dictionary of National Biography*, Oxford University Press, Oxford.

Witten, E. (1995) String theory dynamics in various dimensions, *Nucl. Phys. B*, 443:85-126.

Witten, E. (1996) Reflections on the fate of spacetime, *Phys. Today*, 49:24-30.

Wiiten, E. (1998) Magic, mystery, and matrix, *Not. AMS*, 45:1124-1129.

Witten, E. (2005) Unravelling string theory, *Nature*, 438:1085.

Wittgenstein, L.J.J. (1922) *Tractatus Logico-Philosophicus*, with an Introduction by B. Russell, Kegan Paul, Trench, Trubner & Co., Ltd., London.

Woit, P. (2006) *Not Even Wrong: The Failure of String Theory and the Continuing Challenge to Unify the Laws of Physics*, Basic Books, New York.

Wollaston, W.H. (1802) A method of examining refractive and dispersive powers, by prismatic reflection, *Philos. Trans. R. S. Lond.*, 92:365-380.

Wood, A. (1962) *The Physics of Music*, Revised by J.M. Bowsher, Methuen & Co. Ltd., London.

Woode, P. (2003) Science, philosophy, and the mind, In *The Cambridge History of Science*, Vol. 4 *Eighteenth-Century Science*, Ed. R. Porter, Cambridge University Press, Cambridge, pp. 800-824.

Woodward, J. (1695) *An Essay Toward a Natural History of the Earth and Terrestrial Bodies, Especially Minerals and also of the Sea, Rivers, and Springs with an Account of the Universal Deluge*, Printed for R. Wilkin, London.

Woolf, V. (1925) *Mrs Dalloway*, Hogarth Press, London.

Woolf, V. (1927) *To the Lighthouse*, Hogarth Press, London.

Woolf, V. (1931) *The Waves*, Hogarth Press, London.

Woolf, V. (1978) *The Diary of Virginia Woolf*, Vol. 2 1920–1924, Ed. A.O. Bell, Harcourt Brace & Company, San Diego.

Wordsworth, W. (1963) *Lyrical Ballads: Wordsworth and Coleridge*, Text of the 1798 edition with the additional 1800 poems and the prefaces, Ed. Brett, R.L, and Jones, A.R., Methuen, London.

Wordsworth, W. (2007) *The Excursion*, Ed. Bushell, S., Butler, J.A., and Jaye, M.C., Cornell University Press, Ithaca, NY.

Worth, C. (1991) Swift's 'flying island': Buttons and bomb-vessels, *Rev. Eng. Stud.*, 42:343-360.

Worthen, J, (2004) Lawrence, David Herbert (1885–1930), *Oxford Dictionary of National Biography*, Oxford University Press, Oxford.

Wright, T. (1750) *An Original Theory or New Hypothesis of the Universe*, A facsimile reproduction together with the first publication of A theory of the Universe 1734, Introduction and transcription Hoskin, M.A., Macdonald, New York. (1971).

Yin, J., et al. (2013) Lower bound on the speed of nonlocal correlations without locality and measurement choice loopholes, *Phys. Rev. Lett.*, 110:260407.

Young, T. (1802) An account of some cases of the production of colours, not hitherto described, *Philos. T. R. Soc. Lond.*, 92:387-397.

Young, T. (1804) Experiments and calculations relative to physical optics, *Philos. T. R. Soc. Lond.*, 94:1-16.

Zahlan, A.R. (1999) Crossing the Border: Lawrence Durrell's Alexandrian Conversion to Postmodernism, *S. Atl. Rev.*, 64:84-99.

Ziff, J. (1964) John Langhorne and Turner's 'Fallacies of Hope', *J. Warburg Courtauld*, 27:340-342.

Zimmer, E. (1936) *The Revolution in Physics*, Faber and Faber, London.

Zukav, *The Dancing Wu Li Masters*, Fontana, London.

Zweig, G. (1964). "An SU3 Model for strong interaction symmetry and its breaking". *CERN Report* No.8182/TH.401.

Index

Abbott, Edwin
Flatland, 433–34
Abro, Abraham d'
popularization modern physics, 379
Adams, Fred
cosmological parameters, 348
Addison, Joseph, 240
Aether, 11, 13, 21, 23, 141, 156, 301–4, 377, 390, 391, 393, 433
Airy, George
stars apparently before the Moon, 90
Alembert, Jean-Baptiste d'
Rameau's music theory, 252
vis viva and momentum, 11
Alexander, Samuel
Space, Time, and Deity, 420
Algarotti, Francesco
Cartesianism and Newtonianism, 126
Locke and freemasonry, 126
Newton's optics, 125
Amis, Kinsley, 417
Amontons, Guillaume
absolute zero temperature, 16
Anaxagoras of Samos
banished for impiety, 61
heliocentricity, 5
Anderson, Carl
positron, 43
Andrade, Edward
Introduction to Science, 364
Structure of the Atom, 374
Anthropic principle, 341–53
conversion mass into energy, 345
electromagnetic and gravtitational forces, 343–45
primodial fluctuations, 346
results learnt, 348–53
Anti-Einsteinians, 389–91
Anti-Newtonians, 86–90, 155–57
Benjamin Taylor Kavanaugh, 157
Hutchinsonians, 87–89
Muggletonians, 87, 89–90
Richard Phillips, 155–56
Thomas Simmons Mackintosh, 157
Aquinas, St Thomas
Aristotlian philosophy, 61
Boethius and eternity, 311
five ways and cosmology, 337
Arbuthnot, John
satirist, 240
Whiston's longitude measurement, 244
Aristarchus
persecuted for heliocentricism, 61
Aristotle
antipluralist, 91
Catholicism, 61–62
chief scientist for Catholic Church, 337
circle, the perfect form, 223
colour and musical harmony, 239
division of knowledge, 455
geocentricism, 302
light/darkness and colours, 269
metaphysics, 164
Arnold, Matthew, 217, 425
The Two Cultures, 219
Astronomy
Brahe's hybrid system, 6
comets, 7, 11, 13, 67, 77, 80–1, 85–6, 94, 180, 222, 235, 242, 244, 255, 277
Copernican helocentric system, 5
elliptical planetary orbits, 5–6, 11, 12, 51
nebulae and galaxies, 28

Ptolemy's geocentric theory, 4–5
spectra of nebulae, 152–54
transit of Venus, 120
vortex theory, 11
Atkinson, Peter
x-rays, 367
Atomic energy, 372
Atoms
Democritus, 21
Newton's view, 13
vortex model, 190
Auden, Wystan Hugh
politicized writers, 414
Augustine, St.
Arianism, 88
miracles, 70
Babbage, Charles
decline in science, 145
Bacon, Francis
aphorisms, 184
conservation of mass, 178
empiricism, 164
experimental method, 239
inductive method, 164
laws of nature, 207, 212
philosophy of science ideas, 182
religion and physics, 94, 108
science, its practical effect, 242
Baker, Thomas, 99
Barberini, Pope Urban VIII, 7
Barrow, Isaac
nominalism, 306
Bassi, Laura
physics' professor Bologna, 125
Bauch, Andreas
local position invariance, 31
Bayes, Thomas, 353
Beckett, Samuel
comment on *Finnegans Wake*, 418
early postmodernist play, 416
modern physics' influence, 416
Becquerel, Henri
radioactivity, 366
Bedford, David
Greenwich Observatory music, 234
Beer, Gillian
modern physics and Woolf, 421
Woolf, illusion and reality, 422
Behn, Aphra, 91
enthusiasm, 203
Bell, John
Einstein-Podolsky-Rosen paper, 316
hidden variables, 316, 323
Bellarmine, Cardinal Roberto
letter to Paolo Foscarini, 7
Bergman, Günter
Harmony of the World, 234
Berkeley, George, Bishop
his *De Motu*, 176
his philosophy, 175–77
idealist, 175
instrumentalist, 176
laws of nature, 175, 176
metaphysical idealist, 211
Bernoulli, Daniel
vibrating string, 253
Berington, Samuel
Hutchinsonianism, 88
Bernoulli, Daniel
electric charge, 14
Birbeck, George
mechanics' institutes, 132–34
Black holes, 31–33
Chandrasekhar's limit, 32
coalescence, 31, 442
Blake, William
antagonistic to science, 258, 265
heroes and villiana, 266
Newton and Nebuchadnezzar, 265
Newton's Pantocrator, 265
Bloch, Ernst
liked by Einstein, 445
Bode, Johann Elert
Plurality of the Worlds, 94
Boethius
eternity, 311
Bohm, David
concept of wholeness, 324
exile in Brazil, 318
influence of Engels, 319
non-locality, 323
politics, 318, 319, 322

realist quantum theory, 315, 319, 322–24
Bohr, Niels
atoms, quantum structure, 39–41, 373–74
Complementarity Principle, 319, 320
Einstein-Podolsky-Rosen, 315
England visit 1911-1912, 373
epistemological problems, 324
free will, 310
Rutherford's atom, 372
USA visit 1923, 375
Bolingbroke, Lord
befriended Voltaire, 122
Bolton, Lyndon
essay on relativity, 388
Boltzmann, Ludwig
Bild theory, 192
black body radiation, 34–35
H theorem, 20
multiverse, 351
thermodynamics, second law, 34
Bondi, Hermann
cosmology a science? 325–26, 395–98
Bonnor, William
cosmological theories, 397
Born, Max
physics is not a Marxist paradigma, 321
quantum wave function, 43
uncertainty principle, 41–42
Børresen, Hakon
Tycho Brahe ballet, 234
Boscovich, Roger
particle theory, 104
quantity of motion, 11
Bowen, Ira
spectra of nebulae, 153
Boyle Lectures
John Harris, 71
Richard Bentley, 67, 71, 72–76, 78, 114
Samuel Clarke, 71, 76–77, 78
William Derham, 71, 77–78
William Whiston, 71
trustees, 71
Boyle, Robert
atoms, 74
Boyle's gas law, 16
Christianity, 64, 69, 71
crystallography, 367
Sceptical Chymist, 127
science, its practical effect, 242
universe, clock model, 104
Bradshaw, David
Huxley's flippancy, 426
Bragg, William Henry
crystallography and X-rays, 367, 369–70
particle wave duality, 377
physics popularization, 360
RI Christmas lectures, 142
Bragg, William Lawrence
Bill Coates, 140
crystallography and X-rays, 367–70
RI Christmas lectures, 140
Brahe, Tycho
hybrid system, 5
Braithwaite, Richard Bevan
indeterminacy, 309
Brande, William
Royal Institution, 137
Braque, Georges
co-creator Cubism, 435
Bravais, Auguste
crystal lattices, 368
Brentano, Franz
thermodynamics and religion, 106
Breton, André
Surrealist manifesto, 441
time/space duality, 441
Brewster, David
advice to popular writers, 145
British Association, 142, 143
Cabinet Cyclopedia, 145
popular books, 144, 158
primary colours, 267
Treatise on Optics, 145
Bridgewater Treatises, 85–86
British Association, 457
Lectures for the operative classes, 143
nineteenth century, 142–44
twentieth century, 361
Broad, Charlie Dunbar
existence demands a cause, 339
indeterminacy, 309
Broglie, Louis de

particle-wave duality, 37
quantum mechanics, relativistic, 42
quantum mechanics' relevance, 378–79
Brougham, Henry
technical education, 134, 135, 151
Bunsen, Robert
Fraunhofer lines, 97
Burnet, Thomas
Great Deluge, 64, 81
miracles, 64
natural and revealed religion, 78
Platonists, 64
plurality of the worlds, 91
Sacred Theory of the Earth, 64
Burno, Giordano
plurality of the worlds, 90
Byron, George Gordon (Lord)
eulogy of Newton, 262
Caccini, Tommaso
denounced Galileo's opinions, 6
Callender, Craig
number of space dimensions, 346–47
Cambridge Platonists, 63–64
Campbell, Norman Robert
Bohr's atom, 374
WEA, 220, 362
Capra, Fritjof
physics and Eastern mysticism, 325
Carlson, Per
cosmic microwave background, 346
Carlyle, Thomas
romantic idealism, 103
Caroline of Ansbach, Princess of Wales
Leibniz, 67
Carroll, Sean
Higgs boson, 413
Cassini, Jacques
Earth prolate shaped, 251
Catcott, Alexander
antipluralist, 99
distance of the stars, 100
Categorical frameworks, 300
Catholicism
cosmology, 5, 6–8, 62–63, 456
revelation and reason, 61
Cavallo, Tiberius
meteor of 1783, 83
Cavendish, Henry
deflection of light passing the Sun, 30
Cavendish, Margaret
poetry and physics, 231
Woolf, dismissive comments, 232
Chalmers, Thomas
plurality of the worlds, 95
Chamberlen, Paul
using Newton in advertising, 114
Chambers, Robert
Vestiges, 103
Charles, Jacques
gas laws, 16
Châtelet, Émilie du
relationship with Voltaire, 124
translation of *Principia*, 124
Chesterton, Gilbert Keith
great poets, 418
Cigoli (Lodovico Cardi)
artist friend of Galileo, 238
Circle image, 223–24, 237
breaking the image, 224, 228, 229
Clapeyron Émile
gas laws, 16
Clarke, Adam
plurality of the worlds, 95
Clarke, Samuel
Newton's Bulldog, 70, 77
universe, clock model, 70
Clausius, Rudolf
conservation of energy, 18
kinetic theory of gases, 20
second thermodynamic law, 19
thermodynamics and religion, 105
thermodynamics and the universe, 20, 105, 325
Clerke, Agnes
popular astronomy, 151–53
Coffee houses, 114–17, 157, 240–41, 243, 245, 456
Cohen, Bernard
de Broglie, 378, 379
Declaration of Independence, 207
Principia, definitive translation, 10
Colour of the sky, 289

Comte, Auguste
metaphysics, 186
philosophy of science, 182
postivism, 186
Colding, Ludwig August
first law of thermodynamics, 17, 18
thermodynamics and religion, 105
Coleridge, Samuel Taylor
millenarian, 259
poet receptive to science, 258–60
Condamine, Charles Marie de la
length of meridian at Equator, 251
Conrad, Joseph
thermodynamics, 282–85
Constable, John
rainbows, 272–74
Copernicus, Nicolaus
heliocentricity, 5, 62
new science, 221, 360
trepidation, 227
Correggio, Antonio da
ellipses in painting, 238
Corti, Alfonso
structure of the ear, 286
Cosmology, 325–40, 391–98
Big Bang, 311, 326–33, 336–42, 346–47, 351
cosmic microwave background radiation, 329
cosmic purpose, 301
cosmological principle, 327
creation, 335–38
dark matter and energy, 333–35
expanding universe, 327
Friedmann universes, 327
Friedmann-Lemaître-Robertson-Walker model, 329
Hubble's constant, 328, 333, 393, 395, 398
inflation, 333
part of science, 325–26
popularity, 452
space dimensions, 346–48
Steady State, 326, 390–93
theism/atheism, 326, 338–40
Cotes, Roger
dispute with Leibniz, 70
Coulomb, Charles-Augustin de
electrostatic forces, 15
Craig, William Lane
existence must have a cause, 338
theist cosmological argument, 338–40
Creti, Donato
paintings of the planets, 255
Cronin, James
violation CP symmetry, 429
Crystallography, 367–70
Cudworth, Ralph
Cambridge Platonists, 63
critic of Hobbes, 66
Cultural Revolution 17th cent., 221–22
Cunningham, Ebenezer
first relativity book in English, 30, 381
Curie, Marie
radioactivity, 366
Curie, Pierre
radioactivity, 366, 370
Dali, Salvador
Camembert watches/relative time, 441
fourth dimension, 441
tesseract *Crucifixion*, 441
Dalton, John
atoms, 21
Dardel, Robert Guillaume
sculpture of Descartes, 256
sculpture of Newton, 257
Darrow, Karl
popular physics writer of distinction, 375
Darwin, Erasmus
Lunar Society, 255
poetry to popularize science, 258
proto-evolutionary ideas, 86
Darwin, Charles
rival influence to physics, 212
Darwin, Charles Galton
possible intuitive reality, 377
Davidson, George
stars apparently before the Moon, 90
Davies, John
Copernican System and poetry, 222–23
Davies, Paul
early popular book on superstrings, 401

popular quantum mechanics, 381
Royal Society of Literature, 411
Davisson, Clinton
electron diffraction, 37
Davy, Humphry
Caloric theory, 139
poetry, 258
Royal Institution, 136–40
Democritus
atomism, 21, 225
atoms and the soul, 102
Derham, William
physico-theological poets, 92–93
Desaguliers, John Theophilus
Course of Experimental Philosophy, 121
feats of strongmen, 121–22
force and energy, 121
lady audiences, 124
popular lectures, 116–18
The Newtonian System of the World, 204
Descartes, Rene
action at a distance, 169
books on epistemology, 165
cogito ergo sum, 166
Kepler ignored, 222
plurality of the worlds, 92
Principia Philosophiae, 9, 64, 66, 165, 166
quantity of motion, 10, 11, 173, 174
religious self-censure, 62
Snell's law, 165
spatial extension, 168
universe, clock model, 67–68
vortex theory, 11, 240
wave theory of light, 37
Dessau, Paul
opera *Einstein*, 446
Dialectic method
Hegel, 212
Marxism, 208–11
Popper, 210
Dick, Thomas
meteor of 1783, 83
people's library, 84
plurality of the worlds, 96
popular physics books, 147
teacher science writers, 146
The Christian Philosopher, 84
Digges, Thomas
Copernican System, 221
Dingle, Herbert
cosmythology, 325
Modern Astrophysics, 392
twins' paradox, 383
Dirac, Paul
magnetic monopole, 43
physics and mathematics, 305
positron, 43
quantum mechanics, relativistic, 43
quantum mechanics, renormalization, 44
Ditton, Humphrey
popular lectures, 116
Whiston's longitude measurement, 244
Doesburg, Theo van
De Stijl, 437–38
Domenico, Giovanni
Earth prolate sphere, 251
Donne, John
new atomism, 225
poetry, astronomy, and physics, 223–28
trepidation, 227
Drouin, Jeffrey
Joyce and Marsden, 420
Joyce's reading, 417
Drury, Elizabeth
Donne's elegy, 226
Drysdale, William
theories of the Earth, 83
Duchamp, Marcel
Arensberg Circle, 437
fourth dimension, 436
Duhem, Pierre
chaos and philosophy, 309
conventionalists, 51
reality of atoms, 107
separation of religion and physics, 105
Dühring, Eugen
criticism of Marxism, 210
Durrell, Lawrence
relativity, 427–28
spacetime, 428
Duszenko, Andrzej
Joyce and physics, 418

Joyce's reading, 417
Ebury, Katherine
Beckett's use of physics, 416
Eddington, Arthur
classical and modern reality, 20
energy of the stars, 392
expanding universe, 393–94
indeterminism, 309
is the world deterministic? 307
modern physics and materialism, 426
Nature of the Physical World, 20, 22, 365, 376–77, 389
physics and religion, 300, 310
physics writer of distinction, 378
quantum mechanics reality, 304
redshifts spectra in galaxies, 326, 327
Space, Time and Gravitation, 387
Stars and Atoms, 392
statistical mechanics, 277
universe's quantum beginning, 328
Egerton, Francis Henry
Bridgewater Treatises, 85
Eggeling, Viking
film technique, 437
Ehrenfest, Paul
early discussion on relativity, 382
number of space dimensions, 346
Einstein, Albert
actions at a distance, 316
aether and general relativity, 303
assessment of Mach, 193
biggest blunder, 28
blackboard, 393
Bohm's quantum theory, 316
celebrity, 457
Copenhagen interpretation, 317
cosmological constant, 326
determinism, 313
difficulty in understandin Bohr, 319
electron wave, 43
free will, 310, 313
general theory of relativity, 26–31, 50, 51
importance of imagination, 414
indeterminacy disturbing, 314
influenced by Spinoza, 313
Lobaschevsky, 420
manifesto for peace, 321
philosophy and religion, 354
photoelectric effect, 36
physics and religion, 314
popular book on relativity, 387
quantum mechanics reality, 314
quotations, 12, 34, 316, 349, 381
relativity in weak gravitational fields, 384
secular opposition, 389
special theory of relativity, 22–25
tensors, 446–47
The Times, 1919, 387
unified theory of everything, 47
visit to USA, 1921, 420
Electrostatic forces
Coulomb's law, 15
Eliot, George
influence of thermodynamics, 280–81
Eliot, Thomas Stearn, 285
influence of Russell and Lowell, 286
modernism, 414
poetry and science, 415
thermodynamics and *Hollow Men*, 285
Emerson, Ralph Waldo
imagination and physics, 413
Ellis, George
multiverse not testable, 352
Ellison, William
amateur telescope maker, 154
Encke, Johann
comet, 85
Engels, Friedrich, 208
Anti-Dühring, 210
dialetics, 208–10, 320
dismissive of Newton, 209
Erasmian humanism, 3, 455
Eve, Arthur
atom early, popular modern account, 373
False analogies, 221
Faraday, Michael
cathode rays, 21
electric force, 15
electromagnetism, 137, 138
herald of second physics' revolution, 366
History of a Candle, 142
lines of force, 157, 189

RI Christmas Lectures, 142
RI Friday evenings, 138
Royal Institution, 137–38, 142
Favaro, Antonio
Galileo's manuscripts, 8
Fedorov, Evgraf
crystal structure, 368
Ferguson, Adam
moral and natural philosophy, 201–02, 205
Ferguson, James, 119
astronomy, 94, 95, 129
popular lectures, 117, 120
Feynman, Richard
arts and science both universe's story, 413
Character of Physical Law, 365
comparison of poetry to physics, 411
Feynman diagrams, 44–45
Lectures on Physics, 38, 364
mathematical culture, 219
Ozma problem, 431
particle wave duality, 39
philosophy of science, 182
QED, 44, 375
quantum entanglement teleportation, 316
relativity and quantum mechanics, 301
superstring stuff crazy, 401
Filmer, Robert
divine right of kings, 204
Fitch, Val
violation CP symmetry, 429
Flammarion, Camille
La Fin du Monde, 281
Folkes, Martin
RS and Christianity, 79
Whiston and RS fellowship, 82
Fontenelle, Bernard le Bovier de
Anciens et les Modernes, 243
Cartesianism, 124
Catholic suppession, 91
plurality of the worlds, 91
satirized by Voltaire, 251
usefulness of pure science, 248
writing style, 125
Fourier series, 183, 253
Franklin, Benjamin
electric charge, 14
tool-making, 52
US Constitution, 206, 207
Fraunhofer lines, 97
Fresnel, Augustin-Jean
wave theory, 141
Frost, Isaac
critique of Newtonianism, 90
Muggletonian, 87
plurality of the worlds, 100
stars apparently before the Moon, 90
Fresnel, Augustin-Jean
wave theory of light, 36
Friedmann, Alexander
cosmological principle, 29, 327
evolution of the universe, 327
Gabo, Naum
Minkowksi diagrams, 440
relativity, 439
Galilei, Galileo
anticipated Newton's first law, 8
Catholic condemnation, 6–8
Catholic reconciliation, 63
Dialogue and Discourse, 6
free fall, 173
Friendship with Cigoli, 238
Galilean transformation, 22–23
Kepler ignored, 222
mass and weight, 18
Milton, 230
Newtonian physics, 8–9
physics is mathematical, 305, 306
telescope, 221, 225
vibrating strings, 235
Gamow, George
Big Bang, 398
cosmic opera, 398
Tompkins' adventures, 378
Gardner, Martin
exchange with Nabokov, 430
influence on Nabokov, 429
Ozma problem, 429, 431
Garnett, Thomas
Royal Institution, 136, 137
Gassendi, Pierre
atoms and Christainity, 102

Gauss, Carl Friedrich
non-Euclidean geometry, 433
tensors, 446
Gehrcke, Ernst, 390–91
Geller, Abraham (Rabbi)
quantum mechanics and determinism, 313
Giacconi, Riccardo
Cygnus X-1, 32
Gibbons, Gary
black holes, 32
Giedion, Sigfried
influence of Doesburg, 438
Gieger, Hans
Gieger counter, 371
Gilbert, Stuart
Joyce, Work in Progress, 418
Glashow, Sheldon
force unification, 46
string theory atheist, 401, 403
Glass, Philip
Einstein on the Beach, 445, 446
Michelson-Morley experiments, 445
opera Galileo Galilei, 235
opera *Kepler*, 234
Glover, Richard
Newton demands the Muse, 240
God
clockmaker, 67, 68, 70, 86, 104
design argument, 67, 71, 77, 78, 79, 86, 92, 94, 99, 103
Goethe, Johann Wolfgang von
prisms, 269, 270
rainbows, 272
romantic idealism, 103
Gödel, Krut
mathematics and axioms, 50
Gold, Thomas
Steady State theory, 395
Goldsmith, Oliver
poem for Newbery, 128
Goldstein, Eugen
cathode rays, 21
Goll, Yvan
Surrealist manifesto, 441
Górecki, Henryk
symphony *The Copernican*, 234
Gardner, Martin
anthropic principle, 341
Graetz, Leo
atom early, popular modern account, 374
Grassi, Orazio
comets, 7
Graves, Charles
Punch, 393
Gravitation
dimensions of space, 15
instant action, 13
Newtonian, 11–12
Great Deluge, 64, 79–81, 108
Moses's Principia, 87
Green, Michael
superstring revolution, 48, 400
Greene, Brian
Elegant Universe, 34, 399, 403
Gregory, David
Newtonian mechanics, 14
miracles, 68
Gregory, Olinthus
physics book for youths, 129
popular astronomy book, 146
semi-popular book, 147
teacher science writers, 146
Gregory, Richard
resented women popular writers, 152
Gribbin, John
popular quantum mechanics, 381
Royal Society of Literature, 411
Guericke, Otto von
air pump, 8
Guth, Alan
inflation, 330–32
Haekel, Ernst
monism, 107
Hafele, Joseph
clock paradox experiment, 383–84
Haldane, John
Marxist apologist, 209
Hale, Mathew
law and physics, 212
Hales, Stephen
ascent of sap, 117, 118
biophysical experiments, 249

Halley, Edmund
comets, 13, 80, 222
Great Deluge, 81
Hamilton, William Rowan
unity of culture, 258
Harcourt, William Vernon
midwife of British Association, 143
Harris, John
coffee-house lectures, 115
used Fontenelle's style, 125
Harris, Moses
three primary colours, 267
Harriot, Thomas
telescope, 225
Harrow, Benjamin
dislike of relativity books, 388
Hass, Arthur
total energy of universe is zero, 336
Hauksbee, Francis
popular lectures, 115–16
Hayles, Katherine
Nabokov, 429
Hawking, Stephen
black holes, 32
Brief History of Time, 394, 401, 405
Grand Design, 50
philosophy is dead, 299, 353
principle of ignorance, 339
quantum cosmology, 338
Hayden, Joseph
comic opera Il Mondo della Luna, 254
Hegel, Georg
dialetic method, 181
influence of Kant, 179, 181
metaphysics, 179
philosophy, 179–81
physics and mathematics, 180
biased against Newton, 180
Heisenberg, Werner
Complementarity Principle, 320
Einstein, dogmatic realist, 314
exact state of an entity, 305
free will, 310
physics and mathematics, 305
popularization of uncertainty, 376
QED, 43
quantum mechanics reality, 307
S-matrix, 4
surrealism and uncertainty, 441
uncertainty principle, 41–42, 43, 317
Helmholtz, Herman von
colour theory, 268
conservation of energy, 17, 18
intutition, 191
music, 286–89
music and painting, 275
North British Group of physicists, 186
optics and physiology, 267
painting, 289–90
physiology, 191
thermodynamics, 275, 278
Henderson, Linda Dalrymple
art and the fourth dimension, 433, 436
Breton time/space duality, 441
Doesburg and relativity, 437
Planes by Colour and X-rays, 436
Herschel, John
Cabinet Cyclopedia, 145
philosopher of science, 182–85
Preliminary Discourse, 182–84
Whewell's Inductive Sciences, 184
Herschel, William
plurality of the worlds, 94
Hertz, Heinrich
Bild images, 191–92
Heyl, Paul
popularization modern physics, 377–78
Higgins, Dick
postmodernism, 410
Higgs boson, 45
Hilbert, David
reality of infinity, 306
Hill, Christopher
physics for poets, 441
Hindemith, Paul
opera Harmony of the World, 233
Hinton, Howard
fourth dimension, 433
tesseract, 434
Hobbes, Thomas
face of atheism, 66
Leviathan, 66, 102

natural philosophy, 164–65
political philosophy, 165
Hodgskin, Thomas
London Mechanics' Institution, 134
Hodgson, James
popular lectures, 115
Hoffman, Banesh
popularization quantum mechanics, 380
Hogarth, William
Whiston's longitude method, 244–45
Hoist, Helge
Bohr's atom, 375
Holmes, Arthur
age of the Earth, 371
Hooke, Robert
crystallography, 367
gravitation, inverse square law, 12
wave theory of light, 13, 239
Hopkins, Gerald Manley
interest in Tyndall, 278
thermodynamics and poetry, 277–79
Horne, George
Newton a Hutchinsonian, 89
Horrocks, Jeremiah
Kepler's work, 6
Housman, Alfred
poetry test, 265
Howson, Colin
Popperism and Bayesianism, 353
Hoyle, Fred
Big Bang and Christianity, 326
Olbers' paradox, 396
Steady State theory, 326, 395, 396, 398
Hubble, Edwin
closed universe's curvature, 327
expanding universe, 28, 393, 395
galaxies, 98
Hudson, James
physics' decline, mechanics' institutes, 135
Huggins, William
astronomical spectroscopy, 97
spectra of nebulae, 152
Hughes, John
poetry and physics, 241
Hume, David
design argument, 79
miracles, 69
moral philosophy, 196–200
Newton of moral philosophy, 197
political science, 205
teleological arguments, 86
Hutcheson, Francis
quantification of benevolence, 195–96
Hutchinson, John
contemptuous of Newtonianism, 88
influenced by Woodward, 87
Old Testament incorrectly read, 88
Huxley, Aldous, 425–27
influence of D.H. Lawrence, 427
influence of Sullivan, 425, 426
Island, 284
Newton and poets, 264
science and arts in his DNA, 285, 425
Two Cultures, 220
Huxley, Julian
Introduction to Science, 364
Huxley, Thomas
Darwin's Bulldog, 70
persistence of force, 106
Two Cultures, 217, 219
X-Club, 103
Huygens, Christiaan
Cosmotheros, 92
distance to Sirius, 78
quantity of motion, 10
wave theory of light, 37, 239
Hyperspace
Riemann, 26–27
Inge, William Ralph, Dean
science and religion, 108
thermodynamics and religion, 107, 108
Invisible rays, 21, 365–73
James, William
multiverse, 350
Ozma problem, 431
Jammer, Max
eternity and temporal moment, 312
Javacheff, Christo and Jeanne-Claude, 432
Jeans, James
atom, Bohr model, 40
early remarks on Bohr's atom, 373

heat death, 274
popularizer of modern physics, 392–94
Jefferson, Thomas
Declaration of Independence, 205
influence of William Small, 207
possible influence of Keill, 207
Johnson, Barton
Nabokov's signal for incest, 429
Johnson, Samuel
influence of the Boyle Lectures, 114
metaphysical poets, 222
usefulness of science, 246
Jones, Stephen
curious children's book, 130
Jordan, Pascual
energy of the universe, 336
matrix equations, 41
uncertainity principle, 41–42
Jouffret, Esprit
fourth dimension, 436
Joule, James Prescott
conservation of energy, 17
mechanical equivalent of heat, 17, 18
North British Group of physicists, 186
thermodynamics and religion, 105
Joyce, James, 417–21
Finnegans Wake, 47, 418, 419
Joycean language, 418
mathematical doodling, 419
modernism, 414
Ulysses and physics, 419
Joyce, Jeremiah
books for the young, 146
Joyce, Stanislaus (brother of James), 417
Kaluza, Theodor
fourth dimension, 49, 448
Kant, Immanuel
analogies, 93
Bild theory, 192
Critique, 177, 458
dimensions of space, 15, 346, 347
epistemological idealist, 211
influence of Hume, 197
modern physics, 305
nebular hypothesis, 75
Ozma problem, 347, 431
philosophy, 177–79
Thomas Wright's influence, 93
three laws of mechanics, 178
Karlholm, Dan
postmodernistic art, 432
Keating Richard
clock paradox experiment, 383–84
Keats, John
antagonistic to science, 258, 264–65
Keill, John
comments on Burnet, 65
comments on Huygens, 92
comments on Whiston, 80, 81
laws of nature, 207
Newtonian mechanics, 14
Kennard, Earle
uncertainty principle, 42
Kepler, Johannes
Astronomia Nova, 222
Catholic condemnation, 62
elliptical orbits, 5–6
Harmony of the World, 232–33
new stars, 226
King, Erasmus
ascent of saps, 117
popular lectures, 117
King-Hele, Desmond
interpretation of Shelley, 262, 263
Kinoshita, Toichiro
Feynman diagrams, 45
Kirchhoff, Gustav
Fraunhofer lines, 97
Klein, Oskar
fourth dimension, 49, 448
Knox, Ronald
Berkeleyan limerick, 1751
Kornblum, Hanns
film on relativity (1922) 385
Koteliansky, Samuel
D.H.Lawrence's friend, 423
Kragh, Helge
cosmology, 325, 392
popularization problems, 361
multiverse, 352, 353
Kramers, Hendrik
Atom and the Bohr Theory, 375

Kristensen, Tom
Vico, 417
Kroupa, Pavel
dark matter and energy, 335
Kupka, František
fourth dimension, 436
Kurosawa, Akira
postmodernist film, 410
Lagrange, Joseph-Louis
velocity of sound, 183
Lalande, Jérôme
plurality of the worlds, 95
Lamarck, Jean-Baptiste
evolution of the universe, 103
Lambert, Johann Heinrich
comets populated, 94
Lange, Friedrich
materialism, 104
Langevin, Paul
twins' paradox, 382–83
Laplace, Pierre-Simon
dark bodies, 33
nebular hypothesis, 75, 94
orbits of Jupiter and Saturn, 181
reply to Napoleon, 75
velocity of sound, 183
Lardner, Dionysius
Cabinet Cyclopedia, 144
Latitudinarians, 63–64, 66, 71, 456
Laud, William, Archbishop
Puritans a schismatic threat, 63
Laue, Max von
History of Physics, 364
x-ray interference, 368
Lawrence, David Herbert, 423–25
dislike of Newtonian physics, 423
Fantasia of the Unconscious and relativity, 424
modernism, 414
poem Relativity, 424
relativity and relativism, 424
relativity not mechanical, 424
Leavis, Frank Raymond
background to the Two Cultures, 217–18
Two Cultures, 410
Leclerc, Georges-Louis, (Comte Buffon)
Burnet's *Sacred Theory* a romance, 65
proto-evolutionary ideas, 86
Lederman, Leon
God Particle, 402, 413
physics for poets, 411
Leibniz, Gottfried
dispute with Newton, 70
letter to Burnet, 172
letter to Princess Caroline, 70
metaphysics and the real world, 172
Monadologie, 171, 175
monads and physics, 174
philosophy, 171–75
principle of sufficient reason, 338
scalars and vectors, 174
universe, clock model, 67, 70
vis viva, 11, 16, 173
Lemaître, Georges
Big Bang, 326–28
influence on the Catholic Church, 329
Pontifical Academy of the Sciences, 329
primeval atom, 328, 393
science and religion, 328–29
Lenin
Mach and Helmholtz criticized, 318
Materialism and Empirio-Criticism, 319
Lewis, Pericles
Introduction to Modernism, 410
Lewis, Wyndham
model for Pursewarden, 428
Time and Western Man, 412
Light
corpuscular theory, 13, 33, 36
invarience in speed, 21
wave theory, 13, 36
Linde, Andrei
inflation, 331
Lissitzky, El
Minkowski diagrams, 438
Suprematism and *Proun*, 438
Literacy
beginning of the twentieth century, 360
Literature
Enlightenment, 240–52
Romantic Age, 258–59
Romantics against science, 264–66

Romantics favourable to science, 259–64
modernism and postmodernism, 414–31
seventeenth century, 222–32
Victorian Age and early 20th cent., 275–86
Livingstone, David
influenced by Thomas Dick, 84
Lobachevsky, Nikolai
non-Euclidean geometry, 433
Pangeometry, 438
Locke, John
action at a distance, 169
comments on Whiston, 81
empiricism, 167
influence of Descartes, 167
moral philosophy, 194
Newton's philosopher, 167
philosophy, 167–71
refutation of Filmer, 204
Locke, Richard
satire on pluralists, 96
Lockyer, Norman
founder editor of *Nature*, 154
physical and social worlds, 275
spectra of nebulae, 152
Lodge, Oliver
aether, 302-04, 390
aether and spiritualism, 303
psychic phenomena, 303
spiritualism, 301
Lorentz, Hendrik
classic theory, 44
Lorentzian transformation, 23
Lowell, Percival
Martian canals, 98
Luther, Martin
reality is rational, 62
Lyell, Charles
Principles of Geology, 101
Lyons, Henry
London Science Museum, 362
Lyttleton, Raymond
cosmological theories, 397
McGilvary, Evander Bradley
praise of Eddington, 425
Mach, Ernst
atoms epistemologically denied, 193
physical laws and the universe, 325
physicists and psychic phenomena, 301
positivism and modern physics, 186
Principia, definitions, 18
reality of unobservables, 192–94
Macintyre, John
first radiology unit, 367
Mackintosh, Ernest
London Science Museum, 363
MacSwinny, Owen
art entrepreneur, 255
Madison, James
USA Constituition, 205
Malevich, Kazimir
Suprematism, 438
Mansbridge, Albert
WEA, 362
Marcet, Jane
popular woman writer, 150
March, Robert
physics fror poets, 411
Marsden, Dora
influence on Joyce, 420
philosophical ramblings, 420
Marsden, Ernest
atom's nucleus, 372
Marsili, Luigi Ferdinando
Bolognese astronomy collector, 254
Martin, Benjamin, 457
astronomical lectures, 95
popular lectures, 118–19
Marx, Karl
dialectic method, 181
dialectical materialist, 208
influence of Hegel, 208
Materialism, 102–4
atomism, 102, 104
Tyndall's Belfast Address, 103
Maupertuis, Moreau de
satirized by Voltaire, 251
Maxwell, James Clerk
colour theory, 268
electromagnetism, 49
kinetic theory of gases, 20
molecular vortices, 188
North British Group of physicists, 186–91

QED, 44
thermodynamics and religion, 105, 106
transition to modern physics, 16
Mayer, Alfred
roughness in music, 288
Mayer, Julius Robert von
conservation of energy, 17
mechanical equivalent of heat, 18
redness of the blood, 17
Mechanics' institutes, 132–36
Anderson's Institution, 132–33
decline in physics, 135
management, 133–34
Mersenne, Marin
frequency of a vibrating string, 235
Michell, John
black holes, 33
earthquakes, 127
Michelson, Albert
velocity of light, 21
Mickel, Karl
opera *Einstein*, 446
Mill, John Stuart
philosophy of science, 182
System of Logic, 185, 186
Miller, Dayton
recording sound waves, 253
Miller, William
spectroscopy of the stars, 97
Millington, John
Royal Institution, 137
Milton, John
knowledge of astronomy, 229–30
Paradise Lost, 228–31
Miracles, 64, 68, 69–70, 80, 82
Mitchel, Ormsby
Cincinnati telescope, 147
Möbius, August
four space dimensions, 347
Monism, 106–7, 192
Moral philosophy, 194–202
More, Henry
Cambridge Platonists, 63
critic of Hobbes, 66
Morgan, Augustus de
comment on Richard Phillips, 156
Morley, Edward
velocity of light, 21
Mossotti, Ottaviano-Fabrizio
aether, 156
Metzinger, Jean
fourth dimension, 436
Moore, George Edward
The Apostles, 421
Woolf and rejection of idealism, 422
Mozart, Wolfgang Amadeus
fundamental bass, 253
Muggleton, Lodowicke
geocentric system, 90
prophet, 87
Murry, Ann
popular physics for young ladies, 131
Murry, John Middleton
Literature and Science, 410
Music
atonal, 444–45
Baroque period and tonality, 252
consonance and dissonance, 232, 235–36, 252–53, 286–89
Enlightenment, 252–54
harmony of the planets, 232–33
influence of modern physics, 440–42
influence of postmodern physics, 427–44
musical intervals, 235–37
seventeenth century, 232–37
Victorian Age and Helmholtz, 286–89
Nabokov, Vladimir, 428–31
Ada and *LATH*, 428
exchange with Gardner, 430
signal for incest, 429
symmetry in cosmology, 430
symmetry in *LATH*, 430
Napier, John
logarithms, 236
Neumann, Carl
hidden variables, 315–16
Newtonian gravitation, 28
Newbery, Elizabeth
deterioration in the Newberys, 130
Newbery, John
outstanding children's physics book, 127
Newman, James

best popular science books, 388
Newton, Isaac
absolute space and time, 22, 165, 168, 310
cause, 176, 194, 200
colour and musical harmony, 239
colour theory, 266, 267, 269
corpuscular theory of light, 37
De Motu, 11
design argument, 67, 301
dispute with Leibniz, 70
duration of creation, 80
experimental method, 239
general relativity and the 4th Rule, 334
influence on popular culture, 126
inverse square law, 346
letters to Bentley, 28, 67, 68, 72–74, 73, 75
letters to Burnet, 65
materialism, 102
method of fluxions (calculus), 27
miracles, 64, 69, 70
new science, 360
nominalism, 306
oblate shape of the Earth, 251
Pantokrator, 69
physics is mathematical, 305
plurality of the worlds, 92
politics, 203
religious beliefs, 66–70, 82
resolution of a refracting telescope, 123
stability of the universe, 68, 241
theory of music, 237
universe, clock model, 104
vortex hypothesis, 11, 65
Nicolson, Marjorie
Donne's circle image, 224
Donne's two Elizabeths, 226
Nielsen, Kristian
popularization problems, 361
North British Group of physicists, 186–91
Ohm, Georg
ear's analysis of music, 286
Olbers' paradox, 263–64, 396–97
Opticks, 3
atomism, 104
effect of atmosphere on telescopes, 90
harmony and dischord of colours, 239
laws of nature, 69
mixing coloured lights and pigments, 267
queries, 10, 13, 67, 74, 92, 104, 118, 194, 241, 302
rainbows, 272–73
slit and prism, 269
spectrum, 97
Orsini, Cardinal Alessandro, 7
Orwell, George
politicized writers, 414
Ostwald, Wilhelm
monism, 107
positivism and modern physics, 186
Ozma problem, 347–48, 431
Paalen, Wolfgang
Dyn, 441
Figure Pandynamique, 442
Page, Don
unobservable entities, 351
Paine, Tom
nature's revelation, 95
plurality of the worlds, 95
Paley, William
watchmaker argument, 86
Panofsky, Erwin
ellipses in painting, 238
Paolozzi, Eduardo
statue of Newton, 266
Parkinson, Gavin
Figure Pandynamique and particle/wave duality, 442, 443
Parra, Hèctor, 446–49
coalescence of black holes, 447
Hypermusic Prologue, 448
interpreting physics in music, 446
Randall-Sundrum model, 448
Stress Tensor, 447
Parnell, Thomas
Scriblerus Club, 244
Pascal, Blaise
satirized by Voltaire, 251
Patrick, Simon, Bishop, 114
latitudinarians, 63
Paul II, John, Pope, 63
Pauli, Wolfgang

exclusion principle, 41
neutrino, 47
QED, 43
quantum mechanics,possibilities, 34
Rosenfeld, 320
Peierls, Rudolf
Laws of Nature, 364
Perry, Stephen
influence on Hopkins, 278
Penrose-Hawking
singularity theorems, 339
Penrose, Roger
final theory? 401
postmodernism in physics, 399
Phillips, Richard
Arago, French physicist, 156
letter from Faraday, 156
notorious publisher, 146
Philosophy of science, 181–86
Picabia, Francis
Arensberg Circle, 437
Picasso, Pablo
Bateau-Lavoir group, 435
co-creator Cubism, 435
Pittoni brothers
painting commemorating Newton, 255
Pius XI, Pope
Pontifical Academy of the Sciences, 329
Pius XII, Pope, 63, 105
Big Bang and Christianity, 326, 329
cosmology and Christianity, 337–38
God and science, 337
Place, Francis
London Mechanics' Institution, 134
Planck, Max
blackbody radiation, 21, 34
quanta of energy, 35
Plurality of the Worlds, 90–102
antipluralists, 99–102
atheists' argument, 108
theory of evolution, 98, 100
Poe, Allan
beauty in science, 258
influenced by Keats? 265
Olbers' Paradox, 263
To Science, 264
Poincaré, Henri
doubts on cause and effect (1911) 379
chaos and philosophy, 309
conventionalists, 51
Poisson, Siméon
unimpressed by light wave theory, 141
Politics
17th and 18th centuries, 202–5
Declaration of Indepedence, 205–8
Ferguson, Political Science, 201
Hobbes, Elements of Law, 165
Marxism, 208–11
modern physics, 354
Newtonianism and monarchy, 211
quantum mechanics, 317–22
US Constitution, 205–8
Polkinghorne, John
cosmology and religion, 340
gee-whiz popular books, 361
popularization modern physics, 380
multiverse and creation, 351
Pope, Alexander
epitaph of Newton, 242
Essay on Man, 241–42
Scriblerus Club, 81
well administered government, 207
Popper, Karl
Big Bang an immature theory, 332
dialetics, 181, 210
falsification, 50, 326
interpretation of quantum mechanics, 317
objective criterion for science, 352
particle wave duality, 50
physical world real, 317
physics and metaphysics, 211
not fashionable, 404
renormalization, 44
we see what we want to see, 335
Popular physics books
cost, 132, 145–46, 457
eighteenth century, 121–31
energy and force, 121
feats of strongmen, 122
for children, 126–31
for ladies, 124–26
lady writers, 149–54

natural optical phenomena, 145
nineteenth century, 144–54
twentieth century, 360–61, 363–65
Popular physics lectures
cost, 119
eighteenth century, 114–21
technology, 121
Popular physics modern media
film, 385–861
London Science Museum, 362–63
radio and television, 361, 370
Popular physics periodicals, 154–55
Pound, Ezra
make it new, 414, 449
modernism, 414
proper subject for poetry, 260
science and modernity, 415
science not unpoetic, 415
Pound, Robert
gravitational redshift, 31
Preti, Giovanni
critical *M/R* ratio, 33
Princet, Maurice, 435
***Principia*, 9–14, 456**
aether, 302
changes to second edition, 170
corpuscular theory and vacuum, 74
definitions, 10, 18, 22
General Scholium, 9, 12–13, 67–69, 76–77, 92, 170
gravitation, 11, 180
laws of motion, 10, 11, 178
life of the solar system, 68
mathematics and assumptions, 306
metaphysics, 164
political impact, 204
rules, 198, 335
System of the World, 67
theology, 108
Principle of sufficient reason, 338
Proctor, Mary, daughter of Richard
popular astronomy writer, 149
Proctor, Richard
founder editor *Knowledge*, 155
popular journals, 154
popular writer, 148–49
Protestantism
cosmology, 62
Ptolemaeus, Claudius
geocentric astronomy, 4, 51
Pugwash Conferences
manifesto for peace, 321
Pynchon, Thomas
postmodernism, 416
Quantum mechanics
atomic structure, 39–41, 372–81
black body radiation, 34–35
Copenhagen interpretation, 304, 307, 309, 317–23, 354
Einstein-Podolsky-Rosen, 314–15, 316
particle wave duality, 36–37
Planck units, 35–36
Planck's constant, 35
political turmoil, 317–18
QED, 43–45
relativistic mechanics, 43–46
standard model particle physics, 46–47
two slit experiment, 38–39
uncertainty principle, 41–42
wave mechanics, 42–43
Rameau, Jean-Phillippe
influence of Castel, 252
principles of harmony, 252–53
Randall, Lisa
five dimensional spacetime, 448
libretto, Hypermusic Prologue, 448
string theory agnostic, 403
Warped Passages, 403
Rankine, Macquorn
entropy, 188
molecular vortices, 187
North Britisn Group of physicists, 186
Ranyard, Arthur Cowper
Proctor's *magnum opus*, 149
Ray, Man
Arensberg Circle, 437
Rayleigh, Lord (John William Strutt)
blue sky, 289
Royal Institution, 141, 142
Reality
atoms, 107
dark matter and energy, 334

deterministic/indeterministic, 22, 34, 301, 304, 307–10, 313–14, 318–20, 323–24, 353–54
God's, 70
idealist's view, 192, 211
language controlled, 324
mathematical, 306, 305–6, 306, 324, 351
model, 306, 307, 347, 352
physical world, 172
postmodern physics, 190
quantum mechanics, 47, 304, 305, 314–17, 323
rationality not religiosity, 62
supersymmetry, 49
theories, 29, 47, 176
thought experiments, 307–8, 315
uunobservables, 194, 307
Rebka, Glen
gravitational redshift, 31
Regnault, Victor, 187
Reid, Thomas
causes and philosophy, 200
first cause, 200
induction, 200
moral philosopher, 196
observable phenomena, 200
philosophy and common sense, 199
Reimann, Bernard
non-Euclidean geometry, 191
Reith, John, 370
Rees, Martin
effect of all significant physical constants, 349
fine tuned universe, 348
just six numbers, 342–48, 350
multiverse, 351
Reeve, John
Muggletonian prophet, 87
Relativity, 22–23
clock paradox, 383–84
Eddington/Dyson expeditions, 29–30, 386
energy/mass equivalence, 25
eternity and temporal moment, 311–13
experimental validation, 29–31
film (1920), 380–81
general theory, 26–31, 386–89
gravitational waves, 31
Minkowski diagrams, 25, 311, 382, 385, 433
relativism, 424, 428
simultaneity, 310–11
spacetime, 310–14
special theory, 22–25, 381–86
twin's paradox, 382–83
Reynolds, Joshua
early British art critics, 254
Richardson, Jonathan
early Bitish art critics, 254
Richter, Hans
film technique, 437
Ritchie, Matthew
M Theory, 443
postmodernist artist, 443
sets, Hypermusic Prologue, 448
Ritchie, William
Royal Institution, 137
Robertson, Joseph
London Mechanics' Institution, 134
Robison, John
lack of British physics treatises, 147
Rodgers, John
Harmony of the World, 233
Röntgen, Wilhelm
X-rays, 366
Rood, Ogden
colour theory, 268
Rosen, Nathan
zero total energy, 395
Rosenberg, Léonce
Cubism, 434
Rosenfeld, Léon, 321
Bohr's bulldog, 320
complementarity a Marxist method, 320
Copenhagen interpretation, 319, 322
Lenin and Stalin's Marxist philosophy, 320
quantum mechanics and politics, 321–22
Roush, Sherrilyn
anthropic principle, 341
Rowney, Thomas and Richard
colourmen to the romantics, 266
Royal Institution, 136–42

Christmas lectures, 140, 142, 370, 457
electromagnetism, 137–38
financial problems, 457
management, 136–37
twentieth century, 361
Ruff, Willie
Harmony of the World, 233
Rumford, Count (Benjamin Thompson)
caloric theory, 139, 140
cannon boring, 139
Royal Institution, 136, 137
Russell, Bertrand
ABC of Atoms, 374
ABC of Relativity, 26, 27, 388
causes and philosophy, 200
fin de siècle pessimism, 283
induction, 199
Leibniz, 171, 172
Mach and neutral monism, 192
manifesto for peace, 321
relativity and relativism, 412
The Apostles, 416
The Fourth Dimension, 433
Woolf and rejection of idealism, 422
Rutherford, Ernest
age of the Earth, 371
alpha and beta rays, 366
atomic structure, 39
radioactivity, 370–71
relativity and the 1919 eclipse, 30
structure of the atom, 372, 373, 375
Sagan, Carl, 361
Sahl, Ibn, 165
Salam, Abdus
force unification, 46
quantum mechanics and politics, 322
Sandby, Thomas and Paul, 83
Sauveur, Joseph
musical intervals and logarithms, 237
overtones, 235
Schiaparelli, Giovanni
canals on Mars, 149
Schlegel, Victor
fourth dimension, 436
Schöenberg, Arnold
atonal music not physics related, 445
tonal revolution, 444
Schoenflies, Arthur
crystal structure, 368
Schönberg, Mario
influence on Bohm, 319
Schrödinger, Erwin
cat paradox, 304
physics and philosophy, 299
quantum wave function, 42–3
time and religion, 311
Schwarz, John Henry
superstring revolution, 48, 400
Schwarzschild, Karl
black holes, 33
Schwinger, Julian
QED, 44, 380
Scientific theory, 51–2
Scott-James, Rolfe Arnold
modernism, 414
Scriblerus Group, 244
Sea depth gauge, 118
Seurat, Georges-Pierre
pointillism, 268
Shakespeare, William
Troilus and Cressida:, 223
Shapiro, Barbara
Erasmian humanism, 3
Shapiro, Irwin
verification general relativity, 334
Shaw, George Bernard
Tyndall's Belfast Address, 103
Shelley, Percy Busshe
interest in science, 262–63
Sim, Lorraine
Woolf and George Moore, 422
Sitter, Willem de
evolution of the universe, 327
Slipher, Vesto
spectrum of the spiral galaxies, 326
Smith, Adam
influence of Newtonian physics, 201–202
Smith, Quentin
atheistic cosmological argument, 338–40
Smolin, Lee
five major physics' problems, 398
string theory atheist, 399, 404

Snow, Charles Percy
background to the Two Cultures, 217–18
Discovery, 378
two cultures, 410
Soldner, Johann von
deflection of light passing the Sun, 30
Soddy, Frederick
atomic energy, 372
crystallography, 369
radioactivity, 370-72
Society for Psychical Research, 301-03
Somerville, Mary
woman popular writer, 150
Spencer, Herbert
friend of George Eliot's, 280
monism, 106, 107
thermodynamics, conception, 107
Spenger, Oswald
New Götterdämerung, 108
Spinoza, Baruch
conservation of mass, 178
face of atheism, 66
influence of Descartes, 166
necessity of God's Nature, 77
pantheist, 66
Spiritualism, 301–4
Stannard, Russell
cosmology and religion, 340
Steinhardt, Paul
eternal inflation, 332
Stenger, Victor, 347
toy universes, 348
Steele, Richard, 116, 240
Stefan, Jožef, 34
local position invariance, 31
Stephenson, Bruce
Harmonices Mundi, 232
Steven, Simon
equal temperament, 236
Stewart, Balfour
physical and social worlds, 275
Stewart, George Walter
Bohr's 1913 atom, 373
String theory, 47–52
M-theory, 49–50, 402
problems, 50–52
string theory atheists, 403–4
superstrings, 48–50, 400–403
Stukeley, William
fall of the apple, 11
Sullivan, John William Navin
Bohr's 1920 atom, 374
early discussion of relativity, 389
modern physics and materialism, 426
Science and Literature, 410, 411
Sundrum, Raman
five dimensional spacetime, 448
Swedenborg, Emanuel
nebular hypothesis, 75
Swift, Jonathan
Battle of the Books, 243
coffee houses, 240
falsity in learning ridiculed, 81
influence of Fontenelle? 248
relationship with science, 245–50
science, its practical effect, 243
scientists satirized in *Laputa*, 249
Scriblerus Club, 243
Symmetry
breaking, 401, 402
charge, parity, time (CPT), 429
Higgs particle, 402
is symmetry fundamental? 401
electricity and magnetism, 43
fundamental forces, 49
gauge theories, 45
Lorentzian transformation, 23
modern physics, 347
Nabokov theme, 429
string theory, 47
supersymmetry, 48, 400
Tait, Peter Guthrie
knot tables, 190
metaphysics, 187
molecular vortices, 190
North British Group of physicists, 186
thermodynamics and life, 275, 280
thermodynamics and religion, 105, 106
Tartini, Giuseppe
Tartini tones, 286
Tausk, Klaus
quantum mechanics and politics, 321–22

Tegmark, Max
multiverses, 351
physics is mathematical, 306
significant physical constants, 349
Temple, William
Anciens et les Modernes, 243
Tennyson, Alfred
non-influence of thermodynamics, 276
Theories of the Earth
Burnet, 64–66
Drysdale, 83
Walsh, 91
Whiston, 79–82
Thermodynamics, 16–20
age of the Earth, 86
caloric theory, 139
first law, 16–19, 105
gas laws, 16
heat death, 108
influence on religion, 105–8
second law, 19–20, 105
Thirring, Hans
Ideas of Einstein's Theory, 388
Thompson, Silvanus
early children's lecture on X-rays, 367
Thomson, George
electron diffraction, 37
Thomson, James brother of William
North British Group, 186
Thomson, James Scottish poet, 204
friend of science, 250
The Seasons, 250, 254, 274
Thomson, Joseph John
cathode rays, 21
electrons, 366
magnitude of relativity effects, 389
plum pudding atom model, 190, 373
Thomson, William (Lord Kelvin)
absolute temperature, 16
aether, 302, 390
age of the Earth, 86, 371
images, 191
science and religion, 106
thermodynamics, 16, 19
thermodynamics, and pessimism, 274
thermodynamics and religion, 105, 106
vortex atom model, 190–91
Thought experiments, 382, 384–85
Tilghman, Shirley
liberal arts courses, 411
Time's arrow, 20, 109
Toland, John
Letters to Serena, 77
matter and motion, 71
Tomonaga, Sin-Itiro
formalized QED, 44
QED, 44, 380
Trimble, Virginia
dark matter, 334
Tryon, Edward
creatio ex nihilo, 395
universe a vacuum fluctuation, 336
Turnbull, George
moral philosophy, 194–95
Turner, Joseph Mallord William
Goethe's colour theory, 268, 270–71
Light and Colour, 271
mixing pigments, 268
Two Cultures Debate, 217–20, 458
Tyndall, John
atomism, 104
Belfast Address, 103, 299, 456
British Association, 143
curiosity, 140
heat as a mode of motion, 139
imagination, 194
influence of Carlyle, 103
materialism, 104, 106
RI Christmas lectures, 142
RI spectacular experiments, 140
Royal Institution, 137–42
thermodynamics and religion, 106
X-Club, 103
Universities
Bodleian Library, 455
matriculants' mathematics, 133
new science, 455
physics 17th and 18th centuries, 14
postmodernism, 399–400
Savilian chairs, 455
Untermeyer, Louis
parody of T.S, Eliot, 415

Rhyme and Relativity, 415
Ussher, James, Archbishop, 73
Vanderburgh, William
alternatives to general relativity, 334, 335
Velocity of Sound, 183
Veneziano, Gabriele, 4
Vico, Giambattista
Scienza Nuova, 417
Vilard, Paul
gamma rays, 366
Vilenkin, Alexander
universe created from nothing, 336
Vinci, Leonardo da
circle image, 224
colour theory, 266
ellipses, 237
Visual arts
Cubism, 432–36
Dada and De Stijl, 437–38
ellipse, 237–38
Emlightenment, 254–57
modernism, 431–42
postmodernism, 442–44
Romantic Age, 266–74
Russian *avante-garde*, 435–40
seventeenth century, 237–39
Surrealism, 440–42
Victorian Age and Helmholtz, 289–90
Voltaire
Cartesianism and Newtonianism, 123
dissemination of Newtonianism, 123–24
exile to England, 123
Leibniz and *Candide*, 78
Micromégas, 250–52
plurality of the worlds, 251
rejection of Cartesianism, 124
Swift, 250
Voss-Andreae, Julian
painting and physics, 443
Quantum Man, 444
Waals, Johannes van der
early discussion of relativity, 382
Walker, Adam
influence on Shelley, 262
Wallace, Alfred
plurality of worlds, 149
Wallis, John
experiments with partial notes, 235
laws of motion, 69
Walpole, Horace, 95
Walsh, Francis
theory of the Earth, 91
legitimate scientific hypotheses, 353
multiverse, 352
universe's cosmic purpose, 301
Warhol, Andy
Moonwalk, 442
populist postmodernism, 432
Weinberg, Stephen
cosmology and a Supreme Being, 340
Dreams of a Final Theory, 401, 404
force unification, 46
Wells, Herbert George
atomic energy, 372
funding of science, 155
heat death, 282
influence of Jeans, 282
science is not entertainment, 360
Time Machine and thermodynamics, 282
Wesley, John, 109
Weyers, Stefan
local postion invariance, 31
Wheeler, John
quantum foam, 47
Whewell, William
antipluralist, 100–102
Bridgewater Treatises, 85
Inductive Sciences, 184
John Herschel's *Discourse*, 184
philosopher of science, 182–86
Whiston, William
Arianism, 82
Button's Coffee House, 240
dissemination of Newtonianism, 13
gravitation a Divine effect, 80
Great Deluge, 79–81
longitude measurement, 244
Memoirs of Martinus Scriblerus, 244
New Theory of the Earth, 79–82, 114
physics and natural religion, 82–83
popular lectures, 116, 241

Whitehead, Alfred
remark on Shelley, 263
Whitehurst, John
theories of the Earth, 83
Whitman, Anne
Principia, definitive translation, 10
Whitrow, Gerald
cosmology a science? 326
Whitworth, Michael
D.H. Lawrence, 423
Wigner, Eugene
physics is mathematical, 305
symmetry, 430
Wilberforce, Samuel, Bishop
ape descent speech, 144
Wilkins, John, 64
Williams, William Carlos
Einstein of the Daffodils, 415
modernism, 415
Wilson, Charles
atomic tracks, 363
Wilson, James
influence of Newton, 205, 206
metaphysics, 164
Wilson, Robert
Einstein on the Beach, 445
Wilson, Woodrow
US Constitution, influence of physics, 208
Witches and portents, 13
Witherspoon, John
influence of Reid 205
Witten, Edward
M-theory, 49–50
second superstring revolution, 401, 402
Wittgenstein, Ludwig
language and reality, 324
physics and reality, 307
space dimensions and symmetry, 347
Woit, Peter
string theory atheist, 399, 404
Workers' Educational Association, 361–62, 457
Wollaston, William
Fraunhofer lines, 97
Woodward, John, 79
Wolff, Christian
satirized by Voltaire, 251
Woolf, Leonard
The Apostles, 421
Woolf, Virginia, 421–22
Berkeleyan idealism, 422
illusion and reality, 422
influence of modern physics, 422
modernism, 414
Wordsworth, William
French Revolution, 257, 258
men of science, 260, 261
millenarian, 259
proper subject for poets, 260
Wren, Christopher, 69, 114
Wright, Georg Henrik von
forgotten, Eddington remembered, 299
Wright, Joseph of Derby
science genre paintings, 255
Wright, Thomas
plurality of the worlds, 93
Wu, Chien-Shung, 429
violation of parity, 431
X-Club, 103
Young, Edward
Night Thoughts, 93
Young, Thomas
colour theory, 268
energy, 17, 276
frequency of the colour spectrum, 239
light, two slit experiment, 38
physics and physiology of colour, 267
Royal Institution, 136, 137, 139
sensitifity optical nerves, 289
spectroscopy, 97
wave theory of light, 36
Zahlan, Anne
Alexandria Quartet, 428
Zarlino, Gioseffo
Western just intonation, 236
Zha, Zai-Yu equal temperament, 236

Zimmer, Ernst
physics, philosophy and religion, 300
popular writer of distnction, 378
Zukav, Gary
physics and Eastern mysticism, 325
Zweig, George
quarks, 47
Zwicky, Fritz
dark matter, 333